Microbiology Monographs

Series Editor
Alexander Steinbüchel
Münster, Germany

The Springer book series ***Microbiology Monographs*** presents carefully refereed volumes on selected microbiological topics. Microbiology is a still rapidly expanding field with significant impact on many areas of basic and applied science. The growth in knowledge of microbial physiology, cell structure, biotechnological capabilities and other aspects of microorganisms is increasing dramatically even in the face of the breakthroughs that already have been made.

Reflecting these most recent achievements, the series' wide scope encompasses such topics as inclusions in prokaryotes, predatory prokaryotes, magnetoreception and magnetosomes in bacteria, uncultivable microorganisms, microbial endosymbionts, bacterial resistance, extremophilic microorganisms, analyses of genome sequences and structures, microorganisms as cell factories for chemicals and fuels, metabolic engineering, gene transfer and expression systems and distinct physiological groups of bacteria.

The volume editors are well-known experts in their particular fields, and each volume offers 10 to 20 comprehensive review articles covering all relevant aspects of the topic in focus.

More information about this series at http://www.springer.com/series/7171

Masayuki Inui • Koichi Toyoda
Editors

Corynebacterium glutamicum

Biology and Biotechnology

Second Edition

Springer

Editors
Masayuki Inui
Molecular Microbiology and Biotechnology
Research Institute of Innovative Technology
Kyoto, Japan

Koichi Toyoda
Molecular Microbiology and Biotechnology
Research Institute of Innovative Technology
Kyoto, Japan

ISSN 1862-5576 ISSN 1862-5584 (electronic)
Microbiology Monographs
ISBN 978-3-030-39266-6 ISBN 978-3-030-39267-3 (eBook)
https://doi.org/10.1007/978-3-030-39267-3

This Springer imprint is published by the registered company Springer Nature Switzerland AG.
The registered company address is: Gewerbestrasse 11, 6330 Cham, Switzerland

Preface to the Second Edition

The first edition covered omics analyses and molecular biological tools of *Corynebacterium glutamicum* and the regulatory network consisting of transcription factors that have been characterized by the techniques. Transcriptional regulation of central metabolic genes, respiratory system, and cell division mechanism were also reviewed. For the application of *C. glutamicum*, protein secretion and the production of biorefinery products as well as the production and export of amino acids were illustrated. While all of the contents are still important and fundamental for understanding of this microorganism, in the seven years since the first edition was published, the knowledge of the physiology and molecular biology of *C. glutamicum* largely expanded. Moreover, recent developments in metabolic engineering and synthetic biology have demonstrated *C. glutamicum* as a promising production platform for not only conventional amino acids but also commodity chemicals, biofuels, cosmetics, and pharmaceuticals.

Considering these advances, we decided to update the book contents by replacing and combining the chapters, not just adding the recent findings to the chapters of the first edition. Despite the large modification, the purpose of the monograph is unchanged: to review and summarize the research on the molecular biology of *C. glutamicum* and to introduce readers to the potential of this bacterium for versatile applications.

In the first three chapters, characteristic cell division mechanism and cell surface structure of corynebacteria and respiratory system characteristic to actinobacteria are described. The following three chapters cover sigma factors and their specific promoters, transcriptional regulators involved in the primary metabolic pathway genes, and post-transcriptional regulatory mechanisms. The chapters for amino acid production and amino acid exporters are largely updated with latest advances in the field. The chapter dealing with metabolic flux analysis is fully rewritten with recent knowledge and technologies. The chapter of metabolic engineering illustrates how *C. glutamicum* is successfully engineered for the production of a variety of products. The last two chapters focus on aromatic compounds' catabolism and production. Probably due to the capability of utilizing a broad spectrum of aromatic compounds as a carbon source, *C. glutamicum* is inherently highly resistant to these toxic compounds. This property makes this bacterium as an attractive host for the production of aromatic compounds, which can be used for the synthesis of polymers, cosmetics, and pharmaceuticals.

We would like to sincerely appreciate all the authors for their time and contribution. We hope that this second edition will be informative for researchers in our fields to update their knowledge and used as a reference for their studies.

Finally, we would like to thank Springer Nature for giving this opportunity to publish the second edition. We would also like to thank Springer's editors, Bibhuti Sharma and Markus Späth, for their great support during the course of this project.

Kyoto, Japan

Masayuki Inui
Koichi Toyoda

Contents

Part I
Characteristics of *Corynebacterium glutamicum*

Chromosome Organization and Cell Growth of *Corynebacterium glutamicum*

Kati Böhm, Giacomo Giacomelli, Fabian Meyer, and Marc Bramkamp

Contents

Abstract *Corynebacterineae* share an unusually complex cell envelope that discriminates species of this group markedly from other actinobacteria. Phylogenetically, they belong to the Gram-positives, but contain additionally to their peptidoglycan cell wall an extra cell envelope layer composed of arabinose and galactose polymers, the arabinogalactan. Covalently linked to the outer arabinose subunits are mycolic acids forming an inner leaflet of a membrane that is capped by trehalose mono- and dimycolates and forms the mycolic acid membrane, a permeability barrier similar to the outer membrane of Gram-negative bacteria. Like all actinobacteria, *Corynebacterium glutamicum* grows apically due to a polarly localized elongasome complex. A scaffold protein, DivIVA, is required for spatial organization of the elongasome. Besides organization of elongation growth, DivIVA is also involved in spatial organization of the chromosome. DivIVA serves as an anchor for the partitioning protein ParB that in turn binds DNA sequences in proximity to the origin of replication, thereby orienting the chromosomes to the cell pole.

K. Böhm
Faculty of Biology, Ludwig-Maximilians-University Munich, Planegg, Germany

G. Giacomelli · F. Meyer · M. Bramkamp (✉)
Faculty of Biology, Ludwig-Maximilians-University Munich, Planegg, Germany

Institute for General Microbiology, Christian-Albrechts-University Kiel, Kiel, Germany
e-mail: marc.bramkamp@lmu.de

M. Inui, K. Toyoda (eds.), *Corynebacterium glutamicum*, Microbiology Monographs 23, https://doi.org/10.1007/978-3-030-39267-3_1

Remarkably, both cell poles contain a stably tethered chromosome, making *C. glutamicum* diploid. At fast growth conditions, DNA replication can be over-initiated and, hence, polyploidy is common under these conditions. Correct chromosome segregation is key for septum placement, while classical division-site-selection systems seem absent. Recent work unraveled that subcellular organization of Corynebacteria is in many aspects different to that in classical model organisms, indicating the importance to study bacterial species in all their various aspects.

1 Introduction

Cell growth and division are coupled to chromosome replication and segregation in all organisms to ensure production of viable offspring. In recent years a wealth of new data became available that helped us to understand the molecular mechanisms used by bacteria to govern these essential processes (Eswara and Ramamurthi 2017). However, most insight into chromosome organization, replication and segregation as well as cytokinesis and cell (wall) growth are based on classical model organisms such as *Bacillus subtilis*, *Escherichia coli*, or *Caulobacter crescentus*. Comparatively little was known in bacteria belonging to the actinobacteria phylum. This has started to change very recently with several groups analyzing cell biological aspects of *Streptomyces*, *Mycobacterium* and *Corynebacterium* (Donovan and Bramkamp 2014; Flärdh 2003b; Noens et al. 2007; Ramijan et al. 2018; Schlimpert et al. 2017; Trojanowski et al. 2018). Understanding of subcellular organization in these species is important, since their mode of growth differs drastically from that of the classical model organisms. All members of the actinobacteria studied so far grow by apical extension of their cell wall (Flärdh 2003a; Kang et al. 2008; Letek et al. 2008b; Umeda and Amako 1983). Elongation growth is governed by an elongasome complex that is spatially organized by a coiled-coil protein termed DivIVA that is conserved also in many low GC Gram positives (Fig. 1). DivIVA localizes to the curved region at the cell pole and at the membrane facing either site of the division septum. In *C. glutamicum*, actin-like proteins such as MreB are not involved in elongation growth (Donovan and Bramkamp 2014; Letek et al. 2008b). Septal cell wall synthesis in *C. glutamicum* and other actinobacteria is governed by the tubulin homolog FtsZ, as in most bacteria (Ramos et al. 2005). A surprising difference is the low number of conserved accessory proteins that are involved in FtsZ membrane binding and FtsZ polymer regulation (Donovan and Bramkamp 2014). Although *C. glutamicum* has a rod shaped morphology, division site selection is not dependent on known FtsZ regulators such as Min proteins. It has therefore been assumed that the cytokinetic apparatus found in *C. glutamicum* can be considered a minimal divisome (Donovan and Bramkamp 2014; Letek et al. 2008a).

Corynebacterium shares with *Mycobacterium*, *Nocardia*, and *Rhodococcus* a complex, multi-layered cell wall architecture as a unique feature among actinobacteria (Dover et al. 2004; Jankute et al. 2015; Laneelle et al. 2013; Mishra

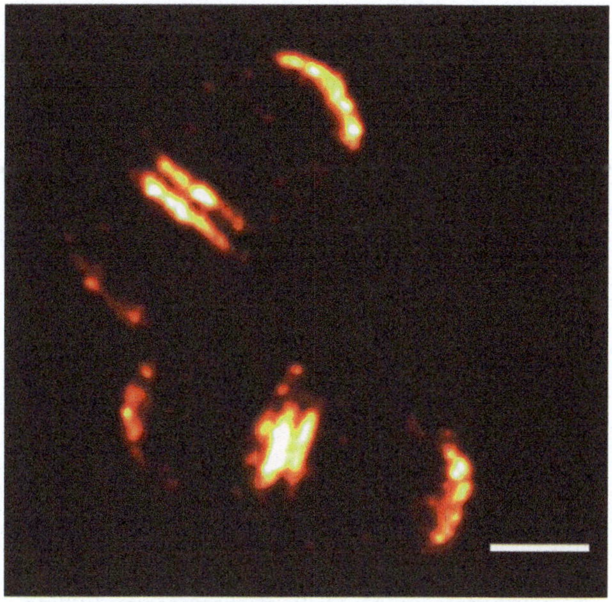

Fig. 1 DivIVA—A polar scaffold for multiple processes. Gaussian rendering of DivIVA localizations in *C. glutamicum* cells imaged with PALM (Photoactivated Localization Microscopy). DivIVA localizes at poles and on the two sides of the septum via its intrinsic ability to recognize negatively curved membranes. Here, it works as a scaffold for diverse cellular functions, such as the elongasome and the ParABS partitioning complex, respectively. Scale-bar: 500 nm

et al. 2011). Despite being phylogenetically a Gram-positive, the cell wall structure differs compared to the conventional cell envelope found for example in Firmicutes. In contrast, *Corynebacterineae* possess a complex, covalently linked hetero-polysaccharide layer that serves as scaffold for an additional outfacing hydrophobic barrier and thus, mimics the function of the outer membrane of Gram-negatives (Jackson 2014).

The subunits of the so called mycomembrane (MM) are referred to as mycolic acids, a condensed structure derived from two fatty acids, forming the α-alkyl, β-hydroxy mycolic motif (Jackson 2014; Laneelle et al. 2013). The length of the fatty acid varies by species and thus serves as taxonomic marker. For *Mycobacterium spp.* the length varies between 60 and 90 carbon atoms, whereas in *Corynebacterium spp.* the aliphatic chain is shorter with roughly between 22 and 36 carbon atoms (Marrakchi et al. 2014). This difference is also observable on a macroscopic scale. Colonies of *Mycobacterium* species show a dry, structured and waxy surface, in contrast to the less structured and moist surface of *C. glutamicum* colonies. The effectivity of the barrier also depends on the length of the mycolic acids, consequently *Mycobacterium spp.* stain acid-fast, while *Corynebacterium* spp. not.

Although, *C. glutamicum* is well understood with respect to its physiology, the cell cycle was only analyzed recently. A major finding was that *C. glutamicum* harbors two chromosome copies that are stably anchored to each cell pole and that the newly replicated chromosomes segregate to midcell positions, where the cell will eventually divide (Böhm et al. 2017). At fast growth rates DNA replication can be over-initiated, a characteristic feature shared with many other bacteria including *Mycobacterium* (Böhm et al. 2017; Trojanowski et al. 2017). Despite many similarities, there are several distinct features that are different between *C. glutamicum* and *Mycobacterium* species, indicating that even closely related species use different strategies for subcellular organization (Trojanowski et al. 2018). *C. glutamicum* is supposed to be a soil bacterium (Kinoshita et al. 1957), but many members of the genus are human commensals found as main components of the human skin microbiome (Cundell 2018). Several species, such as C. *ulcerans*, *C. striatum*, or *C. jeikeium* are emerging opportunistic pathogens (Bernard 2012). Therefore, a more detailed understanding of the cell biology of *Corynebacterium* is of broad interest.

2 Chromosome Structure

Bacterial chromosomes are dynamically folded into highly compacted nucleoids, which refold in the course of each cell cycle in order to allow for processes like gene expression, replication and segregation. Nucleoids are further shaped by DNA-supercoiling, condensin complexes and by an interactive system of nucleoid associated proteins. A variety of chromosome-folding strategies exist across different model organisms that differ in composition of DNA-binding proteins and in spatial domain organizations.

C. glutamicum harbors, similar to most bacteria, a circular chromosome sized 3.28 megabases, where replication is initiated at one single origin of replication (*oriC*) and proceeds bidirectionally towards the opposing terminus of replication (*terC*) (Kalinowski et al. 2003). Here, the *oriC* domain appears to be the hub of chromosome organization in *C. glutamicum*, being organized by the partition protein ParB (cg3426) that binds *oriC*-proximal *parS* sites sequence-specific (Donovan et al. 2010). ParB further interacts with several kilo bases of non-specific DNA surrounding *parS* by spreading along the DNA after nucleating at the *parS* sites (Breier and Grossman 2007; Murray et al. 2006; Rodionov et al. 1999). Spreading, in combination with 3D-bridging events between ParB dimers, induces *oriC*-domain condensation, eventually leading to large nucleoprotein complexes formation (Broedersz et al. 2014; Graham et al. 2014; Sanchez et al. 2015).

Further, *C. glutamicum* stably tethers *oriC*s to the cell poles, similarly to the related actinobacteria *Mycobacterium smegmatis* and *Streptomyces coelicolor* (Ditkowski et al. 2013; Donovan et al. 2012; Ginda et al. 2013). To this end, *C. glutamicum* ParB-*parS* complexes are recruited to polar positions via ParB-interactions with the coiled-coil landmark protein DivIVA (cg2361) (Donovan et al. 2012; Letek et al. 2008b). Notably, different to classical model species, *C. glutamicum* is diploid at various growth rates, where at least two fully replicated genome copies are maintained throughout generations (Böhm et al. 2017). Therefore, both cell poles are permanently occupied by ParB-*oriC* complexes, while chromosomal *terC* domains position at midcell, indicating a longitudinal *ori-ter-ter-ori* chromosome configuration (Böhm et al. 2017).

In addition to ParB, *C. glutamicum* harbors a SMC/ScpAB (structural maintenance of chromosomes) complex (cg2265, *smc*; cg1611, *scpA*; cg1614, *scpB*), which is ubiquitously present in all branches of life (Cobbe and Heck 2004). The mechanism of DNA condensation is strongly conserved amongst these complexes, where ATP-dependent conformational changes drive binding as well as translocation along DNA by extrusion of DNA loops (Minnen et al. 2016; Wilhelm et al. 2015). Different to *B. subtilis*, SMC is not an essential factor for DNA segregation in *C. glutamicum*, but localizes ParB-proximal within cells (Britton et al. 1998, unpublished data), therefore, it likely mediates pairing of chromosomal arms by establishing inter-arm contacts starting from loading sites at ParB-*parS* complexes, as described in *B. subtilis* and *C. crescentus* (Le et al. 2013; Wang et al. 2015). Interestingly, unlike in most other model bacteria, the *C. glutamicum* SMC/ScpAB coexists with the distantly related MksBEFG condensin complex (MukB-like SMC, cg3103-cg3106) (Böhm et al. 2019), yet its precise function in genome organization remains to be characterized. Moreover, nucleoid associated proteins, like MrgA, Lrp and Lsr2 (cg3327, cg0313, cg1966), presumably contribute to chromosomal contacts in *C. glutamicum*, as shown for *E. coli* transcriptional regulators FIS, H-NS and HU (Lioy et al. 2018) or for the mycobacterial HupB (Holowka et al. 2017).

Notably, many corynebacterial genomes contain prophages. This holds true also for the *C. glutamicum* chromosome, which harbors three cryptic prophage regions: CGP1-3 (Kalinowski et al. 2003). Here, CGP3 alone accounts for roughly 6% of the genome size. Viral DNA of this prophage region has been shown to spontaneously excise from its host genome in 1–3% of cells within a population (Frunzke et al. 2008; Nanda et al. 2014). Furthermore, CGP3 encodes its own system for spatio-temporal organization of viral DNA, that comprises actin-like AlpC (cg1890) and the viral DNA-binding adapter protein AlpA (cg1891), which act independently of *C. glutamicum* chromosome organizing proteins. AlpC polymerization is regulated by AlpA, establishing a diffusion-capture mechanism of viral DNA segregation (Donovan et al. 2015; Forde et al. 2017).

3 Cell Cycle

Chromosome replication and cell division are interdependent processes, which are prerequisites for cell growth (Arjes et al. 2014; Bates and Kleckner 2005). Both define cell cycle parameters and, in turn, are timely regulated by the bacterial metabolism (reviewed in Willis and Huang 2017). The bacterial cell cycle commonly consists of three phases, which are coupled to nutrient-availability. The time frame of chromosome replication is defined as C-period, while the subsequent D-period covers the interval between replication termination and completion of cell division (Cooper and Helmstetter 1968). In general, the length of C- and D-periods remains relatively constant amongst different nutrient-rich growth conditions, however, they can be prolonged in nutrient-poor media. Particularly at slow growth, replication is not synchronized with cell birth and is instead preceded by an

Table 1 Overview of species-specific cell cycle parameters at distinct growth rates

Organism[a]	Doubling time (T_d)	Genome size (Mb)	Repl. speed[b] (bases/s)	C	D	References
Cgb	63 min	3.21	340	78 min	20 min	Böhm et al. (2017)
Cgb	130 min	3.21	280	97 min	26 min	Böhm et al. (2017)
Mtu	24 h	4.42	55	11 h	≤13 h	Nair et al. (2009)
Msm	180 min	6.98	400	110 min	25 min	Santi et al. (2013)
Eco	20 min	4.64	820	47 min	25 min	Kubitschek and Freedman (1971)
Bsu	30 min	4.22	700	52 min	21 min	Sharpe et al. (1998)

[a]Cgb *C. glutamicum* RES167; Mtu *M. tuberculosis*; Msm *M. smegmatis*; Eco *E. coli*; Bsu *B. subtilis*
[b]*Repl. speed* replication speed

additional B-period that precedes replication initiation (Wang and Levin 2009). At high growth rates overlapping C-periods abolish the B-period.

C. glutamicum performs one single round of replication per cell cycle at slow growth rates (lower than 0.3 h^{-1}), similar to *Mycobacterium tuberculosis* (for cell cycle parameters, see Table 1). However, *C. glutamicum* cells can adapt their cell cycles to fast growth conditions (0.66 h^{-1}) by overlapping C-periods, thereby reducing generation times. To this end, new rounds of replication are initiated before the termination of previous ones, leading to multiply nested replication forks (Böhm et al. 2017). Brain-heart-infusion (BHI)-grown *C. glutamicum* cells can for instance double almost every 60 min, even though a C-period takes roughly one third longer. Subsequent D-periods range in *C. glutamicum* from 20 to 26 min, equally depending on growth rates (Table 1).

C. glutamicum harbors high DNA contents (Neumeyer et al. 2013), due to di- and polyploidy, where at least two fully replicated genome copies per cell are maintained throughout generations (Böhm et al. 2017). Several evolutionary advantages, like reduced mutation rates, desiccation resistance or growth ability in phosphorous-depleted environments, have been attributed to this growth strategy (reviewed in Soppa 2014). Growth rate-dependent cell cycle models of *C. glutamicum* illustrate DNA contents per cell, where mainly four up to eight fully replicated chromosomes are present per cell prior to cell division (Fig. 2). Accordingly, many replication events are ongoing throughout the *C. glutamicum* life cycle, where fast growth in BHI medium allows for up to six replisomes per cell (Böhm et al. 2017). However, at any cell cycle mode, even in the presence of overlapping C-periods, only one round of genome replication is completed per life cycle (Bremer and Churchward 1977).

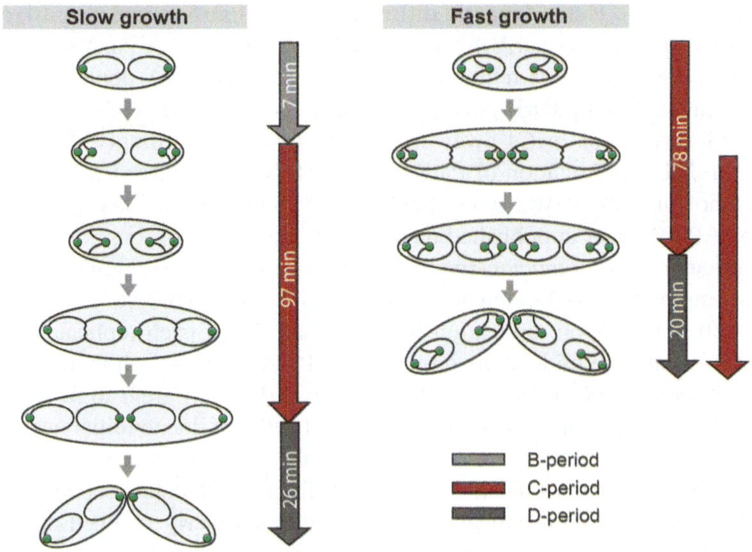

Fig. 2 Spatiotemporal chromosome organization of *C. glutamicum*. Cell cycle modes for short and long generation times. Chromosomes are depicted as black lines with *oriCs* as green circles. Notably, stages with single chromosomes per cell are absent at any growth rate and replication cycles of at least two sister chromosomes take place simultaneously. Left: Cell cycle of slow growing cells in MMI medium supplemented with 4% glucose ($T_d = 130$ min). A short B-period precedes replication initiation, where C-periods are completed within one generation. Right: Life cycle of BHI-grown cells ($T_d = 63$ min). Multifork replication takes place with C-periods starting around 15 min prior to termination of the previous one

4 Chromosome Segregation

Within every cell cycle, duplicated genomes must segregate to opposing cell halves, which will eventually turn into two separate daughter cells. Bacteria evolved different strategies to segregate chromosomes. *E. coli* and related species separate duplicated chromosomes by a self-organizing process involving entropic forces of DNA polymers (Jun and Mulder 2006). However, the vast majority of bacteria including *C. glutamicum* additionally possess a mitotic-like type I ParABS system, which actively segregates duplicated sister chromosomes (Böhm et al. 2017; Donovan et al. 2010; Livny et al. 2007).

In *C. glutamicum* ten chromosomal *parS* sequences contribute to ParB recruitment and nucleoprotein complex formation close to *oriC* (Donovan et al. 2010). The partner protein of ParB is the P-loop ATPase ParA (cg3427), which belongs to the diverse ParA/MinD family (Leipe et al. 2002). ParA binds upon ATP-dependent dimerization via conserved arginine residues nonspecifically to the bacterial nucleoid, which fills large parts of the cell volume (Hester and Lutkenhaus 2007; Leonard et al. 2005). In the course of chromosome segregation, ParAB act together as a self-organizing Brownian ratchet (Hwang et al. 2013; Vecchiarelli et al. 2013; Zhang and

Schumacher 2017). Here, dynamic diffusion of ParB-*parS* complexes drives interactions with surrounding ParA that result in ATP-hydrolysis of ParA and its release from DNA. Delayed DNA-rebinding of ParA creates a ParA depletion zone that trails behind the moving ParB-*parS* complexes (Vecchiarelli et al. 2013) mediating a directed movement of *oriC*s away from each other. Accordingly, ParA deletion abolishes reliable segregation of ParB-*oriC* clusters in *C. glutamicum* (Böhm et al. 2017; Donovan et al. 2010), while ParA-gradients do not form along nucleoids in absence of ParB (own unpublished data).

In particular, *C. glutamicum*-specific features of DNA separation derive from its genetic complexity. Firstly, newborn cells localize one ParB-*oriC* complex to each of their cell poles. Upon replication initiation one out of each replicated *oriC* pair segregates toward the midcell position along a ParA gradient, while the other sister remains tethered to polar DivIVA (Böhm et al. 2017). Different to *M. smegmatis*, *C. glutamicum oriC*s remain within their original cell half due to initial occupancy of the opposing pole by a second *oriC* complex (Santi and McKinney 2015; Trojanowski et al. 2015). To this end, ParA likely establishes separate dynamic gradients within each cell half and a fraction is particularly retained at cell poles interacting with ParB (Donovan et al. 2010). Since the majority of cells at fast growth reinitiate replication, new rounds of chromosome replication can occur at polar, as well as at midcell positions. Notably, sister *oriC* cohesion periods appear to be long and highly variable in *C. glutamicum*. Since diploidy postpones the separation of replicated chromosomes in sister cells for one generation (Böhm et al. 2017), DNA-separation processes may be very flexible in this organism, lacking a tight regulation with the cell cycle. After polar *oriC* splitting, chromosome segregation continues behind replication forks that passively translocate towards mid-cell positions in the course of polar cell elongation (Böhm et al. 2017). Replicated *oriC* regions are segregated across the existing nucleoid towards the cell center (Böhm et al. 2019). Disturbing chromosome segregation in *C. glutamicum* by ParAB deletions causes DNA mis-segregation defects that lead to the formation of up to 16 % of DNA-free cells, underlining the essentiality of the ParABS system in chromosome homeostasis (Donovan et al. 2010). Monoploid Mycobacteria, on the other hand, localize their *oriC* domains via an interaction between ParA and the DivIVA homologue Wag31 to one cell pole, similar to Streptomyces (Ditkowski et al. 2013; Ginda et al. 2013). Despite mechanistic differences and deviant genetic features, fundamental principles of *oriC* tethering to cell poles and of ParABS-mediated chromosome segregation processes are conserved amongst those actinobacteria members (Trojanowski et al. 2018).

In addition to ParABS, *C. glutamicum* harbors the ubiquitous recombinases XerCD (cg2224, cg1608) that mediate decatenation of sister chromosomes at *terC*-proximal *dif* sites after replication termination (Blakely et al. 1993). Further, a homologue of the DNA translocase FtsK exists in *C. glutamicum* (*cg2158*), which aids in the final step of chromosome partitioning to daughter cells by directly pumping chromosomal DNA across the closing septum shortly before completion of cell division. FtsK-powered DNA transport is directed by distinct DNA motifs (KOPS) (Aussel et al. 2002; Bigot et al. 2005; Massey et al. 2006).

5 Division Site Selection

Defects in the ParABS-mediated chromosome partitioning process of *C. glutamicum* are, apart from chromosome mis-segregation, further associated with cell division phenotypes, supporting the notion that both cellular processes are tightly linked (Donovan et al. 2013). In general, division site selection is relatively flexible in *C. glutamicum* and related actinobacteria, where septum-placement may slightly deviate from midcell positions (Donovan et al. 2013; Santi et al. 2013). However, ParAB mutants exhibit strongly aberrant septum positioning and cell elongation rates (Donovan et al. 2013). Here, septa often form at polar positions and frequently constrict over chromosomes, yielding high numbers of non-viable cells that are to large part anucleoid. Consequently, spatiotemporal division site selection is governed in *C. glutamicum* by chromosome organization.

The huge impact of DNA segregation on cell division may, in part, explain why *C. glutamicum* and related actinobacteria lack common Min systems that mediate precise Z-ring positioning at midcell and accessory nucleoid occlusion systems (Donovan and Bramkamp 2014). Consequently, bacteria like *C. glutamicum*, in which these protein machineries are absent, needed to evolve alternative regulators dedicated to placing their Z-rings. Examples of alternative regulators that use the chromosome as topological factor for division site selection include the ParA family member MipZ in *C. crescentus* (Thanbichler and Shapiro 2006). In *C. glutamicum*, another ParA-like ATPase, termed PldP (ParA-like division protein, cg1610), has been implicated in division site selection (Donovan et al. 2010). PldP localizes to the site of future cell division, while its deletion causes altered placement of division septa. However, the underlying mechanism has not yet been investigated.

6 Cytokinesis

C. glutamicum cells divide by binary fission and similar to most bacterial cells, the first protein known to assemble at the nascent division site is the bacterial tubulin homologue FtsZ (cg2366) (Adams and Errington 2009). FtsZ assembles into polymeric structures that subsequently recruit other proteins to the division site. The fully assembled multi-enzyme complex is termed divisome. The FtsZ polymers have been shown to be highly dynamic and recent super resolution studies have shown treadmilling of FtsZ that seems to power the direction of cytokinetic cell wall synthesis (Bisson-Filho et al. 2017; Yang et al. 2017). The level of FtsZ is differently regulated in various bacteria. Overexpression of FtsZ is causing division phenotypes and artificially increased FtsZ levels lead to severe phenotypic consequences in *C. glutamicum* (Letek et al. 2007; Ramos et al. 2005). Localization of FtsZ in *C. glutamicum* was shown by immunofluorescence and revealed that FtsZ not only localizes as a ring-like structure at the future division site, but also assembles close to the cell poles (Ramos et al. 2005). It was speculated that these polar FtsZ assemblies

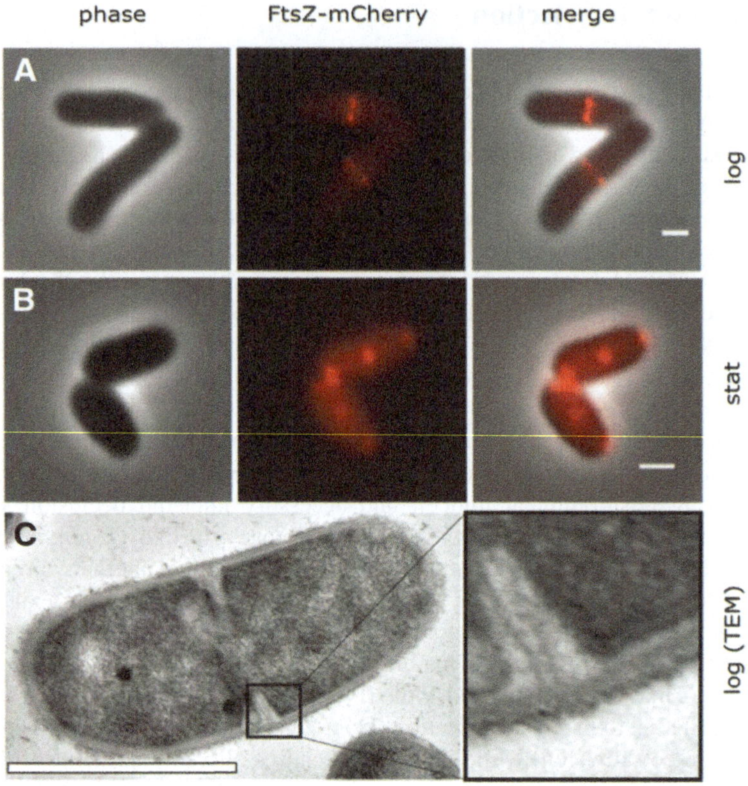

Fig. 3 Division septum formation in *C. glutamicum*. Localization of FtsZ-mCherry in the logarithmic (**a**) and in the stationary growth phase (**b**). In the stationarity phase FtsZ-mCherry shows additional polar localized patches. Transmission electron micrograph of the septal invagination in the logarithmic phase (**c**). Scale-bar: 1 μm

are remnants of previous division machineries. Using a FtsZ-mCherry fusion these polar localizations have been confirmed (own, unpublished Data) (Fig. 3a, b). However, polar FtsZ assemblies are mainly observed in stationary phase cells.

Divisome assembly seems to follow a two-step process (Gamba et al. 2009). First, a proto-ring is assembled on the cytoplasmic site. In *B. subtilis* this complex includes the accessory proteins FtsA, ZapA, EzrA, SepF and ClpX. *C. glutamicum* lacks genes encoding FtsA and ZapA, but encodes a SepF (cg2363) homolog. The so far uncharacterized protein Cg3203 contains a domain associated with cytokinesis (PFAM ID: PF13845) (Slayden et al. 2006), but this protein lacks a transmembrane domain and, hence, might be an unlikely candidate to tether the Z-ring to the cell membrane. Likely, the enzyme complex synthesizing the cell wall precursor lipid II composed of Mra and Mur proteins is also localized at the division site. In a second, timely distinct step, the membrane components of the divisome are assembled. The membrane integral proteins include division specific penicillin binding proteins. The division-specific transpeptidase FtsI (cg2375) and the regulatory proteins FtsL (cg2376) and FtsQ (cg2367) are essential for division. Recently, it was shown that the SEDS (shape, elongation, division and sporulation) proteins RodA (cg0061) and FtsW (cg2370) act as major cell wall transglycosidases (Meeske et al. 2016; Taguchi et al. 2019). Before, it was speculated that RodA and FtsW proteins might rather act as lipid II flippase, a function that has now be assigned to MurJ (Ruiz 2008; Sham et al. 2014). The SEDS transglycosidases seem to act only in intimate cooperation with their cognate PBP (penicillin binding protein). FtsW (cg2370) is essential for cytokinesis and localizes specifically at the division site (Sieger et al. 2013).

The corynebacterial peptidoglycan (PG) belongs to the peptidoglycan type A1γ with meso-diaminopimelic acid (mDpm) in position 3 and a linkage between positions 3 and 4, similar to PG crosslinks found in many Gram-negative bacteria (Schleifer and Kandler 1972). The single building blocks are synthesized in the cytosol. Subsequent reactions, carried out by homologues of the Mur and Mra family result in the coupling of a PG precursor to the membrane-integrated carrier molecule decaprenylphosphate (DP) (Scheffers and Tol 2015). The resulting lipid II is subsequently translocated across the cell membrane by the flippase MurJ (Ruiz 2008, 2015; Sham et al. 2014). *C. glutamicum* encodes a homolog of the putative lipid II flippase MurJ (cg3419). The gene product has not been characterized in detail, but localization of MurJ to the cell poles and septa in *C. glutamicum* is in line with the predicted function (own, unpublished data).

On the extracellular site of the membrane, transglycosylases release the PG precursor from the DP molecule and integrate it via the GlcNAc-residue into a growing strand of PG (Typas et al. 2011). The liberated DP is recycled by back-flipping. In order to integrate the linear polymer into the three-dimensional PG mesh, the cross-linking of the existing sacculus must be degraded in a confined area (Typas et al. 2011).

Despite encoding a core set of conserved cytokinetic genes, *C. glutamicum* lacks several associated factors found in other bacteria (Donovan and Bramkamp 2014). Furthermore, little is known about the stabilization of the Z-ring and the mechanism by which FtsZ is tethered to the membrane. SepF, which is supposed to function in Z-ring tethering in *B. subtilis* (Hamoen et al. 2006), is not essential in *C. glutamicum* (Honrubia et al. 2001), however, a described protein-protein interaction between FtsW and FtsZ might serve this function (Valbuena et al. 2007).

Septal PG cross-walls can be visualized in transmission electron micrographs and reveal an unusual emergence of a Π-shaped invagination that is flanked by the de-novo synthesized PG (Zhou et al. 2019) (Fig. 3c). On top of the opposing PG surfaces a new arabino-galactan (AG) layer starts to form. This new AG-layer is connected to the AG/mycolic membrane (MM)-complex of the mother cell. After division, cell wall hydrolases cleave the connecting cell wall between the daughter cells and the turgor creates a force pushing the daughter cells apart. Before reaching the limits of the tensile strength, the connecting outer layers become leaky and allow the MM to propagate along the newly formed poles. The turgor further inflates the new poles until a breaking point in the MM/AG-complex that quickly rips apart along the division plane, leaving behind a minimal connecting hinge. This characteristic V-snapping occurs within 10 ms and was shown to be sensitive to osmotic changes (Zhou et al. 2016, 2019). Hence, the ring-like surface structures discovered by scanning electron microscopy derive from the primary MM/AG-complex and are birth scars rather than excess cell wall material synthesized at the onset of cytokinesis. After completion of cytokinesis, daughter cell separation is governed by the activity of cell wall hydrolases. The major peptidoglycan DL-endopeptidase RipA (cg1735; cgR_1596 in *C. glutamicum* R) was shown to be essential for V-snapping (Zhou et al. 2019). *C. glutamicum* encodes three cell wall-associated hydrolases (cg0784, cg2401, and NlpC (cg2402)). However, these have not been analyzed in detail. The homolog of NlpC in *C. glutamicum* R, cgR_2070, has been shown to be involved in cell separation (Tsuge et al. 2008), but there seems to be redundancy between the PG hydrolases. In *E. coli* the cell wall hydrolase EnvC is controlled by action of an ABC-transporter-like protein pair, FtsEX (Yang et al. 2011). Similarly, in *B. subtilis* FtsEX regulate the activity of the murein hydrolase CwlO (Meisner et al. 2013). Given the fact that FtsE (cg0914) and FtsX (cg0915) are conserved also in *C. glutamicum*, it seems likely that cell wall hydrolases are, also in this actinobacterium, regulated similar to *B. subtilis* and *E. coli*. For *C. glutamicum* R it has been shown that RNAse III regulates MraZ and RipA transcript levels, thereby increasing *ftsEX* transcription. Consequently, deletion of RNAse III leads to cell separation defects (Maeda et al. 2016).

7 Apical Growth

Cell elongation machineries in bacteria can be spatially localized differently. Spatial localization of cell elongation in bacteria can be divided roughly into two sub-categories: those that are mainly FtsZ-dependent and occur at midcell (as found in cocci) and those that are elsewhere in the cell. It is important to make this distinction, because those organisms that do not rely on FtsZ for their elongasome localization still rely on it for the placement of the machinery responsible for cell wall synthesis during septation as discussed above. The first category can be further divided between bacteria showing septal (i.e.: *Staphylococcus aureus*) (Monteiro et al. 2015; Pinho and Errington 2003) and unipolar (i.e.: *Agrobacterium tumefaciens*) (Brown et al. 2012; Cameron et al. 2014) cell wall synthesis, while the second group includes bacteria that place their elongation machineries at the lateral site (i.e.: *B. subtilis, E. coli, C. crescentus*) (Defeu Soufo and Graumann 2005; Doi et al. 1988; Figge et al. 2004) and those that rely on apical cell growth (i.e.: *C. glutamicum*) (Daniel and Errington 2003; Flärdh 2003b; Letek et al. 2008b). The different elongasome localizations observed amongst bacteria depend on the molecular scaffold used for PG synthesis. Actin-like MreB-homologues support lateral elongation in many rod-shape bacteria while DivIVA mediates apical growth in actinobacteria.

The latter PG synthesis mechanism was originally proposed for *Corynebacterium diphteriae* (Umeda and Amako 1983) and later visualized via fluorescently labelled

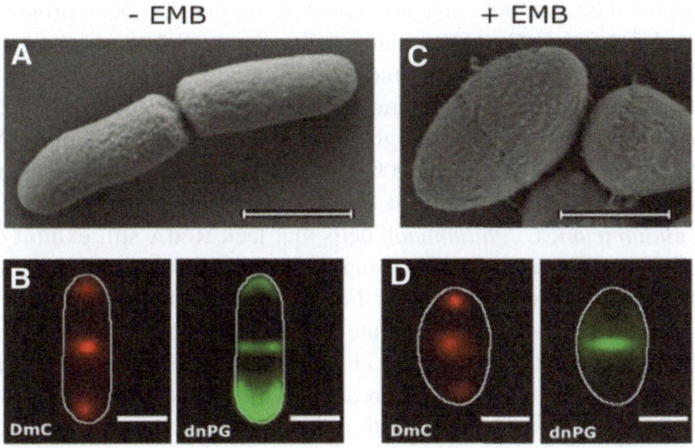

Fig. 4 Importance of the apical elongation machinery in *C. glutamicum*. Comparison of untreated *C. glutamicum* cells (− EMB; **a**, **b**) and cells treated with the antituberculosis drug ethambutol (+ EMB; **c**, **d**). The upper panel shows scanning electron micrographs. In the lower panel, fluorescent micrographs show localization of DivIVA-mCherry (DmC) and the sites of de novo PG synthesis (dnPG), visualized by biorthogonal labelling with azido-D-alanine. The comparison shows the impaired function of the apical synthesis machinery upon treatment with EMB. Scale-bar: 1 μm

vancomycin in three members of the Actinobacteria phylum (*S. coelicolor, M. smegmatis*, *C. glutamicum*) (Daniel and Errington 2003; Flärdh 2003a; Letek et al. 2008b; Nguyen et al. 2007). In these organisms, DivIVA monomers polymerise via a mechanism known as molecular bridging, forming a relatively uniform superstructure that is able to autonomously recognise membrane characterized by a high degree of negative curvature such as the one found at the poles and the septum (Lenarcic et al. 2009). Here, DivIVA works as a scaffold for a multitude of processes that often differ depending on the organism. Members of the DivIVA super-family share a conserved short N-terminal domain responsible for membrane interaction, and at least two coiled-coil regions essential for the protein oligomerization (Letek et al. 2009; Muchova et al. 2002). The protein-protein interaction regions that allow for DivIVA to interact with different partners are not conserved and are the reason why members of this family differ in function across different Gram-positive bacteria (Lin and Thanbichler 2013).

The hypothesis that DivIVA is involved in cell elongation in *C. glutamicum* is further supported by the analysis of cells that are partially depleted in DivIVA. This mutant exhibits a lack of polar PG synthesis and adopts a spherical cell shape (Letek et al. 2008b) similar to what happens in presence of ethambutol (EMB), an antibiotic that has been shown to specifically inhibit apical growth in *Corynebacterium* and *Mycobacterium* (Schubert et al. 2017) (Fig. 4). In contrast, over-expression of DivIVA results in accumulation of protein at one cell pole, which in turn leads to enhanced PG synthesis at this position and aberrant rod shapes in *C. glutamicum* and *M. tuberculosis* (Letek et al. 2008b; Nguyen et al. 2007), or massively enlarged hyphae in filamentous bacteria such as *S. coelicolor* (Flärdh 2003a). This interdependence between pole size and DivIVA amount has been proposed to be a reason for the typical *C. glutamicum* club shape (Schubert et al. 2017). In *C. glutamicum,* DivIVA has been found to interact with the cell elongation machinery via the putative glycosyltransferase RodA (cg0061) (Sieger and Bramkamp 2014) and the high-molecular-weight penicillin-binding protein (HMW-PBP) PBP1a (cg0336) (Valbuena et al. 2007). Notably, viable *rodA* deletion mutants have been described for many model organisms including *B. subtilis*, *E. coli*, but also in *C. glutamicum*. *C. glutamicum* cells that lack RodA still exhibit elongation growth (Sieger and Bramkamp 2014), suggesting that there is redundancy of RodA with bifunctional PBPs with respect to the transglycosylation reaction.

Genes encoding RodA and its cognate PBP, PbpA (cg0060), are not encoded in the dcw cluster (division cell wall gene cluster), but in a separate genome island that also encodes two Hanks-type serine/threonine kinases (STPK), PknA (cg0059) and PknB (cg0057) (Fiuza et al. 2008). Both protein kinases have been implicated with posttranslational modification of cell division proteins such as FtsZ (Fiuza et al. 2008; Kang et al. 2005; Schultz et al. 2009). PknB contains four PASTA domains that are involved in binding of PG components (Shah et al. 2008; Yeats et al. 2002). *C. glutamicum* encodes a third STPK, PknL (cg2388) that also contains five PASTA domains and also plays a role in regulation of cell growth and division (Schultz et al. 2009). Clean genetic deletions of PknA, PknB, or PknL show that single deletions lead to a mild cell elongation phenotype. However, there seems to be redundancy and a complete knock-out of PknABL has not been achieved, suggesting synthetic lethality. *C. glutamicum* encodes only a single protein phosphatase Ppp (cg0062) and deletion of the corresponding gene leads to a severe growth and division phenotype with short, bulged cells and occasional filamentation (Schultz et al. 2009). The existence of PASTA domains in PknB strongly hints that the signal perceived by PknB and PknL might be coupled with cell wall synthesis or turn over, linking growth to genetic regulation. However, detailed analyses describing the control of cytokinesis in *C. glutamicum* are so far lacking. The osmo-sensitive two component system MtrAB (cg0862, cg0864) was shown to be involved in the process of cell elongation (Möker et al. 2004, 2007a, b). Upon osmotic stress the response-regulator MtrA downregulates the putative cell wall peptidase MepA (cg2747), the putative secreted cell wall peptidases MepB (cg0980) and NlpC (cg2402), as well as the putative membrane-bound protease modulator PpmA (cg3138). In *M. tuberculosis*, the homologues of MtrAB (TBMG_03294, TBMG_03293) were shown to be essential (Fol et al. 2006), whereas in *C. glutamicum* deletions lead to cell-length phenotypes. For *M. tuberculosis* it was reported that MtrA also regulates the expression of DnaA, which could explain essentiality in this organism (Fol et al. 2006).

Divisome and elongasome complexes use different subsets of enzymes in a spatio-temporal regulation to generate the three connected structural elements of the cell envelope. These structures are built through different metabolic pathways (Alderwick et al. 2006; Gande et al. 2004; Letek et al. 2008a). To maintain structural integrity of their cell envelope, *Corynebacterineae* couple PG synthesis, the assembly of the connecting arabinogalactan (AG) and the synthesis of the mycolic acid layer (MM) during cell elongation. So far little is known about the spatial organization of the complex that synthesizes the AG and MM layer. For *M. smegmatis* a copurification of DivIVA with two enzymes involved in the early mycolate synthesis (AccA3 and AccD5) was shown, indicating a putative spatial association of MM-precursor synthesis with sites of cell elongation (Meniche et al. 2014). In *M. tuberculosis* a crucial protein for arabinose production, DprE (cg0238), has been shown to localize to the cell pole, suggesting that the AG synthesis is part of the elongasome. This is further supported by the localization of LcpA (cg0847) to the cell poles and late septa in *C. glutamicum*. LcpA was shown to act as transglycosylase connecting the AG layer to the PG polymer (Baumgart et al. 2016). LcpB (cg3210) was also shown to localize to sites of nascent cell wall synthesis, but a null allele had only a mild phenotype (Baumgart et al. 2016).

8 Outlook

In recent years quite some progress was made in the understanding of the cell biology of *C. glutamicum* and its closest relatives. Several labs started to address questions about cytokinesis, cell growth and chromosome organization in this organism, making it arguably one of the better understood actinobacteria. However, there are still many open questions and often new data show that molecular details, even between closely related bacteria, can be quite different. Several important aspects of the *C. glutamicum* cell organization will be revealed in near future. The function of bacterial condensin in *C. glutamicum* is currently investigated and there is accumulating evidence that the two condensin complexes (SMC-ScpAB and MksBEFG) might function in non-related pathways. While SMC likely governs chromosome organization, MksBEFG is assumingly involved in plasmid defense (Böhm et al. 2019; Doron et al. 2018). Although, evidence for this in *C. glutamicum* is still lacking, results *from M. tuberculosis* using a loss-of-function mutation *eptC*, encoding a MksB homolog, show that this condensin restricts plasmid maintenance (Panas et al. 2014).

Future research will also need to address again the regulation of cytokinesis by the two component system MtrAB and the STKPs PknABL. Further, the mode of growth is still unclear and while several reports exist describing an adder-like growth behavior of many bacteria including *M. tuberculosis*, apically growing, pleomorphic cells still remain a challenge for a very clean growth analysis that can be addressed with mathematical modelling. Earlier observation with *C. glutamicum* revealed that growth does not follow sizer or timer regimes (Donovan et al. 2013), but more accurate growth descriptions are needed.

In summary, *C. glutamicum* has developed into a new model system to understand bacterial apical growth and subcellular organization, and despite the lack of sophisticated cell differentiation, *C. glutamicum* remains a fascinating bacterium with many surprises.

Acknowledgements The authors thank the Deutsche Forschungsgemeinschaft (DFG) for financial support (Grants BR 2915/6-1, BR 2915/6-2 and TRR174 to M.B.).

References

Adams DW, Errington J (2009) Bacterial cell division: assembly, maintenance and disassembly of the Z ring. Nat Rev Microbiol 7:642–653

Alderwick LJ, Dover LG, Seidel M, Gande R, Sahm H, Eggeling L, Besra GS (2006) Arabinan-deficient mutants of *Corynebacterium glutamicum* and the consequent flux in decaprenylmonophosphoryl-D-arabinose metabolism. Glycobiology 16:1073–1081

Arjes H, Kriel A, Sorto NA, Shaw JT, Wang JD, Levin PA (2014) Failsafe mechanisms couple division and DNA replication in bacteria. Curr Biol 24:2149–2155

Aussel L, Barre F-X, Aroyo M, Stasiak A, Stasiak AZ, Sherratt D (2002) FtsK is a DNA motor protein that activates chromosome dimer resolution by switching the catalytic state of the XerC and XerD recombinases. Cell 108:195–205

Bates D, Kleckner N (2005) Chromosome and replisome dynamics in *E. coli*: loss of sister cohesion triggers global chromosome movement and mediates chromosome segregation. Cell 121:899–911

Baumgart M, Schubert K, Bramkamp M, Frunzke J (2016) Impact of LytR-CpsA-Psr proteins on cell wall biosynthesis in *Corynebacterium glutamicum*. J Bacteriol 198:3045–3059

Bernard K (2012) The genus *Corynebacterium* and other medically relevant coryneform-like bacteria. J Clin Microbiol 50:3152–3158

Bigot S, Saleh OA, Lesterlin C, Pages C, El Karoui M, Dennis C, Grigoriev M, Allemand J-F, Barre F-X, Cornet F (2005) KOPS: DNA motifs that control *E. coli* chromosome segregation by orienting the FtsK translocase. EMBO J 24:3770–3780

Bisson-Filho AW, Hsu Y-P, Squyres GR, Kuru E, Wu F, Jukes C, Sun Y, Dekker C, Holden S, VanNieuwenhze MS, Brun YV, Garner EC (2017) Treadmilling by FtsZ filaments drives peptidoglycan synthesis and bacterial cell division. Science 355:739–743

Blakely G, May G, McCulloch R, Arciszewska LK, Burke M, Lovett ST, Sherratt DJ (1993) Two related recombinases are required for site-specific recombination at *dif* and *cer* in *E. coli* K12. Cell 75:351–361

Böhm K, Meyer F, Rhomberg A, Kalinowski J, Donovan C, Bramkamp M (2017) Novel chromosome organization pattern in *Actinomycetales* - overlapping replication cycles combined with diploidy. MBio 8:e00511–e00517

Böhm K, Giacomelli G, Schmidt A, Imhof A, Koszul R, Marbouty M, Bramkamp M (2019) Chromosome organization by a conserved condensin-ParB system in the actinobacterium *Corynebacterium glutamicum*. BioRxiv 10. https://doi.org/10.1101/649749

Breier AM, Grossman AD (2007) Whole-genome analysis of the chromosome partitioning and sporulation protein Spo0J (ParB) reveals spreading and origin-distal sites on the *Bacillus subtilis* chromosome. Mol Microbiol 64:703–718

Bremer H, Churchward G (1977) An examination of the Cooper-Helmstetter theory of DNA replication in bacteria and its underlying assumptions. J Theor Biol 69:645–654

Britton RA, Lin DC, Grossman AD (1998) Characterization of a prokaryotic SMC protein involved in chromosome partitioning. Gene Dev 12:1254–1259

Broedersz CP, Wang X, Meir Y, Loparo JJ, Rudner DZ, Wingreen NS (2014) Condensation and localization of the partitioning protein ParB on the bacterial chromosome. Proc Natl Acad Sci USA 111:8809–8814

Brown PJ, de Pedro MA, Kysela DT, van der Henst C, Kim J, De Bolle X, Fuqua C, Brun YV (2012) Polar growth in the Alphaproteobacterial order Rhizobiales. Proc Natl Acad Sci USA 109:1697–1701

Cameron TA, Anderson-Furgeson J, Zupan JR, Zik JJ, Zambryski PC (2014) Peptidoglycan synthesis machinery in *Agrobacterium tumefaciens* during unipolar growth and cell division. MBio 5:e01219–e01214

Cobbe N, Heck MM (2004) The evolution of SMC proteins: phylogenetic analysis and structural implications. Mol Biol Evol 21:332–347

Cooper S, Helmstetter CE (1968) Chromosome replication and the division cycle of *Escherichia coli* Br. J Mol Biol 31:519–540

Cundell AM (2018) Microbial ecology of the human skin. Microb Ecol 76:113–120

Daniel RA, Errington J (2003) Control of cell morphogenesis in bacteria: two distinct ways to make a rod-shaped cell. Cell 113:767–776

Defeu Soufo HJ, Graumann PL (2005) *Bacillus subtilis* actin-like protein MreB influences the positioning of the replication machinery and requires membrane proteins MreC/D and other actin-like proteins for proper localization. BMC Cell Biol 6:10

Ditkowski B, Holmes N, Rydzak J, Donczew M, Bezulska M, Ginda K, Kędzierski P, Zakrzewska-Czerwińska J, Gabriella H. Kelemen G, Jakimowicz D (2013) Dynamic interplay of ParA with the polarity protein, Scy, coordinates the growth with chromosome segregation in *Streptomyces coelicolor*. Open Biol 3:130006

Doi M, Wachi M, Ishino F, Tomioka S, Ito M, Sakagami Y, Suzuki A, Matsuhashi M (1988) Determinations of the DNA-Sequence of the *mreB*-gene and of the geneproducts of the *mre*-Region that function in formation of the rod shape of *Escherichia coli* cells. J Bacteriol 170:4619–4624

Donovan C, Bramkamp M (2014) Cell division in *Corynebacterineae*. Front Microbiol 5:132

Donovan C, Schwaiger A, Krämer R, Bramkamp M (2010) Subcellular localization and characterization of the ParAB system from *Corynebacterium glutamicum*. J Bacteriol 192:3441–3451

Donovan C, Sieger B, Krämer R, Bramkamp M (2012) A synthetic *Escherichia coli* system identifies a conserved origin tethering factor in Actinobacteria. Mol Microbiol 84:105–116

Donovan C, Schauss A, Krämer R, Bramkamp M (2013) Chromosome segregation impacts on cell growth and division site selection in *Corynebacterium glutamicum*. PloS One 8:e55078

Donovan C, Heyer A, Pfeifer E, Polen T, Wittmann A, Krämer R, Frunzke J, Bramkamp M (2015) A prophage-encoded actin-like protein required for efficient viral DNA replication in bacteria. Nucleic Acids Res 43:5002–5016

Doron S, Melamed S, Ofir G, Leavitt A, Lopatina A, Keren M, Amitai G, Sorek R (2018) Systematic discovery of antiphage defense systems in the microbial pangenome. Science 359. https://doi.org/10.1126/science.aar4120

Dover LG, Cerdeno-Tarraga AM, Pallen MJ, Parkhill J, Besra GS (2004) Comparative cell wall core biosynthesis in the mycolated pathogens, *Mycobacterium tuberculosis* and *Corynebacterium diphtheriae*. FEMS Microbiol Rev 28:225–250

Eswara PJ, Ramamurthi KS (2017) Bacterial cell division: nonmodels poised to take the spotlight. Annu Rev Microbiol 71:393–411

Figge RM, Divakaruni AV, Gober JW (2004) MreB, the cell shape-determining bacterial actin homologue, co-ordinates cell wall morphogenesis in *Caulobacter crescentus*. Mol Microbiol 51:1321–1332

Fiuza M, Canova MJ, Zanella-Cléon I, Becchi M, Cozzone AJ, Mateos LM, Kremer L, Gil JA, Molle V (2008) From the characterization of the four serine/threonine protein kinases (PknA/B/G/L) of *Corynebacterium glutamicum* toward the role of PknA and PknB in cell division. J Biol Chem 283:18099–18112

Flärdh K (2003a) Essential role of DivIVA in polar growth and morphogenesis in *Streptomyces coelicolor* A3(2). Mol Microbiol 49:1523–1536

Flärdh K (2003b) Growth polarity and cell division in *Streptomyces*. Curr Opin Microbiol 6:564–571

Fol M, Chauhan A, Nair NK, Maloney E, Moomey M, Jagannath C, Madiraju MVVS, Rajagopalan M (2006) Modulation of *Mycobacterium tuberculosis* proliferation by MtrA, an essential two-component response regulator. Mol Microbiol 60:643–657

Forde AJ, Albrecht N, Klingl A, Donovan C, Bramkamp M (2017) Polymerization dynamics of the prophage-encoded actin-like protein AlpC is influenced by the DNA-Binding adapter AlpA. Front Microbiol 8:1429–1429

Frunzke J, Bramkamp M, Schweitzer J-E, Bott M (2008) Population heterogeneity in *Corynebacterium glutamicum* ATCC 13032 caused by prophage CGP3. J Bacteriol 190:5111–5119

Gamba P, Veening JW, Saunders NJ, Hamoen LW, Daniel RA (2009) Two-step assembly dynamics of the *Bacillus subtilis* divisome. J Bacteriol 191:4186–4194

Gande R, Gibson KJC, Brown AK, Krumbach K, Dover LG, Sahm H, Shioyama S, Oikawa T, Besra GS, Eggeling L (2004) Acyl-CoA carboxylases (*accD2* and *accD3*), together with a unique polyketide synthase (*Cg-pks*), are key to mycolic acid biosynthesis in *Corynebacterianeae* such as *Corynebacterium glutamicum* and *Mycobacterium tuberculosis*. J Biol Chem 279:44847–44857

Ginda K, Bezulska M, Ziolkiewicz M, Dziadek J, Zakrzewska-Czerwinska J, Jakimowicz D (2013) ParA of *Mycobacterium smegmatis* co-ordinates chromosome segregation with the cell cycle and interacts with the polar growth determinant DivIVA. Mol Microbiol 87:998–1012

Graham TGW, Wang X, Song D, Etson CM, van Oijen AM, Rudner DZ, Loparo JJ (2014) ParB spreading requires DNA bridging. Genes Dev 28:1228–1238

Hamoen LW, Meile JC, de Jong W, Noirot P, Errington J (2006) SepF, a novel FtsZ-interacting protein required for a late step in cell division. Mol Microbiol 59:989–999

Hester CM, Lutkenhaus J (2007) Soj (ParA) DNA binding is mediated by conserved arginines and is essential for plasmid segregation. Proc Natl Acad Sci USA 104:20326–20331

Holowka J, Trojanowski D, Ginda K, Wojtas B, Gielniewski B, Jakimowicz D, Zakrzewska-Czerwinska J (2017) HupB is a bacterial nucleoid-associated protein with an indispensable eukaryotic-like tail. MBio 8:e01272–e01217

Honrubia MP, Ramos A, Gil JA (2001) The cell division genes *ftsQ* and *ftsZ*, but not the three downstream open reading frames YFIH, ORF5 and ORF6, are essential for growth and viability in *Brevibacterium lactofermentum* ATCC 13869. Mol Genet Genomics 265:1022–1030

Hwang LC, Vecchiarelli AG, Han YW, Mizuuchi M, Harada Y, Funnell BE, Mizuuchi K (2013) ParA-mediated plasmid partition driven by protein pattern self-organization. EMBO J 32:1238–1249

Jackson M (2014) The mycobacterial cell envelope-lipids. CSH Perspect Med 4:a021105

Jankute M, Cox JA, Harrison J, Besra GS (2015) Assembly of the mycobacterial cell wall. Annu Rev Microbiol 69:405–423

Jun S, Mulder B (2006) Entropy-driven spatial organization of highly confined polymers: Lessons for the bacterial chromosome. Proc Natl Acad Sci USA 103:12388–12393

Kalinowski J, Bathe B, Bartels D, Bischoff N, Bott M, Burkovski A, Dusch N, Eggeling L, Eikmanns BJ, Gaigalat L, Goesmann A, Hartmann M, Huthmacher K, Krämer R, Linke B, McHardy AC, Meyer F, Möckel B, Pfefferle W, Pühler A, Rey DA, Rückert C, Rupp O, Sahm H, Wendisch VF, Wiegräbe I, Tauch A (2003) The complete *Corynebacterium glutamicum* ATCC 13032 genome sequence and its impact on the production of L-aspartate-derived amino acids and vitamins. J Biotechnol 104:5–25

Kang CM, Abbott DW, Park ST, Dascher CC, Cantley LC, Husson RN (2005) The *Mycobacterium tuberculosis* serine/threonine kinases PknA and PknB: substrate identification and regulation of cell shape. Genes Dev 19:1692–1704

Kang CM, Nyayapathy S, Lee JY, Suh JW, Husson RN (2008) Wag31, a homologue of the cell division protein DivIVA, regulates growth, morphology and polar cell wall synthesis in mycobacteria. Microbiology 154:725–735

Kinoshita S, Udaka S, Shimono M (1957) Studies on the amino acid fermentation. Part 1. Production of L-glutamic acid by various microorganisms. J Gen Appl Microbiol 3:193–205

Kubitschek HE, Freedman ML (1971) Chromosome replication and the division cycle of *Escherichia coli* Br. J Bacteriol 107:95–99

Laneelle MA, Tropis M, Daffe M (2013) Current knowledge on mycolic acids in *Corynebacterium glutamicum* and their relevance for biotechnological processes. Appl Microbiol Biotechnol 97:9923–9930

Le TB, Imakaev MV, Mirny LA, Laub MT (2013) High-resolution mapping of the spatial organization of a bacterial chromosome. Science 342:731–734

Leipe DD, Wolf YI, Koonin EV, Aravind L (2002) Classification and evolution of P-loop GTPases and related ATPases. J Mol Biol 317:41–72

Lenarcic R, Halbedel S, Visser L, Shaw M, Wu LJ, Errington J, Marenduzzo D, Hamoen LW (2009) Localisation of DivIVA by targeting to negatively curved membranes. EMBO J 28:2272–2282

Leonard TA, Butler PJ, Löwe J (2005) Bacterial chromosome segregation: structure and DNA binding of the Soj dimer – a conserved biological switch. EMBO J 24:270–282

Letek M, Ordóñez E, Fiuza M, Honrubia-Marcos P, Vaquera J, Gil JA, Mateos LM (2007) Characterization of the promoter region of *ftsZ* from *Corynebacterium glutamicum* and controlled overexpression of FtsZ. Int Microbiol 10:271–282

Letek M, Fiuza M, Ordonez E, Villadangos AF, Ramos A, Mateos LM, Gil JA (2008a) Cell growth and cell division in the rod-shaped actinomycete *Corynebacterium glutamicum*. A Van Leeuw J Microb 94:99–109

Letek M, Ordonez E, Vaquera J, Margolin W, Flardh K, Mateos LM, Gil JA (2008b) DivIVA is required for polar growth in the MreB-lacking rod-shaped actinomycete *Corynebacterium glutamicum*. J Bacteriol 190:3283–3292

Letek M, Fiuza M, Ordonez E, Villadangos AF, Flardh K, Mateos LM, Gil JA (2009) DivIVA uses an N-terminal conserved region and two coiled-coil domains to localize and sustain the polar growth in *Corynebacterium glutamicum*. FEMS Microbiol Lett 297:110–116

Lin L, Thanbichler M (2013) Nucleotide-independent cytoskeletal scaffolds in bacteria. Cytoskeleton 70:409–423

Lioy VS, Cournac A, Marbouty M, Duigou S, Mozziconacci J, Espéli O, Boccard F, Koszul R (2018) Multiscale structuring of the *E. coli* chromosome by nucleoid-associated and condensin proteins. Cell 172:771–783.e718

Livny J, Yamaichi Y, Waldor MK (2007) Distribution of centromere-like *parS* sites in bacteria: insights from comparative genomics. J Bacteriol 189:8693–8703

Maeda T, Tanaka Y, Takemoto N, Hamamoto N, Inui M (2016) RNase III mediated cleavage of the coding region of *mraZ* mRNA is required for efficient cell division in *Corynebacterium glutamicum*. Mol Microbiol 99:1149–1166

Marrakchi H, Laneelle MA, Daffe M (2014) Mycolic acids: structures, biosynthesis, and beyond. Chem Biol 21:67–85

Massey TH, Mercogliano CP, Yates J, Sherratt DJ, Löwe J (2006) Double-stranded DNA translocation: structure and mechanism of hexameric FtsK. Mol Cell 23:457–469

Meeske AJ, Riley EP, Robins WP, Uehara T, Mekalanos JJ, Kahne D, Walker S, Kruse AC, Bernhardt TG, Rudner DZ (2016) SEDS proteins are a widespread family of bacterial cell wall polymerases. Nature 537:634–638

Meisner J, Montero Llopis P, Sham LT, Garner E, Bernhardt TG, Rudner DZ (2013) FtsEX is required for CwlO peptidoglycan hydrolase activity during cell wall elongation in *Bacillus subtilis*. Mol Microbiol 89:1069–1083

Meniche X, Otten R, Siegrist MS, Baer CE, Murphy KC, Bertozzi CR, Sassetti CM (2014) Subpolar addition of new cell wall is directed by DivIVA in mycobacteria. Proc Natl Acad Sci USA 111:E3243–E3251

Minnen A, Bürmann F, Wilhelm L, Anchimiuk A, Diebold-Durand M-L, Gruber S (2016) Control of Smc coiled coil architecture by the ATPase heads facilitates targeting to chromosomal ParB/*parS* and release onto flanking DNA. Cell Rep 14:2003–2016

Mishra AK, Driessen NN, Appelmelk BJ, Besra GS (2011) Lipoarabinomannan and related glycoconjugates: structure, biogenesis and role in *Mycobacterium tuberculosis* physiology and host-pathogen interaction. FEMS Microbiol Rev 35:1126–1157

Möker N, Brocker M, Schaffer S, Krämer R, Morbach S, Bott M (2004) Deletion of the genes encoding the MtrA-MtrB two-component system of *Corynebacterium glutamicum* has a strong influence on cell morphology, antibiotics susceptibility and expression of genes involved in osmoprotection. Mol Microbiol 54:420–438

Möker N, Kramer J, Unden G, Krämer R, Morbach S (2007a) In vitro analysis of the two-component system MtrB-MtrA from *Corynebacterium glutamicum*. J Bacteriol 189:3645–3649

Möker N, Reihlen P, Krämer R, Morbach S (2007b) Osmosensing properties of the histidine protein kinase MtrB from *Corynebacterium glutamicum*. J Biol Chem 282:27666–27677

Monteiro JM, Fernandes PB, Vaz F, Pereira AR, Tavares AC, Ferreira MT, Pereira PM, Veiga H, Kuru E, VanNieuwenhze MS, Brun YV, Filipe SR, Pinho MG (2015) Cell shape dynamics during the staphylococcal cell cycle. Nat Commun 6:8055

Muchova K, Kutejova E, Scott DJ, Brannigan JA, Lewis RJ, Wilkinson AJ, Barak I (2002) Oligomerization of the *Bacillus subtilis* division protein DivIVA. Microbiology 148:807–813

Murray H, Ferreira H, Errington J (2006) The bacterial chromosome segregation protein Spo0J spreads along DNA from *parS* nucleation sites. Mol Microbiol 61:1352–1361

Nair N, Dziedzic R, Greendyke R, Muniruzzaman S, Rajagopalan M, Madiraju MV (2009) Synchronous replication initiation in novel *Mycobacterium tuberculosis dnaA* cold-sensitive mutants. Mol Microbiol 71:291–304

Nanda AM, Heyer A, Kramer C, Grunberger A, Kohlheyer D, Frunzke J (2014) Analysis of SOS-induced spontaneous prophage induction in *Corynebacterium glutamicum* at the single-cell level. J Bacteriol 196:180–188

Neumeyer A, Hübschmann T, Müller S, Frunzke J (2013) Monitoring of population dynamics of *Corynebacterium glutamicum* by multiparameter flow cytometry. Microb Biotechnol 6:157–167

Nguyen L, Scherr N, Gatfield J, Walburger A, Pieters J, Thompson CJ (2007) Antigen 84, an effector of pleiomorphism in *Mycobacterium smegmatis*. J Bacteriol 189:7896–7910

Noens EE, Mersinias V, Willemse J, Traag BA, Laing E, Chater KF, Smith CP, Koerten HK, Van Wezel GP (2007) Loss of the controlled localization of growth stage-specific cell-wall synthesis pleiotropically affects developmental gene expression in an *ssgA* mutant of *Streptomyces coelicolor*. Mol Microbiol 64:1244–1259

Panas MW, Jain P, Yang H, Mitra S, Biswas D, Wattam AR, Letvin NL, Jacobs WRJ (2014) Noncanonical SMC protein in *Mycobacterium smegmatis* restricts maintenance of *Mycobacterium fortuitum* plasmids. Proc Natl Acad Sci USA 111:13264–13271

Pinho MG, Errington J (2003) Dispersed mode of *Staphylococcus aureus* cell wall synthesis in the absence of the division machinery. Mol Microbiol 50:871–881

Ramijan K, Ultee E, Willemse J, Zhang Z, Wondergem JAJ, van der Meij A, Heinrich D, Briegel A, van Wezel GP, Claessen D (2018) Stress-induced formation of cell wall-deficient cells in filamentous actinomycetes. Nat Commun 9:5164

Ramos A, Letek M, Campelo AB, Vaquera J, Mateos LM, Gil JA (2005) Altered morphology produced by *ftsZ* expression in *Corynebacterium glutamicum* ATCC 13869. Microbiology 151:2563–2572

Rodionov O, Lobocka M, Yarmolinsky M (1999) Silencing of genes flanking the P1 plasmid centromere. Science 283:546–549

Ruiz N (2008) Bioinformatics identification of MurJ (MviN) as the peptidoglycan lipid II flippase in *Escherichia coli*. Proc Natl Acad Sci USA 105:15553–15557

Ruiz N (2015) Lipid Flippases for bacterial peptidoglycan biosynthesis. Lipid Insights 8:21–31

Sanchez A, Cattoni DI, Walter JC, Rech J, Parmeggiani A, Nollmann M, Bouet JY (2015) Stochastic self-assembly of ParB proteins builds the bacterial DNA segregation apparatus. Cell Syst 1:163–173

Santi I, McKinney JD (2015) Chromosome organization and replisome dynamics in *Mycobacterium smegmatis*. MBio 6:e01999–e01914

Santi I, Dhar N, Bousbaine D, Wakamoto Y, McKinney JD (2013) Single-cell dynamics of the chromosome replication and cell division cycles in mycobacteria. Nat Commun 4:2470

Scheffers DJ, Tol MB (2015) LipidII: just another brick in the wall? PLoS Pathog 11:e1005213

Schleifer KH, Kandler O (1972) Peptidoglycan types of bacterial cell walls and their taxonomic implications. Bacteriol Rev 36:407–477

Schlimpert S, Wasserstrom S, Chandra G, Bibb MJ, Findlay KC, Flärdh K, Buttner MJ (2017) Two dynamin-like proteins stabilize FtsZ rings during *Streptomyces* sporulation. Proc Natl Acad Sci USA 114:E6176–E6183

Schubert K, Sieger B, Meyer F, Giacomelli G, Böhm K, Rieblinger A, Lindenthal L, Sachs N, Wanner G, Bramkamp M (2017) The antituberculosis drug ethambutol selectively blocks apical growth in CMN group bacteria. MBio 8:e02213–e02216

Schultz C, Niebisch A, Schwaiger A, Viets U, Metzger S, Bramkamp M, Bott M (2009) Genetic and biochemical analysis of the serine/threonine protein kinases PknA, PknB, PknG and PknL of *Corynebacterium glutamicum*: evidence for non-essentiality and for phosphorylation of OdhI and FtsZ by multiple kinases. Mol Microbiol 74:724–741

Shah IM, Laaberki MH, Popham DL, Dworkin J (2008) A eukaryotic-like Ser/Thr kinase signals bacteria to exit dormancy in response to peptidoglycan fragments. Cell 135:486–496

Sham LT, Butler EK, Lebar MD, Kahne D, Bernhardt TG, Ruiz N (2014) Bacterial cell wall. MurJ is the flippase of lipid-linked precursors for peptidoglycan biogenesis. Science 345:220–222

Sharpe ME, Hauser PM, Sharpe RG, Errington J (1998) Bacillus subtilis cell cycle as studied by fluorescence microscopy: constancy of cell length at initiation of DNA replication and evidence for active nucleoid partitioning. J Bacteriol 180:547–555

Sieger B, Bramkamp M (2014) Interaction sites of DivIVA and RodA from *Corynebacterium glutamicum*. Front Microbiol 5:738

Sieger B, Schubert K, Donovan C, Bramkamp M (2013) The lipid II flippase RodA determines morphology and growth in *Corynebacterium glutamicum*. Mol Microbiol 90:966–982

Slayden RA, Knudson DL, Belisle JT (2006) Identification of cell cycle regulators in *Mycobacterium tuberculosis* by inhibition of septum formation and global transcriptional analysis. Microbiology 152:1789–1797

Soppa J (2014) Polyploidy in archaea and bacteria: about desiccation resistance, giant cell size, long-term survival, enforcement by a eukaryotic host and additional aspects. J Mol Microbiol Biotechnol 24:409–419

Taguchi A, Welsh MA, Marmont LS, Lee W, Sjodt M, Kruse AC, Kahne D, Bernhardt TG, Walker S (2019) FtsW is a peptidoglycan polymerase that is functional only in complex with its cognate penicillin-binding protein. Nat Microbiol 4:587

Thanbichler M, Shapiro L (2006) MipZ, a spatial regulator coordinating chromosome segregation with cell division in *Caulobacter*. Cell 126:147–162

Trojanowski D, Ginda K, Pióro M, Hołówka J, Skut P, Jakimowicz D, Zakrzewska-Czerwińska J (2015) Choreography of the *Mycobacterium* replication machinery during the cell cycle. MBio 6:e02125–e02114

Trojanowski D, Holowka J, Ginda K, Jakimowicz D, Zakrzewska-Czerwinska J (2017) Multifork chromosome replication in slow-growing bacteria. Sci Rep 7:43836

Trojanowski D, Hołówka J, Zakrzewska-Czerwińska J (2018) Where and when bacterial chromosome replication starts: a single cell perspective. Front Microbiol 9:2819–2819

Tsuge Y, Ogino H, Teramoto H, Inui M, Yukawa H (2008) Deletion of *cgR_1596* and *cgR_2070*, encoding NlpC/P60 proteins, causes a defect in cell separation in *Corynebacterium glutamicum* R. J Bacteriol 190:8204–8214

Typas A, Banzhaf M, Gross CA, Vollmer W (2011) From the regulation of peptidoglycan synthesis to bacterial growth and morphology. Nat Rev Microbiol 10:123–136

Umeda A, Amako K (1983) Growth of the surface of *Corynebacterium diphtheriae*. Microbiol Immunol 27:663–671

Valbuena N, Letek M, Ordóñez E, Ayala J, Daniel RA, Gil JA, Mateos LM (2007) Characterization of HMW-PBPs from the rod-shaped actinomycete *Corynebacterium glutamicum*: peptidoglycan synthesis in cells lacking actin-like cytoskeletal structures. Mol Microbiol 66:643–657

Vecchiarelli AG, Hwang LC, Mizuuchi K (2013) Cell-free study of F plasmid partition provides evidence for cargo transport by a diffusion-ratchet mechanism. Proc Natl Acad Sci USA 110: E1390–E1397

Wang JD, Levin PA (2009) Metabolism, cell growth and the bacterial cell cycle. Nat Rev Microbiol 7:822–827

Wang X, Le TBK, Lajoie BR, Dekker J, Laub MT, Rudner DZ (2015) Condensin promotes the juxtaposition of DNA flanking its loading site in *Bacillus subtilis*. Genes Dev 29:1661–1675

Wilhelm L, Bürmann F, Minnen A, Shin HC, Toseland CP, Oh BH, Gruber S (2015) SMC condensin entraps chromosomal DNA by an ATP hydrolysis dependent loading mechanism in *Bacillus subtilis*. eLife 4:e06659

Willis L, Huang KC (2017) Sizing up the bacterial cell cycle. Nat Rev Microbiol 15:606

Yang DC, Peters NT, Parzych KR, Uehara T, Markovski M, Bernhardt TG (2011) An ATP-binding cassette transporter-like complex governs cell-wall hydrolysis at the bacterial cytokinetic ring. Proc Natl Acad Sci USA 108:E1052–E1060

Yang X, Lyu Z, Miguel A, McQuillen R, Huang KC, Xiao J (2017) GTPase activity-coupled treadmilling of the bacterial tubulin FtsZ organizes septal cell wall synthesis. Science 355:744–747

Yeats C, Finn RD, Bateman A (2002) The PASTA domain: a beta-lactam-binding domain. Trends Biochem Sci 27:438

Zhang H, Schumacher MA (2017) Structures of partition protein ParA with nonspecific DNA and ParB effector reveal molecular insights into principles governing Walker-box DNA segregation. Genes Dev 31:481–492

Zhou X, Halladin DK, Theriot JA (2016) Fast mechanically driven daughter cell separation is widespread in actinobacteria. MBio 7:e00952–e00916

Zhou X, Rodriguez-Rivera FP, Lim HC, Bell JC, Bernhardt TG, Bertozzi CR, Theriot JA (2019) Sequential assembly of the septal cell envelope prior to V snapping in *Corynebacterium glutamicum*. Nat Chem Biol 15:221–231

Architecture and Biogenesis of the Cell Envelope of *Corynebacterium glutamicum*

Christine Houssin, Célia de Sousa d'Auria, Florence Constantinesco, Christiane Dietrich, Cécile Labarre, and Nicolas Bayan

Contents

Abstract *Corynebacteriales* cells are surrounded by a multilayered cell envelope with unique architecture and composition. *Corynebacterium glutamicum* has become an increasingly used model to understand both biosynthesis and assembly of this complex compartment of the cell. The specific core of this envelope consists of a huge mycoloyl-arabinogalactan-peptidoglycan (mAGP) complex that constitutes the building support for an outer hydrophobic barrier mainly composed of mycolates containing glycolipids. Besides this basic specific core, the envelope contains other polymers such as lipomannan (LM) and lipoarabinomannan (LAM) but their localization and their functional role is still elusive. Finally, the more

C. Houssin · C. de Sousa d'Auria · F. Constantinesco · C. Dietrich · C. Labarre · N. Bayan (✉)
Institute for Integrative Biology of the Cell (I2BC), CEA, CNRS, Université Paris-Saclay, Gif-sur-Yvette cedex, France
e-mail: nicolas.bayan@u-psud.fr

M. Inui, K. Toyoda (eds.), *Corynebacterium glutamicum*, Microbiology Monographs 23, https://doi.org/10.1007/978-3-030-39267-3_2

external layers encompass some strains of *C. glutamicum*: the outer layer and the S-layer which may have important roles in specific environments. The biosynthetic pathways involved in the synthesis of peptidoglycan, arabinogalactan and mycolates have been quite well characterized and are exhaustively described. In the frame of this review, we focused exclusively on what has been clearly demonstrated in *Corynebaterium glutamicum* and, when relevant, we compare to what is known in *Mycobacterium* species. The precise assembly of the whole envelope as a function of time and space is still to be discovered and represent an exciting challenge for the next decade.

1 Introduction: The General Features of the Bacterial Cell Envelope of *C. glutamicum*

The bacterial cell envelope represents all the structures that enclose the cytoplasm, from the plasma membrane to the most peripheral surface layers (Silhavy et al. 2010). These structures are chemically complex and essential for interactions and adaptation of bacteria to their specific environment. It is a cell compartment as its own, containing up to 30% of total bacterial proteins, many saccharidic polymers and most cellular lipids. Its biogenesis represents a major challenge with considerable energy costs. The study of bacterial envelopes has implications in different fields of biology, for the study of fundamental mechanisms of life (translocation of proteins through membranes, cell division, morphology), the study of interactions between bacteria and their hosts (notably interaction with cells of the immune system) and finally in several domains of biotechnology (research of new targets for antibiotics, production and secretion of metabolites of interest . . .).

Lots of bacteria are non-cultivable and therefore biochemical analysis of their envelope is impossible. Despite this, the experimental data coming from studies of a limited number of bacterial genera highlight the considerable diversity and molecular complexity of bacterial cell envelopes. These findings were greatly improved the last decade thanks to the use of efficient analytical techniques. This includes the routine use of (i) mass-spectrometry coupled with HPLC, (ii) up to date microscopy techniques (cryo-electron microscopy techniques and super-resolution microscopy) or (iii) the systematic use of bio-orthogonal chemistry approaches. Although extremely diverse, the overall architecture of bacterial envelopes is surprisingly well preserved. The minimal configuration consists of a phospholipidic cytoplasmic membrane surrounded by a large peptidoglycan exoskeleton, as found for most genera in *Firmicutes* and *Actinobacteria* phyla, which therefore have a monoderm organization. In other phyla (mostly *Cyanobacteria* and *Proteobacteria*), this peptidoglycan polymer is often thinner but covalently associated with a second lipid barrier (the outer membrane) of very specific lipid and protein composition (porins and lipo-polysaccharide (LPS)). This diderm-LPS organization seems to be the most largely represented in the bacterial kingdom as suggested by the study of Sutcliffe in his survey of genes encoding key specific outer membrane biosynthetic enzymes in all metagenomic sequences available (Sutcliffe 2010). In most bacteria, several additional structures can also be superimposed on these basic elements that form the core of the bacterial shell such as secondary polymers, capsules, S-layer and other various optional elements.

C. glutamicum is a non-pathogenic species, widely used for biotechnology applications. It belongs to *Corynebacteriales* order of the *Actinobacteria* phylum (Goodfellow and Jones 2015), an order of paramount importance in human health since it includes numerous pathogens, such as the agents of tuberculosis, leprosy, diphtheria, as well as emergent pathogens causing opportunistic diseases (*Mycobacterium abscessus, Corynebacterium jeikeium* ...). Although sharing very peculiar cell wall properties such as the presence of specific cell wall complex polysaccharides and a large variety of non-conventional lipids, all members of this bacterial order have long been considered to have, like all *Actinobacteria*, a monoderm cell wall architecture. With the improvement of biochemical and microscopy analysis, it has become progressively clear that they rather had an atypical diderm organization: a phospholipidic cytoplasmic membrane surrounded by a complex unusual exoskeleton, itself covalently linked to a mycolic acid-containing outer membrane (the so-called mycomembrane) (Hoffmann et al. 2008; Zuber et al. 2008). These mycolic acids, the hallmark of *Corynebacteriales*, are unique long chain (C_{30}–C_{90}) α-branched β-hydroxylated fatty acids esterifying trehalose or arabinose moieties of a polymer of arabinose and galactose (the arabinogalactan or AG) (Lanéelle et al. 2013; Marrakchi et al. 2014). Most of our knowledge on this very intriguing architecture comes from numerous studies performed on a limited set of species, namely *M. tuberculosis*, *Mycobacterium smegmatis* and *C. glutamicum* (Puech et al. 2001; Bayan et al. 2003; Dover et al. 2004; Daffe 2005; Burkovski 2013; Angala et al. 2014; Alderwick et al. 2015), the latter having proven to be an increasingly important model to depict the intimate organization of the *Corynebacteriales* cell envelope. An updated speculative model organization of *C. glutamicum* cell envelope is shown in Fig. 1. The specific core of this envelope consists of a thick peptidoglycan network on which a rather limited number of mycolate-esterified arabinogalactan chains are associated by a covalent rhamnose-glucosamine linker. In turn, these lipo-polysaccharidic chains provide basic platforms for the insertion of trehalose mycolates (trehalose mono- or di-mycolate) together with conventional phospholipids to assemble into the mycomembrane. Other complex glycolipids are found in mycobacterial outer membrane but are absent in *C. glutamicum* (Layre et al. 2014; Jackson 2014; Klatt et al. 2018). Secondary polymers such as lipoglycans are essential in this structure although their exact function is unknown. Most of them are anchored in the cytoplasmic membrane and may stabilize the overall structure as proposed for teichoic acids in monoderm architectures (Swoboda et al. 2010). Although proteomic analysis of different fractions of the cell envelope has identified a wide range of proteins present in this compartment (Hansmeier et al. 2006b), only a few of them, namely porins, were functionally characterized in vivo or in vitro by their ability to form hydrophilic pores in lipid bilayers (Costa-Riu et al. 2003a, b;

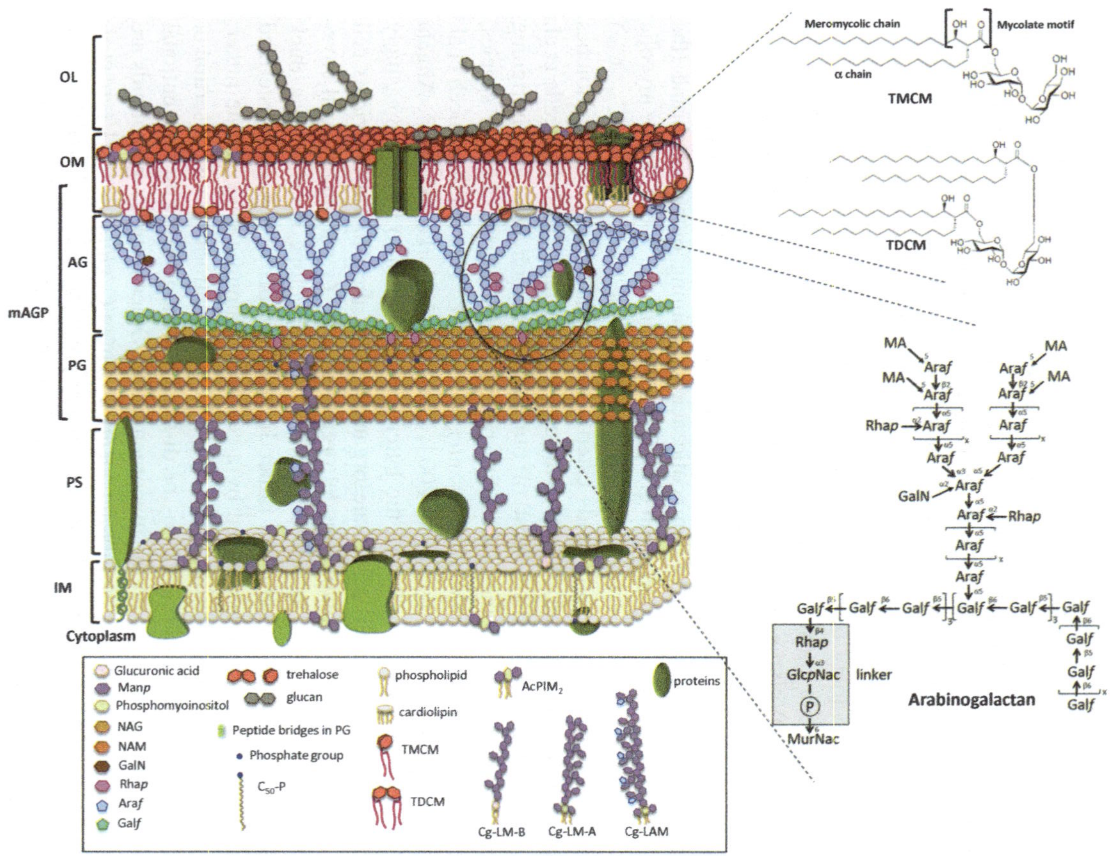

Fig. 1 A model of the cell envelope of *C. glutamicum* depicting the organization of its key components. From the cytoplasm to the external medium this envelope contains: (i) an inner membrane (IM), composed of proteins and phospholipids (mainly phosphatidyl inositol, phosphatidyl glycerol and cardiolipin).

Fig. 1 (continued) Also shown in this bilayer some molecules of decaprenyl phosphate (C_{50}-P), the lipid carrier of different precursors of the synthesis of envelope components. (ii) A periplasmic space (PS) which appears in CEMOVIS (Cryo-Electron Microscopy Of Vitreous Sections) as a zone of low density containing a granular layer next to the IM but which is not represented in this model because the molecular nature of this layer is not clear (Zuber et al. 2008). (iii) A cell wall skeleton: the mycoloyl-arabinogalactan-peptidoglycan (mAGP) made up of a peptidoglycan network (PG) bound to arabinogalactan (AG) via a covalent phosphoryl-glucosamine rhamnose linker, with some of the AG terminations esterified by mycolic acids. Detailed structure of mAGP is shown in the zoom at the bottom right of the figure. The precise number of Gal*f*, Ara*f* and Rha*p* residues is not known but according to Birch et al. the relative molar ratio of Rha/Ara/Gal is 21:71:31 (Birch et al. 2009). (iv) An outer membrane (OM also named mycomembrane) constituted of AG bound mycolates forming the basis of the inner leaflet and trehalose monocorynomycolate (TMCM) and dicorynomycolates (TDCM) forming the outer leaflet (detailed structures of TMCM and TDCM are shown in the zoom at the top right of the figure). Also shown in the inner leaflet some molecules of cardiolipin as suggested by Bansal-Mutalik and Nikaido (2011), TMCM and TDCM. Some porins have been shown to localize in this OM and are drawn as oligomers. (v) An outer layer (OL) mainly composed of glucan. (vi) Different lipoglycans: mono-acylated phosphatidylinositol di-mannosides (AcPIM$_2$), Cg-lipomannan A (Cg-LM-A), Cg-lipomannan B (Cg-LM-B) and Cg-lipoarabinomannan (Cg-LAM). As discussed in the text, lipoglycans are found essentially associated with the inner membrane. The surface layer (which is only present in a limited number of species) is not represented in this picture

Hünten et al. 2005). More recently, thanks to improved cell fractionation methods, more than 40 different proteins (including predicted lipoproteins) have been shown to specifically localize in the mycomembrane of *C. glutamicum* (Marchand et al. 2012). Their function remains to be determined. *Corynebacteriales* envelope, because of its specific organization and the presence of long chain mycolic acids, is predicted to be highly impermeable to small hydrophilic solutes and antimicrobial compounds. This has been clearly demonstrated in *M. smegmatis* (Mailaender et al. 2004), but is not necessarily true for all members of *Corynebacteriales*. In that respect, it may be important to note that engineered *C. glutamicum* cells are highly effective in secreting a wide range of metabolites and proteins pointing that its envelope may have evolved with its own specificities and could be much more permeable than the envelope of other *Corynebacteriales* genera (Lee et al. 2016).

In this chapter we review current knowledge on the composition, architecture and biosynthesis of *C. glutamicum* envelope components and their assembly. The names of the proteins and/or the genes (or their locus tags) involved in these processes are systematically given either in the text or in the figures and refer to those of the reference strain ATCC13032 (Kalinowski et al. 2003) (RefSeq: NC_006958).

2 The PG-AG Complex: A Specific Rigid Cocoon

2.1 Peptidoglycan

The peptidoglycan is an essential polymer involved in cell morphology and cell protection against osmotic stress. In *C. glutamicum* its general structure and biosynthesis is quite similar to that of other bacteria since most biosynthetic genes are conserved. Still, several peculiar specificities were identified and are reported below. They are important to understand the general architecture of the cell envelope but also its susceptibility to different anti-bacterial agents.

2.1.1 Peptidoglycan Structure and Properties

C. glutamicum peptidoglycan (PG) belongs to the A1γ-type, in which *m*-DAP and D-Ala (at position 4) from adjacent stems are directly cross-linked (Schleifer and Kandler 1972). As conserved in all PG, glycan strands are composed of alternating β-(1→4)-linked N-acetylglucosamine (GlcNAc) and N-acetylmuramic acid (MurNAc) units and are cross-linked via short MurNAc-linked peptides that are initially synthesized as L-Ala-γ-D-*iso*Glu-*m*-DAP-D-Ala-D-Ala (DAP, diaminopimelic acid). Measurement of the overall peptide cross-linking for the PG of the ATCC13032 strain in exponential growth phase, gave a value of 32.5%, very close to that determined in parallel for the PG of *E. coli* (29%) (Levefaudes et al. 2015). The precise structure of the *C. glutamicum* PG in terms of inter-peptide linkage is not available. However, results obtained on *C. jeikeium* PG, showed that, in addition to the regular 4→3 linkages, 38%

of the cross-links are of the non-classical 3→3 type (*m*-DAP→*m*-DAP crosslink) (Lavollay et al. 2009), suggesting that *C. glutamicum* PG could also contain a significant part of 3→3 crosslinks.

One major modification of the *C. glutamicum* PG, which is a unique feature of *Corynebacteriales*, is the covalent attachment to some muramyl units of a large heteropolysaccharide, the arabinogalactan. Structure of AG and its linker will be discussed later in the text. Two other modifications were identified in the stem peptides of the *C. glutamicum* PG: amidation of the α-carboxyl group of D-Glu and of the ε-carboxyl group of *m*-DAP (100% and 80% in ATCC13032 strain respectively) (Levefaudes et al. 2015). A total absence of m-DAP amidation leads to a PG that exhibits a huge decrease in lysozyme resistance both in vivo and in vitro, providing evidence that DAP amidation is an essential mechanism for *C. glutamicum* adaptation to lytic enzymes (Hirasawa et al. 2000; Levefaudes et al. 2015). It should be noted that the use of a mutant strain unable to amidate DAP (Δ*ltsA*, see below) can provide a particularly interesting tool in protein or DNA purification processes because the hydrolysis of the very resistant *C. glutamicum* cell wall becomes much easier to achieve in this mutant (Hirasawa et al. 2000; Levefaudes et al. 2015). It has also been shown that mutations or disruption of the *ltsA* gene in the *C. glutamicum* KY9611 and ATCC13032 strains lead to overproduction of L-glutamate specially at 37 °C (Hirasawa et al. 2000, 2001). Another property conferred by DAP amidation is a significant increase in the resistance of the bacteria towards β-lactam antibiotics (Levefaudes et al. 2015).

2.1.2 Peptidoglycan Biosynthesis

As compared to mycobacteria, only very few studies have been carried out on corynebacterial PG biosynthesis. Nevertheless, the biosynthetic pathway of PG is well conserved along evolution and in fact *C. glutamicum* possesses most of the orthologous genes known to code PG biosynthetic enzymes in *E. coli, B. subtilis or M. tuberculosis* (Barreteau et al. 2008; Vollmer and Bertsche 2008; Pavelka et al. 2014; Raghavendra et al. 2018). Classically, PG synthesis is divided in three stages: (i) synthesis of a nucleotide sugar-pentapeptide precursor in the cytoplasm; (ii) transfer of this precursor to a polyprenyl-phosphate (Pol-P) on the inner face of the cytoplasmic membrane to form the membrane-bound precursors Lipid I and then Lipid II; (iii) after translocation of the lipid II to the periplasmic side of the membrane, incorporation of the PG-precursor into the expanding cell-wall by transglycosylation and transpeptidation reactions. A schematic view of PG biosynthesis in *C. glutamicum* is shown in Fig. 2.

Synthesis of the nucleotide sugar-pentapeptide precursor (UDP-MurNAc/L-Ala-γ-D-*iso*Glu-*m*-DAP-D-Ala-D-Ala also named the Park's nucleotide) begins with the formation of UDP-GlcNAc from fructose-6-phosphate (F6-P) which is subsequently converted to glucosamine-6-phosphate then, glucosamine-1-phosphate and finally uridine-diphospho-N-acetylglucosamine (UDP-GlcNAc) by GlmS (*cg2492*), GlmM (*cg0675*) and GlmU (*cg1076*) respectively. UDP-GlcNAc is then

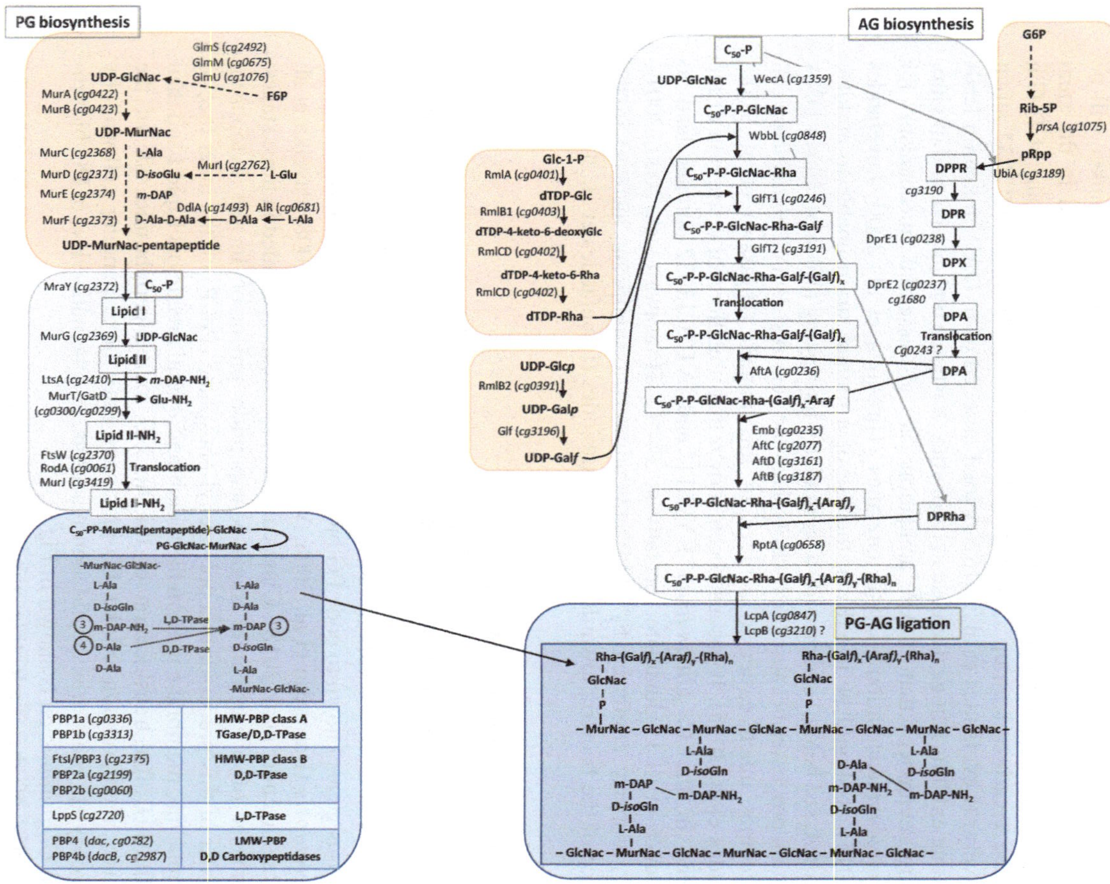

Fig. 2 Biosynthesis pathways of peptidoglycan, arabinogalactan and their assembly in the cell envelope of *C. glutamicum*. See text for all the details. The colors refer to the cellular compartment in which the reactions take place: cytoplasm (orange), IM (grey) and cell wall (blue). In the PG biosynthesis, amidation steps (by LtsA and MurT/GatD) are indicated to occur at the level of the lipid II molecule but it is not excluded that it may take place at the level of lipid I as specified in the text

converted to UDP-MurNAc by MurA and B, on which the different aminoacids residues of the pentapeptide are added sequentially through a series of ATP-dependent reactions (by MurC for L-Ala, MurD for D-*iso*Glu, MurE for *m*-DAP and MurF for D-Ala-D-Ala). This step requires the previous conversion of L-Ala and L-Glu into their corresponding isomer D-Ala and D-*iso*Glu (by Alr and MurI respectively) as well as the ligation of two D-Ala (by DdlA) and the synthesis of *m*-DAP which, in *C. glutamicum*, can result from tetrahydrodipicolinate either by the succinylase or by the dehydrogenase pathways (Wehrmann et al. 1998).

The MurNAc-pentapeptide is then transferred onto a polyprenyl-phosphate (Pol-P) from the Park's nucleotide on the inner side of the cytoplasmic membrane by MraY to form the lipid precursor Lipid I. In *Corynebacteriales*, Pol-P is a decaprenyl-phosphate (C_{50}-P) which is obtained after dephosphorylation of a decaprenyl-diphosphate (C_{50}-PP) by an unknown phosphatase. C_{50}-PP is produced by UppS2 (a Z-decaprenyl-diphosphate synthase encoding by the *cg2508* gene) from isopentenyl-diphosphate precursors which are synthetized by the non-mevalonate pathway (Kaur et al. 2004; Heider et al. 2014; Grover et al. 2014). The subsequent attachment of a GlcNAc (from UDP-GlcNAc) to the MurNAc carbohydrate of Lipid I through a β-(1→4) bond is catalyzed by MurG to form the peptidoglycan precursor Lipid II.

Modifications of the stem peptides are most likely to occur at the level of these membrane precursors in the cytoplasmic compartment. In *C. glutamicum*, a glutamine amidotransferase (LtsA) was shown to catalyze the transfer of an amine group from glutamine to the ε-carboxyl group of *m*-DAP in Lipid I or Lipid II (Levefaudes et al. 2015). In *Staphylococcus aureus* (Münch et al. 2012) and *Streptococcus pneumoniae* (Zapun et al. 2013), glutamate amidation of Lipid II is catalyzed by a two-protein complex composed of a glutamine amidotransferase-like protein (GatD) and a Muramyl (Mur) ligase homologue (MurT). The *murT* and *gatD* genes are conserved in many gram positives including *C. glutamicum* (Figueiredo et al. 2012), suggesting that the complex is also responsible for the D-Glu amidation in this species.

The mechanism responsible for the flipping of Lipid II across the cytoplasmic membrane is still a matter of debate (Ruiz 2015). In *C. glutamicum*, two proteins that belong to the SEDS (Shape, Elongation, Division and Sporulation) protein family, FtsW and RodA, have been reported to play a role in growth and division and proposed to translocate Lipid II (Sieger et al. 2013; Sieger and Bramkamp 2015). FtsW was shown to be essential, located at the septum and assumed to be responsible for the transport of lipid II during the cytokinesis, while RodA was shown to localize exclusively to the cell poles where it interacts with DivIVA. Although not essential, it has been proposed that RodA should ensure the transport of lipid precursors during apical growth of *C. glutamicum* (Sieger et al. 2013). Recent evidences obtained in *E. coli* and *B. subtilis*, suggest that SEDS family members could not act as flippases but rather have a PG glycosyltransferase activity and that the translocation of lipid II would be ensured by proteins of another family (MurJ-like proteins) (Sham et al. 2014; Meeske et al. 2015, 2016). A search for a MurJ orthologue in *C. glutamicum* ATCC13032 retrieves one gene: *cg3419*; the corresponding protein has not been studied so far.

Incorporation of PG precursors from lipid II to pre-existing cell wall is catalyzed by membrane bound enzymes called Penicillin-Binding Proteins (PBPs) that elongate the glycan strands and cross link the peptide stems together (4→3 linkage). Analysis of the genome sequence of *C. glutamicum* indicates nine putative proteins with a domain characteristic of PBPs, five of which being High Molecular Weight (HMW)-PBPs and four Low Molecular Weight (LMW)-PBPs (Valbuena et al. 2007). Of the five HMW-PBPs, two are class A PBPs, i.e. possessing both transglycosidase (TGase) and D,D-transpeptidase (TPase) domains (PBP1a (*cg0336*) and PBP1b (*cg3313*)) while three are class B PBPs, i.e. possessing only a TPase domain (FtsI/PBP3 (*cg2375*), PBP2a (*cg2199*) and PBP2b (*cg0060*)). FtsI, as in other bacteria is required for cell division and is the only essential PBP (Valbuena et al. 2006). In their study of the characterization of HMW-PBPs of *C. glutamicum*, Valbuena et al. (2007) showed that except for a PBP2b mutant which presented division and morphological defects, loss of individual PBP1a, PBP1b and PBP2a did not lead to any particular phenotype. They proposed that bifunctional PBP1a and PBP1b would be involved in polar elongation while monofunctional FtsI, PBP2a and PBP2b would rather be involved in cell division. None of the four putative LMW-PBPs (PBP4, PBP4b, PBP5 and PBP6) have been characterized so far, but PBP4 and PBP4b are probably D,D carboxypeptidases while PBP5 (*cg2649*) and PBP6 (*cg2478*) are most likely β-lactamases (Valbuena et al. 2007). In addition to these PBPs, the *C. glutamicum* genome encodes two L,D transpeptidases (Cg0650 and LppS, (3→3 linkage)). Although enzymatic activity of these proteins has not been investigated in *C. glutamicum*, it has been shown that *lppS* deletion confers a strong lysozyme sensitivity to the bacterial cell, while its overexpression enhances lysozyme resistance (Toyoda and Inui 2018). A study of the activity of recombinant *C. jeikeium* orthologues (Ldt_{cjk1} and Ldt_{cjk2}) showed that only the LppS orthologue (Ldt_{cjk1}) has a PG cross-linking activity in vitro and, unlike PBPs, is not inhibited by ampicillin (Lavollay et al. 2009). In the same study Lavollay et al. identified an ampicillin-sensitive D,D carboxypeptidase ($Pbp4_{Cjk}$) that catalyzes the hydrolysis of the C-terminal D-Ala which is a prerequisite for Ldt_{cjk1} cross-linking reaction to occur. The corresponding *C. glutamicum* orthologue is PBP4b.

2.1.3 Peptidoglycan Hydrolases

PG degrading enzymes are mainly composed of glycosidases that cleave the glycan backbone and peptidases that cleave the peptide side-chain (Frirdich and Gaynor 2013). It has been reported that *C. glutamicum* produces a β-N-acetylglucosaminidase encoded by the *nagA2* gene (Matano et al. 2016). Although the NagA2 lipoprotein is the orthologue of the *E. coli* enzyme NagZ that acts in PG recycling by cleaving the GlcNAc-(1→4)-MurNAc bonds, the exact role of NagA2 in vivo is not known. Based on the analysis of the repertoire of PG hydrolases in mycobacteria (Machowski et al. 2014), search for other putative PG glycosidase encoding genes in the *C. glutamicum* genome, retrieves three genes possessing a lytic transglycosylase domain (that are predicted to cleave the

MurNAc-(1→4)-GlcNAc bond): *rpf1* (*cg0936*), *rpf2* (*cg1037*) and *cg1110*. *rpf1* and *rpf2* are not essential neither individually nor collectively and encode proteins that belong to the Resuscitation-Promoting Factor family (Hartmann et al. 2004; Sexton et al. 2015). The absence of clear phenotype associated with mutants in which these genes have been deleted did not allow a specific function to be assigned to the Rpf1 and Rpf2 proteins (Hartmann et al. 2004). Concerning peptidases enzymes: in addition to the putative D,D carboxypeptidases PBP4b and PBP4 (removal of D-Ala at position 5 of the stem peptide), four genes encoding putative NlpC/P60-type PG endopeptidases (cleavage within the stem peptide) are also found in the *C. glutamicum* genome (*cg0784*, *cg1735*, *cg2401* and *cg2402*). Of these four genes, two have been inactivated in the R strain: *cgR_1596* and *cgR_2070* (the *cg1735* and *cg2402* orthologues respectively). Single disruption of *cgR_1596*, but not that of *cgR_2070* caused cell elongation with multiple septa and morphological changes. These defects were more pronounced in a double *cgR_1596-cgR_2070* disruptant, suggesting that both enzymes are involved in cell division and cell elongation (Tsuge et al. 2008). Finally, one gene encoding a putative N-acetylmuramoyl-L-alanine amidase activity is also present in *C. glutamicum* (*cg3424*). Cg3424 is the orthologue of *M. tuberculosis* CwlM protein, the PG hydrolyzing activity of which has been demonstrated (Deng et al. 2005). However, recent results showed that the function of CwlM is much more complex and that the protein is rather an essential regulator of PG synthesis in *M. tuberculosis* (Boutte et al. 2016; Turapov et al. 2018). The role of Cg3424 has not been investigated in *C. glutamicum*.

2.2 *Arabinogalactan*

Arabinogalactan chains represent the main specific modification of PG in *Corynebacteriales* and disclosure of their biosynthetic pathway is an attractive approach to find new targets for antibiotics. Because *C. glutamicum* is a very accommodating model of *Corynebacteriales*, a lot of studies were performed on this bacterium and led to a quite exhaustive knowledge of AG biosynthesis which is described below.

2.2.1 Arabinogalactan and Linker Structures

C. glutamicum AG is mainly composed, as in all *Corynebacteriales*, of galactose and arabinose residues in their furanoid ring form (β-D-Gal*f* and α-D-Ara*f*), organized in two separated but linked domains (Daffe et al. 1990, 1993; Daffe 2005). *C. glutamicum* AG also contains a large amount of rhamnose (Rha*p*) (Alderwick et al. 2005; Birch et al. 2009). No complete structure of the *C. glutamicum* AG has been determined to date. However, detailed glycosyl-linkage composition analysis of arabinogalactans from *C. glutamicum* indicated the presence of t-Gal*f*, 5-Gal*f*, 6-Gal*f*, 5,6-Gal*f*, t-Ara*f*, 2-Ara*f*, 5-Ara*f* and 3,5-Ara*f* as in *M. tuberculosis* but also of

2,5-Ara*f* and t-Rha*p* that are only found in *C. glutamicum* (Alderwick et al. 2005, 2018). The galactan domain is a linear chain made of alternating β(1→5) and β(1→6) linked Gal*f* residues, which is probably shorter in *C. glutamicum* than the 20–40 sugar units found in the galactan domain of mycobacterial AG (Wesener et al. 2017). Galactan polymers are anchored to the PG through a α-L-Rha*p*-(1→3)-α-D-GlcNAc-1-phosphate linkage unit. This linker is connected on one side to the first Gal*f* of the galactan chain by a β(1→4) bond with the Rha and on the other side, the GlcNAc is esterified to position 6 of some of the Mur residues of the PG (McNeil et al. 1990). Three arabinan domains are linked to each galactan chain at the 8th, 10th and 12th Gal*f* residues. They are made of linear α(1→5) linked Ara*f* residues with branching introduced by 3,5-Ara*f* residues (Alderwick et al. 2005). The number of Ara*f* residues constituting an arabinan domain is not known in *C. glutamicum* but it was determined to be 23 in *M. tuberculosis* (Besra et al. 1995). One particular feature of *C. glutamicum* AG is the presence of a large quantity of t-Rha*p* (approximately 7 t-Rha*p* for 23 Ara*f*) which are attached to the C2 position of α-(1→5)-linked Ara*f* forming branched 2,5-linked Ara*f* motifs (Birch et al. 2009). The presence of galactosamine (GalN) residues in the AG of *C. glutamicum* has also been reported but their exact position within the polymer remains unknown (Marchand et al. 2012). It should be noted that in AG of slow-growing mycobacteria a single GalN residue was identified on the C2 position of 3,5-Ara*f* residues (Lee et al. 2006; Peng et al. 2012). Based on ^{13}C-NMR analyses of purified AG, the structure of the non-reducing termini of *C. glutamicum* AG was proposed to be a tri-arabinofuranosyl instead of the classical penta-arabinosyl termini of mycobacterial AG (Daffe 2005). Importantly, the terminal β(1→2)-Ara*f* was shown to be esterified by mycolic acids, although it is not the only attachment site since in its absence a significant amount of corynomycolic acids are still found covalently linked to AG, most likely to the 5-OH of the 2-Ara*f* (Seidel et al. 2007). In mycobacteria, approximately two-thirds of the pentasaccharide terminal units are esterified with mycolic acids (McNeil et al. 1991). Given the amount of bound mycolates in *C. glutamicum*, which represent only a small percentage of dried corynebacterial mass (about 1%) compared to mycobacteria (about 10%), it is reasonable to think that only a fraction of the AG termini are esterified by these lipids in this species (Puech et al. 2001; Meniche et al. 2008).

2.2.2 Biosynthesis of Arabinogalactan

The biosynthesis of AG has been the subject of intense research over the past decade and *C. glutamicum* has proven to be an excellent model for studying this biosynthetic pathway. Indeed, while AG is essential for mycobacteria, *C. glutamicum* can tolerate large modifications of its AG as important as the absence of the arabinan domain or the loss of some links between AG and PG (Alderwick et al. 2005; Baumgart et al. 2016; Grzegorzewicz et al. 2016; Jankute et al. 2018). A scheme of the biosynthesis of AG and its precursors is given in Fig. 2.

The first step of AG biosynthesis is the synthesis of the linker unit which is initiated on the lipid carrier C_{50}-P by the transfer of GlcNac from UDP-GlcNac to form C_{50}-P-P-GlcNac. The enzyme catalyzing this reaction was shown to be GlcNAc-1-phosphate transferase WecA in *M. tuberculosis* (Ishizaki et al. 2013), the orthologue of which is Cg1359 in *C. glutamicum*. A Rhamnose unit is then attached to C_{50}-P-P-GlcNac from dTDP-Rha by the α-3-L-rhamnosyltransferase WbbL (Mikusová et al. 1996; Mills et al. 2004). dTDP-Rha is synthetized from Glc1-P by a series of reactions catalyzed by RmlA, RmlB1 and RmlCD in mycobacteria (Ma et al. 1997, 2001), the orthologous genes of which form an operon in *C. glutamicum*.

The galactan domain is polymerized on the linker unit by two galactofuranosyltransferases GlfT1 and GlfT2, using UDP-Gal*f* as precursor. GlfT1 is a bifunctional β-(1→4) and β-(1→5) galactofuranosyl transferase that initiates galactan synthesis by adding one or two Gal*f* on the C_{50}-P-P-GlcNac-Rha (Mikusova et al. 2006; Belanova et al. 2008; Wesener et al. 2017). GlfT2 is also a bifunctional enzyme that sequentially adds β-(1→5) and β-(1→6) Gal*f*, generating the galactan polymer and probably controlling its length (Kremer et al. 2001; Belanova et al. 2008; May et al. 2009; Wesener et al. 2017). Unlike the arabinan domain of AG, which is dispensable for *C. glutamicum* viability, the galactan domain is not. Indeed, inhibition of Glf (the enzyme UDP-galactopyranose mutase, see Fig. 2) prevents the cell growth (Wesener et al. 2017). Because Gal*f* additions take place in the cytosolic compartment, the lipid-linked galactan chain must be translocated to the periplasmic side of the plasma membrane where subsequent arabinosylation steps take place. The protein responsible for this transport is not known.

The synthesis of the arabinan domain results from the sequential addition of Ara*f* residues to the galactan domain by specialized arabinosyltransferases that act on the periplasmic face of the inner membrane and therefore use the decaprenylphosphoryl-D-arabinose (DPA) as a substrate (Alderwick et al. 2005). DPA is synthesized from phosphoribose-diphosphate (pRpp) through an unusual series of three successive reactions on the inner face of the plasma membrane (Mikusová et al. 2005; Meniche et al. 2008). The first one is the transfer of ribose 5-P from pRpp to C_{50}-P to form decaprenylphosphoryl-5-phosphoribose (DPPR) catalyzed by UbiA. While the protein is essential in *M. tuberculosis*, the *ubiA* gene can be inactivated in *C. glutamicum* resulting in a slow growing but viable mutant totally devoid of arabinan (Alderwick et al. 2005). DPPR is then dephosphorylated to decaprenylphosphoryl-ribose (DPR) by a phosphatase, probably Cg3190, the mycobacterial Rv3807c orthologue (Cai et al. 2014). The third step is epimerization of DPR to DPA which in *C. glutamicum* involves three enzymes that catalyze two distinct steps, each being essential for the viability of the cell (Meniche et al. 2008). Firstly, DPR is oxidized to decaprenylphosphoryl-2-keto-β-D-erythro-pentofuranose by Cg0238 (DprE1 in mycobacteria) which is subsequently reduced to DPA by Cg0237 (DprE2 in mycobacteria) or a protein with redundant function, Cg1680. DPA is then translocated from the cytoplasmic to the periplasmic side of the inner membrane by a flippase whose identity is not clearly established. Rv3789 was first proposed to fulfill this role in mycobacteria, but this hypothesis was recently challenged by another study suggesting that Rv3789 would rather be involved in the recruitment of the priming arabinosyltransferase AftA (Larrouy-Maumus et al. 2012; Kolly et al. 2015). In *C. glutamicum*, the role of Cg0243 (the orthologue of Rv3789) has not yet been investigated. A set of five arabinofuranosyltransferases (AraTs) is needed to synthesize the entire arabinan domain of AG in *C. glutamicum*. These AraTs are large membrane proteins belonging to the GT-C family of glycosyltransferases that use polyprenylphospho-sugars as substrates (Berg et al. 2007; Alderwick et al. 2015). None of them are essential. The priming AraT that link the first Ara*f* with the C5-OH of a β-(1→6) Gal*f* in the galactan domain is Cg-AftA (Alderwick et al. 2006b). This single Ara*f* is further elongated in a processive manner by Cg-Emb resulting in a α-(1→5) arabinan chain (Alderwick et al. 2005). α-(1→3) branching are introduced in this linear chain by Cg-AftC (Birch et al. 2008). The α-(1→3) primed bifurcation is then elongated by AftD, another α-(1→5) AraT (Alderwick et al. 2018). Finally, non-reducing ends of AG are formed by the linkage of β-(1→2) Ara*f* onto α-(1→5) Ara*f* (or α-(1→3) Ara*f* in mycobacteria) by the capping enzyme Cg-AftB (Seidel et al. 2007; Bou Raad et al. 2010). Transfer of Rha*p* onto the AG is catalyzed by another non-essential glycosyltransferase of the GT-C family: the rhamnopyranosyltransferase A (RptA), which uses decaprenyl-monophosphorylrhamnose as a substrate (Birch et al. 2009). A recent study using bacterial two-hybrid system identified an important network of interactions between most of the proteins involved in AG assembly in *C. glutamicum* (Jankute et al. 2014) probably enhancing the overall efficiency of biosynthesis of this polymer.

2.2.3 AG Attachment to PG

The ligation of AG to PG involves the trans-phosphorylation of the GlcNac-1P of the linker unit onto the C-6 hydroxyl group of a MurNac in the PG. One membrane protein, belonging to the LCP (LytR, CpsA, Psr) family, named LcpA, was recently identified to carry out this function in *C. glutamicum* (Baumgart et al. 2016; Grzegorzewicz et al. 2016). LcpA (*cg0847*) is essential and its depletion leads to bacteria with many growth and morphological defects, part of them displaying a disrupted mycomembrane with large fragments detaching from the cell surface (Grzegorzewicz et al. 2016). *C. glutamicum* possesses a second non-essential LCP protein (LcpB, *cg3210*), the inactivation of which, in contrast to LcpA, does not produce any particular phenotype. Nevertheless LcpB like LcpA has been shown to localize at regions of nascent cell wall synthesis (Baumgart et al. 2016). The biosynthetic stage at which this ligation occurs is not known. However, previous experiments on *C. glutamicum* mutants deficient in AG synthesis seems to indicate that neither the arabinosylation of AG nor its mycoloylation are prerequisites for its attachment to PG (Alderwick et al. 2005, 2006a).

3 Mycolic Acids and Mycomembrane: Original Glycolipids for an Intriguing Membrane

The presence of mycolic acids is the hallmark of *Corynebateriales*. These lipids are first synthetized in the cytoplasm and further exported in the envelope where they are used to build the outer membrane. In this part, we focus on all these essential steps but also on recent advances concerning the regulation of their synthesis.

3.1 Mycolic Acids Structure and Composition

Mycolic acids (MAs) are α-branched and β-hydroxylated fatty acids that are only found in *Corynebacteriales* envelopes. Although all MAs have a common pattern (the mycolic acid motif, see Fig. 1), their chain length and the structure of the β-chain (also known as the meromycolate chain) vary with genus and species, the shortest and the simplest (absence of specific functions as cyclopropanes or oxygenated groups) being found in corynebacteria (Lanéelle et al. 2013). MAs mainly exist in two forms in the envelope (Fig. 1): either bound to AG-PG and forming the mycoloyl-arabinogalactan-peptidoglycan core of the cell wall or attached to trehalose forming trehalose mono- (TMM) and di-mycolates (TDM), these later species being generally referred as "free mycolates". *C. glutamicum* contains corynomycolates with chain lengths ranging from $C_{22:0}$ to $C_{38:3}$ with a majority of $C_{32:0}$, $C_{34:1}$ and $C_{36:2}$, the relative proportion of which depends on the strain but also on the culture medium and growth temperature (Yang et al. 2012; Klatt et al. 2018). In a very recent detailed study of the lipid composition of the ATCC13032 strain grown in exponential phase in a BHI medium, Klatt et al. showed that trehalose mono- and di-corynomycolates (TMCM and TDCM) are the most abundant lipids in this bacteria (Klatt et al. 2018). They were able to identify 125 molecular species associated with corynomycolates, the majority of which containing saturated and mono-unsaturated chains. These species were divided into subclasses, the major one containing the classical hydroxylated mature TMCM and TDCM. Among the minor subclasses were keto-TMCM (non-reduced form of TMCM), acetylated TMCM (AcTMCM) or TDCM (AcTDCM), acylated TMCM (Acyl-TMCM, acylation of the non mycoloylated glucose moiety of the trehalose) and acylated AcTMCM (Acyl-AcTMCM). They also found some glucosylmonocorynomycolate (GMCM) associated with the outer membrane.

3.2 Mycolic Acids Biosynthesis

Because all *Corynebacteriales* synthetize their mycolic acids by a common cytoplasmic pathway and because, unlike mycobacteria, corynebacteria can grow without mycolates, *C. glutamicum* has proven to be a model of choice for elucidating the different steps of MA biosynthesis (Fig. 3). MAs are synthetized from two fatty acids which in *C. glutamicum* are produced by two non-redundant type I fatty acid synthases: FAS-IA is the main one, is essential and produces stearic acid ($C_{18:0}$) and oleic acid ($C_{18:1}$) with a small amount of palmitic acid ($C_{16:0}$), while FAS-IB is secondary, non-essential and synthetizes primarily $C_{16:0}$ (Radmacher et al. 2005). It should be noted that FAS-IB is also involved in the biosynthesis of octanoic acid, the precursor of lipoic acid (Ikeda et al. 2017). Activation of fatty acids is necessary prior to their future condensation and this is achieved on one hand by the fatty acyl-AMP ligase Cg-FadD2 (the orthologue of the mycobacterial FadD32) for the meromycolic chain and on the other hand by the acyl-CoA carboxylase complex for the α-chain (Gande et al. 2004; Portevin et al. 2005). This latter complex consists of the biotinylated α-subunit AccBC, the β-subunits AccD2/AccD3 and the small peptide AccE (Gande et al. 2007). The activated fatty acid precursors are used by the Cg-Pks polyketide synthase (the orthologue of the mycobacterial Pks13) that catalyzes their condensation by a Claisen reaction, forming an α-alkyl β-ketoester (Portevin et al. 2004; Gande et al. 2004). For this, Cg-Pks has to be previously activated (attachment of a 4′-phosphopantetheine (P-pant) group) by the specific enzyme PptT (Chalut et al. 2006). The final step of MA synthesis is the reduction of the β-ketoester by CmrA to yield the final mycolic motif, a reduction that proved to be significant with regard to the subsequent transfer of MA to the AG but not for the formation of TMCM or TDCM (Lea-Smith et al. 2007).

Subsequent steps that follow MAs synthesis and leading to the production of TMCM in the cell envelope are much less understood in *C. glutamicum*. In mycobacteria, most if not all the TMM, is synthesized on the inner leaflet of the cytoplasmic membrane by Pks13, that in addition to the condensation reaction, catalyzes the transfer of the MA onto a trehalose by its thio-esterase domain before its reduction (Gavalda et al. 2014). Whether Cg-Pks, like mycobacterial Pks13, is responsible for the synthesis of TMM in vivo is still to demonstrate. Indeed, in apparent contradiction to this hypothesis, by using mutants unable to synthesize trehalose, Tropis et al. proposed that the transfer of a mycoloyl residue onto a trehalose (or another α-glucosyl-containing sugar present in the medium) occurs in the envelope compartment (Tropis et al. 2005). In mycobacteria, cytoplasmic TMM is transported to the periplasmic side of the membrane by MmpL3 (for Mycobacterial membrane protein Large 3), an essential protein of the RND transporter family (Grzegorzewicz et al. 2012; Varela et al. 2012; Xu et al. 2017). A low expression of MmpL3 or its inhibition by a specific drug leads to a decrease in TDM and AG-bound mycolate content concomitant with TMM accumulation. Inactivation of its orthologue in *C. glutamicum* (Cg3174, also named CmpL1 for Corynebacterial membrane protein Large 1) had no effect on the level of bound or free mycolates (Varela et al. 2012; Yang et al. 2014). In fact, Varela et al. identified three other genes in addition to *cmpL1*: *cg1054*, *cg0722 and cg0284* (also named respectively *cmpL2*, *cmpL3* and *cmpL4*) (Varela et al. 2012). By a single and double gene deletion approach, they showed that transport of mycolates in *C. glutamicum* is achieved by two fully redundant CmpLs: CmpL1 and CmpL4. Surprisingly, loss of

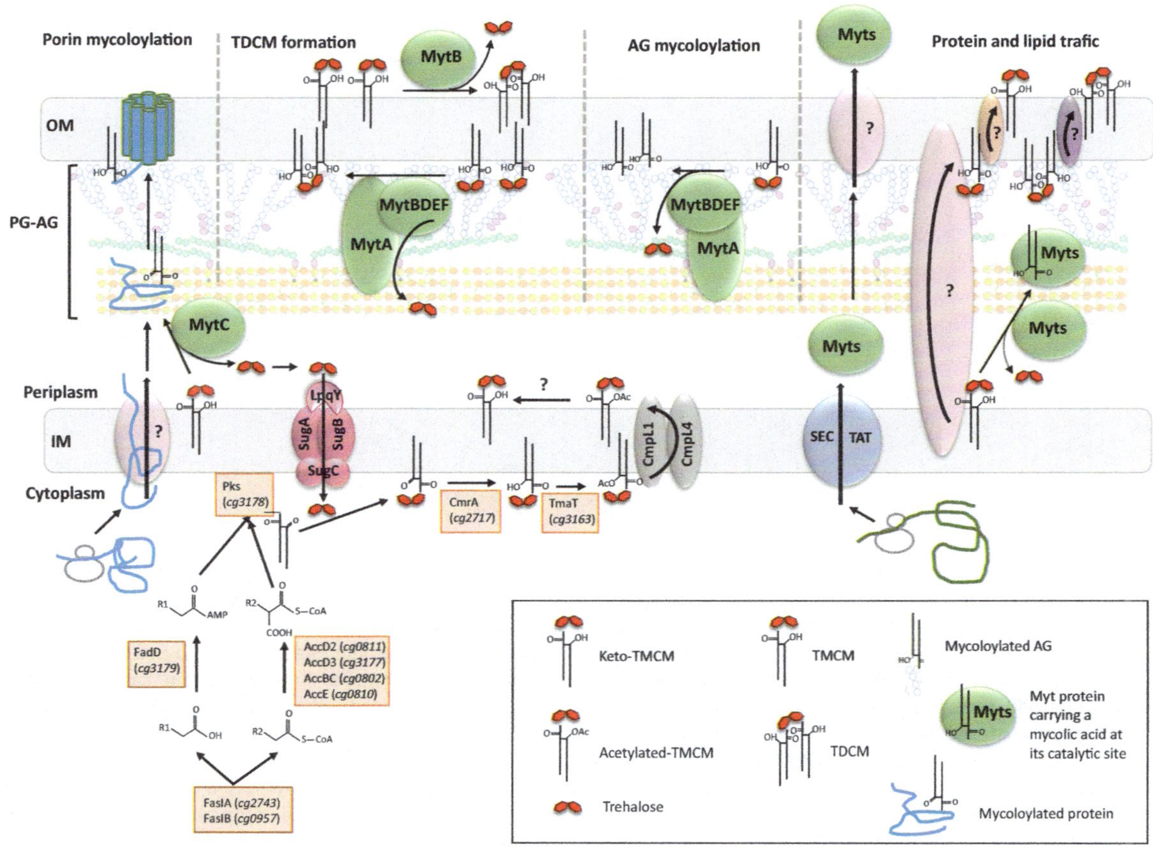

Fig. 3 Biosynthesis of mycolic acids and their transfer on the different terminal acceptors by mycoloyltransferases: porin mycoloylation, TDCM formation and AG mycoloylation. Details of the biosynthesis of TMCM are given in the text. Biosynthesis and transport of mycolic acids are represented in accordance with

Fig. 3 (continued) what has been shown recently for mycobacteria but not still demonstrated in *C. glutamicum*. In particular, we do not represent on this scheme the possibility that forms of mycolates different from TMCM could be transported and that transfer on trehalose could occur in the periplasm as suggested by Tropis et al. (2005). *C. glutamicum* possesses all the genes encoding the transport system that recycles the extracellular trehalose released from trehalose-containing mycolates in mycobacteria: the lipoprotein LpqY (*cg0834*) and the ABC-transporter SugABC, (*cg0831*, *cg0832* and *cg0833*) (Kalscheuer et al. 2010). However, the functionality of this transport system has not been proven in this bacterium. The TAT and SEC abbreviations refer to the Twin Arginine Translocation and the general protein translocation pathways. Porins like PorA/PorH do not have TAT or SEC secretion signals, therefore, they may be translocated through the inner membrane by a specific complex (tagged with a question mark on the figure), then mycoloylated and inserted in the mycomembrane. In the right part of the figure different hypothesis concerning mycolic acid traffic from inner membrane to outer membrane are represented: i.e. (i) a protein complex spanning the envelope (associated or not to CmpL1/4) that transport the mycolate (TMCM) from the outer leaflet of the IM to the inner or the outer leaflet of the OM, (ii) flippases in the OM that translocate TMCM or TDCM from one leaflet to the other or (iii) transport of MA by the Myts themselves. Trafficking of Myt proteins is also shown: across the IM by the SEC or TAT pathway and across the OM by an unknown transporter

these two CmpL proteins led to a complete abrogation of any form of corynomycolates, a result that suggests the possibility of a tight coupling between the Cg-Pks biosynthesis complex and the transporters that could regulate the mycolate synthesis. In a recent study, Yamaryo-Botte et al. provided evidences that the mycolate species recognized and transported by the CmpL1/CmpL4 complex could be an acetylated form of TMCM (AcTMCM). The enzyme responsible for this modification is the acetyltransferase TmaT (for TMCM mycolyl acetyltransferase), essential in mycobacteria but not in *C. glutamicum* (Yamaryo-Botte et al. 2015). Once in the periplasm, AcTMCM would be deacetylated by an unknown enzyme to form mature TMCM which is the periplasmic donor of mycolates to final cell wall acceptors. These steps which are crucial for the construction of the mycomembrane are performed by a family of enzymes named mycoloyltransferases (Myts) (Dautin et al. 2017).

3.3 Assembly and Composition of the Mycomembrane

C. glutamicum possesses six Myts (cMyts A-F, (Table 1)) that display 26–42% sequence identity between each other except for cMytE and F, which are almost identical. All proteins are between 341 and 483 aa-long, with the notable exception of cMytA which possesses a C-terminal domain containing four 54 amino acids-long LGFP tandem repeats. These repeats have been proposed to mediate interactions with the cell wall (either AG or PG) (Adindla et al. 2004a, b). *cmyt* genes are dispersed on the chromosome, except for *cmytA* and *cmytB*, which are only separated by one open reading frame of unknown function but very likely unrelated to the Myt activity (Kacem et al. 2004).

All cMyts proteins have a signal sequence and are predicted, by *in silico* analysis, to be exported across the cytoplasmic membrane. All have been found experimentally associated to the cell wall or secreted in the culture medium by proteomic analysis (Hansmeier et al. 2006b). Interestingly, in *C. glutamicum* R, of the six putative Myts, one (CgR_1023) is predicted to contain an atypical Twin-Arginine Translocation system (TAT) signal (lacking one of the two characteristic arginines) instead of a classical Sec signal. When this N-terminal signal is fused to α-amylase, secretion of the recombinant enzyme is abolished in a *ΔtatC* mutant, strongly

Table 1 Mycoloyl transferases of *C. glutamicum* reference strain ATCC13032

Protein name	Gene name	Locus tag	aa number
cMytA/PS1	*cmytA/csp1/cop1*	*cg3182*	657
cMytB/Cmt2	*cmytB/cmt2*	*cg3186*	341
cMytC/Cmt1	*cmytC/cmt1*	*cg0413*	365
cMytD/Cmt5	*cmytD/cmt5*	*cg1170*	411
cMytE/Cmt3	*cmytE/cmt3*	*cg1052*	483
cMytF/Cmt4	*cmytF/cmt4*	*cg2394*	483

suggesting that at least CgR_1023 could follow the TAT pathway (Watanabe et al. 2009). Similarly, in *M. tuberculosis*, mycoloyltransferases FbpA and FbpC are proposed to use the TAT machinery while FbpB would rather use the Sec pathway (McDonough et al. 2008; Marrichi et al. 2008). The reason for the coexistence of two distinct mechanisms of translocation for the transport of Myts in the cell envelope is not clear but may be related to their folding properties and/or their ability to bind a substrate before translocation.

The apparent redundancy of Myts in *C. glutamicum* is not completely understood but each enzyme could have in vivo its own specific function. Gene inactivation and overexpression studies indicated that, except for cMytC, all the other cMyts (A-F) may contribute to the transfer of mycoloyl residues onto TMCM and/or onto cell wall AG to build the backbone of the mycomembrane with mycoloylated ends of the AG constituting the internal leaflet and TMCM/TDCM the outer leaflet (Figs. 1 and 3) (Brand et al. 2003; De Sousa-D'Auria et al. 2003; Kacem et al. 2004). cMytA and cMytB are the main contributors for TDCM synthesis and can compensate for each other. cMytD and cMytF (which are weakly expressed in *C. glutamicum* ATCC13032 reference strain) can replace cMytA to produce TDCM but not mycoloylated-AG (De Sousa-D'Auria et al. 2003). A *C. glutamicum* mutant lacking all mycoloyltransferases has not been constructed yet but, in *Mycobacterium* species, the use of specific Myt inhibitors has proved to be very effective to inhibit in vitro growth (Rose et al. 2002; Wang et al. 2004; Annamala et al. 2007; Warrier et al. 2012).

Due to the difficulty to biochemically fractionate the cell envelope of *Corynebacteriales*, the exact lipid composition of the mycomembrane has long been unknown. In 2011, Bansal-Mutalik adapted the use of reverse surfactant micelles for specifically extracting the lipids of the mycomembrane of *C. glutamicum* (Bansal-Mutalik and Nikaido 2011). By this way, the authors suggested that, in minimal medium, the outer leaflet of the mycomembrane is mainly composed of TDCM and to a lesser extent of TMCM. No significative amounts of phospholipids were found in the mycomembrane except for the presence of non-extractable cardiolipin which may be anchored to the PG-AG complex. In another study, Marchand et al. (2012) purified the mycomembrane-linked heteropolymer of PG and AG from the cytoplasmic membrane and observed a very sharp segregation of lipids between both membranes: phospholipids being mainly associated with the cytoplasmic membrane while glycolipids (TDCM and TMCM) were totally associated with the mycomembrane. In terms of proteins, this latter study allowed the identification of more than 50 proteins proposed to be associated with the mycomembrane. These include some characterized porins, mycoloyltransferases, but also a long list of unknown proteins. Among them, several putative lipoproteins (typified by their lipobox) are thus predicted to be anchored in the mycomembrane. A functional acylation machinery has indeed been identified in *C. glutamicum* (composed of Lgt: Cg2292 and Lnt also named Ppm2: Cg1673) allowing tri-acylation of a set of exported proteins as found in other species (Mohiman et al. 2012). Besides this classical acylation, a specific post-translational O-mycoloylation has also been discovered in *C. glutamicum* (Fig. 3). Indeed, small proteins PorA, PorH, PorB, PorC and ProtX have been shown to be mycoloylated on serine residues (Huc et al. 2013; Issa et al. 2017). This very unique modification which is absent in a *mytC* gene-disrupted mutant (Huc et al. 2013) is important for anchoring PorB and PorC in the mycomembrane (Carel et al. 2017; Issa et al. 2017) (and, in the case of the porins PorAH, essential for their in vitro activity (Rath et al. 2011)). Whether this modification is also involved in folding, oligomerization, insertion or sorting is still to discover. In any case, the coexistence of acylated and mycoloylated proteins in the mycomembrane is intriguing and may reflect specific functions or subcellular localization of both family of lipidated proteins.

3.4 Regulation of Mycolic Acid Metabolism

The presence of a mycolate barrier confers a unique protection for corynebacteria and mycobacteria and is probably submitted to a tight regulation according to environmental stress. Two different regulators of the mycolate biosynthesis pathway have been identified in *C. glutamicum*. First, the envelope lipids regulation factor (ElrF) which is involved in the reorganization of outer membrane lipid composition following heat stress, by affecting the balance between corynomycolate lipids and phospholipids (Meniche et al. 2009). In vitro, ElrF has an esterase/thioesterase activity but the mechanism by which it increases MA synthesis when bacteria are exposed to a moderate heat shock is still unknown. The second is the Extra Cytoplasmic Function (ECF) sigma factor σ^D (Cg0696) which has been shown to be a key regulator of the mycolate synthesis and transfer (Taniguchi et al. 2017; Toyoda and Inui 2018). Indeed, the σ^D regulon includes five genes of the corynomycolate metabolism: *accD2*, *accD3*, *accE* (acyl-CoA carboxylase subunits), *fadD2* (acyl-AMP ligase) and cg-*pks* (condensase), four of the *cmyt* genes (*cmytA*, *cmytB*, *cmytC* and *cmytE*), as well as *elrF*. Other genes encoding mycoloylated proteins are also found in the σ^D regulon, such as *porA*, *porH* and *cg2875* encoding ProtX. This very specific pattern of genes suggests a role of σ^D on the mycomembrane synthesis (mycolic acids and mycoproteins) but probably more broadly on the cell envelope biosynthesis as indicated by the presence of an L,D-Transpeptidase encoding gene in the regulon (Toyoda and Inui 2018). The cognate anti-sigma factor has been identified as RsdA, a membrane protein with one transmembrane segment (Cg0697). The σ^D-RsdA signaling pathway including the environmental signals that allowed σ^D regulon induction are not currently known.

4 Lipoglycans: Phosphatidyl-myo-Inositol Mannosides (PIM) and LM/LAM

In addition to the core AG-PG polymer, *C. glutamicum* envelope contains other important glycans whose structure has been elucidated in the last decade. PIM, LM and LAM are all composed of mannan chains and their biosynthetic pathways very well characterized. However, their localization and function are still elusive.

4.1 Composition and Structure of PIM and Lipoglycans

As in all *Corynebacteriales,* the cell wall of *C. glutamicum* contains an array of non-covalently associated glycolipids of the phosphatidyl-myo-inositol mannosides (PIM) type and lipoglycans of the lipomannan (LM) and lipoarabinomannan (LAM) types (Crellin et al. 2013). Different PIMs have been found in *C. glutamicum,* essentially in the inner membrane: phosphatidylinositol mono and di-mannosides (PIM_1 and PIM_2) and acylated di-, tri- and tetramannosides ($AcPIM_{2\text{-}4}$) with $AcPIM_2$ being the most important in term of quantity and diversity (Klatt et al. 2018). A recent LC/MS-ESI-TOF analysis of the LM/LAM content of ATCC13032 strain, showed that this bacterium contains 117 clearly resolved LM species with masses ranging from 2712 to 9190 Da (Cashmore et al. 2017). *C. glutamicum* LMs (Cg-LM) can be separated in two distinct populations according to their lipid anchor but sharing a similar mannan backbone. Cg-LM-A, is derived from PIM and is similar to mycobacterial LM, while the other, termed Cg-LM-B, is specific to the species and has a α-D-glucopyranosyluronic acid-(1→3)-glycerol (GlcAGroAc2) based anchor (Tatituri et al. 2007a, b; Lea-Smith et al. 2008; Mishra et al. 2008b). The mannan domain of both Cg-LM-A and B is composed of 10–50 mannopyranosyl (Man*p*) residues organized in an α(1→6)-Man*p* backbone with α(1→2)-Man*p* branching, a number which appears to be reduced in Cg-LM-A as compared to Cg-LM-B (Tatituri et al. 2007b). Cg-LM-B constitutes the major part of the Cg-LM pool (about 95%) but only Cg-LM-A is elaborated to Cg-LAM (Tatituri et al. 2007a, b). Although Cg-LAM is reminiscent to mycobacterial LAM, its structure and composition are much simpler as the arabinan domain is reduced to t-Ara*f* residues capping some Man*p* residues (Tatituri et al. 2007a, b). Cell envelope fractionation experiments performed on bacteria in stationary phase, have shown that Cg-LM/LAM localize essentially in the inner membrane, with small amounts of LM (but not LAM) found in the mycomembrane (Marchand et al. 2012). However, it cannot be excluded that content and localization of the lipoglycans may vary in function of the growth phase. Although the precise function of Cg-LM/LAM is not known, it should be noted that like other actinobacterial lipoglycans, both Cg-LM-A and Cg-LM-B display pro-inflammatory properties (Mishra et al. 2008b).

4.2 Biosynthesis of Lipoglycans

The biosynthetic pathway of Cg-AcPIM$_2$ (Fig. 4a) is similar to that found in mycobacteria and involves two GDP-Man*p*-dependent mannopyranosyltransferases that act sequentially and an acyltransferase (Sancho-Vaello et al. 2017). The first step is catalyzed by PimA and corresponds to the transfer of a Man*p* residue from GDP-Man*p* to the O-2 position of the inositol ring of PI to give a phosphatidylinositol monomannoside (PIM$_1$) (Kordulákova et al. 2002). PIM$_1$ is then acylated on the 6-position of the Man*p* residue to give Ac$_1$PIM$_1$. In mycobacteria the acyltransferase involved in this process is PatA (Rv2611c) the orthologue of which is Cg1877 in *C. glutamicum* (Kordulákova et al. 2003). The second Man*p* is transferred by PimB' on the O-6 of the inositol ring to give Ac$_1$PIM$_2$ (Lea-Smith et al. 2008; Mishra et al. 2008b; Batt et al. 2010). After being translocated across the inner membrane by an unknown flippase, Ac$_1$PIM$_2$ serves as a precursor for the biosynthesis of Cg-LM-A (Fig. 4a). LM-A is synthetized by the action of four non-essential polyprenol-monophosphomannose (PPM)-dependent mannosyltransferases of the GT-C family, MptA-D (Mishra et al. 2007, 2008a, 2011). PPM is the periplasmic mannosyl donor for both lipoglycan biosynthesis and protein glycosylation and is synthetized in the cytosol from GDP-mannose by Cg-Ppm1 (Cg1672) a polyprenol-monophosphomannose synthase (Gibson et al. 2003; Mohiman et al. 2012). After being translocated to the periplasmic face of the inner membrane, PPM serves as a substrate for MptB and MptA that respectively add proximal and distal α(1→6)-Man*p* as well as for MptC and MptD that catalyse α(1→2)-Man*p* branching. Two proteins with regulatory functions are also involved in the biosynthesis of mannan domain: LpqW, a lipoprotein that was proposed to activate MptB (Rainczuk et al. 2012) and Cg3164 a membrane protein with unknown function but whose inactivation results in truncated Cg-LM/LAM (Cashmore et al. 2017). Further transfer of single t-Ara*f* residues from DPA (which is also the AG arabinose donor) onto the mannan backbone of Cg-LM-A leads to Cg-LAM (Fig. 4a) (Tatituri et al. 2007a). The AraT responsible for this transfer in *C. glutamicum* has not been identified so far but does not appear to be neither Emb nor AftA, since mutants that no longer express either glycosyltransferase still produce LM/LAM identical to those of the WT strain (Tatituri et al. 2007a). Surprisingly it has been reported that a *C. glutamicum*::*ubiA* mutant, unable to synthetize DPA, produces a LAM that still contains small amounts of arabinose. This result suggests that in contrast to AG biosynthesis where DPA is the only Ara*f* donor, another source or Ara*f* could exist for Cg-LAM synthesis (Tatituri et al. 2007a).

The second pathway of Cg-LM biosynthesis (Fig. 4b) begins with the synthesis of 1,2-di-O-C_{16}/$C_{18:1}$-(α-D-glucopyranosyluronic acid)-(1→3)-glycerol (GlcAGroAc$_2$ also termed Gl-A) that originated from the transfer of glucuronic acid (GlcA) from CDP-GlcA onto a diacylglycerol (GroAc$_2$) (Tatituri et al. 2007b). A Man*p* residue is then transferred on Gl-A by an α-mannosyl-glucopyranosyluronic acid-transferase A (MgtA, previously named PimB) to give 1,2-di-O-(α-D-mannopyranosyl)-(1→4)-(α-D-glucopyranosyluronic acid)-(1→3)-glycerol (ManGlcAGroAc$_2$ also termed

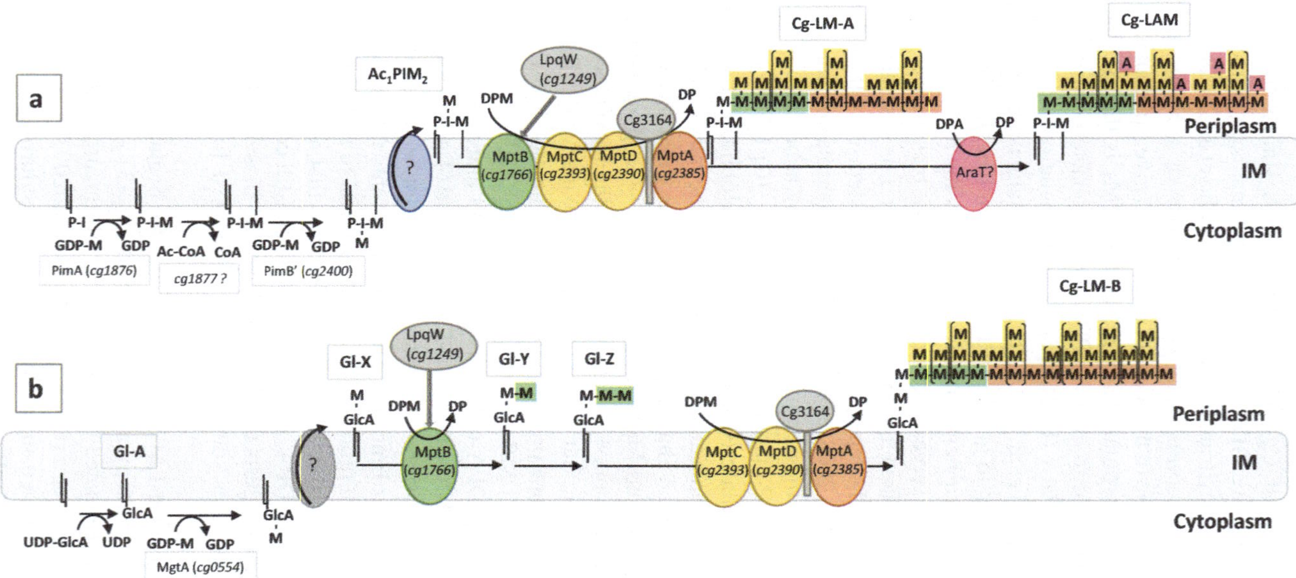

Fig. 4 Biosynthesis pathways of lipoglycans of *C. glutamicum*. (**a**) Cg-LM-A/LAM pathway. (**b**) Cg-LM-B pathway. Gl-A, Gl-X Gl-Y and Gl-Z represent the different intermediates that have been detected in recent studies (Klatt et al. 2018). *M* mannose, *CoA* coenzyme A, *Ac-CoA* acyl-Coenzyme A, *P-I* phosphatidylinositol, *DPM* decaprenylphosphomannose (PPM in the text), *DP* decaprenylphosphate. Other abbreviations are given in the text. The color of each mannose in the Cg-LM-A/LAM and Cg-LM-B corresponds to the color of the protein which add this mannose in the biosynthetic pathway

Gl-X) (Tatituri et al. 2007b; Mishra et al. 2009). After translocation of Gl-X from the cytoplasmic face of the inner membrane to the periplasmic one, elongation with Man*p* residues is performed in the same way as for Cg-LM-A to give the mature Cg-LM-B (Mishra et al. 2007, 2008a, 2011; Rainczuk et al. 2012; Cashmore et al. 2017).

5 Other Peripheral Components of the Envelope

The general organization of the cell envelope of *C. glutamicum* is highly conserved but an important variability may be observed among different strains. It may concern specific modifications of the core AG-PG or alternatively correspond to the presence of additional layers at the surface of the bacterium. The presence of these specific components is probably related to specific niches and may therefore reflect a localized evolution driven by the environment. Two important surface structures are described below.

5.1 The Outer Layer

Ultrastructural analysis of the cell envelope of *C. glutamicum* by electron microscopic observation of ultrathin sections reveals the presence of a non-ordered diffuse outer layer (OL) around the cell, covering the mycomembrane (Marienfeld et al. 1997; Puech et al. 2001). The thickness of this OL is not known. Its apparent value range between 2 and 40 nm according to the technique used to visualize the envelope (freeze-substitution, conventional fixation with or without ruthenium red staining). Chemical analysis of this outermost material showed that it is almost entirely composed of carbohydrate containing mainly Glc but also some Ara and Man which probably derived from LAM (Puech et al. 2001). It is very likely that Glc comes from a high molecular mass glucan (>100 kDa) as it has been described at the cell surface of *Corynebacterium xerosis* or in the capsule of some pathogenic mycobacteria (Puech et al. 2001; Dinadayala et al. 2008; Sambou et al. 2008). The structure of this glucan has not been resolved to date.

5.2 The Surface Layer

Most of the *C. glutamicum* strains (but not the reference strain ATCC13032) possess an S-layer formed by a single protein named PS2 encoded by the *cspB* gene (Peyret et al. 1993; Chami et al. 1995; Hansmeier et al. 2004). This protein assembles into a highly stable two-dimensional crystalline array that covered the cell surface. The layer is resistant to proteases and detergent and could only be dissociated at high temperature (up to 80 °C), indicating uncommon strong interactions between PS2 subunits (Chami et al. 1997; Scheuring et al. 2002). A systematic comparison of PS2 proteins from 28 *C. glutamicum* isolates showed apparent molecular masses ranging from 55 to 66 kDa. Their sequences display highly conserved and highly variable regions and a percent of identity varying between 69 and 98% (Hansmeier et al. 2004). Among the best conserved region is the C-terminal end of the protein which contains a stretch of 21 hydrophobic amino acids that anchors PS2 in the mycomembrane (Chami et al. 1997). Based principally on atomic force microscopy analysis, it was shown that the overall structure of the S-layer is well conserved among the different PS2 and exhibits an hexameric organization within an hexagonal lattice with the C-terminal domains of the hexamer forming a hydrophobic ring structure that might constitute a pore (Chami et al. 1995; Scheuring et al. 2002; Hansmeier et al. 2004). Each protein of an hexamer is connected to two other protomers from two other hexamers giving a M_6C_3 layer in the classification established by Saxton and Baumeister for the assembly of S-layer proteins into massive cores (M) and connecting structures (C) (Saxton and Baumeister 1986).

What is the precise function of the S-layer in the natural biotope of the bacterium (i.e. the soil) is a question currently unanswered but it is obvious that given the energy expenditure required to produce the huge amount of PS2 necessary to cover the cell (about 100,000 copies per cell), an accurate regulation of the protein expression related to the physiological status of the bacteria is expected. Indeed, it has been shown that PS2 expression depends on the nature of the carbon source suggesting a relationship between carbohydrate metabolism and S-layer formation (Soual-Hoebeke et al. 1999). As a confirmation of such hypothesis, a LuxR-type transcriptional activator that binds to the *cspB* promoter region and regulates the gene expression was identified (Hansmeier et al. 2006a). This regulator is RamA (Cg2831), a global regulator of carbon metabolism (Shah et al. 2018).

6 Conclusion

The cell envelope of *Corynebacteriales* is remarkably unique in the bacterial kingdom with a strong original heteropolysaccharidic cell wall and an atypical mycolate-containing outer membrane. To date, very important milestones have been achieved concerning the identification of specific individual components but the elucidation of the complete choreography of cell envelope biosynthetic enzymes remains an exciting challenge for the next decades. The evolutionary origin of this peculiar architecture is not known but the very specific composition of the mycomembrane suggests that the LPS-outer membrane and the mycolate-outer membrane diderm species may have evolved independently from a common monoderm ancestor. Indeed, recent analysis of the distribution of the major genes involved in the synthesis of the mycomembrane-linked heteropolymer of PG and AG (Vincent et al. 2018) supports a scenario in which *Corynebacteriales* species evolved from an actinobacterial ancestor by a stepwise cell envelope remodeling involving gradual acquisition of new enzymatic functions. In this scenario, one can imagine that final evolution of corynebacteria and mycobacteria may differ according to their very different habitat. In fact, both groups of bacteria do not strictly have the same envelope composition and organization. Particularly noteworthy, mycolates are essential for *M. tuberculosis* and dispensable for *C. glutamicum*, a characteristic that makes *C. glutamicum* a very suitable model to study the envelope of *Corynebacteriales* and also a very versatile cell for biotechnological applications. Indeed, *C. glutamicum* has long been used as an industrial workhorse for the production of amino acids and more recently as a protein expression cell and a protein platform display (Lee and Kim 2018). The precise knowledge of the architecture of the cell envelope is important to guide biotechnological experiments. In that respect, the presence of an outer membrane in *C. glutamicum* raises several interesting questions and may help to resolve specific constrains regarding protein secretion. In LPS-diderm bacteria numerous secretion systems have been described so far but none of them are found in *Corynebacteriales*. Identification of such specific mycomembrane machineries involved in protein secretion is thus an interesting research direction for the future.

References

Adindla S, Guruprasad K, Guruprasad L (2004a) Three-dimensional models and structure analysis of corynemycolyltransferases in *Corynebacterium glutamicum* and *Corynebacterium efficiens*. Int J Biol Macromol 34:181–189. https://doi.org/10.1016/j.ijbiomac.2004.03.008

Adindla S, Inampudi KK, Guruprasad K, Guruprasad L (2004b) Identification and analysis of novel tandem repeats in the cell surface proteins of archaeal and bacterial genomes using computational tools. Comp Funct Genomics 5:2–16. https://doi.org/10.1002/cfg.358

Alderwick LJ, Radmacher E, Seidel M et al (2005) Deletion of Cg-emb in *Corynebacterianeae* leads to a novel truncated cell wall arabinogalactan, whereas inactivation of Cg-ubiA results in an arabinan-deficient mutant with a cell wall galactan core. J Biol Chem 280:32362–32371. https://doi.org/10.1074/jbc.M506339200

Alderwick LJ, Dover LG, Seidel M et al (2006a) Arabinan-deficient mutants of *Corynebacterium glutamicum* and the consequent flux in decaprenylmonophosphoryl-D-arabinose metabolism. Glycobiology 16:1073–1081. https://doi.org/10.1093/glycob/cwl030

Alderwick LJ, Seidel M, Sahm H et al (2006b) Identification of a novel arabinofuranosyltransferase (AftA) involved in cell wall arabinan biosynthesis in *Mycobacterium tuberculosis*. J Biol Chem 281:15653–15661. https://doi.org/10.1074/jbc.M600045200

Alderwick LJ, Harrison J, Lloyd GS, Birch HL (2015) The mycobacterial cell wall–peptidoglycan and arabinogalactan. Cold Spring Harb Perspect Med 5:a021113. https://doi.org/10.1101/cshperspect.a021113

Alderwick LJ, Birch HL, Krumbach K et al (2018) AftD functions as an α1 → 5 arabinofuranosyltransferase involved in the biosynthesis of the mycobacterial cell wall core. Cell Surf (Amsterdam, Netherlands) 1:2–14. https://doi.org/10.1016/j.tcsw.2017.10.001

Angala SK, Belardinelli JM, Huc-Claustre E et al (2014) The cell envelope glycoconjugates of *Mycobacterium tuberculosis*. Crit Rev Biochem Mol Biol 49:361–399. https://doi.org/10.3109/10409238.2014.925420

Annamala MK, Inampudi KK, Guruprasad L (2007) Docking of phosphonate and trehalose analog inhibitors into *M. tuberculosis* mycolyltransferase Ag85C: comparison of the two scoring fitness functions GoldScore and ChemScore, in the GOLD software. Bioinformation 1:339–350

Bansal-Mutalik R, Nikaido H (2011) Quantitative lipid composition of cell envelopes of *Corynebacterium glutamicum* elucidated through reverse micelle extraction. Proc Natl Acad Sci U S A 108:15360–15365. https://doi.org/10.1073/pnas.1112572108

Barreteau H, Kovač A, Boniface A et al (2008) Cytoplasmic steps of peptidoglycan biosynthesis. FEMS Microbiol Rev 32:168–207. https://doi.org/10.1111/j.1574-6976.2008.00104.x

Batt SM, Jabeen T, Mishra AK et al (2010) Acceptor substrate discrimination in phosphatidyl-myo-inositol mannoside synthesis: structural and mutational analysis of mannosyltransferase *Corynebacterium glutamicum* PimB′. J Biol Chem 285:37741–37752. https://doi.org/10.1074/jbc.M110.165407

Baumgart M, Schubert K, Bramkamp M, Frunzke J (2016) Impact of LytR-CpsA-Psr proteins on cell wall biosynthesis in *Corynebacterium glutamicum*. J Bacteriol 198:3045–3059. https://doi.org/10.1128/JB.00406-16

Bayan N, Houssin C, Chami M, Leblon G (2003) Mycomembrane and S-layer: two important structures of *Corynebacterium glutamicum* cell envelope with promising biotechnology applications. J Biotechnol 104:55–67

Belanova M, Dianiskova P, Brennan PJ et al (2008) Galactosyl transferases in mycobacterial cell wall synthesis. J Bacteriol 190:1141–1145. https://doi.org/10.1128/JB.01326-07

Berg S, Kaur D, Jackson M, Brennan PJ (2007) The glycosyltransferases of *Mycobacterium tuberculosis* – roles in the synthesis of arabinogalactan, lipoarabinomannan, and other glycoconjugates. Glycobiology 17:35–56R. https://doi.org/10.1093/glycob/cwm010

Besra GS, Khoo KH, McNeil MR et al (1995) A new interpretation of the structure of the mycolyl-arabinogalactan complex of *Mycobacterium tuberculosis* as revealed through characterization of oligoglycosylalditol fragments by fast-atom bombardment mass spectrometry and 1H nuclear magnetic res. Biochemistry 34:4257–4266. https://doi.org/10.1021/bi00013a015

Birch HL, Alderwick LJ, Bhatt A et al (2008) Biosynthesis of mycobacterial arabinogalactan: identification of a novel α(1→3) arabinofuranosyltransferase. Mol Microbiol 69:1191–1206. https://doi.org/10.1111/j.1365-2958.2008.06354.x

Birch HL, Alderwick LJ, Rittmann D et al (2009) Identification of a terminal rhamnopyranosyltransferase (RptA) involved in *Corynebacterium glutamicum* cell wall biosynthesis. J Bacteriol 191:4879–4887. https://doi.org/10.1128/JB.00296-09

Bou Raad R, Méniche X, de Sousa-d'Auria C et al (2010) A deficiency in arabinogalactan biosynthesis affects *Corynebacterium glutamicum* mycolate outer membrane stability. J Bacteriol 192:2691–2700. https://doi.org/10.1128/JB.00009-10

Boutte CC, Baer CE, Papavinasasundaram K et al (2016) A cytoplasmic peptidoglycan amidase homologue controls mycobacterial cell wall synthesis. Elife 5:1–22. https://doi.org/10.7554/eLife.14590

Brand S, Niehaus K, Pühler A, Kalinowski J (2003) Identification and functional analysis of six mycolyltransferase genes of *Corynebacterium glutamicum* ATCC 13032: the genes cop1, cmt1, and cmt2 can replace each other in the synthesis of trehalose dicorynomycolate, a component of the mycolic acid l. Arch Microbiol 180:33–44. https://doi.org/10.1007/s00203-003-0556-1

Burkovski A (2013) Cell envelope of corynebacteria: structure and influence on pathogenicity. ISRN Microbiol 2013:935736. https://doi.org/10.1155/2013/935736

Cai L, Zhao X, Jiang T et al (2014) Prokaryotic expression, identification and bioinformatics analysis of the *Mycobacterium tuberculosis* Rv3807c gene encoding the putative enzyme committed to decaprenylphosphoryl-d-arabinose synthesis. Indian J Microbiol 54:46–51. https://doi.org/10.1007/s12088-013-0418-8

Carel C, Marcoux J, Réat V et al (2017) Identification of specific posttranslational O-mycoloylations mediating protein targeting to the mycomembrane. Proc Natl Acad Sci U S A 114:4231–4236. https://doi.org/10.1073/pnas.1617888114

Cashmore TJ, Klatt S, Yamaryo-Botte Y et al (2017) Identification of a membrane protein required for lipomannan maturation and lipoarabinomannan synthesis in *Corynebacterineae*. J Biol Chem 292:4976–4986. https://doi.org/10.1074/jbc.M116.772202

Chalut C, Botella L, de Sousa-D'Auria C et al (2006) The nonredundant roles of two 4'--phosphopantetheinyl transferases in vital processes of Mycobacteria. Proc Natl Acad Sci U S A 103:8511–8516. https://doi.org/10.1073/pnas.0511129103

Chami M, Bayan N, Dedieu J et al (1995) Organization of the outer layers of the cell envelope of *Corynebacterium glutamicum*: a combined freeze-etch electron microscopy and biochemical study. Biol cell 83:219–229

Chami M, Bayan N, Peyret JL et al (1997) The S-layer protein of *Corynebacterium glutamicum* is anchored to the cell wall by its C-terminal hydrophobic domain. Mol Microbiol 23:483–492

Costa-Riu N, Burkovski A, Krämer R, Benz R (2003a) PorA represents the major cell wall channel of the Gram-positive bacterium *Corynebacterium glutamicum*. J Bacteriol 185:4779–4786

Costa-Riu N, Maier E, Burkovski A et al (2003b) Identification of an anion-specific channel in the cell wall of the Gram-positive bacterium *Corynebacterium glutamicum*. Mol Microbiol 50:1295–1308

Crellin PK, Luo C-Y, Morita YS (2013) Metabolism of plasma membrane lipids in mycobacteria and corynebacteria. In: Valenzuela Baez R (ed) Lipid metabolism. InTech, p 460

Daffe M (2005) The cell envelope of Corynebacteria. In: Eggeling L, Bott M (eds) Handbook of *Corynebacterium glutamicum*. CRC press, pp 121–148

Daffe M, Brennan PJ, McNeil M (1990) Predominant structural features of the cell wall arabinogalactan of *Mycobacterium tuberculosis* as revealed through characterization of oligoglycosyl alditol fragments by gas chromatography/mass spectrometry and by 1H and 13C NMR analyses. J Biol Chem 265:6734–6743

Daffe M, McNeil M, Brennan PJ (1993) Major structural features of the cell wall arabinogalactans of *Mycobacterium*, *Rhodococcus*, and *Nocardia spp*. Carbohydr Res 249:383–398

Dautin N, de Sousa-d'Auria C, Constantinesco-Becker F et al (2017) Mycoloyltransferases: a large and major family of enzymes shaping the cell envelope of *Corynebacteriales*. Biochim Biophys Acta – Gen Subj 1861:3581–3592. https://doi.org/10.1016/j.bbagen.2016.06.020

de Sousa-d'Auria C, Kacem R, Puech V et al (2003) New insights into the biogenesis of the cell envelope of corynebacteria: identification and functional characterization of five new mycoloyltransferase genes in *Corynebacterium glutamicum*. FEMS Microbiol Lett 224:35–44. https://doi.org/10.1016/S0378-1097(03)00396-3

Deng LL, Humphries DE, Arbeit RD et al (2005) Identification of a novel peptidoglycan hydrolase CwlM in *Mycobacterium tuberculosis*. Biochim Biophys Acta 1747:57–66. https://doi.org/10.1016/j.bbapap.2004.09.021

Dinadayala P, Sambou T, Daffé M, Lemassu A (2008) Comparative structural analyses of the alpha-glucan and glycogen from *Mycobacterium bovis*. Glycobiology 18:502–508. https://doi.org/10.1093/glycob/cwn031

Dover LG, Cerdeño-Tárraga AM, Pallen MJ et al (2004) Comparative cell wall core biosynthesis in the mycolated pathogens, *Mycobacterium tuberculosis* and *Corynebacterium diphtheriae*. FEMS Microbiol Rev 28:225–250. https://doi.org/10.1016/j.femsre.2003.10.001

Figueiredo T, Sobral RG, Ludovice AM et al (2012) Identification of genetic determinants and enzymes involved with the amidation of glutamic acid residues in the peptidoglycan of *Staphylococcus aureus*. PLoS Pathog 8:e1002508. https://doi.org/10.1371/journal.ppat.1002508

Frirdich E, Gaynor EC (2013) Peptidoglycan hydrolases, bacterial shape, and pathogenesis. Curr Opin Microbiol 16:767–778. https://doi.org/10.1016/j.mib.2013.09.005

Gande R, Gibson KJC, Brown AK et al (2004) Acyl-CoA carboxylases (accD2 and accD3), together with a unique polyketide synthase (Cg-pks), are key to mycolic acid biosynthesis in *Corynebacterianeae* such as *Corynebacterium glutamicum* and *Mycobacterium tuberculosis*. J Biol Chem 279:44847–44857. https://doi.org/10.1074/jbc.M408648200

Gande R, Dover LG, Krumbach K et al (2007) The two carboxylases of *Corynebacterium glutamicum* essential for fatty acid and mycolic acid synthesis. J Bacteriol 189:5257–5264. https://doi.org/10.1128/JB.00254-07

Gavalda S, Bardou F, Laval F et al (2014) The polyketide synthase Pks13 catalyzes a novel mechanism of lipid transfer in mycobacteria. Chem Biol 32:1–10. https://doi.org/10.1016/j.chembiol.2014.10.011

Gibson KJC, Eggeling L, Maughan WN et al (2003) Disruption of Cg-Ppm1, a polyprenyl monophosphomannose synthase, and the generation of lipoglycan-less mutants in *Corynebacterium glutamicum*. J Biol Chem 278:40842–40850. https://doi.org/10.1074/jbc.M307988200

Goodfellow M, Jones AL (2015) *Corynebacteriales* ord. nov. In: Bergey's manual of systematics of archaea and bacteria. Wiley, Chichester, pp 1–14

Grover S, Alderwick LJ, Mishra AK et al (2014) Benzothiazinones mediate killing of *Corynebacterineae* by blocking decaprenyl phosphate recycling involved in cell wall biosynthesis. J Biol Chem 289:6177–6187. https://doi.org/10.1074/jbc.M113.522623

Grzegorzewicz AE, Pham H, Gundi Va KB et al (2012) Inhibition of mycolic acid transport across the *Mycobacterium tuberculosis* plasma membrane. Nat Chem Biol 8:334–341. https://doi.org/10.1038/nchembio.794

Grzegorzewicz AE, de Sousa-d'Auria C, McNeil MR et al (2016) Assembling of the *Mycobacterium tuberculosis* cell wall core. J Biol Chem 291:18867–18879. https://doi.org/10.1074/jbc.M116.739227

Hansmeier N, Bartels FW, Ros R et al (2004) Classification of hyper-variable *Corynebacterium glutamicum* surface-layer proteins by sequence analyses and atomic force microscopy. J Biotechnol 112:177–193. https://doi.org/10.1016/j.jbiotec.2004.03.020

Hansmeier N, Albersmeier A, Tauch A et al (2006a) The surface (S)-layer gene cspB of *Corynebacterium glutamicum* is transcriptionally activated by a LuxR-type regulator and located on a 6 kb genomic island absent from the type strain ATCC 13032. Microbiology 152:923–935. https://doi.org/10.1099/mic.0.28673-0

Hansmeier N, Chao T-C, Pühler A et al (2006b) The cytosolic, cell surface and extracellular proteomes of the biotechnologically important soil bacterium *Corynebacterium efficiens* YS-314 in comparison to those of *Corynebacterium glutamicum* ATCC 13032. Proteomics 6:233–250. https://doi.org/10.1002/pmic.200500144

Hartmann M, Barsch A, Niehaus K et al (2004) The glycosylated cell surface protein Rpf2, containing a resuscitation-promoting factor motif, is involved in intercellular communication of *Corynebacterium glutamicum*. Arch Microbiol 182:299–312. https://doi.org/10.1007/s00203-004-0713-1

Heider SAE, Wolf N, Hofemeier A et al (2014) Optimization of the IPP precursor supply for the production of lycopene, decaprenoxanthin and astaxanthin by *Corynebacterium glutamicum*. Front Bioeng Biotechnol 2:1–13. https://doi.org/10.3389/fbioe.2014.00028

Hirasawa T, Wachi M, Nagai K (2000) A mutation in the *Corynebacterium glutamicum ltsA* gene causes susceptibility to lysozyme, temperature-sensitive growth, and L-glutamate production. J Bacteriol 182:2696–2701. https://doi.org/10.1128/jb.182.10.2696-2701.2000

Hirasawa T, Wachi M, Nagai K (2001) L-glutamate production by lysozyme-sensitive *Corynebacterium glutamicum ltsA* mutant strains. BMC Biotechnol 1:9

Hoffmann C, Leis A, Niederweis M et al (2008) Disclosure of the mycobacterial outer membrane: cryo-electron tomography and vitreous sections reveal the lipid bilayer structure. Proc Natl Acad Sci U S A 105:3963–3967. https://doi.org/10.1073/pnas.0709530105

Huc E, de Sousa-D'Auria C, de la Sierra-Gallay IL et al (2013) Identification of a mycoloyl transferase selectively involved in o-acylation of polypeptides in *Corynebacteriales*. J Bacteriol 195:4121–4128. https://doi.org/10.1128/JB.00285-13

Hünten P, Costa-Riu N, Palm D et al (2005) Identification and characterization of PorH, a new cell wall channel of *Corynebacterium glutamicum*. Biochim Biophys Acta 1715:25–36. https://doi.org/10.1016/j.bbamem.2005.07.011

Ikeda M, Nagashima T, Nakamura E et al (2017) In vivo roles of fatty acid biosynthesis enzymes in biosynthesis of biotin and α-lipoic acid in *Corynebacterium glutamicum*. Appl Environ Microbiol 83:1–15. https://doi.org/10.1128/AEM.01322-17

Ishizaki Y, Hayashi C, Inoue K et al (2013) Inhibition of the first step in synthesis of the mycobacterial cell wall core, catalyzed by the GlcNAc-1-phosphate transferase WecA, by the novel caprazamycin derivative CPZEN-45. J Biol Chem 288:30309–30319. https://doi.org/10.1074/jbc.M113.492173

Issa H, Huc-Claustre E, Reddad T et al (2017) Click-chemistry approach to study mycoloylated proteins: evidence for PorB and PorC porins mycoloylation in *Corynebacterium glutamicum*. PLoS One 12:e0171955. https://doi.org/10.1371/journal.pone.0171955

Jackson M (2014) The mycobacterial cell envelope-lipids. Cold Spring Harb Perspect Med 4: a021105–a021105. https://doi.org/10.1101/cshperspect.a021105

Jankute M, Byng CV, Alderwick LJ, Besra GS (2014) Elucidation of a protein-protein interaction network involved in *Corynebacterium glutamicum* cell wall biosynthesis as determined by bacterial two-hybrid analysis. Glycoconj J 31:475–483. https://doi.org/10.1007/s10719-014-9549-3

Jankute M, Alderwick LJ, Moorey AR et al (2018) The singular *Corynebacterium glutamicum* Emb arabinofuranosyltransferase polymerises the $\alpha(1 \rightarrow 5)$ arabinan backbone in the early stages of cell wall arabinan biosynthesis. Cell Surf 2:38–53. https://doi.org/10.1016/j.tcsw.2018.06.003

Kacem R, de Sousa-D'Auria C, Tropis M et al (2004) Importance of mycoloyltransferases on the physiology of *Corynebacterium glutamicum*. Microbiology 150:73–84

Kalinowski J, Bathe B, Bartels D et al (2003) The complete *Corynebacterium glutamicum* ATCC 13032 genome sequence and its impact on the production of L-aspartate-derived amino acids and vitamins. J Biotechnol 104:5–25

Kalscheuer R, Weinrick B, Veeraraghavan U et al (2010) Trehalose-recycling ABC transporter LpqY-SugA-SugB-SugC is essential for virulence of *Mycobacterium tuberculosis*. Proc Natl Acad Sci U S A 107:21761–21766. https://doi.org/10.1073/pnas.1014642108

Kaur D, Brennan PJ, Crick DC (2004) Decaprenyl diphosphate synthesis in *Mycobacterium tuberculosis*. J Bacteriol 186:7564–7570

Klatt S, Brammananth R, O'Callaghan S et al (2018) Identification of novel lipid modifications and intermembrane dynamics in *Corynebacterium glutamicum* using high-resolution mass spectrometry. J Lipid Res 59:1190–1204. https://doi.org/10.1194/jlr.M082784

Kolly GS, Mukherjee R, Kilacsková E et al (2015) GtrA protein Rv3789 is required for arabinosylation of arabinogalactan in *Mycobacterium tuberculosis*. J Bacteriol 197:3686–3697. https://doi.org/10.1128/JB.00628-15

Korduláková J, Gilleron M, Mikusova K et al (2002) Definition of the first mannosylation step in phosphatidylinositol mannoside synthesis. PimA is essential for growth of mycobacteria. J Biol Chem 277:31335–31344. https://doi.org/10.1074/jbc.M204060200

Korduláková J, Gilleron M, Puzo G et al (2003) Identification of the required acyltransferase step in the biosynthesis of the phosphatidylinositol mannosides of *Mycobacterium* species. J Biol Chem 278:36285–36295. https://doi.org/10.1074/jbc.M303639200

Kremer L, Dover LG, Morehouse C et al (2001) Galactan biosynthesis in *Mycobacterium tuberculosis*. Identification of a bifunctional UDP-galactofuranosyltransferase. J Biol Chem 276:26430–26440. https://doi.org/10.1074/jbc.M102022200

Lanéelle MA, Tropis M, Daffé M (2013) Current knowledge on mycolic acids in *Corynebacterium glutamicum* and their relevance for biotechnological processes. Appl Microbiol Biotechnol 97:9923–9930. https://doi.org/10.1007/s00253-013-5265-3

Larrouy-Maumus G, Škovierová H, Dhouib R et al (2012) A small multidrug resistance-like transporter involved in the arabinosylation of arabinogalactan and lipoarabinomannan in mycobacteria. J Biol Chem 287:39933–39941. https://doi.org/10.1074/jbc.M112.400986

Lavollay M, Arthur M, Fourgeaud M et al (2009) The beta-lactam-sensitive D,D-carboxypeptidase activity of Pbp4 controls the L,D and D,D transpeptidation pathways in *Corynebacterium jeikeium*. Mol Microbiol 74:650–661. https://doi.org/10.1111/j.1365-2958.2009.06887.x

Layre E, Al-Mubarak R, Belisle JT, Branch Moody D (2014) Mycobacterial lipidomics. Microbiol Spectr 2:1–19. https://doi.org/10.1128/microbiolspec.MGM2-0033-2013

Lea-Smith DJ, Pyke JS, Tull D et al (2007) The reductase that catalyzes mycolic motif synthesis is required for efficient attachment of mycolic acids to arabinogalactan. J Biol Chem 282:11000–11008. https://doi.org/10.1074/jbc.M608686200
Lea-Smith DJ, Martin KL, Pyke JS et al (2008) Analysis of a new mannosyltransferase required for the synthesis of phosphatidylinositol mannosides and lipoarabinomannan reveals two lipomannan pools in *Corynebacterineae*. J Biol Chem 283:6773–6782. https://doi.org/10.1074/jbc.M707139200
Lee MJ, Kim P (2018) Recombinant protein expression system in *Corynebacterium glutamicum* and its application. Front Microbiol 9:2523. https://doi.org/10.3389/fmicb.2018.02523
Lee A, Wu S-W, Scherman MS et al (2006) Sequencing of oligoarabinosyl units released from mycobacterial arabinogalactan by endogenous arabinanase: identification of distinctive and novel structural motifs. Biochemistry 45:15817–15828. https://doi.org/10.1021/bi060688d
Lee J-Y, Na Y-A, Kim E et al (2016) The actinobacterium *Corynebacterium glutamicum*, an industrial workhorse. J Microbiol Biotechnol 26:807–822. https://doi.org/10.4014/jmb.1601.01053
Levefaudes M, Patin D, de Sousa-d'Auria C et al (2015) Diaminopimelic acid amidation in *Corynebacteriales*: new insights into the role of *ltsA* in peptidoglycan modification. J Biol Chem 290:13079–13094. https://doi.org/10.1074/jbc.M115.642843
Ma Y, Mills JA, Belisle JT et al (1997) Determination of the pathway for rhamnose biosynthesis in mycobacteria: cloning, sequencing and expression of the *Mycobacterium tuberculosis* gene encoding alpha-D-glucose-1-phosphate thymidylyltransferase. Microbiology 143:937–945. https://doi.org/10.1099/00221287-143-3-937
Ma Y, Stern RJ, Scherman MS et al (2001) Drug targeting *Mycobacterium tuberculosis* cell wall synthesis: genetics of dTDP-rhamnose synthetic enzymes and development of a microtiter plate-based screen for inhibitors of conversion of dTDP-glucose to dTDP-rhamnose. Antimicrob Agents Chemother 45:1407–1416
Machowski EE, Senzani S, Ealand C, Kana BD (2014) Comparative genomics for mycobacterial peptidoglycan remodelling enzymes reveals extensive genetic multiplicity. BMC Microbiol 14:75. https://doi.org/10.1186/1471-2180-14-75
Mailaender C, Reiling N, Engelhardt H et al (2004) The MspA porin promotes growth and increases antibiotic susceptibility of both *Mycobacterium bovis* BCG and *Mycobacterium tuberculosis*. Microbiology 150:853–864. https://doi.org/10.1099/mic.0.26902-0
Marchand CH, Salmeron C, Raad RB et al (2012) Biochemical disclosure of the mycolate outer membrane of *Corynebacterium glutamicum*. J Bacteriol 194:587–597. https://doi.org/10.1128/JB.06138-11
Marienfeld S, Uhlemann EM, Schmid R et al (1997) Ultrastructure of the *Corynebacterium glutamicum* cell wall. Antonie Van Leeuwenhoek 72:291–297
Marrakchi H, Lanéelle M-A, Daffé M (2014) Mycolic acids: structures, biosynthesis, and beyond. Chem Biol 21:67–85. https://doi.org/10.1016/j.chembiol.2013.11.011
Marrichi M, Camacho L, Russell DG, DeLisa MP (2008) Genetic toggling of alkaline phosphatase folding reveals signal peptides for all major modes of transport across the inner membrane of bacteria. J Biol Chem 283:35223–35235. https://doi.org/10.1074/jbc.M802660200
Matano C, Kolkenbrock S, Hamer SN et al (2016) *Corynebacterium glutamicum* possesses β-N-acetylglucosaminidase. BMC Microbiol 16:1–12. https://doi.org/10.1186/s12866-016-0795-3
May JF, Splain RA, Brotschi C, Kiessling LL (2009) A tethering mechanism for length control in a processive carbohydrate polymerization. Proc Natl Acad Sci 106:11851–11856. https://doi.org/10.1073/pnas.0901407106
McDonough JA, McCann JR, Tekippe EM et al (2008) Identification of functional Tat signal sequences in *Mycobacterium tuberculosis* proteins. J Bacteriol 190:6428–6438. https://doi.org/10.1128/JB.00749-08
McNeil M, Daffe M, Brennan PJ (1990) Evidence for the nature of the link between the arabinogalactan and peptidoglycan of mycobacterial cell walls. J Biol Chem 265:18200–18206

McNeil M, Daffe M, Brennan PJ (1991) Location of the mycolyl ester substituents in the cell walls of mycobacteria. J Biol Chem 266:13217–13223

Meeske AJ, Sham L-T, Kimsey H et al (2015) MurJ and a novel lipid II flippase are required for cell wall biogenesis in *Bacillus subtilis*. Proc Natl Acad Sci 112:6437–6442. https://doi.org/10.1073/pnas.1504967112

Meeske AJ, Riley EP, Robins WP et al (2016) SEDS proteins are a widespread family of bacterial cell wall polymerases. Nature 537:634–638. https://doi.org/10.1038/nature19331

Meniche X, de Sousa-d'Auria C, Van-der-Rest B et al (2008) Partial redundancy in the synthesis of the D-arabinose incorporated in the cell wall arabinan of *Corynebacterineae*. Microbiology 154:2315–2326. https://doi.org/10.1099/mic.0.2008/016378-0

Meniche X, Labarre C, de Sousa-d'Auria C et al (2009) Identification of a stress-induced factor of *Corynebacterineae* that is involved in the regulation of the outer membrane lipid composition. J Bacteriol 191:7323–7332. https://doi.org/10.1128/JB.01042-09

Mikusová K, Mikus M, Besra GS et al (1996) Biosynthesis of the linkage region of the mycobacterial cell wall. J Biol Chem 271:7820–7828

Mikusová K, Huang H, Yagi T et al (2005) Decaprenylphosphoryl arabinofuranose, the donor of the D-arabinofuranosyl residues of mycobacterial arabinan, is formed via a two-step epimerization of decaprenylphosphoryl ribose. J Bacteriol 187:8020–8025. https://doi.org/10.1128/JB.187.23.8020-8025.2005

Mikusova K, Belanova M, Kordulakova J et al (2006) Identification of a novel galactosyl transferase involved in biosynthesis of the mycobacterial cell wall. J Bacteriol 188:6592–6598

Mills JA, Motichka K, Jucker M et al (2004) Inactivation of the mycobacterial rhamnosyltransferase, which is needed for the formation of the arabinogalactan-peptidoglycan linker, leads to irreversible loss of viability. J Biol Chem 279:43540–43546. https://doi.org/10.1074/jbc.M407782200

Mishra AK, Alderwick LJ, Rittmann D et al (2007) Identification of an alpha(1-->6) mannopyranosyltransferase (MptA), involved in *Corynebacterium glutamicum* lipomanann biosynthesis, and identification of its orthologue in Mycobacterium tuberculosis. Mol Microbiol 65:1503–1517. https://doi.org/10.1111/j.1365-2958.2007.05884.x

Mishra AK, Alderwick LJ, Rittmann D et al (2008a) Identification of a novel alpha(1-->6) mannopyranosyltransferase MptB from *Corynebacterium glutamicum* by deletion of a conserved gene, NCgl1505, affords a lipomannan- and lipoarabinomannan-deficient mutant. Mol Microbiol 68:1595–1613. https://doi.org/10.1111/j.1365-2958.2008.06265.x

Mishra AK, Klein C, Gurcha SS et al (2008b) Structural characterization and functional properties of a novel lipomannan variant isolated from a *Corynebacterium glutamicum* pimB' mutant. Antonie Van Leeuwenhoek 94:277–287. https://doi.org/10.1007/s10482-008-9243-1

Mishra AK, Batt S, Krumbach K et al (2009) Characterization of the *Corynebacterium glutamicum* deltapimB' deltamgtA double deletion mutant and the role of *Mycobacterium tuberculosis* orthologues Rv2188c and Rv0557 in glycolipid biosynthesis. J Bacteriol 191:4465–4472. https://doi.org/10.1128/JB.01729-08

Mishra AK, Krumbach K, Rittmann D et al (2011) Lipoarabinomannan biosynthesis in *Corynebacterineae*: the interplay of two α(1→2)-mannopyranosyltransferases MptC and MptD in mannan branching. Mol Microbiol 80:1241–1259. https://doi.org/10.1111/j.1365-2958.2011.07640.x

Mohiman N, Argentini M, Batt SM et al (2012) The ppm operon is essential for acylation and glycosylation of lipoproteins in *Corynebacterium glutamicum*. PLoS One 7:e46225. https://doi.org/10.1371/journal.pone.0046225

Münch D, Roemer T, Lee SH et al (2012) Identification and in vitro analysis of the GatD/MurT enzyme-complex catalyzing lipid II amidation in *Staphylococcus aureus*. PLoS Pathog 8:e1002509. https://doi.org/10.1371/journal.ppat.1002509

Pavelka MS, Mahapatra S, Crick DC et al (2014) Genetics of peptidoglycan biosynthesis. Microbiol Spectr 2:1–20. https://doi.org/10.1128/microbiolspec.MGM2-0034-2013

Peng W, Zou L, Bhamidi S et al (2012) The galactosamine residue in mycobacterial arabinogalactan is α-linked. J Org Chem 77:9826–9832. https://doi.org/10.1021/jo301393s

Peyret JL, Bayan N, Joliff G et al (1993) Characterization of the cspB gene encoding PS2, an ordered surface-layer protein in *Corynebacterium glutamicum*. Mol Microbiol 9:97–109. https://doi.org/10.1111/j.1365-2958.1993.tb01672.x

Portevin D, de Sousa-D'Auria C, Houssin C et al (2004) A polyketide synthase catalyzes the last condensation step of mycolic acid biosynthesis in mycobacteria and related organisms. Proc Natl Acad Sci U S A 101:314–319. https://doi.org/10.1073/pnas.0305439101

Portevin D, de Sousa-D'Auria C, Montrozier H et al (2005) The acyl-AMP ligase FadD32 and AccD4-containing acyl-CoA carboxylase are required for the synthesis of mycolic acids and essential for mycobacterial growth: identification of the carboxylation product and determination of the acyl-CoA carboxylase component. J Biol Chem 280:8862–8874. https://doi.org/10.1074/jbc.M408578200

Puech V, Chami M, Lemassu A et al (2001) Structure of the cell envelope of corynebacteria: importance of the non-covalently bound lipids in the formation of the cell wall permeability barrier and fracture plane. Microbiology 147:1365–1382. https://doi.org/10.1099/00221287-147-5-1365

Radmacher E, Alderwick LJ, Besra GS et al (2005) Two functional FAS-I type fatty acid synthases in *Corynebacterium glutamicum*. Microbiology 151:2421–2427. https://doi.org/10.1099/mic.0.28012-0

Raghavendra T, Patil S, Mukherjee R (2018) Peptidoglycan in mycobacteria: chemistry, biology and intervention. Glycoconj J 35:421–432. https://doi.org/10.1007/s10719-018-9842-7

Rainczuk AK, Yamaryo-Botte Y, Brammananth R et al (2012) The lipoprotein LpqW is essential for the mannosylation of periplasmic glycolipids in corynebacteria. J Biol Chem 287:42726–42738. https://doi.org/10.1074/jbc.M112.373415

Rath P, Demange P, Saurel O et al (2011) Functional expression of the PorAH channel from *Corynebacterium glutamicum* in cell-free expression systems: implications for the role of the naturally occurring mycolic acid modification. J Biol Chem 286:32525–32532. https://doi.org/10.1074/jbc.M111.276956

Rose JD, Maddry JA, Comber RN et al (2002) Synthesis and biological evaluation of trehalose analogs as potential inhibitors of mycobacterial cell wall biosynthesis. Carbohydr Res 337:105–120

Ruiz N (2015) Lipid flippases for bacterial peptidoglycan biosynthesis. Lipid Insights 8:21–31. https://doi.org/10.4137/LPI.S31783

Sambou T, Dinadayala P, Stadthagen G et al (2008) Capsular glucan and intracellular glycogen of *Mycobacterium tuberculosis*: biosynthesis and impact on the persistence in mice. Mol Microbiol 70:762–774. https://doi.org/10.1111/j.1365-2958.2008.06445.x

Sancho-Vaello E, Albesa-Jové D, Rodrigo-Unzueta A, Guerin ME (2017) Structural basis of phosphatidyl-myo-inositol mannosides biosynthesis in mycobacteria. Biochim Biophys Acta – Mol Cell Biol Lipids 1862:1355–1367. https://doi.org/10.1016/j.bbalip.2016.11.002

Saxton WO, Baumeister W (1986) Principles of organization in S layers. J Mol Biol 187:251–253

Scheuring S, Stahlberg H, Chami M et al (2002) Charting and unzipping the surface layer of *Corynebacterium glutamicum* with the atomic force microscope. Mol Microbiol 44:675–684

Schleifer KH, Kandler O (1972) Peptidoglycan types of bacterial cell walls and their taxonomic implications. Bacteriol Rev 36:407–477

Seidel M, Alderwick LJ, Birch HL et al (2007) Identification of a novel arabinofuranosyltransferase AftB involved in a terminal step of cell wall arabinan biosynthesis in *Corynebacterianeae*, such as *Corynebacterium glutamicum* and *Mycobacterium tuberculosis*. J Biol Chem 282:14729–14740. https://doi.org/10.1074/jbc.M700271200

Sexton DL, St-Onge RJ, Haiser HJ et al (2015) Resuscitation-promoting factors are cell wall-lytic enzymes with important roles in the germination and growth of *Streptomyces coelicolor*. J Bacteriol 197:848–860. https://doi.org/10.1128/JB.02464-14

Shah A, Blombach B, Gauttam R, Eikmanns BJ (2018) The RamA regulon: complex regulatory interactions in relation to central metabolism in *Corynebacterium glutamicum*. Appl Microbiol Biotechnol 102:5901–5910. https://doi.org/10.1007/s00253-018-9085-3

Sham L-T, Butler EK, Lebar MD et al (2014) Bacterial cell wall. MurJ is the flippase of lipid-linked precursors for peptidoglycan biogenesis. Science 345:220–222. https://doi.org/10.1126/science.1254522

Sieger B, Bramkamp M (2015) Interaction sites of DivIVA and RodA from *Corynebacterium glutamicum*. Front Microbiol 6:1–11. https://doi.org/10.3389/fmicb.2014.00738

Sieger B, Schubert K, Donovan C, Bramkamp M (2013) The lipid II flippase RodA determines morphology and growth in *Corynebacterium glutamicum*. Mol Microbiol 90:966–982. https://doi.org/10.1111/mmi.12411

Silhavy TJ, Kahne D, Walker S (2010) The bacterial cell envelope. Cold Spring Harb Perspect Biol 2:a000414. https://doi.org/10.1101/cshperspect.a000414

Soual-Hoebeke E, de Sousa-D'Auria C, Chami M et al (1999) S-layer protein production by *Corynebacterium* strains is dependent on the carbon source. Microbiology 145:3399–3408. https://doi.org/10.1099/00221287-145-12-3399

Sutcliffe IC (2010) A phylum level perspective on bacterial cell envelope architecture. Trends Microbiol 18:464–470. https://doi.org/10.1016/j.tim.2010.06.005

Swoboda JG, Campbell J, Meredith TC, Walker S (2010) Wall teichoic acid function, biosynthesis, and inhibition. ChemBioChem 11:35–45. https://doi.org/10.1002/cbic.200900557

Taniguchi H, Busche T, Patschkowski T et al (2017) Physiological roles of sigma factor SigD in *Corynebacterium glutamicum*. BMC Microbiol 17:158. https://doi.org/10.1186/s12866-017-1067-6

Tatituri RVV, Alderwick LJ, Mishra AK et al (2007a) Structural characterization of a partially arabinosylated lipoarabinomannan variant isolated from a *Corynebacterium glutamicum* ubiA mutant. Microbiology 153:2621–2629. https://doi.org/10.1099/mic.0.2007/008078-0

Tatituri RVV, Illarionov PA, Dover LG et al (2007b) Inactivation of *Corynebacterium glutamicum* NCgl0452 and the role of MgtA in the biosynthesis of a novel mannosylated glycolipid involved in lipomannan biosynthesis. J Biol Chem 282:4561–4572. https://doi.org/10.1074/jbc.M608695200

Toyoda K, Inui M (2018) Extracytoplasmic function sigma factor σD confers resistance to environmental stress by enhancing mycolate synthesis and modifying peptidoglycan structures in *Corynebacterium glutamicum*. Mol Microbiol 107:312–329. https://doi.org/10.1111/mmi.13883

Tropis M, Meniche X, Wolf A et al (2005) The crucial role of trehalose and structurally related oligosaccharides in the biosynthesis and transfer of mycolic acids in *Corynebacterineae*. J Biol Chem 280:26573–26585. https://doi.org/10.1074/jbc.M502104200

Tsuge Y, Ogino H, Teramoto H et al (2008) Deletion of cgR_1596 and cgR_2070, encoding NlpC/P60 proteins, causes a defect in cell separation in *Corynebacterium glutamicum* R. J Bacteriol 190:8204–8214. https://doi.org/10.1128/JB.00752-08

Turapov O, Forti F, Kadhim B et al (2018) Two faces of CwlM, an essential PknB substrate, in *Mycobacterium tuberculosis*. Cell Rep 25:57–67.e5. https://doi.org/10.1016/j.celrep.2018.09.004

Valbuena N, Letek M, Ramos A et al (2006) Morphological changes and proteome response of *Corynebacterium glutamicum* to a partial depletion of FtsI. Microbiology 152:2491–2503. https://doi.org/10.1099/mic.0.28773-0

Valbuena N, Letek M, Ordóñez E et al (2007) Characterization of HMW-PBPs from the rod-shaped actinomycete *Corynebacterium glutamicum*: peptidoglycan synthesis in cells lacking actin-like cytoskeletal structures. Mol Microbiol 66:643–657. https://doi.org/10.1111/j.1365-2958.2004.05943.x

Varela C, Rittmann D, Singh A et al (2012) MmpL genes are associated with mycolic acid metabolism in mycobacteria and corynebacteria. Chem Biol 19:498–506. https://doi.org/10.1016/j.chembiol.2012.03.006

Vincent AT, Nyongesa S, Morneau I et al (2018) The mycobacterial cell envelope: a relict from the past or the result of recent evolution? Front Microbiol 9:2341. https://doi.org/10.3389/fmicb.2018.02341

Vollmer W, Bertsche U (2008) Murein (peptidoglycan) structure, architecture and biosynthesis in *Escherichia coli*. Biochim Biophys Acta 1778:1714–1734. https://doi.org/10.1016/j.bbamem.2007.06.007

Wang J, Elchert B, Hui Y et al (2004) Synthesis of trehalose-based compounds and their inhibitory activities against *Mycobacterium smegmatis*. Bioorg Med Chem 12:6397–6413. https://doi.org/10.1016/j.bmc.2004.09.033

Warrier T, Tropis M, Werngren J et al (2012) Antigen 85C inhibition restricts *Mycobacterium tuberculosis* growth through disruption of cord factor biosynthesis. Antimicrob Agents Chemother 56:1735–1743. https://doi.org/10.1128/AAC.05742-11

Watanabe K, Tsuchida Y, Okibe N et al (2009) Scanning the *Corynebacterium glutamicum* R genome for high-efficiency secretion signal sequences. Microbiology 155:741–750. https://doi.org/10.1099/mic.0.024075-0

Wehrmann A, Phillipp B, Sahm H, Eggeling L (1998) Different modes of diaminopimelate synthesis and their role in cell wall integrity: a study with *Corynebacterium glutamicum*. J Bacteriol 180:3159–3165

Wesener DA, Levengood MR, Kiessling LL (2017) Comparing galactan biosynthesis in *Mycobacterium tuberculosis* and *Corynebacterium diphtheriae*. J Biol Chem 292:2944–2955. https://doi.org/10.1074/jbc.M116.759340

Xu Z, Meshcheryakov VA, Poce G, Chng S-S (2017) MmpL3 is the flippase for mycolic acids in mycobacteria. Proc Natl Acad Sci 114:7993–7998. https://doi.org/10.1073/pnas.1700062114

Yamaryo-Botte Y, Rainczuk AK, Lea-Smith DJ et al (2015) Acetylation of trehalose mycolates is required for efficient MmpL-mediated membrane transport in *Corynebacterineae*. ACS Chem Biol 10:734–746. https://doi.org/10.1021/cb5007689

Yang Y, Shi F, Tao G, Wang X (2012) Purification and structure analysis of mycolic acids in *Corynebacterium glutamicum*. J Microbiol 50:235–240. https://doi.org/10.1007/s12275-012-1459-0

Yang L, Lu S, Belardinelli J et al (2014) RND transporters protect *Corynebacterium glutamicum* from antibiotics by assembling the outer membrane. Microbiologyopen 3:484–496. https://doi.org/10.1002/mbo3.182

Zapun A, Philippe J, Abrahams KA et al (2013) In vitro reconstitution of peptidoglycan assembly from the gram-positive pathogen *Streptococcus pneumoniae*. ACS Chem Biol 8:2688–2696. https://doi.org/10.1021/cb400575t

Zuber B, Chami M, Houssin C et al (2008) Direct visualization of the outer membrane of mycobacteria and corynebacteria in their native state. J Bacteriol 190:5672–5680. https://doi.org/10.1128/JB.01919-07

Respiratory Chain and Energy Metabolism of *Corynebacterium glutamicum*

Naoya Kataoka, Minenosuke Matsutani, and Kazunobu Matsushita

Contents

Abstract The production of glutamic acid as well as amino acids of the aspartic acid family from carbohydrates is carried out industrially by a group of bacteria represented by *Corynebacterium glutamicum. C. glutamicum* is a Gram-positive facultative aerobe with a thick cell wall comprising, besides the peptidoglycan layer, a layer of mycolic acid and arabinogalactan. Industrially, glutamic acid production is induced by various stresses such as biotin limitation or the addition of surfactants or

N. Kataoka · M. Matsutani · K. Matsushita (✉)
Graduate School of Science and Technology for Innovation, Yamaguchi University, Yamaguchi, Japan

Research Center for Thermotolerant Microbial Resources, Yamaguchi University, Yamaguchi, Japan
e-mail: kazunobu@yamaguchi-u.ac.jp

M. Inui, K. Toyoda (eds.), *Corynebacterium glutamicum*, Microbiology Monographs 23, https://doi.org/10.1007/978-3-030-39267-3_3

antibiotics, which may alter the cell membrane tension or opening of the mechanosensitive channel as well as lead to the reduction of 2-oxoglutarate dehydrogenase activity that in turn affects the flow of the TCA cycle. Albeit the molecular events associated with these phenotypes are not fully understood, glutamate production in *C. glutamicum* is related to the cell surface structure (cell membrane) and the metabolic flux (especially through the TCA cycle) since these are deeply involved in cellular energetics. In the aerobic respiratory chain of *C. glutamicum*, several primary dehydrogenases, including type II NADH dehydrogenase, and malate: quinone oxidoreductase- and lactate dehydrogenase-dependent NADH reoxidizing systems, function donating electrons from each substrates to menaquinone, and the resulting menaquinol is oxidized by cytochrome $bcc\text{-}aa_3$ supercomplex and cytochrome *bd* oxidase. The $bcc\text{-}aa_3$ supercomplex and *bd* oxidase generate proton-motive force, H^+/O ratio of 6 and 2, respectively. In this chapter, molecular characters and energetics of these respiratory components are summarized and also some metabolic engineering of *C. glutamicum* to produce valuable chemicals are described related to the respiratory chain.

1 Menaquinone Reduction in the Respiratory Chain of *C. glutamicum*

The respiratory chain of *C. glutamicum* consists of many primary dehydrogenases and four cytochrome complexes, cytochrome bc_1c (*bcc*) complex (uniquely this bc_1 complex contains an additional heme *c*, so that it is called as bc_1c or *bcc* separately from reference to reference. In this chapter, *bcc* is used only for general description), cytochrome aa_3 oxidase, cytochrome *bd* oxidase, and nitrate reductase. The primary dehydrogenases transfer the reducing equivalents obtained by the oxidation of various substrates to menaquinone (MQ) to produce menaquinol (MQH_2). MQH_2 is further oxidized by three terminal branches, two oxygen reductases and one nitrate reductase, which couple the electron transfer to the generation of an electrochemical proton gradient, namely, a proton motive force (Δp), across the cytoplasmic membrane (Fig. 1).

1.1 Primary Dehydrogenases of the Respiratory Chain

Biochemical analyses, together with information obtained from the genome sequence, have shown that *C. glutamicum* possesses at least eight different primary dehydrogenases in the entry (primary) site of the respiratory chain (Bott and Neibisch 2005), of which six enzymes relatively well characterized are shown in Fig. 1. These primary dehydrogenases are single subunit peripheral flavoproteins located at the inner surface

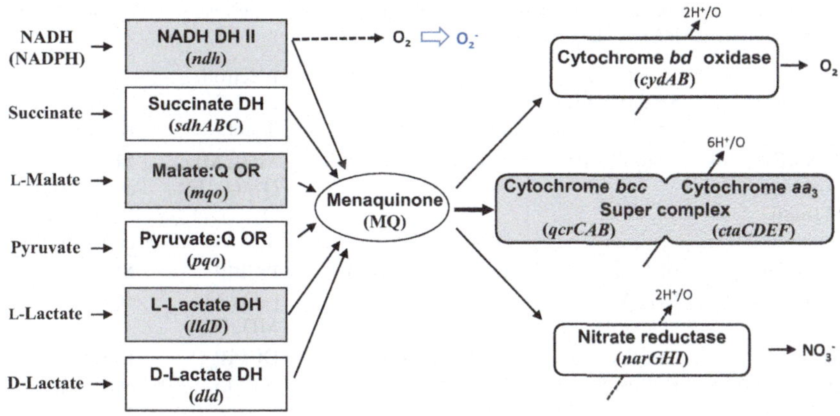

Fig. 1 Known respiratory components of the respiratory chain of *C. glutamicum*. Abbreviation: *DH* dehydrogenase, *Q* quinone, *OR* oxido-reductase

of the cytoplasmic membrane, except for succinate dehydrogenase or succinate: menaquinone oxidoreductase (SDH or SQR) consisting of three subunits, SdhA, SdhB, and SdhC. Five of these primary dehydrogenases have been purified and characterized (Table 1), of which SDH, type II NADH dehydrogenase (NDH-2), D-lactate dehydrogenase, and L-lactate dehydrogenase (LldD) have been solubilized and purified with detergents, such as dodecylmaltoside, Triton X-100, or *N,N*-dimethyldodecylamine *N*-oxide. The need to use strong detergents to solubilize these proteins indicates that these enzymes are tightly bound to the membrane. However, malate:quinone oxidoreductase (MQO) and pyruvate:quinone oxidoreductase (PQO) seem to be only weakly attached to the membrane since they are released into the supernatant by, respectively, the mere addition of some chelators or by mechanical cell disruption without solubilization. Of these dehydrogenases, SDH is unique in having an anchoring subunit, SdhC, which is a cytochrome *b* having five transmembrane helices (Fig. 2). However, the exact membrane-anchoring sites have not been identified in any other single subunit flavoproteins, of which only NDH-2 is suggested to be a membrane protein by the secondary structure prediction, or by information with the crystal structures from other microbes, as described later. All other enzymes, including loosely attached MQO and PQO, are not expected to be membrane proteins by such the prediction as HHMTOP (http://www.enzim.hu/hmmtop/index.php), SOSUI (http://harrier.nagahama-i-bio.ac.jp/sosui/), or TMHMM (http://www.cbs.dtu.dk/services/TMHMM/).

Nevertheless, all these dehydrogenases seem to donate electrons directly to membranous MQ since the reactivity to quinone analogues has been shown with purified preparations of these peripheral flavoproteins. SDH has been demonstrated to reduce dichlorophenolindophenol (DCPIP) in a reaction mediated by decylquinone (DQ) as well as phenazine methosulfate (Kurokawa and Sakamoto 2005). Likewise, MQO has been shown to reduce DCPIP via ubiquinone-1 (Q_1) or menadione (MD) (Molenaar et al. 1998). In NDH-2, quinone reduction has been

Table 1 Primary dehydrogenases and cytochrome oxidases in the *C. glutamicum* respiratory chain that have been characterized at least partly

Enzymes (gene name)	Solubilization with	Subunit (cofactors)	Electron acceptors or donors	References
Type II NADH dehydrogenase; NDH-2 (*ndh*)	Triton X-100	55 kDa (FAD)	MQ_2, MD, Q_1, Q_2, FR, O_2	Matsushita et al. (2001) Nantapong et al. (2005)
Malate:quinone oxidoreductase (*mqo*)	EDTA/ EGTA/ glycerol	56-57 kDa (FAD[a])	DCPIP (+Q_0, Q_1, MD, DQ)	Molenaar et al. (1998)
Succinate: menaquinone oxidoreductase or Succinate dehydrogenase (*sdhABC*)	Dodecyl maltoside	67, 29, 23 kDa (2 Heme B, FAD[b], FeS[b])	DCPIP (+PMS, DQ)	Kurokawa and Sakamoto (2005)
Pyruvate:quinone oxidoreductase (*pqo*)	Soluble	62 kDa (tetramer) (FAD, TPP[a], divalent cation[a])	DCPIP, MD, INT	Schreiner and Eikmanns (2005)
D-Lactate dehydrogenase (*dld*)	1) LDAO	66 kDa (FAD[a])	DCPIP, DCPIP (+PMS)	Bott and Niebisch (2003)
	2) Triton X-100/ 300 mM NaCl	–	DCPIP	Kato et al. (2010)
L-Lactate dehydrogenase (*lldD*)[c]	Triton X-100/ glycerol	46 kDa (FMN[b])	DCPIP, DCPIP (+PMS)	Matsushita et al. (unpublished)
Cytochrome aa_3	Dodecyl maltoside	64.3, 40.5, 21 kDa (geranyl-geranyl Heme A, Cu)	TMPD, yeast cyt *c*, horse-heart cyt *c*	Sakamoto et al. (2001)
Cytochrome bc_1c/ cytochrome aa_3 supercomplex (*qcrCAB, ctaCDEF*)	Dodecyl maltoside/ 100 mM NaCl	60, 45, 30 kDa/65, 40, 22, 16 kDa/P29, P24, P20, P17 (Heme A, Heme B, Heme C, Cu)	$DMNQH_2$, TMPD, bovine-heart cyt *c*, yeast cyt *c*	Niebisch and Bott (2003) Bott and Niebisch (2005) Kao et al. (2016)
Cytochrome *bd* (*cydAB*)	MEGA 9+10	56.4, 41.5 kDa (Heme B, Heme D)	$DMNQH_2$ (+ MQ_1, MQ_2, MQ_3)	Kusumoto et al. (2000)

MQ_n menaquinone-n, *MD* menadione, *$DMNQH_2$* dimethylnaphthoquinol, *Q_n* ubiquinone-n, *DQ* decylquinone, *FR* ferricyanide, *DCPIP* 2,6-dichlorophenolindophenol, *PMS* phenazine methosulfate, *INT* *p*-indonitrotetrazolium violet, *TMPD* *N,N,N′,N′*-tetramethyl phenylene diamine, *cyt* c cytochrome *c*, *LDAO* lauryldimethylamine oxide, *MEGA* alkyl-N-methylglucamides

[a]Activated by the addition

[b]Not determined in this sample but detected in other sources

[c]Not purified yet

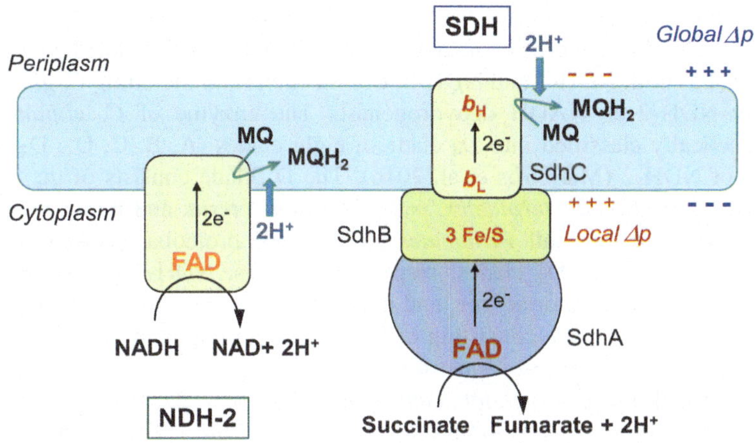

Fig. 2 Succinate dehydrogenase and Δp-dependent reaction. Schematic representation of the electron transfer route model in the succinate dehydrogenase complex (SDH) of *C. glutamicum*. The electrons from succinate are transferred to MQ via FAD, three iron-sulfur clusters (Fe/S), and low and high potential heme B (b_L and b_H, respectively). During the electron transfer from succinate to MQ, $2H^+$ are released inside and taken up outside, and an inside positive membrane potential (local Δp) is generated. Type II NADH dehydrogenase (NDH-2) is also shown as a Δp-independent control. Global Δp is generated by MQH_2 oxidation systems of cytochrome *bcc-aa*$_3$ oxidase or cytochrome *bd* oxidase, and used to drive the electron transfer in SDH from succinate to MQ

shown to occur more directly, as described later. Thus, all primary dehydrogenases present in *C. glutamicum* respiratory chain seems to donate electrons directly to MQ.

1.2 NADH Dehydrogenase

None of these primary dehydrogenases as described above should contribute to Δp generation, except for SDH as described below. A single subunit NDH-2 is a non-energy-generating NADH dehydrogenase that is different from type I NADH dehydrogenase (Complex I or NDH-1) (Matsushita et al. 1987). NDH-1 is a large molecular complex consisting of 14 conserved core subunits from bacteria to mammals, and has been proposed to pumps $4H^+$ per one molecule of NADH oxidation on the basis of the X-ray crystal structures from bacteria (Efremov et al. 2010; Baradaran et al. 2013) to eukaryotic mitochondria (Zickermann et al. 2015; Zhu et al. 2016; Fiedorczuk et al. 2016), and also by biochemical evidence (Jones et al. 2017). NDH-1 is present in almost all of eukaryotic mitochondria, while its occurrence is limited in bacteria, of which 52% equip the enzyme (Spero et al. 2015). More than half of *Actinobacteria*, including the genus *Mycobacterium*, also have NDH-1 (Spero et al. 2015), but the genus *Corynebacterium*, one of *Actinobacteria*, has no NDH-1 except for some minor strains of this genus (unpublished data). Whereas, NDH-2 is predicted to be present in 61% of the microbial genomes, and

89% of *Actinobacteria*, including genus *Corynebacterium*, contain at least one gene encoding NDH-2 (Marreiros et al. 2016). Thus, though closely related genus *Mycobacterium* has both NDH-1 and NDH-2, *Corynebacterium* including *C. glutamicum* has only NDH-2 as NADH dehydrogenase. The enzyme of *C. glutamicum* is phylogenetically classified into D_4 clade of eight clades (A, B, C, D_1, D_2, D_3, D_4 and D_5) of NDH-2 (Marreiros et al. 2016). The D_4 clade consists of the enzymes from *Actinobacteria*, *Bacteroidetes*, part of *Cyanobacteria*, and minor numbers of *Proteobacteria*. Almost all *Firmicutes* NDH-2 and proteobacterial enzymes are classified into D_1 and D_2 clades, respectively. Thus, NDH-2 phylogeny is not congruent with the taxonomic tree, and interestingly the D_4 clade is more phylogenetically close to D_5 clade consisting of NDH-2 from Fungi and Protists, including *Saccharomyces cerevisiae* or *Neurospora crassa*.

As described above, *C. glutamicum* NDH-2 has been purified from the membranes of the over-producing strain (Nantapong et al. 2005). The enzyme is tightly bound to the membranes so that it is solubilized with Triton X-100 (Table 1). Actually, the secondary structure prediction of NDH-2 suggests that it exhibits a single putative transmembrane helix in the C-terminal (the residue numbers from 386 to 405 by HHMTOP, 383 to 405 by TMHMM, or 384 to 406 by SOSUI). In *Escherichia coli* NDH-2, it has been shown that C-terminal domain including the same transmembrane helix as the *C. glutamicum* one (considered as an amphipathic helix in *E. coli*) is indispensable for the membrane binding (Villegas et al. 2011). Later, the structure of NDH-2 was determined in two different enzymes from *S. cerevisiae* (Ndi1: Feng et al. 2012) and *Caldoalkalibacillus thermarum* (Heikal et al. 2014) that conferred more substantial information for the membrane binding or anchoring role of the C-terminal domain. The domain has a β-sheet structure (three anti-parallel β-strands; β19~β21 in Ndi1) and two amphipathic α-helices (α15 and α16 in Ndi1) (Fig. 3b), which seem not to be transmembraneous, instead anchoring to the membrane surface. When *C. glutamicum* enzyme is aligned with these known enzymes of *S. cerevisiae*, *C. thermarum*, *Mycobacterium tuberculosis, E. coli*, similar conserved domains were detected (Fig. 3b). Furthermore, the modeling structure of *C. glutamicum* enzyme also shows a similarly oriented β-sheet structure and two amphipathic α-helicies (Fig. 3a). Thus, *C. glutamicum* NDH-2 seems to be anchored into the membrane surface based on the similar C-terminal a β-sheet and two α-helicies like NDH-2s of *C. thermarum* or *S. cerevisiae* (Fig. 3a).

This enzyme, like other NDH-2, is able to donate electrons to quinone, ubiquinone (UQ) or MQ. It has been shown to be able to reduce Q_1, ubiquinone-2 (Q_2), and MD directly, and also ferricyanide via menaquinone-2 (MQ_2) (Nantapong et al. 2005). The reactivity with these quinone analogues is likely to be physiologically relevant because it has been shown to have a high affinity (Km values are 41, 32, 31, and 5 μM for Q_1, Q_2, MD, and MQ_2, respectively) and a high reactivity (Vmax values of 82~841 units/mg protein) *in vitro* at pH 7.0. Thus, it can be inferred that NDH-2 in *C. glutamicum* is likely to oxidize NADH in a process that is tightly coupled with the natural electron acceptor MQ located within the membrane. The quinone binding portion has been found near the same C terminal domain in yeast Ndi1 structure (Feng et al. 2012), where one UQ was located around the β-sheet and

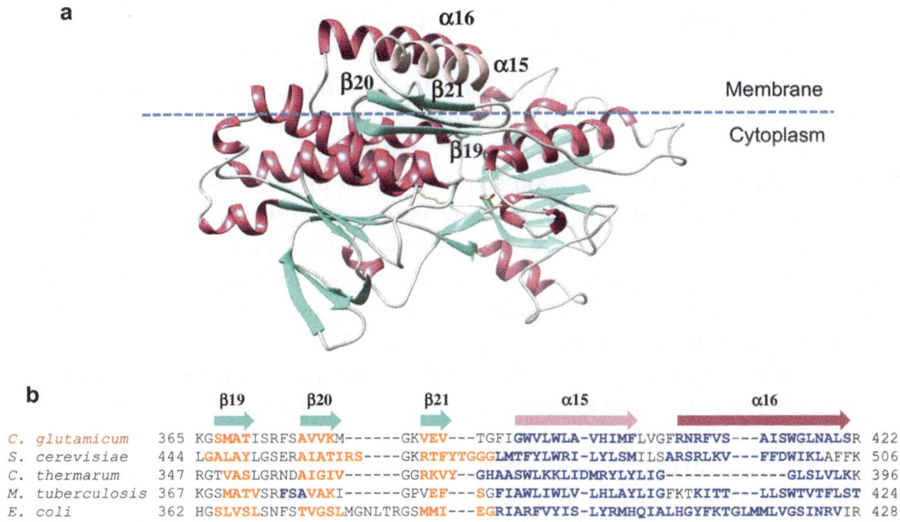

Fig. 3 Membrane anchoring domain of NDH-2. (**a**) Comparative modeling structure of Ndh protein from *C. glutamicum*. Model was constructed by the SWISS-MODEL (PMID: 29788355; Waterhouse et al. 2018). Crystal structure of NADH dehydrogenase from *Plasmodium falciparum* (PDB code: 5jwa.1A) was used as a template structure. (**b**) Amino acid sequence alignment of the membrane anchoring domains of NDH-2s from *C. glutamicum*, *Saccharomyces cerevisiae*, *Caldalkalibacillus thermarum*, *Mycobacterium tuberculosis*, and *E. coli*. Sequence alignment was performed by using MUSCLE v.3.8.31 (PMID: 15034147; Edgar 2004). Sequence data from NCBI GenBank and Reference Sequence databases, *C. glutamicum* ATCC 13032 (NP_600682.1), *S. cerevisiae* S288C (NP_013586.1), *C. thermarum* TA2.A1 (EGL84165.1), *M. tuberculosis* H37Rv (NP_216370.1), and *E. coli* K-12 (NP_415627.1), were used for the sequence alignment construction. Bold orange and blue colors indicate β-sheet and α-helix, respectively. Secondary structure of *C. glutamicum* was calculated from the modeling structure as shown in (**a**). Secondary structures of *C. thermarum* and *S. cerevisiae*, were calculated from their crystal structures, PDB accession No. 4NWZ and 4G6G, respectively, by using DSSP program (PMID: 6667333; Kabsch and Sander 1983). Secondary structure of two sequences, *M. tuberculosis*, and *E. coli*, were predicted using PredictProtein (Yachdav et al. 2014). The three arrows (cyan) for β-strands in the β-sheet (β19, β20, β21) and the two arrows (purple) for α-helices (α15, α16) exhibit the sequences of *C. glutamicum* Ndh modeling structure (**a**)

one α-helix (α15). Almost the same quinone location has been expected in *C. thermarum* enzyme (Heikal et al. 2014), where two conserved glutamine residues (Q394 and Q398 in Ndi1) occupy near UQ site, and "AQXAXQ motif" is proposed as the quinone binding site. Later, Marreiros et al. (2016) indicated that the AQ(P)xAxQ(R) motif is widely spread in many NDH-2s. *C. glutamicum* enzyme also have the same motif AQVAIQ (A340 to Q346).

In the catalytic site, NDH-2 has two conserved nucleotide binding domains; flavin adenine dinucleotide (FAD) binding domains with a glycine-rich consensus sequence (GXGXXG) along with an NADH binding domain containing a negatively charged residue (D or E) or neutral residue (Q or N) at the end of the second β-sheet, which determines the preference for NADH or NADPH, respectively (Desplats et al. 2007). These domains correspond to GAGPTG (G175 to G180) and LLDGA (L211 to A215) in *C. glutamicum* NDH-2. Komati et al. (2015) have found suppressor mutants that are able to oxidize NADPH better and thus grow faster than a parental (mutant) strain having NADPH-generating glycolytic pathway. The suppressor mutants had a single point mutation leading to the amino acid change D213G in NDH-2, and did NADPH oxidation activity even at pH 7.5. They also constructed the mutants, D213N and D213Q, to show similar NADPH oxidation activity with the reactivity of 36–64% of the NADH oxidation, and with relatively high affinity for NADPH from 103 to 180 μM. Thus, *C. glutamicum* enzyme is a typical NADH-oxidizing enzyme having the negatively charged residue (D213) and thus has a preference for NADH. However, besides the NADH oxidation, the NDH-2 of *C. glutamicum* has a relatively high NADPH oxidation (Q_1 reduction) activity at acidic pH but not at neutral pH (20.5 and ~0 units/mg protein for NADPH; 95 and 120 units/mg for NADH, at pH 5.0 and 7.0, respectively), although the affinity for the nucleotides is 40 times lower in NADPH than NADH (Km values of 394 μM for NADPH at pH 5.0, and 11.6 μM for NADH at pH 7.0) (Nantapong et al. 2005). Actually, D213G mutant also shows ~6 times higher NADPH oxidation activity at pH 5.5 than the wild-type enzyme (Komati et al. 2015).

1.3 Other Aspects of NADH Dehydrogenase: Cyanide-Resistant Oxidase Activity and Superoxide Generation

As described above, the respiratory chain of *C. glutamicum* has branched terminal oxidases, which show different cyanide sensitivities that depend on growth phase (Kusumoto et al. 2000) and respiratory substrate (Matsushita et al. 1998). In the stationary phase cells, NADH oxidase activity in the membrane exhibits three different cyanide sensitivities corresponding to three distinct enzymes: the first sensitivity level is observed at low cyanide concentrations (~0.5 μM), the second at intermediate concentrations (~3 mM), and the third at concentrations up to 30 mM (Kusumoto et al. 2000). The former two oxidase activities could be respectively assigned to cytochrome aa_3 oxidase and cytochrome *bd* oxidase as described below, but the most cyanide-resistant oxidase activity seems to be attributed to a primary (flavoprotein) dehydrogenase-dependent activity (electron leak), where the flavin moiety in the proximity of the substrate binding site is expected to easily react with oxygen by subtle change of the substrate-binding site conformation, as shown in L-aspartate oxidase (Tedeschi et al. 1997) or carbohydrate oxidase (Xu et al. 2001). It has been widely known that aerobic respiration generates reactive oxygen species (ROS) as byproducts, mainly via flavoproteins, especially NADH dehydrogenase, that react directly with oxygen to generate superoxide (O_2^-) or hydrogen peroxide

(H_2O_2), as shown in many microbes, such as yeast (Davidson and Schiestl 2001), and *E. coli* (Messner and Imlay 1999, 2002).

In *C. glutamicum*, the cyanide-resistant oxidase activity was observed in the membranes with all the substrate oxidations examined, more critical in NADPH and L-lactate oxidations than in NADH and succinate oxidations (Matsushita et al. 1998). Furthermore, NDH-2 purified from *C. glutamicum* has been shown to react directly with oxygen to generate primarily O_2^- then H_2O_2, especially when it oxidizes NADPH (Nantapong et al. 2005). The *C. glutamicum* NDH-2 may thus mainly function as the primary dehydrogenase of the NADH oxidase respiratory chain, but partly function as an NADH oxidase or an NADPH oxidase in a process that is not coupled with the respiratory chain to release oxides (Fig. 1), which in turn could be detoxified by SOD, catalase, and/or several different types of peroxidase. In *C. glutamicum*, Mn-SOD and catalase, has been shown to function to reduce ROS; either SOD or catalase-disruption increased the ROS level, and inversely their overexpression dramatically decreased ROS level (Nantapong et al. 2019). When the growth temperature was increased in *C. glutamicum*, endogenous ROS level was very much increased concomitant with the defect in cell growth (Nantapong et al. 2019). Thus, ROS is probably generated from the respiratory enzymes including NDH-2 by increasing membrane fluidity when grown at higher temperature.

1.4 Succinate Dehydrogenase and Δp-Dependent MQ Reduction

SDH or SQR, including QFR (quinol:fumarate oxidoreductase), have been classified into five different types based on the structural information (subunit composition and number of heme B) (Ohnishi et al. 2000). The *C. glutamicum* SDH falls in the group of *Bacillus* dehydrogenase enzymes that have a single transmembrane cytochrome *b* subunit equipped with two hemes B: namely, high potential (b_H) and low potential (b_L) hemes; notably, this classification is also supported by phylogenetic analyses (Kurokawa and Sakamoto 2005). The two hemes B moieties are expected to be localized close to each other but on opposite sides of the membrane (Fig. 2). This location of the electron carriers prompts us to speculate about the generation of Δp with the electron transfer reaction, which has been shown clearly in the cytochrome *b* subunit of the cytochrome bc_1 complex (see Fig. 4 and a review: Brandt and Trumpower 1994). *Bacillus* SDH has been proposed to generate Δp in the opposite direction from that of cytochrome *b* in the bc_1 complex (compare Figs. 2 and 4). Specifically, it has been shown in *Bacillus subtilis* cells and derived membrane vesicles that the electron transfer from succinate to MQ is inhibited by protonophores and that the reverse electron transfer from MQH_2 to fumarate generates Δp (Schirawski and Unden 1998; Schnorpfeil et al. 2001). More unequivocally, it has been shown in proteoliposomes with purified *Bacillus licheniformis* SDH (Madej et al. 2006) that succinate oxidation by quinone is Δp (inside negative)-

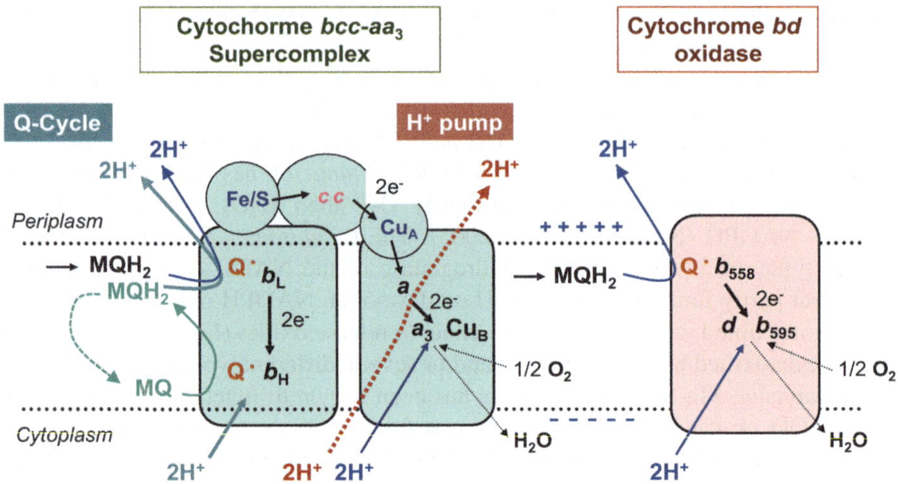

Fig. 4 Schematic representation of H^+ translocation mechanism in cytochrome *bcc*-cytochrome *aa*$_3$ oxidase supercomplex and cytochrome *bd* oxidase. H^+ translocation due to quinone (Q)-cycle is shown by light green arrows, Pump-dependent H^+ translocation by red dashed arrow. Other scalar reactions of H^+ translocation are shown by blue solid arrows. $Q^{\cdot}$ represents semiquinone tightly bound to enzymes

dependent and a reverse electron flow occurs from quinol to fumarate that generates Δp (inside negative). This endergonic reaction with *Bacillus* or *Corynebacterium* enzyme is thermodynamically possible because the redox potential (−80 mV) of the electron acceptor MQ is more negative than that of the electron donor succinate (+31 mV) (Fig. 5), which is in contrast to the exergonic succinate oxidation by UQ (+90 mV) carried out by the typical SDH of mitochondrial Complex II or by the *E. coli* SDH. These latter enzymes comprise a single heme B and, unlike the *C. glutamicum* SDH, do not have two hemes B transmembranously separated.

Thus, the molecular function of the *C. glutamicum* SDH is likely to be similar to that of the *Bacillus* system, since the corynebacterial SDH is structurally classified as being of the *Bacillus*-type and functions in a respiratory chain that contains MQ as the natural quinone, as observed in *Bacillus*. This view is further promoted by the observation that the succinate oxidation in *C. glutamicum* cells is sensitive to a protonophore, CCCP, similarly to what is observed in *B. subtilis* and *B. licheniformis* (Schirawski and Unden 1998). In order to describe this phenomenon more clearly, the Δp-dependent electron transfer of SDH in succinate oxidase system was examined with inside-out membrane (ISO) vesicles prepared from *C. glutamicum* KY9714 (Matsushita 2013). In the presence of 15 mM $MgCl_2$ and 100 μM DCCD (for repressing the leakiness), ISO vesicles generated a sufficiently high membrane potential (inside positive) by the addition of succinate or NADH as the respiratory substrate. Both the succinate- and NADH-dependent membrane potential generations are completely abolished by CCCP less than 1 μM. While, succinate oxidase activity was inhibited completely with 5 μM CCCP but NADH

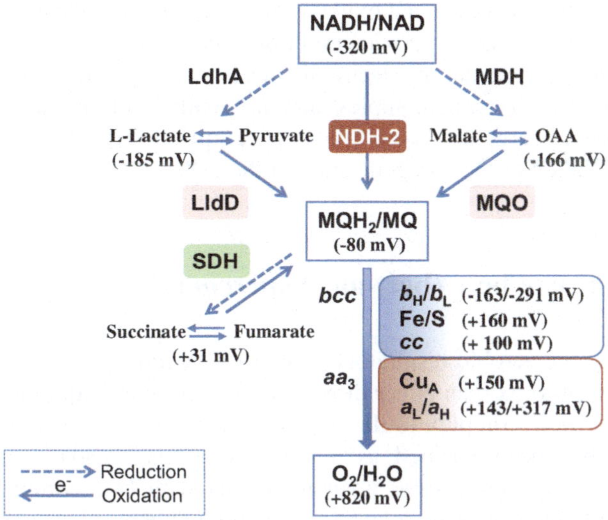

Fig. 5 Redox potential-based schematic representation of NAD oxidation systems, succinate dehydrogenase, and cytochrome *bcc-aa*$_3$ supercomplex. Enzymes represented by their names with colored background are MQ-linked respiratory enzymes, NDH-2, malate-quinone oxidoreductase (MQO), L-lactate dehydrogenase (LldD), and succinate dehydrogenase (SDH), while enzymes represented by their names without any colored background are NAD(P)-dependent enzymes, lactate dehydrogenase (LdhA) and malate dehydrogenase (MDH). This figure also includes the prosthetic groups and their redox potentials (Kao et al. 2016) of cytochrome *bcc* and cytochrome *aa*$_3$ oxidase to oxidize MQH_2

oxidase activity not at all up to the 10 μM. Thus, like the *Bacillus* enzyme, the *C. glutamicum* SDH is expected to generate an opposite Δp (inside positive) (Fig. 2), which could be dissipated by the much higher Δp (inside negative) generated by the quinol oxidase portion or cytochrome *bc*$_1$*c*/cytochrome *c* oxidase as described later (see Fig. 4). Thus, in *C. glutamicum*, electron transfer occurs from succinate to MQ or further to oxygen. At the same time, the opposite electron transfer from fumarate to succinate (fumarate reduction) could occur to generate Δp (inside negative), which have been shown to possibly occur in a mutant strain having a defect of *bc*$_1$*aa*$_3$ respiratory chain or a *ndh*-overproduced pyruvate producer strain, as described later.

2 Menaquinol Oxidation in the Respiratory Chain of *C. glutamicum*

In the respiratory chain of *C. glutamicum*, MQH_2 generated by the several primary dehydrogenases described in the previous section is further oxidized by two oxygen reductases, cytochrome *bcc-aa*$_3$ supercomplex or cytochrome *bd* ubiquinol oxidase, and also by an additional terminal oxidase, nitrate reductase, when nitrate is present in the culture medium under anoxic condition. These terminal oxidases couple the electron transfer to generate an electrochemical proton gradient with the different efficiency between the supercomplex and *bd* oxidase or nitrate reductase. The difference in their energetics would affect the cell physiology and thus their expression is tightly linked to the energetic state of the cells.

2.1 *Cytochrome* bc$_1$c (bcc)-aa$_3$ *Supercomplex*

Downstream of the aerobic respiratory chain of *C. glutamicum*, the electrons from MQH_2 are passed to oxygen either directly to the *bd*-type ubiquinol oxidase or via the cytochrome bc_1c complex to the aa_3-type cytochrome *c* oxidase (Fig. 1). The latter branch has been purified as a cytochrome aa_3 oxidase complex (Cu_A, 2 heme A, and Cu_B) and also as a supercomplex with bc_1c complex (2 heme B, Fe-S, 2 heme C) that is unique in having an additional heme C besides the heme for cytochrome c_1 (Sone et al. 2001), and cytochrome aa_3 oxidase. The aa_3 oxidase was first purified as a three-subunit cytochrome aa_3 oxidase consisting of CtaD, CtaC, and CtaE, and shown to have a unique heme A moiety having a geranyl-geranyl side chain instead of the normal farnesyl group, and also to have an extra-charged amino acid cluster in CtaC expected to have a role in the interaction of the diheme cytochrome *c* subunit of bc_1c complex to the oxidase (Sakamoto et al. 2001). Later, a four-subunits aa_3 oxidase complex (CtaD, CtaC, CtaE, and CtaF) and a super complex, consisting of a total of 11 subunits (of which 4 subunits come from cytochrome aa_3 oxidase, 3 from cytochrome bc_1c complex: QcrB, QcrA, and QcrC, and an additional 4 subunits: P29, P24, P20, and P17) were subsequently separately isolated by affinity chromatography (Niebisch and Bott 2003). Since the three-subunit oxidase and the four-subunit oxidase exhibit in vitro similar turnover numbers for the oxidation of TMPD or horse heart cytochrome *c*, CtaF may not be essential for enzymatic activity despite of being essential for the formation of the active complex (Niebisch and Bott 2003). Recently, the *bcc*-aa_3 supercomplex (now called as *bcc* instead of bc_1c) were shown to be a stable dimeric $(bcc)_2$-$(aa_3)_2$ having a whole molecular weight of ~1100 kDa estimated in Blue-Native gel (Kao et al. 2016). Additionally, more recently, it has been reported that the supercomplex formation is impaired when CtiP (Cg2699) or CopC (Cg1884), possible chaperone for copper insertion into Cu_A or Cu_B sites of the aa_3, was deleted, suggesting these copper centers in aa_3 complex are necessary for the formation of a stable supercomplex (Morosov et al. 2018). Furthermore, only recently, the structure of mycobacterial *bcc*-aa_3 supercomplex that is homologous to the corynebacterial *bcc*-aa_3 supercomplex, was determined by cryo-electron microscopy (Gong et al. 2018). This structure shows aa_3-$(bcc)_2$-aa_3 dimer as expected in corynebacterial supercomplex (Kao et al. 2016), but surprisingly an additional component, SOD, is located at the periplasmic side in the center of the complex. The SOD was expected to work for scavenging O_2^- generated from the *bcc* complex.

The corynebacterial supercomplex has all the prosthetic groups for cytochromes *bcc* and aa_3 (Fig. 4), of which the redox potentials (see Fig. 5) were estimated to be −163, −291, +160, +100, +150, and +143/+317 mV, respectively, for heme b_H, heme b_L, Fe/S, dihemes *c*, Cu_A, and heme a/a_3 (Kao et al. 2016). These redox potentials are much lower than the mitochondrial counterpart to which UQH_2 (+90 mV) donates electrons instead of MQH_2 (−80 mV); highest difference at the heme *b* (b_H/b_L: 280 mV lower), medium difference at the Fe/S to Cu_A (110–160 mV lower), and minimum difference at the aa_3 (77 and 43 mV lower). Thus, corynebacterial supercomplex is a low or mixed potential complex having the lowest potential *bcc* complex reacting with low potential MQ and the relatively high potential aa_3 binuclear center reacting with oxygen. Thus, the *bcc*-aa_3 supercomplex exhibits quinol oxidase activity able to transfer electrons from MQH_2 (2,3-dimethylnaphthoquinol: $DMNQH_2$) directly to oxygen (Niebisch and Bott 2003). The electron transfer rate in the complex from $DMNQH_2$ was determined to be ~210 e^-/s/aa_3, while soluble cytochrome *c* (cyt. *c*)-dependent activity (with TMPD + cyt. *c*: 130 e^-/s/aa_3) was lower than the MQH_2 oxidase activity (Graf et al. 2016). Since the isolated cytochrome aa_3 exhibited much higher cyt. *c* oxidase activity (440 e^-/s/aa_3), soluble cyt. *c* may not be easily accessible to the electron entry (near Cu_A) site by tight binding of the cytochrome *bcc* to the aa_3 in the supercomplex.

2.2 *Cytochrome* bd *Oxidase and Nitrate Reductase*

Cytochrome *bd* oxidase, widely spread in Bacteria and Archaea, is another branch for the MQH_2 oxidation to transfer electrons to oxygen in *C. glutamicum*. There is one more another branch for MQH_2 oxidation in this organism, which is nitrate reductase transferring electrons to nitrate instead of oxygen (Fig. 1). The *bd* oxidase has been purified and shown to consist of large and small subunits, corresponding to CydA and CydB, respectively (Kusumoto et al. 2000). The purified enzyme exhibits a redox difference spectrum having peaks at 560, 595, and 627 nm, typical of cytochrome *bd* such as in *E. coli* enzyme (a multisubunit enzyme comprising 2 heme B, b_{558} and b_{595}, and 1 heme D) (Miller and Gennis 1983; Kita et al. 1984). The *bd* oxidase has MQH_2 oxidase activity, which could be measured as MQ_2-dependent $DMNQH_2$ oxidation activity, via the quinone MQ bound to the enzyme (Kusumoto et al. 2000). This is similar to what is observed in *E. coli* where the cytochrome bo_3 oxidase is associated with a bound quinone, Q_8, which functions for bulk Q reduction (Sato-Watanabe et al. 1994). The *bd* oxidase has a relatively high *K*i value to cyanide of 5.3 mM (Kusumoto et al. 2000). The structure of this enzyme was reported from *Geobacillus thermodenitrificans* (Safarian et al. 2016), and only recently from *E. coli* (Safarian et al. 2019). Both the structures showed the presence of an additional third subunit of ~4 kDa, and also a unique arrangement of the three heme groups, where low spin (hexacoordinated) b_{558}, and two high-spin b_{595} and *d* form a triangular arrangement and the b_{558} is located in close to the Q loop (quinone-binding domain). However, the position of *d* and b_{595} of *Geobacillus* enzyme was shown to be interchanged in *E. coli* one. *C. glutamicum bd* oxidase was expected to be *E. coli*-type enzyme (Safarian et al. 2019).

C. glutamicum has been shown to grow anaerobically in the presence of nitrate, and to consume nitrate to produce nitrite as the reaction product (Nishimura et al. 2007; Takeno et al. 2007). The nitrate-dependent growth seems to be due to one of the anaerobic respirations, nitrate respiration, where nitrate acts as the terminal electron acceptor for the respiratory chain to generate Δp, instead of oxygen as is the case in the aerobic respiration. The nitrate reduction occurs via a membrane-bound respiratory nitrate reductase, called NAR, as demonstrated by the abolition of nitrate-dependent growth upon disruption of the *narG* or *narH* genes of the *C. glutamicum nar* operon (Nishimura et al. 2007), or of the *narG* or *narJ* genes of the same operon (Takeno et al. 2007). The *C. glutamicum* genome has a single *narKGHJI* operon that is very similar to the *narK* gene and *narGHJI* operon of *E. coli*, which is different from the periplasmic nitrate reductase called NAP (see Zumft 1997 for a review). Notably, nitrate may not be utilized as an electron acceptor for nitrate respiration as long as oxygen is available because the *nar* operon expression is repressed by the transcriptional regulator ArnR under aerobic conditions (Nishimura et al. 2008).

2.3 Energetics of the Aerobic Menaquinol Oxidations

As shown in Figs. 1 and 4, the energy coupling efficiencies of the cytochrome *bcc*-aa_3 supercomplex and cytochrome *bd* oxidase (that is, the H^+ translocation ability), are expected to be a H^+/O ratio of 6 and 2, or a H^+/e^- ratio of 3 and 1, respectively. For a two-electron transfer (O reduction), in the *bcc*-aa_3 supercomplex, $2H^+$ are translocated by a Q cycle mechanism (light green arrows in Fig. 4) of the *bcc* complex that does not include H^+ pumping but rather a pure scalar reaction: $2H^+$ are translocated by the H^+ pumping (red broken arrow in Fig. 4) of aa_3 oxidase. Additional scalar $2H^+$ translocation (blue arrows in Fig. 4) can be performed by $2H^+$ release outside from MQH_2 generated in the substrate oxidation by primary dehydrogenase ($MQH_2 \rightarrow MQ + 2H^+_{out}$) and $2H^+$ uptake inside with O_2 reduction to water by aa_3 oxidase ($O_2 + 2H^+_{in} \rightarrow H_2O$). Whereas, MQH_2 oxidation with cytochrome *bd* oxidase could result in only the scalar $2H^+$ translocation (blue arrows in Fig. 4) by releasing H^+ outside from MQH_2 ($MQH_2 \rightarrow MQ + 2H^+_{out}$) and H^+ uptake from inside to reduce oxygen to produce water ($O_2 + 2H^+_{in} \rightarrow H_2O$) during the two-electron transfer. In nitrate reductase (not shown in Fig. 4), the same scalar reaction generates Δp, inside negative potential and alkaline pH, as in the case of *bd* oxidase. Only the difference is the nitrate reduction to produce nitrite ($NO_3^- + 2H^+_{in} \rightarrow NO_2^- + H_2O$), instead of O_2 reduction to H_2O.

Phenotypic evidence attained with several *C. glutamicum* mutants exhibiting various respiratory component defects supports the energy coupling efficiency described above (Sone et al. 2004). Particularly, the growth phenotype of *qcrCAB*-

or *ctaD*-deficient mutant (Δ*qcrCAB* or Δ*ctaD*) has revealed that the *bcc-aa*$_3$ branch is of major importance for energy generation in this organism (Niebisch and Bott 2003; Koch-Koerfges et al. 2013). However, the defective mutant of another branch, that of cytochrome *bd* oxidase, exhibits a normal cell growth during the exponential growth phase, but a relatively large defect in the late growth phase (Bott and Niebisch 2005; Kabus et al. 2007). This growth defect is presumably due to the requirement of *bd* oxidase in the late growth phase where the dissolved oxygen concentration is typically lower. What is more, relatively high expression of *bd* oxidase has previously been evidenced in *C. glutamicum* cells at the stationary phase of growth (Kusumoto et al. 2000). However, the overexpression of *bd* oxidase leads to reduced cell growth that is concomitant with increased absorption spectra at 560 nm and 630 nm corresponding to cytochromes *b* and *d*, respectively, and also with an increased respiration rate (Kabus et al. 2007). The same growth defect has been shown more recently to occur by the upregulation of *cydABCD* operon which is caused by a transcriptional regulator *sigC* overexpression (Toyoda and Inui 2016). These observations are in agreement with expectations because the *bd* oxidase branch has an overall energy coupling efficiency that is much lower than that of the *bcc-aa*$_3$ oxidase branch described below.

Kabashima et al. (2009) have analyzed the H^+/O ratio of *C. glutamicum* respiratory chain using *bd* oxidase- and *aa*$_3$ oxidase-deficient mutants (Δ*cydAB* and Δ*ctaD*). In these experiments, Δ*cydAB* exhibited a H^+/O ratio of 5.23, which is close to the expected value when only the *bcc-aa*$_3$ branch functions in the respiratory chain. Although the Δ*ctaD* showed a H^+/O ratio of 2.76, which is higher than the expected value when only *bd* oxidase functions in the respiratory chain, the H^+/O ratio decreases from 2.76 to 2.29, which is close to the expected value of 2, when the *bd* oxidase is overexpressed in the Δ*ctaD* mutant. Thus, the H^+ translocation ability (H^+/O ratio of 2 and 6) expected for both terminal oxidase branches appear congruent with experimental data, and both branches function at least in the growth conditions tested, since the proton translocation ability of the wild-type strain (that has both branches) in the exponential phase reaches a H^+/O ratio of 3.94. Later, Koch-Koerfges et al. (2013) have done more detailed analysis of Δp with Δ*qcrCAB*, Δ*cydAB*, and also with double mutant (called as DOOR strain). Δ*cydAB* shows almost the same Δp of −228 mV as the wild strain (−224 mV) at pH 7.0, while Δ*qcrCAB* does 12% lower Δp (−197 mV). The Δp value is decreased in DOOR strain by 31% (−154 mV). Thus, it is likely that cytochrome *bd* oxidase has a minor role and essentially complements the cytochrome *bcc-aa*$_3$ supercomplex activity under conditions where this enzymatic complex does not work properly (Kabus et al. 2007), although it remains important, according to the nutrients, for growth even under aerobic conditions (Kabashima et al. 2009).

3 Physiological and Biotechnological Aspects of Respiratory Chain Enzymes of *C. glutamicum*

C. glutamicum synthesizes ATP both by substrate-level phosphorylation, typified by glycolysis, and by oxidative phosphorylation in respiratory chain which directly links to TCA cycle. In the respiratory chain of *C. glutamicum*, NADH formed through the central carbon metabolism is re-oxidized mainly by NDH-2, reducing MQ to MQH_2. Subsequently, the electrons from MQH_2 are passed to oxygen molecules either via the cytochrome *bcc*-aa_3 supercomplex or by cytochrome *bd* oxidase, unless nitrate is added to the culture medium. The Δp generated by this process is used to drive F_1F_o-ATPase to synthesize ATP.

3.1 Metabolic Changes Caused by Overproduction of NDH-2

Numerous kinds of valuable compounds have been produced from sugar substrates by microbial fermentation. In these studies, the development of host strains that can rapidly consume sugar substrate is important. Sugar substrate is commonly metabolized through the central carbon metabolism, which is tightly subjected to regulation responsive to the redox state. The redox state is deeply depending on aeration level of the culture. For a robust metabolism, NADH generated by the central carbon metabolism has to be re-oxidized to NAD^+ for maintaining the intracellular redox balance. Re-oxidation of NADH is carried out either by NDH-2 of the respiratory chain or by reductive coupling reactions from pyruvate or oxaloacetate to L-lactate or malate, respectively, in *C. glutamicum* (Figs. 5 and 6).

Tsuge et al. (2015) investigated the effects of *ndh* overexpression on glucose consumption rate as well as intracellular NADH/NAD^+ ratio in *C. glutamicum* strain R cultured under oxygen-deprived conditions, and found that glucose consumption rate is increased by 14–26%, while NADH/NAD^+ ratio is decreased by 44–59% compared to the control strain. Additionally, applicability of the *ndh* overexpression for metabolic engineering was examined in the several mutants derived from *C. glutamicum* strain R. Among them, the Δ*ppc* (encoding phosphoenolpyruvate carboxylase; PEPC in Fig. 6) and Δ*mdh* (encoding NAD-dependent malate dehydrogenase) mutants, both of which are useful production host strains with the ability to accumulate less succinate, and the engineered D-lactate producer strain, which lacks *ppc* and *ldhA* (encoding NAD-dependent lactate dehydrogenase) but expresses *D-ldhA* from *Lactobacillus delbrueckii*, showed similar results with the wild-type strain, i.e., increased glucose consumption rate and decreased NADH/NAD^+ ratio compared to the control strain. Since NADH is mainly generated by glyceraldehyde-3-phosphate dehydrogenase (GAPDH) in the glycolysis due to the decreased activity of TCA cycle under oxygen-deprived conditions and the GAPDH activity has already been shown to be inhibited under higher NADH/NAD^+ ratio (Dominguez et al. 1998), these results can be explained that overproduction of NDH-2 decreases

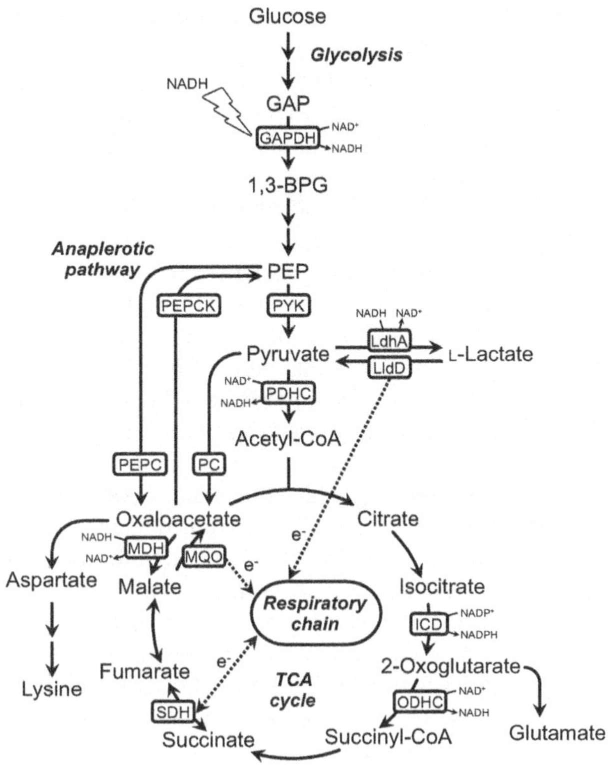

Fig. 6 Central carbon metabolism of *C. glutamicum*. The zigzag arrowhead indicates inhibition by NADH. Abbreviations: *1,3-BPG* 1,3-bisphosphoglycerate; *GAP* glyceraldehyde-3-phosphate; *GAPDH* GAP dehydrogenase; *ICD* isocitrate dehydrogenase; *LdhA* NAD-dependent lactate dehydrogenase; *LldD*, L-lactate dehydrogenase; *MDH* NAD-dependent malate dehydrogenase; *MQO* malate:quinone oxidoreductase; *ODHC* 2-oxoglutarate dehydrogenase complex; *PC* pyruvate carboxylase; *PDHC* pyruvate dehydrogenase complex; *PEP* phosphoenolpyruvate; *PEPC* PEP carboxylase; *PEPCK* PEP carboxykinase; *PYK* pyruvate kinase; *SDH* succinate:quinone oxidoreductase

$NADH/NAD^+$ ratio, relieving the inhibition of GAPDH activity, and consequently enhances the glucose consumption.

However, in the Δ*ldhA* mutant with a higher $NADH/NAD^+$ ratio than other strains, overexpression of *ndh* does not affect glucose consumption rate despite resulting in decreased $NADH/NAD^+$ ratio, suggesting that the $NADH/NAD^+$ ratio in Δ*ldhA* is still too high and thus the $NADH/NAD^+$ ratio is not decreased enough to the level that relieves the inhibition of GAPDH activity and can enhance the glucose consumption rate.

The potential of enhanced glucose consumption, triggered by NDH-2 overproduction, for metabolic engineering was further investigated in the culture under aerobic growth conditions. In the engineered pyruvate producer strain (Δ*ramA* Δ*ldhA* Δ*pqo* Δ*cat* Δ*pta-ackA*), the *ndh* overexpression increased the biomass formation but had no positive effect on glucose consumption rate under aerobic growth conditions compared to the control strain (Kataoka et al. unpublished data), which already exhibited increased glucose consumption rate as explained below. The pyruvate producer strain was derived from the strain ATCC 13032 lacking all known L-lactate- and acetate-pathways by inactivating RamA, a global regulator responsible for activating the oxidative TCA cycle (Kataoka et al. 2019). The *ramA*-

deficient strain has been shown to perform fermentative metabolism even under aerobic growth conditions and thus consume glucose rapidly (Kataoka et al. 2019). In this *ndh*-overexpressed pyruvate producer strain, the NADH/NAD^+ ratio was low enough to maintain the GAPDH activity under aerobic growth conditions, but more succinate and fumarate, and less 2-oxoglutarate were produced than in the control strain, the reason of which could be explained as described below (Fig. 6). *C. glutamicum* possesses MQO and LldD in the respiratory chain, which catalyze the highly exergonic one-way oxidation of malate to oxaloacetate, and of L-lactate to pyruvate, respectively, and transfer electrons to MQ (Figs. 5 and 6). Whereas MDH or LdhA catalyzes a reversible oxido-reduction reaction, in which the reaction favors reduction of oxaloacetate or of pyruvate, respectively. Thus, MQO and LldD have been shown to function as alternative NADH oxidation systems by coupling with the reverse reactions catalyzed by MDH and LdhA, respectively (Molenaar et al. 2000; Nantapong et al. 2004; Sawada et al. 2012) (Figs. 5 and 6). The LdhA/LldD system was previously reported to compensate for the ability of re-oxidation of NADH in the *ndh*-disrupted mutant (Nantapong et al. 2004). Additionally, the increased MQO activity was observed in the *ndh*-disrupted mutant (Nantapong et al. 2004), indicating the involvement of the MDH/MQO system for the completion of weakened NADH re-oxidation system. The metabolic change observed in the *ndh*-overexpressed pyruvate producer, i.e., increased production of fumarate and succinate and decreased 2-oxoglutarate accumulation, could be explained by the activation of reductive TCA cycle. Overproduction of NDH-2 accelerates the re-oxidation of NADH and results in over-accumulation of MQH_2, which is used to drive the fumarate reduction by SDH as mentioned in the Sect. 1.4.

3.2 *Effects of Alteration of Menaquinol-Oxidizing Enzymes on Glucose Consumption Rate and Amino and Organic Acids Production*

A mutant lacking the *cydAB* genes (encoding subunits I and II of cytochrome *bd* oxidase) of *C. glutamicum* strain ATCC 13032 shows similar growth to that of the wild type during the logarithmic growth phase, but strongly impairs growth thereafter, in which the dissolved oxygen levels are almost 0% of air saturation, indicating the importance of cytochrome *bd* oxidase under oxygen-limited conditions (Kabus et al. 2007; Koch-Koerfges et al. 2013). Meanwhile, the *cydABDC*-overexpressed strain exhibits a strongly reduced growth rate and biomass formation (Kabus et al. 2007). This result might be explained by the difference in bioenergetic efficiency between the cytochrome *bcc*-aa_3 supercomplex and cytochrome *bd* oxidase. The shift of the terminal oxidase from the *bcc*-aa_3 supercomplex to *bd* oxidase in the menaquinol oxidation system reduces Δp, and results in reduced ATP synthesis by F_1F_o-ATPase. To compensate this reduced ATP generation by oxidative phosphorylation, glucose required for catabolism is increased while the amount of glucose for biomass formation (glucose assimilation) is decreased.

The effects of absence or overproduction of cytochrome *bd* oxidase on amino acid production were also analyzed in the lysine producer, strain MH20-22B. The deletion of *cydAB* in strain MH20-22B increases the lysine production but does not affect the growth and glucose consumption (Kabus et al. 2007), which may reflect the less ability to flow the carbon into biomass formation in strain MH20-22B than in strain ATCC 13032. As discussed above, this result also can be explained by the difference in bioenergetic efficiency between two MQH_2-oxidizing enzymes, i.e., the inactivation of the *bd* oxidase increased the dependency on the *bcc-aa*$_3$ supercomplex to produce more Δp and thus increased ATP synthesis, and resulted in the glucose availability for lysine synthesis. On the other hand, overproduction of the *bd* oxidase decreased the production of lysine, growth rate and biomass formation (Kabus et al. 2007), showing a consistency with the profile of the *cydABDC*-overexpressed strain derived from ATCC 13032.

The strategy to alter the *bd* oxidase was also applied for the pyruvate producer strain because pyruvate is the end product of glucose catabolism, glycolysis, which will be enhanced by the reduced Δp. Overexpression of *cydABDC* in the pyruvate producer strain (Δ*ramA* Δ*ldhA* Δ*pqo* Δ*cat* Δ*pta-ackA*) expectedly further enhanced the pyruvate production and glucose consumption during the stationary phase, exhibiting the decreased biomass formation as observed in other strains (Kataoka et al. unpublished data). Interestingly, the *cydABDC*-overexpressed pyruvate producer strain accumulated less fumarate, succinate, and 2-oxoglutarate compared to the control strain (Fig. 6). Although the reason for the decreased production of 2-oxoglurate is still uncertain, the reduced accumulation of fumarate and succinate could be explained by the decreased carbon flow to the reductive TCA cycle. Overproduction of *bd* oxidase accelerates the re-oxidation of MQH_2 and results in decreased MQH_2 availability to drive the fumarate reduction by SDH, which is an adverse phenomenon observed in the *ndh*-overexpressed pyruvate producer strain.

As described in the Sect. 2.3, both Δ*qcrCAB* (encoding *bcc* complex) and Δ*ctaD* (encoding subunit I of cytochrome *aa*$_3$ oxidase) mutants show impaired growth in glucose minimal medium (Koch-Koerfges et al. 2013). When the Δ*qcrCAB* and Δ*ctaD* mutants were cultured in a biotin-limited glutamic acid fermentation medium, interestingly, both mutants showed 36–59% increased glucose consumption rate per cell but no glutamic acid production (Kataoka et al. unpublished data). Instead of glutamic acid, both the mutants accumulated more pyruvate and pyruvate-derived metabolites than wild type. Although the reason for this fermentation profile of mutants of cytochrome *bcc-aa*$_3$ supercomplex is still unclear, this would be due to the deficiency of the oxidative phosphorylation by respiration.

3.3 Effects of Decreased F_1F_o-ATPase Activity on Glucose Consumption, Respiration, and Amino Acid Production

The Δp generated by the function of primary dehydrogenases and MQH_2-oxidizing enzymes is finally converted to ATP by F_1F_o-ATPase in *C. glutamicum*.

A mutant of *C. glutamicum* strain ATCC 14067 with a 75% reduced F_1F_o-ATPase activity, strain F172-8, was firstly reported by Sekine et al. (2001). This mutant shows a 70% enhanced glucose consumption per cell and two-times higher respiration rate but much less glutamic acid production compared to the wild type when cultured in a standard biotin-limited glutamic acid fermentation medium. In the metabolic point of view, the formation of lactate and pyruvate-family amino acids such as alanine and valine is increased in strain F172-8, whereas the accumulation of 2-oxoglutarate and glutamate-family amino acids such as glutamate and proline is inhibited. The increased glucose catabolism in strain F172-8 might be due to a response of the cells to compensate the shortage of intracellular ATP caused by the defect in F_1F_o-ATPase. This change consequently leads to high intracellular pyruvate concentration, and results in the enhanced production of pyruvate-derived compounds. Later, the mechanism of increased respiration rate in the F_1F_o-ATPase-defective mutants was examined by proteomic and transcriptional analyses and enzyme activity measurement. It was found that the MDH/MQO and LdhA/LldD systems, SDH, and cytochrome *bd* oxidase were up-regulated and enhanced in the mutants compared to the wild-type strains, but NDH-2 and cytochrome *bcc-aa*$_3$ supercomplex were not (Li et al. 2007; Sawada et al. 2012).

A strain A-1, which is derived from strain ATCC 13032 and possesses the same mutation as strain F172-8, was engineered to improve the production of valine, a pyruvate-family amino acid, by overexpressing the genes for a feedback-insensitive acetohydroxyacid synthase and acetohydroxyacid isomeroreductase from *C. glutamicum*. The engineered strain produced 2.5-times higher amount of valine than the control strain (Wada et al. 2008), indicating the superiority of F_1F_o-ATPase-defective mutant as a host strain for the production of pyruvate-derived metabolites. Additionally, even though the detailed mechanism has not been clarified so far, when strain F172-8 is cultured under exhaustive biotin limitation with much smaller inoculum size or under Tween 40-triggered production conditions, a 22–54% increased glutamic acid production is observed compared with wild type, strain ATCC 14067 (Aoki et al. 2005).

4 Conclusions and Future Directions

C. glutamicum has a rather simple respiratory chain in terms of the terminal menaquinol oxidases, a higher energy-generating cytochrome *bcc-aa*$_3$ supercomplex and a lower energy-generating cytochrome *bd* oxidase or nitrate reductase, although these menaquinol oxidase systems could be regulated depending on the cellular situation, growth phase, nutrient sources, aeration and so on. On the other hand, this bacterium has unique and rather sophisticated respiratory chain in the menaquinone reduction part, comprising many different flavoprotein dehydrogenases. Of these primary dehydrogenases, MQO and LldD could participate in NADH re-oxidation coupled with their respective cytoplasmic MDH and LdhA. The unique NADH oxidation systems, together with NDH-2-dependent NADH oxidase, seems to be critically regulated by the energetic state of the cells, which may be related to the cellular $NADH/NAD^+$ level. As a result, the energetic state of the corynebacterial cell is deeply related to its metabolism; this could be exploited further as critically linked to the amino acid production in this bacterium.

Since, as described herein, changes of the energetic efficiency and of the related metabolic flux caused by qualitative changes of the respiratory chain or by a defective F_1F_0-ATP synthase have strong effects on the metabolism and amino acid production, the respiratory chain and the energy metabolism are an attractive target for improving the amino acid productivity of *C. glutamicum*. The modulation of the respiratory chain composition in favor of either an increased or decreased bioenergetic efficiency, or of alternating NADH re-oxidation manner possibly relating to O_2^- generation ability, could thus be implemented as a strategy to improve the amino acid productivity of *C. glutamicum*-based processes.

Acknowledgement The author thanks Prof. Toshiharu Yakushi, Yamaguchi University, for his critical reading and suggestions for this chapter.

References

Aoki R, Wada M, Takesue N, Tanaka K, Yokota A (2005) Enhanced glutamic acid production by H^+-ATPase-defective mutant of *Corynebacterium glutamicum*. Biosci Biotechnol Biochem 69:1466–1472

Baradaran R, Berrisford JM, Minhas GS, Sazanov LA (2013) Crystal structure of the entire respiratory complex I. Nature 494(7438):443–448

Bott M, Niebisch A (2003) The respiratory chain of *Corynebacterium glutamicum*. J Biotechnol 104:129–153

Bott M, Niebisch A (2005) Respiratory energy metabolism. In: Eggeling L, Bott M (eds) Handbook of *Corynebacterium glutamicum*. Taylor & Francis, New York, pp 305–332

Brandt U, Trumpower B (1994) The proton motive Q cycle in mitochondria and bacteria. Crit Rev Biochem Mol Biol 29:165–197

Davidson JF, Schiestl RH (2001) Mitochondrial respiratory electron carriers are involved in oxidative stress during heat stress in *Saccharomyces cerevisiae*. Mol Cell Biol 21 (24):8483–8489

Desplats C, Beyly A, Cuiné S, Bernard L, Cournac L, Peltier G (2007) Modification of substrate specificity in single point mutants of *Agrobacterium tumefaciens* type II NADH dehydrogenase. FEBS Lett 581(21):4017–4022

Dominguez H, Rokkin C, Guyonvarch A, Guerquin-Kern JL, Cocaign-Bousquet M, Lindley ND (1998) Carbon-flux distribution in the central metabolic pathways of *Corynebacterium glutamicum* during growth on fructose. Eur J Biochem 254:96–102

Edgar RC (2004) MUSCLE: multiple sequence alignment with high accuracy and high throughput. Nucleic Acids Res 32(5):1792–1797

Efremov RG, Baradaran R, Sazanov LA (2010) The architecture of respiratory complex I. Nature 465:441–445

Feng Y, Li W, Li J, Wang J, Ge J, Xu D, Liu Y, Wu K, Zeng Q, Wu JW, Tian C, Zhou B, Yang M (2012) Structural insight into the type-II mitochondrial NADH dehydrogenases. Nature 491 (7424):478–482

Fiedorczuk K, Letts JA, Degliesposti G, Kaszuba K, Skehel M, Sazanov LA (2016) Atomic structure of the entire mammalian mitochondrial complex I. Nature 538(7625):406–410

Gong H, Li J, Xu A, Tang Y, Ji W, Gao R, Wang S, Yu L, Tian C, Li J, Yen HY, Man Lam S, Shui G, Yang X, Sun Y, Li X, Jia M, Yang C, Jiang B, Lou Z, Robinson CV, Wong LL, Guddat LW, Sun F, Wang Q, Rao Z (2018) An electron transfer path connects subunits of a mycobacterial respiratory supercomplex. Science 362(1020):eaat8923

Graf S, Fedotovskaya O, Kao WC, Hunte C, Ädelroth P, Bott M, von Ballmoos C, Brzezinski P (2016) Rapid electron transfer within the III–IV supercomplex in *Corynebacterium glutamicum*. Sci Rep 6:34098

Heikal A, Nakatani Y, Dunn E, Weimar MR, Day CL, Baker EN, Lott JS, Sazanov LA, Cook GM (2014) Structure of the bacterial type II NADH dehydrogenase: a monotopic membrane protein with an essential role in energy generation. Mol Microbiol 91(5):950–964

Jones AJ, Blaza JN, Varghese F, Hirst J (2017) Respiratory complex I in *Bos taurus* and *Paracoccus denitrificans* pumps four protons across the membrane for every NADH oxidized. J Biol Chem 292(12):4987–4995

Kabashima Y, Kishikawa J, Kurokawa T, Sakamoto J (2009) Correlation between proton translocation and growth: genetic analysis of the respiratory chain of *Corynebacterium glutamicum*. J Biochem 146:845–855

Kabsch W, Sander C (1983) Dictionary of protein secondary structure: pattern recognition of hydrogen-bonded and geometrical features. Biopolymers 22(12):2577–2637

Kabus A, Niebisch A, Bott M (2007) Role of cytochrome *bd* oxidase from *Corynebacterium glutamicum* in growth and lysine production. Appl Environ Microbiol 73:861–868

Kao WC, Kleinschroth T, Nitschke W, Baymann F, Neehaul Y, Hellwig P, Richers S, Vonck J, Bott M, Hunte C (2016) The obligate respiratory supercomplex from Actinobacteria. Biochim Biophys Acta 1857(10):1705–1714

Kataoka N, Vangnai AS, Pongtharangkul T, Yakushi T, Wada M, Yokota A, Matsushita K (2019) Engineering of *Corynebacterium glutamicum* as a prototrophic pyruvate-producing strain: characterization of *ramA*-deficient mutant and its application for metabolic engineering. Biosci Biotechnol Biochem 83:372–380

Kato O, Youn J-W, Stansen KC, Matsui D, Oikawa T, Wendisch VF (2010) Quinone-dependent D-lactate dehydrogenase Dld (Cg1027) is essential for growth of *Corynebacterium glutamicum* on D-lactate. BMC Microbiol 10(1):321

Kita K, Konishi K, Anraku Y (1984) Terminal oxidases of *Escherichia coli* aerobic respiratory chain. II. Purification and properties of cytochrome b_{558}-*d* complex from cells grown with limited oxygen and evidence of branched electron-carrying systems. J Biol Chem 259:3375–3381

Koch-Koerfges A, Pfelzer N, Platzen L, Oldiges M, Bott M (2013) Conversion of *Corynebacterium glutamicum* from an aerobic respiring to an aerobic fermenting bacterium by inactivation of the respiratory chain. Biochim Biophys Acta 1827(6):699–708

Komati RG, Lindner SN, Wendisch VF (2015) Metabolic engineering of an ATP-neutral Embden-Meyerhof-Parnas pathway in *Corynebacterium glutamicum*: growth restoration by an adaptive point mutation in NADH dehydrogenase. Appl Environ Microbiol 81(6):1996–2005

Kurokawa T, Sakamoto J (2005) Purification and characterization of succinate: menaquinone oxidoreductase from *Corynebacterium glutamicum*. Arch Microbiol 183:317–324

Kusumoto K, Sakiyama M, Sakamoto J, Noguchi S, Sone N (2000) Menaquinol oxidase activity and primary structure of cytochrome *bd* from the amino-acid fermenting bacterium *Corynebacterium glutamicum*. Arch Microbiol 173:390–397

Li L, Wada M, Yokota A (2007) A comparative proteomic approach to understand the adaptations of an H^+-ATPase-defective mutant of *Corynebacterium glutamicum* ATCC14067 to energy deficiencies. Proteomics 7:3348–3357

Madej MG, Nasiri HR, Hilgendorff NS, Schwalbe H, Unden G, Lancaster CR (2006) Experimental evidence for proton motive force-dependent catalysis by the diheme-containing succinate: menaquinone oxidoreductase from the Gram-positive bacterium Bacillus licheniformis. Biochemistry 45:15049–15055

Marreiros BC, Sena FV, Sousa FM, Batista AP, Pereira MM (2016) Type II NADH:quinone oxidoreductase family: phylogenetic distribution, structural diversity and evolutionary divergences. Environ Microbiol 18(12):4697–4709

Matsushita K (2013) *Corynebacterium glutamicum*. In: Yukawa H, Inui M (eds) Microbiology monographs 23. Springer, pp 315–334

Matsushita K, Ohnishi T, Kaback R (1987) NADH-ubiquinone oxidoreductase of the *Escherichia coli* aerobic respiratory chain. Biochemistry 26:7732–7737

Matsushita K, Yamamoto T, Toyama H, Adachi O (1998) NADPH oxidase system works as superoxide-generating cyanide-resistant pathway in the respiratory chain of *Corynebacterium glutamicum*. Biosci Biotechnol Biochem 62:1968–1977

Matsushita K, Otofuji A, Iwahashi M, Toyama H, Adachi O (2001) NADH dehydrogenase of *Corynebacterium glutamicum*. Purification of NADH dehydrogenase II homologue able to oxidize NADPH. FEMS Microbiol Lett 204:271–276

Messner KR, Imlay JA (1999) The identification of primary sites of superoxide and hydrogen peroxide formation in the aerobic respiratory chain and sulfite reductase complex of *Escherichia coli*. J Biol Chem 274(15):10119–10128

Messner KR, Imlay JA (2002) Mechanism of superoxide and hydrogen peroxide formation by fumarate reductase, succinate dehydrogenase, and aspartate oxidase. J Biol Chem 277 (45):42563–42571

Miller MJ, Gennis RB (1983) The purification and characterization of the cytochrome d terminal oxidase complex of the *Escherichia coli* aerobic respiratory chain. J Biol Chem 258:9159–9165

Molenaar D, van der Rest ME, Petrović S (1998) Biochemical and genetic characterization of the membrane-associated malate dehydrogenase (acceptor) from *Corynebacterium glutamicum*. Eur J Biochem 254:395–403

Molenaar D, van der Rest ME, Drysch A, Yucel R (2000) Functions of the membrane-associated and cytoplasmic malate dehydogenases in the citric acid cycle of *Corynebacterium glutamicum*. J Bacteriol 182:6884–6891

Morosov X, Davoudi CF, Baumgart M, Brocker M, Bott M (2018) The copper-deprivation stimulon of *Corynebacterium glutamicum* comprises proteins for biogenesis of the actinobacterial cytochrome bc_1-aa_3 supercomplex. J Biol Chem 293(40):15628–15640

Nantapong N, Kugimiya Y, Toyama H, Adachi O, Matsushita K (2004) Effect of NADH dehydrogenase-disruption and over-expression on the respiratory-related metabolism in *Corynebacterium glutamicum* KY9714. Appl Microbiol Biotechnol 66:187–193

Nantapong N, Otofuji A, Migita CT, Adachi O, Toyama H, Matsushita K (2005) Electron transfer ability from NADH to menaquinone and from NADPH to oxygen of type II NADH dehydrogenase of *Corynebacterium glutamicum*. Biosci Biotechnol Biochem 69:149–159

Nantapong N, Murata R, Trakulnaleamsai S, Kataoka N, Yakushi T, Matsushita K (2019) The effect of reactive oxygen species (ROS) and ROS-scavenging enzymes, superoxide dismutase and catalase, on the thermotolerant ability of *Corynebacterium glutamicum*. Appl Microbiol Biotechnol 103(13):5355–5366

Niebisch A, Bott M (2003) Purification of a cytochrome bc-aa_3 supercomplex with quinol oxidase activity from *Corynebacterium glutamicum*. Identification of a fourth subunity of cytochrome aa_3 oxidase and mutational analysis of diheme cytochrome c_1. J Biol Chem 278:4339–4346

Nishimura T, Vertès AA, Shinoda Y, Inui M, Yukawa H (2007) Anaerobic growth of *Corynebacterium glutamicum* using nitrate as a terminal electron acceptor. Appl Microbiol Biotechnol 75:889–897

Nishimura T, Teramoto H, Vertès AA, Inui M, Yukawa H (2008) ArnR, a novel transcriptional regulator, represses expression of the *narKGHJI* operon in *Corynebacterium glutamicum*. J Bacteriol 190:3264–3273

Ohnishi T, Moser CC, Page CC, Dutton PL, Yano T (2000) Simple redox-linked proton-transfer design: new insights from structures of quinol-fumarate reductase. Structure 8:R23–R32

Safarian S, Rajendran C, Müller H, Preu J, Langer JD, Ovchinnikov S, Hirose T, Kusumoto T, Sakamoto J, Michel H (2016) Structure of a *bd* oxidase indicates similar mechanisms for membrane-integrated oxygen reductases. Science 352(6285):583–586

Safarian S, Hahn A, Mills DJ, Radloff M, Eisinger ML, Nikolaev A, Meier-Credo J, Melin F, Miyoshi H, Gennis RB, Sakamoto J, Langer JD, Hellwig P, Kühlbrandt W, Michel H (2019) Active site rearrangement and structural divergence in prokaryotic respiratory oxidases. Science 366(6461):100–104

Sakamoto J, Shibata T, Mine T, Miyahara R, Torigoe T, Noguchi S, Matsushita K, Sone N (2001) Cytochrome *c* oxidase contains an extra charged amino acid cluster in a new type of respiratory chain in the amino-acid-producing Gram-positive bacterium *Corynebacterium glutamicum*. Microbiology 147:2865–2871

Sato-Watanabe M, Mogi T, Ogura T, Kitagawa T, Miyoshi H, Iwamura H, Anraku Y (1994) Identification of a novel quinone-binding site in the cytochrome *bo* complex from *Escherichia coli*. J Biol Chem 269:28908–28912

Sawada K, Kato Y, Imai K, Li L, Wada M, Matsushita K, Yokota A (2012) Mechanism of increased respiration in an H^+-ATPase-defective mutant of *Corynebacterium glutamicum*. J Biosci Bioeng 113:467–473

Schirawski J, Unden G (1998) Menaquinone-dependent succinate dehydrogenase of bacteria catalyzes reversed electron transport driven by the proton potential. Eur J Biochem 257:210–215

Schnorpfeil M, Janausch IG, Biel S, Kröger A, Unden G (2001) Generation of a proton potential by succinate dehydrogenase of *Bacillus subtilis* functioning as a fumarate reductase. Eur J Biochem 268:3069–3074

Schreiner ME, Eikmanns BJ (2005) Pyruvate:quinone oxidoreductase from *Corynebacterium glutamicum*: purification and biochemical characterization. J Bacteriol 187:862–871

Sekine H, Shimada T, Hayashi C, Ishiguro A, Tomita F, Yokota A (2001) H^+-ATPase defect in *Corynebacterium glutamicum* abolishes glutamic acid production with enhancement of glucose consumption rate. Appl Microbiol Biotechnol 57:534–540

Sone N, Nagata K, Kojima H, Tajima J, Kodera Y, Kanamaru T, Noguchi S, Sakamoto J (2001) A novel hydrophobic diheme *c*-type cytochrome. Purification from *Corynebacterium glutamicum* and analysis of the QcrCBA operon encoding three subunit proteins of a putative cytochrome reductase complex. Biochim Biophys Acta 1503:279–290

Sone N, Hägerhäll C, Sakamoto J (2004) Aerobic respiration in the gram-positive bacteria. In: Zannoni D (ed) Respiration in archaea and bacteria, vol 2: Diversity of prokaryotic respiratory systems. Springer, The Netherlands, pp 35–62

Spero MA, Aylward FO, Currie CR, Donohue TJ (2015) Phylogenomic analysis and predicted physiological role of the proton-translocating NADH:quinone oxidoreductase (complex I) across bacteria. MBio 6(2):e00389-15

Takeno S, Ohnishi J, Komatsu T, Masaki T, Sen K, Ikeda M (2007) Anaerobic growth and potential for amino acid production by nitrate respiration in *Corynebacterium glutamicum*. Appl Microbiol Biotechnol 75:1173–1182

Tedeschi G, Zetta L, Negri A, Mortarino M, Ceciliani F, Ronchi S (1997) Redox potentials and quinone reductase activity of L-aspartate oxidase from *Escherichia coli*. Biochemistry 36:16221–16230

Toyoda K, Inui M (2016) The extracytoplasmic function σ factor σ(C) regulates expression of a branched quinol oxidation pathway in *Corynebacterium glutamicum*. Mol Microbiol 100:486–509

Tsuge Y, Uematsu K, Yamamoto S, Suda M, Yukawa H, Inui M (2015) Glucose consumption rate critically depends on redox state in *Corynebacterium glutamicum* under oxygen deprivation. Appl Micobiol Biotechnol 99:5573–5582

Villegas JM, Volentini SI, Rintoul MR, Rapisarda VA (2011) Amphipathic C-terminal region of *Escherichia coli* NADH dehydrogenase-2 mediates membrane localization. Arch Biochem Biophys 505(2):155–159

Wada M, Hijikata N, Aoki R, Takesue N, Yokota A (2008) Enhanced valine production in *Corynebacterium glutamicum* with H^+-ATPase and C-terminal truncated acetohydroxyacid synthase. Biosci Biotechnol Biochem 72:2959–2965

Waterhouse A, Bertoni M, Bienert S, Studer G, Tauriello G, Gumienny R, Heer FT, de Beer TAP, Rempfer C, Bordoli L, Lepore R, Schwede T (2018) SWISS-MODEL: homology modelling of protein structures and complexes. Nucleic Acids Res 46(W1):W296–W303

Xu F, Golightly EJ, Fuglsang CC, Schneider P, Duke KR, Lam L, Christensen S, Brown KM, Jørgensen CT, Brown SH (2001) A novel carbohydrate:acceptor oxidoreductase from *Microdochium nivale*. Eur J Biochem 268:1136–1142

Yachdav G, Kloppmann E, Kajan L, Hecht M, Goldberg T, Hamp T, Hönigschmid P, Schafferhans A, Roos M, Bernhofer M, Richter L, Ashkenazy H, Punta M, Schlessinger A, Bromberg Y, Schneider R, Vriend G, Sander C, Ben-Tal N, Rost B (2014) PredictProtein – an open resource for online prediction of protein structural and functional features. Nucleic Acids Res 42(Web Server issue):W337–W343

Zhu J, Vinothkumar KR, Hirst J (2016) Structure of mammalian respiratory complex I. Nature 536 (7616):354–358

Zickermann V, Wirth C, Nasiri H, Siegmund K, Schwalbe H, Hunte C, Brandt U (2015) Structural biology. Mechanistic insight from the crystal structure of mitochondrial complex I. Science 347 (6217):44–49

Zumft WG (1997) Cell biology and molecular basis of denitrification. Microbiol Mol Biol Rev 61:533–616

Part II
Regulation at Various Levels

Sigma Factors of RNA Polymerase in *Corynebacterium glutamicum*

Miroslav Pátek, Hana Dostálová, and Jan Nešvera

Contents

Abstract *Corynebacterium glutamicum* is an important biotechnological organism as well as a model organism for other corynebacteria including pathogenic species. *C. glutamicum* also seems suitable as a model organism for corynebacteria in respect of studies of regulatory networks controlled by sigma factors of RNA polymerase, because its sigma factors represent a common subset, which was found in most *Corynebacterium* species. The *C. glutamicum* genome encodes seven σ factors: A primary σ^A, a primary-like σ^B and five σ factors of the extracytoplasmic function (ECF) group (σ^C, σ^D, σ^E, σ^H and σ^M) that are involved in various stress responses. Activities of σ^D, σ^E and σ^H are controlled by the cognate anti-sigma factors.

Activities of ECF sigma factors in response to heat, cold, cell surface and oxidative stresses, DNA damage, growth phases, nutritional limitations and chemostresses caused by various harmful substances often overlap. Most of the consensus sequences of different promoter classes recognized by individual sigma factors have been defined. *C. glutamicum* sigma factors σ^A and σ^B initiate transcription from nearly identical promoters and their recognition specificity probably depends largely on physiological conditions. Comparison of consensus sequences of promoters recognized by ECF sigma factors indicates that they are mutually similar to various extent and recognition specificities of these sigma factors may also overlap.

M. Pátek (✉) · H. Dostálová · J. Nešvera
Institute of Microbiology, Academy of Sciences of the Czech Republic, Prague 4, Czech Republic
e-mail: patek@biomed.cas.cz

M. Inui, K. Toyoda (eds.), *Corynebacterium glutamicum*, Microbiology Monographs 23, https://doi.org/10.1007/978-3-030-39267-3_4

1 Introduction

The transcription of bacterial genes driven by DNA-dependent RNA polymerase (RNAP) is the first step in gene expression. The RNAP core is a large complex enzyme consisting of the subunits $\alpha_2\beta\beta'\omega$. The RNAP core is sufficient for transcription elongation and termination, however, an additional dissociable subunit, sigma, is necessary for the initiation of specific transcription. The sigma subunit (factor), which binds to the RNAP core to form the RNAP holoenzyme, is responsible for promoter recognition and subsequent transcription initiation. Thus σ factors are the regulators that are necessary for the transcription initiation of each bacterial gene. Bacteria contain several different σ factors. Most σ factors belong to the σ^{70}-family and all eubacteria encode at least one such σ factor, while not all of them contain a σ factor of the σ^{54}-family. Only a single σ factor was found in *Mycoplasma genitalium*, whereas 7 σ factors were reported in *Escherichia coli*, 13 in *Mycobacterium tuberculosis*, 17 in *Bacillus subtilis*, 34 in *Rhodococcus jostii* and over 60 in *Streptomyces* species. The σ^{70}-type factors were classified into four groups (Gruber and Gross 2003): Group 1, essential primary σ factors (4 domains) that drive the transcription of housekeeping genes; Group 2, the primary-like σ factors (4 domains) that are not essential; Group 3, σ factors (3 domains) which control the genes involved in specific functions such as sporulation and the biosynthesis of flagella and Group 4, σ factors (called also extracytoplasmic function, ECF σ; 2 domains) which are mostly active in response to various stresses. Sigma factors thus function as regulators which can switch the expression of gene groups (regulons) depending on growth conditions and environmental stimuli. The σ factors of the σ^{70}-family consist of two to four conserved domains. The domain structure of *Corynebacterium glutamicum* σ factors is shown in Fig. 1.

The activity of many ECF σ factors is regulated by anti-σ factors, which are sensory proteins mostly encoded by genes located nearby the *sig* genes (Paget 2015). The anti-σ factors inhibit the function of the σ factors by direct protein-protein interactions. In the absence of stress stimuli, the anti-σ factors inactivate the σ factors by binding and occluding their RNAP-binding determinants. Sigma factors become active under inducing conditions when anti-σ factors are degraded or released from σ factors in response to extracellular stimuli (Paget 2015). The mechanisms which regulate the activity of σ factors by anti-σ factors in *C. glutamicum* have not yet been described.

2 Sigma Factors and Anti-sigma Factors in Corynebacteria

Searching for the genes encoding σ factors of RNA polymerase in the complete genome sequences of corynebacterial strains deposited in the KEGG Genes Database (in total 95 strains classified into 47 species) revealed that only sigma factors of the σ^{70}-family (and not of the σ^{54}-family) are present in corynebacteria. The essential

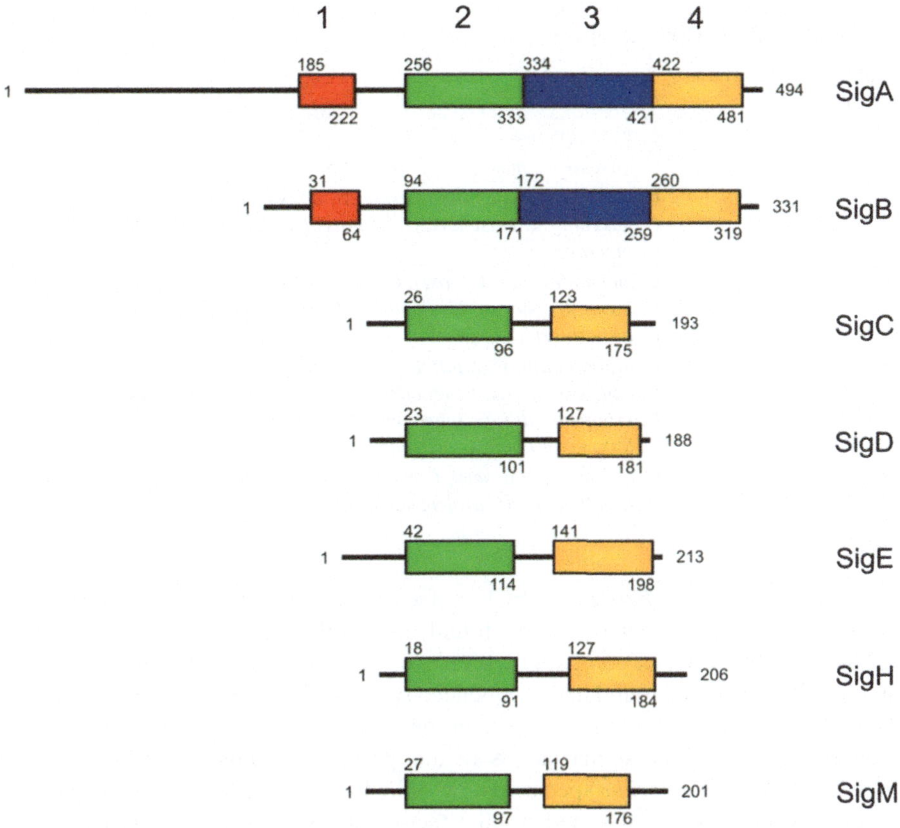

Fig. 1 Domain structure of *C. glutamicum* σ factors. The boundaries of the domains are marked with the numbers of amino acids within the proteins. Numbers assigned to the conserved regions (domains) are shown on the top

primary sigma factor σ^A (Group 1) as well as the primary-like sigma factor σ^B (Group 2) were found to be present in each of the sequenced species of the genus *Corynebacterium*. On the other hand, no σ factor of Group 3 has been found in corynebacteria so far. As shown in Table 1, the number of ECF σ factors (Group 4) varies from four to nine in the individual species of the genus *Corynebacterium*.

Among actinobacteria, σ factors were mostly studied in *M. tuberculosis*, encoding 13 different σ factors. The nomenclature of the actinobacterial σ factors is therefore based on their sequence similarity to the related well-characterized *M. tuberculosis* σ factors. A high level of identity (>45%) to the *M. tuberculosis* ECF σ factors was found in five corynebacterial ECF σ factors (designated σ^C, σ^D, σ^E, σ^G and σ^H according to their *M. tuberculosis* homologs). Sigma factors σ^C and σ^E are present in each of the 47 sequenced corynebacterial species. Sigma factor σ^H is missing in two species (*Corynebacterium sphenisci* and *C. ureicelerivorans*) and σ^D in three species (*C. lactis*, *C. stationis* and *C. urealyticum*). On the other hand, σ^G was only found in one species of the genus *Corynebacterium* (*C. variabile*).

Table 1 Numbers of ECF sigma factors in 47 sequenced genomes of *Corynebacterium* species (according to KEGG Genes Database)

Number of ECF sigma factors	*Corynebacterium* species
9	*C. glyciniphilum*
8	*C. falsenii, C. mustelae, C. resistens*
7	*C. aurimucosum, C. diphtheriae, C. doosanense, C. epidermidicanis, C. jeikeium, C. kutscheri, C. minutissimum, C. pseudotuberculosis, C. ulcerans*
6	*C. ammoniagenes, C. casei, C. crudilactis, C. glaucum, C. halotolerans, C. lactis, C. phocae, C. simulans, C. singulare, C. striatum, C. testudinoris, C. variabile, C. vitaeruminis*
5	***C. glutamicum***, *C. aquilae, C. argentoratense, C. callunae, C. camporealensis, C. efficiens, C. humireducens, C. imitans, C. kroppenstedtii, C. marinum, C. maris, C. stationis, C. terpenotabidum, C. uterequi*
4	*C. atypicum, C. deserti, C. flavescens, C. frankenforstense, C. sphenisci, C. urealyticum, C. ureicelerivorans*

As for other *M. tuberculosis* ECF σ factors (σ^J, σ^K, σ^L, σ^M), their putative homologs in corynebacteria were found to exhibit lower levels of identity (25–41%) to them when compared with σ^C, σ^D, σ^E, σ^G and σ^H. The most frequent are the σ^M-like factors found in 41 sequenced species of the genus *Corynebacterium*. The σ^M-like factors seem to form a heterogenous group, since only a low level of identity (<45%) of the sigma factors of this group was found, even between the individual *Corynebacterium* species. The σ^K-like factors were found in 15 sequenced species, whereas the σ^J-like and σ^L-like factors are present in six corynebacterial species each.

It was found that another σ^E-like sigma factor is present in 14 species of the genus *Corynebacterium* in addition to σ^E. Even two different σ^E-like sigma factors were found in *C. resistens* in addition to σ^E. Similarly, a σ^D-like sigma factor was found in five corynebacterial species in addition to σ^D.

In *M. tuberculosis*, genes encoding anti-σ factors were found downstream of the genes encoding the corresponding ECF sigma factors σ^D, σ^E, σ^H, σ^K, σ^L and σ^M. Similarly, genes encoding anti-σ factors pertinent to σ^D, σ^E, σ^H and σ^K are present downstream of the respective *sig* genes in the vast majority of corynebacterial species. On the other hand, no genes encoding anti-σ factors were found downstream of the genes encoding σ^L-like and σ^M-like factors in any of the sequenced species of the genus *Corynebacterium*. Similarly, no genes encoding anti-σ factors are present downstream of the genes encoding accessory σ^E-like and σ^D-like sigma factors.

In conclusion, σ factors and anti-σ factors in corynebacteria are very variable. Sigma factors designated with the same letter (e.g. σ^M) are not necessarily highly similar. On the other hand, similar homologous σ factors can fulfil different functions e.g. in pathogenic and in non-pathogenic corynebacteria. Cells of the pathogenic species are able to overcome stresses produced by the host, e.g. the influence of acid, bile, lysozyme or other antimicrobial agents. Since most of these damaging substances endanger the integrity of the cell envelope, the pathogens may respond by activating the same σ factor regulons which direct the response to cell envelope disturbance by the action of detergents or toxic aromatic compounds in non-pathogenic soil species. Some σ factors are directly involved in pathogenicity functions, e.g. σ^E and σ^K in *C. pseudotuberculosis* (Pacheco et al. 2012; Ruiz et al. 2011). *C. glutamicum* seems to be a suitable model organism for actinobacteria in terms of σ factors, because its seven σ factors represent a common subset found in most actinobacteria.

3 Sigma and Anti-sigma Factors in *Corynebacterium glutamicum*

The *C. glutamicum* genome encodes seven σ factors: A primary σ^A, a primary-like σ^B and five σ factors of the ECF group (σ^C, σ^D, σ^E, σ^H and σ^M) that are involved in various stress responses. The activity of three of them is regulated by anti-sigma factors: RshA (anti-σ^H), CseE (anti-σ^E) and RsdA (anti-σ^D). The description of *C. glutamicum* σ factors is arranged here according to their supposed impact on cell activities.

3.1 Sigma Factors of Group 1 and 2

3.1.1 σ^A, a Primary Sigma Factor

The *C. glutamicum sigA* gene encodes a primary sigma factor σ^A (belonging to Group 1 of σ factors) that controls the expression of housekeeping genes representing the majority of the genes of the genome. The *sigA* gene is essential, considering the fact that all attempts to delete it failed. The amino acid sequences of *C. glutamicum* σ^A DNA-binding regions are highly similar to the corresponding sequences of sigma factors of Group 1 in related actinobacteria and also in *E. coli*. This suggests that the function of the major essential sigma factor is conserved in the evolution of bacteria and the respective promoters might also be similar in various bacteria. The *C. glutamicum sigA* gene is transcribed from two promoters, vegetative P1 and stress responsive P2 (Toyoda et al. 2015). The *sigA* gene is expressed during the exponential growth phase and the expression decreases at the beginning of the transition phase (Larisch et al. 2007). During the transition period, the σ^A is replaced by the primary-like sigma σ^B. However, the *sigA* gene is still weakly expressed during the stationary phase and also during certain stress responses, probably due to recognition of its promoter P2 by the stress responsive σ^H. The σ^A level seems to be high throughout the whole exponential phase, since the *sigA* transcript is almost constantly present during the exponential growth phase in the cell (Larisch et al.

2007). Overproduction of σ^A using an expression vector with cloned *sigA* in the recombinant *C. glutamicum* producers of lycopene, β-carotene or linear C50 carotenoid substantially increased production of the desired substances in all cases, particularly in stationary phase (Taniguchi et al. 2017). Thus, *sigA* overexpression may lead to increased transcription of some genes even in the stationary growth phase. Determination of global gene expression changes due to *sigA* overexpression showed that relatively small number of genes were up-regulated: 49 genes in exponential phase and 64 genes in stationary phase. If we consider that probably thousands of genes are under control of σ^A in the cell, the proportion exhibiting increased activity due to high level of σ^A is very low. In agreement with this finding, a weak increase in activity of several σ^A-dependent promoters was observed in most two-plasmid assays with overexpressed *sigA* (Dostálová et al. 2017). The permanently high level of σ^A in the cell is probably the reason why the deliberate overexpression of *sigA* does not further increase the activity of these tested σ^A-dependent promoters.

Sequences of bacterial promoters of housekeeping genes were generally found to be similar to the *E. coli* consensus, which is TTGACA (−35 region) and TATAAT (−10 region) (Lisser and Margalit 1993). In the initial analysis of *C. glutamicum* promoters (33 sequences), the −10 element sequence was defined as TANAAT, whereas the sequences of the −35 element were much less conserved (Pátek et al. 1996). The −35 element consensus sequence based on the analysis of 159 *C. glutamicum* promoter sequences was defined as TTGNCA. However, most of the positions are conserved to a lower extent than those in the −10 hexamer (Pátek and Nešvera 2011). An extensive analysis of the *C. glutamicum* transcriptome obtained by RNA sequencing provided 2454 transcription start sites (TSSs) which are most likely under the control of σ^A-dependent promoters (Albersmeier et al. 2017). The analysis of all deduced promoter sequences resulted in the −10 conserved motif GNTANNNTNG (−10 hexamer is underlined). The trimer TTG only appeared with the highest frequency within the −35 region, but no clear consensus sequence was detected as a −35 hexamer (Albersmeier et al. 2017).

The activity of promoters of bacterial housekeeping genes is also influenced by some conserved nucleotides in the upstream regions of −10 and −35 hexamers. A TG dimer located one base upstream of the −10 hexamer was found specifically in *C. glutamicum* promoters with sequences less similar to the consensus. This motif is a part of the extended −10 region in other bacteria, especially in Gram-positive (e.g. *B. subtilis*), but also in *E. coli* (Burr et al. 2000). The TG dimer extending the −10 region to TGNTATAATNG strongly increased promoter activity in *C. glutamicum* in mutagenesis studies (Holátko et al. 2008; Vašicová et al. 1999). Another feature of the σ^A-dependent promoters could be connected to the AT-richness of the region upstream of the −35 element. The analysis of trimers occurring at positions between −35 and −150 showed a preference for AAA and TTT, particularly in short-leadered and leaderless transcripts (63 and 47%, respectively) (Albersmeier et al. 2017). An AT-rich region called the UP-element, characterized as a low-conserved consensus sequence, was defined in *E. coli* promoters upstream of the −35 element (Gourse et al. 2000). Although the statistical analysis showed that the A + T content and particularly AAA and TTT trimers exhibit a periodicity with maxima 11 nt apart (helical periodicity) within the region −20 to −90 in *C. glutamicum* promoter sequences, no particular consensus of the UP element could be defined (Albersmeier et al. 2017). The key sequences of the strongest σ^A-dependent *C. glutamicum* promoters could be found in the synthetic promoter libraries: –35 TTGACA and −10 TGTGTTATAATGG (Rytter et al. 2014) and −35 TTGACA and −10 TGAGATATAATGG (Zhang et al. 2018). These libraries could serve as sources of promoters with the required activity levels for the purposes of overproducing specific proteins or metabolic engineering in *C. glutamicum*.

3.1.2 σ^B, a Primary-like Sigma Factor

The *C. glutamicum sigB* gene encodes a primary-like sigma factor σ^B (belonging to Group 2 of σ factors) that mainly activates the genes in the transition period between the exponential and stationary growth phase (Larisch et al. 2007). The expression of *sigB* is also upregulated by ethanol, salt, heat and cold stress (Halgasova et al. 2002) and during lactic acid adaptation (Jakob et al. 2007). In addition to the genes expressed in the transition phase and in response to various stresses, σ^B was found to be involved in the transcription of several genes determining glucose utilization under conditions of oxygen limitation as well as during exponential aerobic growth (Ehira et al. 2008). However, σ^B is non-essential and knocking out the *sigB* gene only results in a slightly slower growth of the σ^B-deficient strain (Larisch et al. 2007). The *sigB* gene is transcribed from a promoter controlled alternatively by σ^E and σ^H (Dostálová et al. 2017). The σ^E-dependent *sigB* expression was found to be stronger than its σ^H-dependent expression by in vitro transcription and particularly in the stationary phase by two-plasmid in vivo assay. The control of *sigB* expression by two ECF σ factors involved in cell surface stress response (σ^E) and in heat and oxidative stress response (σ^H) clearly classifies σ^B as a stress response regulator. Similarly, the *M. tuberculosis sigB* gene was found to be under the control of σ^E and σ^H (Rodrigue et al. 2006). In addition to the stress-responsive promoter just upstream of *sigB*, a σ^A- and/or σ^B-dependent promoter driving transcription of the *cg2099-cg2100-cg2101-sigB* operon was detected by RNA sequencing (Pfeifer-Sancar et al. 2013). This might explain why some glucose metabolism genes which are under the control of σ^B are also expressed during aerobic exponential growth (Ehira et al. 2008). Thus, σ^B fulfils several functions: it activates genes during the transition growth phase, promotes the expression of various stress response genes and participates in the regulation of cell energy management during exponential growth. σ^B could therefore be considered to be a general stress response σ factor as well as another vegetative σ factor.

The σ^A and σ^B domains 2.4, 3.0 and 4.2 that interact with promoter DNA, exhibit a high amino acid sequence similarity (Fig. 2). This high degree of similarity suggests that the structure of σ^A- and σ^B-controlled promoters may also be closely similar. Indeed, the difference between the consensus sequences of the −35 and −10

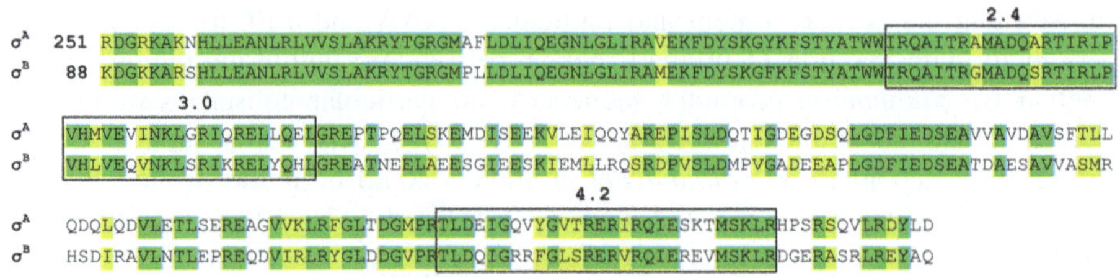

Fig. 2 σ^A and σ^B sequence similarities. Identical amino acids are shown in green, similar amino acids in yellow. Domains 2.4, 3.0 and 4.2, that bind the −10 region, extended −10 region and −35 region of a promoter, respectively, are boxed

elements of these promoter classes has been difficult to reliably define. The suggested motifs −35 cgGCaa and −10 tAnAAT based on 7 determined promoters (Ehira et al. 2008) and the extended −10 region tgngnTAtaaTgg based on 6 promoters (Larisch et al. 2007) are not reliable due to a low number of experimentally proven promoter sequences.

The situation in *C. glutamicum* is reminiscent of the relationship between the primary (vegetative) sigma factor σ^{70} and the general stress sigma factor σ^{38} (σ^S) in *E. coli*. It is notoriously difficult to distinguish between the consensus sequences of σ^{70}- and σ^{38}-controlled promoters (Weber et al. 2005). The genes which are dependent on σ^{70} in vivo can often be transcribed in vitro with σ^{38} and vice versa. Apparently, a number of genes can switch from to σ^{70} to σ^{38} dependency under specific physiological conditions (Weber et al. 2005). Such switching depends on the physiological status of the cell and the activity of a number of regulatory elements in *E. coli* cells. It was found that the recognition specificities of *C. glutamicum* σ^A and σ^B overlap in their transcription from several promoters. The promoters of two σ^B-controlled genes (Ehira et al. 2008), *fba* (fructose 1,6-bisphosphate aldolase) which is active mainly during exponential growth and *pqo* (pyruvate:quinone oxidoreductase) which is active mainly in the transition phase, were active with both σ^A and σ^B when tested in vivo and in vitro (Dostálová et al. 2017; Šilar et al. 2016). In addition, the typical vegetative *C. glutamicum* promoter P*per* (of the gene encoding positive effector of replication from the plasmid pGA1; Nesvera et al. 1997) provided strong signals with both σ^A and σ^B when analyzed by in vitro transcription (Šilar et al. 2016). Another strong vegetative promoter, P2*sigA*, which drives transcription of the *sigA* gene, also provided specific transcripts with both σ^A and σ^B in vitro. However, when the *sigA* and *sigB* genes were alternatively overexpressed in a *C. glutamicum* cell, only σ^A induced a clear increase in P2*sigA* activity (Dostálová et al. 2017). It is still not clear whether the different activity of P2*sigA* with σ^B in vitro and in vivo is due to the absence of regulatory elements (e.g. repressors or activators) and/or the less strict specificity of RNA polymerase in vitro.

The analysis of the *C. glutamicum* transcriptome by RNA sequencing (Albersmeier et al. 2017) looking for differences in the consensus sequences of σ^A- and σ^B-dependent promoters detected a conserved G located 2 nucleotides (nt) downstream of the −10 region (position −5) in approximately 75% out of

114 promoters located upstream of 82 σ^B-dependent genes. Besides this single G, the whole G-content of the region between the TSS and −10 hexamer TANNNT was found to be higher (43%) in the supposedly σ^B-dependent promoters than in other analyzed promoters (less than 30%) (Albersmeier et al. 2017). To test the possible importance of this G-rich spacer for sigma specificity, the whole spacer sequences (−6 to +1 relative to the TSS) within several typical σ^A- and σ^B-dependent promoters were replaced by mutagenesis and their sigma dependence was determined. However, a shift from σ^A- to σ^B-dependency or vice versa could not be generally demonstrated (Pátek, unpublished results). The G-rich spacer between the −10 element and the TSS thus probably only has a statistical significance and does not define individual σ^B-controlled promoters.

There is still insufficient data on the structure of *C. glutamicum* σ^B-dependent promoters, the regulatory mechanisms controlling their function, the basis for σ^A- and σ^B-promoter selectivity and physiological principles for the possible overlap of their recognition by σ^A- and σ^B. We hypothesize that most σ^A- and σ^B-dependent promoters could be recognized to some degree by both σ factors depending on the growth phase, environmental conditions and physiological status of the *C. glutamicum* cells.

3.2 *Sigma Factors of Group 4 (ECF)*

3.2.1 Sequence Similarities of *C. glutamicum* ECF Sigma Factors

Amino acid sequences of *C. glutamicum* ECF sigma factors show significant levels of overall mutual similarity (with the exception of σ^D; Fig. 3). When only comparing the regions of *C. glutamicum* ECF sigma factors which are supposed to interact with the −35 and −10 promoter elements (Fig. 4), the level of their amino acid similarity between all ECF σ factors is still more obvious. This suggests that the different *C. glutamicum* ECF σ factors might recognize similar promoter sequences. The subtle differences, especially in the amino acids that are supposed to interact directly with the specific nucleotides within promoter sequences (Li et al. 2019), between the individual ECF σ factors might result in recognition of individual promoters by various combinations of the ECF σ factors.

Fig. 3 Similarities of *C. glutamicum* ECF sigma factors (% identity)

	SigH	SigE	SigM	SigC	SigD
SigH	-----	38%	25%	26%	< 25%
SigE		-----	31%	31%	< 25%
SigM			-----	28%	< 25%
SigC				-----	< 25%
SigD					-----

a) σ region interacting with -35 element

```
SigH 151  DLAYKEIAEIMDVPLGTVMSRLHRGRKQLRGMLKEVAKE
SigE 162  GMSYDEIAETLGVKMGTVRSRIHRGRSQLRASLEAAAMT
SigD 149  GLSAEETAEMVGSTPGAVRVAQHRALTTLRSTLEQQENK
SigC 145  GYTYEEAAKIADVRVGTIRSRVARARADLIAATATGDSS
SigM 151  GYTVEDVAEIEGIKVGTVKSRRGRARKALRALLHADFFG
```

b) σ region interacting with -10 element

```
SigH 38   NPADAEDLVQDTYIKAYQAFASFKPGTNLKAWL-YRIMTNTYINMYRKKQRQPSQTSADEITDYQLV
SigE 60   NQHDAEDLTQETFMRVFRSLKSYQPGT-FEGWL-HRITTNLFLDMVRHRGKIRMEAEDYERVPGNDI
SigD 45   RQPTAEDVAQEICLAVATSIRNFVDQGRPFMAFVYGIASNKVADAHRAMSRDKSTPIEEVPETSPDT
SigC 45   GHEIADDLTQETYLRVMSALPRFAARSSARTWL-LSLARRVWVDNIRHDMARPRKSIVEYEDTGATD
SigM 47   KPEDAQDILQEALFRASRNMHLYRAEAALGTWL-HKLVLNSGFDWATHRSQVEFPILNEPTIDLEKD
```

Fig. 4 Similarities of *C. glutamicum* ECF sigma factor regions interacting with the −35 (**a**) and −10 (**b**) promoter elements. The AAs identical in all *C. glutamicum* ECF σ factors are shown in purple, those identical in 2, 3 or 4 ECF σ factors are shown in cyan or green and similar amino acids are in yellow. The AAs of σ^H, identical or similar with the AAs of *M. tuberculosis* σ^H whose direct interactions with specific nucleotides within promoter sequences have been proven (Li et al. 2019), are in bold and underlined

3.2.2 σ^H, a Sigma Factor with a Function of Global Regulator of Stress Responses

The *C. glutamicum sigH* gene encodes a Group 4 sigma factor σ^H that mainly controls heat and oxidative stress responses (Engels et al. 2004; Kim et al. 2005). The *rshA* gene, encoding an anti-σ factor, is located immediately downstream of *sigH* (Busche et al. 2012). The same gene organization was found in mycobacteria, in which the genes *sigH* and *rshA* form an operon encoding analogous σ and anti-σ factors (Song et al. 2003). *C. glutamicum* σ^H was found to share 69% identity with the σ^H from *M. tuberculosis*, whereas the respective anti-σ^H factors only share 32% identity. Both *C. glutamicum sigH* and *rshA* genes could be deleted and such deletion strains (Δ*sigH* and Δ*rshA*) were used to discover the σ^H regulon (Busche et al. 2012; Ehira et al. 2009).

The *sigH* gene is expressed from a σ^H-dependent promoter in many bacteria (e.g. *M. tuberculosis*). In contrast, *C. glutamicum sigH* was found to be transcribed from three vegetative promoters, which are under the control of σ^A and/or σ^B (Busche et al. 2012). The measurement of activity of the σ^H-dependent promoter P*trxB1* showed that σ^H is produced at a low level during exponential growth (Dostálová et al. 2017), whereas its production and activity grow markedly after a heat shock (Ehira et al. 2009).

The *C. glutamicum* anti-σ-encoding *rshA* gene was found to be expressed from two promoters: a σ^H-controlled promoter located closely upstream of *rshA* (Busche et al. 2012) and a vegetative promoter driving the expression of *rshA* from the TSS that was located 113 nt upstream (Pfeifer-Sancar et al. 2013). The σ^A- and/or σ^B-controlled *rshA* promoter probably ensures a sufficient level of anti-σ^H in the cell to keep σ^H inactive during exponential growth and under non-stressed conditions, whereas the σ^H-controlled promoter ensures the production of anti-σ^H and the rapid ending of the stress response when the stress conditions are over. The separate control of anti-σ gene expression in *C. glutamicum* differs from the situation in most other actinobacteria, where anti-σ genes are only transcribed as part of the operon from the upstream *sig* promoters (Song et al. 2003). Similar arrangement was only found in some actinobacteria carrying homologs of *rsrA* encoding an anti-sigma factor (Kim et al. 2012).

Using *sigH* deletion or overexpression, 45 genes organized in 29 operons were found to be members of the σ^H regulon in *C. glutamicum* strain R (Ehira et al. 2009). Another study which employed deletion of the anti-σ^H gene *rshA* revealed 85 genes directly or indirectly regulated by σ^H (Busche et al. 2012). These genes belong to various gene groups related among others to proteasome components, disulfide and heat stress response and DNA repair. Using chromatin immunoprecipitation combined with a microarray (ChIP-chip) technique, 74 genes directly regulated by σ^H were identified (Toyoda et al. 2015). This approach uncovered further σ^H-regulated genes which are involved in the pentose phosphate pathway, riboflavin synthesis and Zn-uptake. The large σ^H-regulon thus contains genes with highly diverse functions including those necessary for NADPH and flavin cofactor synthesis under specific conditions (Taniguchi and Wendisch 2015). Deliberate overexpression of *sigH* resulted in an increased expression of riboflavin biosynthesis genes and consequently, riboflavin and flavin mononucleotide production. The study showed that graded overexpression of *sigH* (and generally any sigma factor) may switch the cell metabolism to produce a useful compound and discovered still untapped biotechnological potential of *C. glutamicum* (Taniguchi and Wendisch 2015).

σ^H was also found to drive the transcription of other sigma genes: *sigM* (Nakunst et al. 2007), *sigB* (Ehira et al. 2009) and *sigA* (Toyoda et al. 2015). Since σ^B is a general stress sigma factor, σ^H is indirectly involved in responses to a variety of other environmental stimuli. Moreover, the genes encoding transcriptional regulators ClgR, SufR and HspR were found to be σ^H-controlled (Busche et al. 2012). Recently, a partially overlapping specificity of σ^H with σ^D was observed which extends the influence of σ^H to also include maintenance of cell wall integrity (Dostálová et al. 2019) In conclusion, a large group of σ^H-regulated genes form a complex regulatory network (Fig. 5) that plays a key role in many kinds of stress responses, and σ^H can thus be considered to be a global transcriptional regulator (Schröder and Tauch 2010; Toyoda and Inui 2016a).

Promoters active with σ^H have been well described: the consensus sequence GGAA—(18–20 nt)—GTT is highly conserved (Busche et al. 2012; Ehira et al. 2009). The key elements are identical to the core sequences which were found in *M. tuberculosis* and *Streptomyces coelicolor* σ^H- and σ^R-controlled promoters, respectively (Paget et al. 2001; Raman et al. 2001). Moreover, σ^H is able to run transcription from σ^E- and to some degree from σ^D-dependent promoters (Dostálová

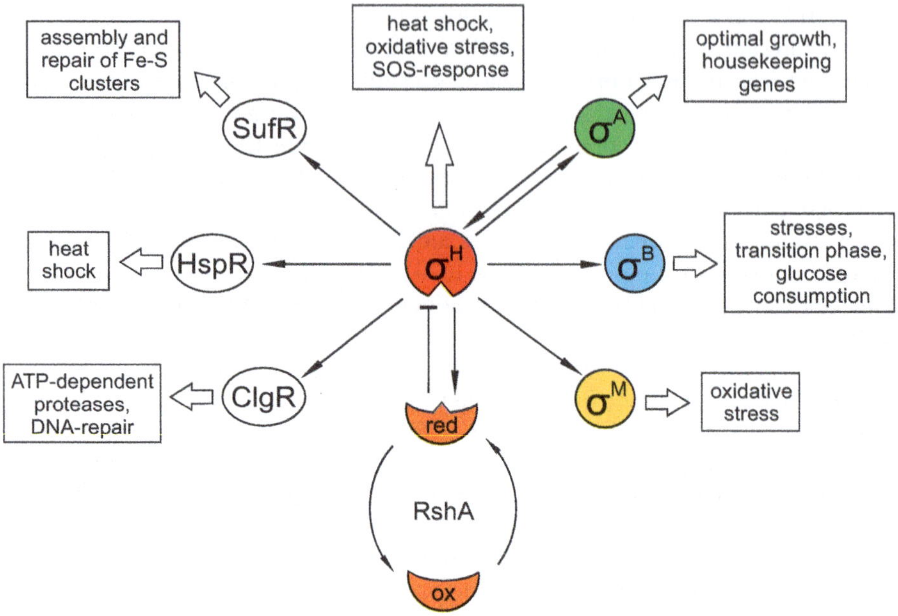

Fig. 5 Scheme of σ^H regulatory network in *C. glutamicum*. σ factors are in circles; anti-σ^H (RshA) is shown as a crescent (in its reduced or oxidized form); transcriptional regulators are in ovals; processes affected by σ factors and transcriptional regulators are in boxes; thin arrows depict a positive effect on transcription; a negative regulation of σ^H activity is shown by a T-bar. Based on Busche et al. (2012) and Toyoda and Inui (2016a)

et al. 2017, 2019; Šilar et al. 2016). The genes which were defined as σ^M-dependent (Nakunst et al. 2007) were found to also be active with σ^H (Busche et al. 2012). These versatile activities of σ^H are in agreement with the finding that various ECF sigma factors can recognize similar promoters (Gaballa et al. 2018; Helmann 2016). The observation that promoters of genes belonging to the regulons of other ECF sigma factors are also recognized by σ^H strengthens the idea that σ^H is a crucial regulator of stress responses in *C. glutamicum* (Fig. 5). The overlapping specificities of the ECF σ factors form the basis of the complex regulatory network of σ factors and contribute to the cell response to complex stresses. Toxic aromatic compounds such as ferulic acid (Chen et al. 2017), vanillin (Chen et al. 2016) and phenol (Dostálová, unpublished results) were found to initiate such complex chemostress responses, which induce the expression of many stress genes regulated by the cooperation of various *C. glutamicum* ECF sigma factors.

3.2.3 σ^E, a Sigma Factor Involved in Cell Surface Stress Response

The *C. glutamicum sigE* gene encodes the Group 4 sigma factor σ^E that mainly controls cell surface stress response (Park et al. 2008). The *cseE* gene, encoding an anti-σ factor, is located downstream of *sigE* (Park et al. 2008). The genes encoding anti-σ factors are located downstream of *sigE* in most other corynebacterial species. The same gene organization was also found in *M. tuberculosis*, in which the genes *sigE* and *rseA* (encoding an anti-σ factor) form an operon (Boldrin et al. 2019). The σ^E activity was found to be specifically inhibited by the anti-σ factor CseE (anti-σ^E). Protein-protein interaction in vitro showed that CseE binds σ^E and in this way mediates the antagonistic activity of anti-σ^E against σ^E (Park et al. 2008).

The function of *C. glutamicum* σ^E seems to be similar to that of the homologous σ^E in *M. tuberculosis*, which is essential for survival under conditions of cell surface stresses and heat shock. *C. glutamicum* σ^E was found to share 71% identity with the σ^E from *M. tuberculosis*.

The *C. glutamicum* Δ*sigE* strain grows at nearly the same rate as the WT strain, whereas the Δ*cseE* strain grows much slower (Park et al. 2008). The strain lacking *sigE* was more sensitive to detergents, lysozyme and some antibiotics, which suggests that *sigE* has a role in cell envelope defense. *sigE* expression is quite low during growth under optimum conditions, whereas it strongly increases after cell surface stress (Park et al. 2008) and heat shock (Barreiro et al. 2004). In addition, the *sigE* transcript level increases during nitrogen starvation (Silberbach et al. 2005), and lactic acid adaptation (Jakob et al. 2007). This indicates that σ^E is also involved in gene expression during the slow growth caused by nutritional limitation. In agreement with this, the *sigE* promoter was most active during the transition and stationary growth phases (our unpublished results).

According to the TSS determination by primer extension technique, the *sigE* gene is transcribed from three σ^A- and/or σ^B-dependent promoters (our unpublished results). The respective −10 promoter elements are conserved in other *Corynebacterium* species closely related to *C. glutamicum* (*C. callunae*, *C. efficiens* and *C. ammoniagenes*) according to their genome sequences. Two main *sigE* TSSs were also detected by transcriptome sequencing (Pfeifer-Sancar et al. 2013). Two promoters were also detected upstream of the *cseE* gene. According to the sequences of their −35 and −10 elements, one is vegetative, whereas the other with the motifs GGAA—(19 nt)—GTT identical to the consensus sequences of the σ^H-specific promoters, is most probably a stress promoter. Since the consensus motif of promoters dependent on related mycobacterium σ^E is GGAAC—(18–19 nt)—GTT (Rodrigue et al. 2006; Sachdeva et al. 2010), the *C. glutamicum cseE* stress promoter might also be σ^E-dependent. Although the consensus sequence of the σ^E-specific promoters has not yet been defined, it seems that it is closely similar to the σ^H consensus sequence. We showed that the *C. glutamicum* promoters P*sigB* (Dostálová et al. 2017), P2*dnaK*, P2*dnaJ2*, and P1*clgR* (Šilar et al. 2016) drove transcription with σ^H or σ^E both in vitro and in vivo. A similar overlap of σ^H and σ^E promoter specificity was observed in *M. tuberculosis*. The *M. tuberculosis sigB* gene is also transcribed with both σ^H and σ^E, and the two sigma factors recognize nearly identical promoters with subtle differences immediately downstream of the −35 element GGAA (Song et al. 2008). The nucleotide at position −31 was also found to be decisive for distinguishing between σ^R- and σ^E-dependent *S. coelicolor* promoters with the main common motif GGAA within the −35 region (Kim et al. 2016).

The participation of σ^E in the formation of the general stress sigma factor σ^B, the chaperons DnaK and DnaJ2 and transcriptional regulator ClgR, which controls the expression of the genes involved in proteolysis and DNA repair (Engels et al. 2004), suggests that this sigma factor may participate in responses to several stress types and complex stress conditions.

3.2.4 σ^D, a Sigma Factor Implicated in the Maintenance of Cell Wall Integrity

The *C. glutamicum sigD* gene encodes a Group 4 sigma factor σ^D that regulates expression of the genes involved in mycolate synthesis and peptidoglycan formation (Taniguchi et al. 2017; Toyoda and Inui 2018). The gene encoding anti-σ factor was found to be located downstream of *sigD* in most other corynebacterial and mycobacterial species. *C. glutamicum* σ^D shares 55% identity with the σ^D from *M. tuberculosis*. In contrast to the other *C. glutamicum* ECF σ factors whose amino acid sequences are mutually similar to various extents (Fig. 3), the overall amino acid sequence of σ^D differs substantially from the sequences of other *C. glutamicum* ECF σ factors (identity <25%).

The *C. glutamicum sigD* gene is transcribed from two vegetative (σ^A- and/or σ^B-dependent) promoters according to the TSS determination (49 and 120 nt upstream of translation initiation site, respectively) by comprehensive transcriptome sequencing and genome-wide TSS determination (Pfeifer-Sancar et al. 2013). The *rsdA* gene, encoding an anti-σ factor, is transcribed from a single promoter with the key sequence motifs −35 GTAAC and−10 CGAT, which were suggested to be consensus sequences of σ^D-dependent promoters (Toyoda and Inui 2018).

The *C. glutamicum sigD* deletion strain grew slightly slower than the WT strain, whereas strong overexpression of *sigD* cloned in the expression vector severely inhibited the growth of the culture (Taniguchi et al. 2017). The *sigD* deletion strain was more sensitive to lysozyme and also to some antibiotics (vancomycin and cephalothin), whereas *sigD* overexpression resulted in resistance to lysozyme (Toyoda and Inui 2018).

Analysis of differential gene expression in the *C. glutamicum* ATCC 13032 WT strain versus the *sigD* overexpression strain using the whole transcriptome data obtained by RNA sequencing provided a set of 29 σ^D-dependent genes (Taniguchi et al. 2017). This group included genes essential for the synthesis of mycolic acids contained in the corynebacterial cell wall, the modification of peptidoglycan and other cell envelope-related functions. Several of these genes which are involved in the regulation of mycolate metabolism are specific for strains of the genera *Corynebacterium*, *Mycobacterium* and *Rhodococcus*, which belong to the Mycolata group of bacteria. TSSs of these σ^D-dependent genes were mapped using a specific cDNA library of primary 5-end transcripts using the same technique as that currently used for *C. diphtheriae* (Wittchen et al. 2018). A search for the conserved sequence motifs resulted in the localization of 11 σ^D-specific promoters (Dostálová et al. 2019). Both in vitro transcription and in vivo two-plasmid assay proved unequivocally that all these promoters initiate the σ^D-specific synthesis of the transcripts (Dostálová et al. 2019). Using another methodical approach (ChIP-chip analysis), a similar set of 26 σ^D-dependent genes were revealed in the *C. glutamicum* strain R (Toyoda and Inui 2018). In total 22 promoter sequences with nearly identical promoter sequence elements −35 and −10, respectively, were defined by determining the TSSs by the 5′RACE technique and additional sequence analysis (Toyoda and Inui 2018). The fairly conserved common consensus sequence GTAAC—(18–19 nt)—GAT of two overlapping promoter sets could be deduced (Toyoda and Inui 2018; Dostálová et al. 2019). This σ^D consensus is different from that of σ^H-specific promoters (GGAA—(18–20 nt)—GTT). However, one of the σ^H consensus motifs −35 GGAA or −10 GTT could be also found in two σ^D-dependent promoters, P*cg0607* and P*lpd*, respectively (Dostálová et al. 2019). Although the σ^H and σ^D promoter consensus sequences are different, it was found that most of the σ^D-dependent promoters tested by in vivo two-plasmid assay were also active to a much lower extent with σ^H (Dostálová et al. 2019). Homology modelling of σ^D and σ^H interactions with the −35 and −10 elements of the analyzed σ^D-dependent promoters showed that the two σ factors bind to the mainly σ^D-dependent promoter in a similar way and only a few amino acids within the σ^D and σ^H sequence are responsible for the different affinity of each σ factor to the promoter DNA (Dostálová et al. 2019). These findings prove that σ^H can partially substitute for σ^D and thus may also be involved in the maintenance of the cell wall integrity in *C. glutamicum*. Overexpression of the *sigD* gene resulted in the accumulation of trehalose dicorynomycolate. This finding may contribute to the development of the biotechnological production of this useful compound (Taniguchi et al. 2017). On the other hand, strains with a lower content of corynomycolate (possibly with a *sigD* deletion) were reported to excrete more L-lysine and L-glutamate, which are large-scale products of *C. glutamicum* biotechnological applications (Gebhardt et al. 2007).

3.2.5 σ^C, a Sigma Factor Regulating the Aerobic Respiratory Chains

The *C. glutamicum sigC* gene encodes the Group 4 sigma factor σ^C, which is involved in the control of two branches of the aerobic respiratory chain (Toyoda and Inui 2016b). *C. glutamicum* σ^C was found to share 49% identity with σ^C from *M. tuberculosis*. The function of *C. glutamicum* σ^C, its regulon and consensus sequence of the σ^C-specific promoters have comprehensively been described in a single publication (Toyoda and Inui 2016b). The *sigC* gene is probably expressed at a low level during cultivation under non-stressed conditions, since its transcript was not detected by transcriptome sequencing (Pfeifer-Sancar et al. 2013) although it was shown that σ^C is essential for growth during exponential growth. No anti-σ gene was detected in the vicinity of *sigC* in *C. glutamicum*.

The promoter of the *sigC* gene was found immediately upstream of the start codon of the gene, which indicates that its transcript is leaderless. The *sigC* promoter is σ^A- and/or σ^B-dependent according to the −35 and −10 regions, which were deduced from the position of TSS determined by 5′-RACE. The *sigC* gene is thus most likely not autoregulated, similarly to the homologous gene in *M. tuberculosis* (Chang et al. 2012). The *sigC* gene is not induced by the impaired aerobic respiration conditions (e.g. copper deficiency and addition of cyanide), whereas the genes of the *sigC* regulon were upregulated under these conditions.

The deletion of *sigC* led to clones with an unstable phenotype and subsequent mutant formation. The detection of clearly upregulated or downregulated genes was therefore not reproducible with the Δ*sigC* strain. Overexpression of *sigC* resulted in the activation of 39 genes forming 8 operons. However the growth of that strain was slowed down and the culture reached a lower final optical density than the wild type.

The *sigC* regulon was determined by ChIP-chip technique and subsequent sequence analysis (Toyoda and Inui 2016b). The regulon mainly contains genes involved in heme and cytochrome synthesis as well as genes encoding parts of ATP-binding transporters. Analysis of the aerobic respiratory circuits finally showed that the σ^C regulon is induced in response to impaired electron transfer by cytochrome aa_3. In general, the σ^C regulon participates in the regulation of the aerobic respiration system in *C. glutamicum* (Toyoda and Inui 2016b).

ChIP-chip analysis revealed eight σ^C-specific promoters. The consensus sequence GGGAACT—(15–16 nt)—CGAC was confirmed by mutagenesis. Although the core of the −35 region is identical to the consensus of this element within the σ^H-dependent promoters, no recognition overlap between σ^C and σ^H specificity has been observed so far.

In conclusion, σ^C is involved in synthesis at terminal oxidases (cytochrome *bd* and cytochrome aa_3 oxidases) which enhance bacterial tolerance to oxidative and nitrosative stress conditions by scavenging reactive oxygen species. The detailed physiological function of the σ^C regulon and the signal transduction pathway from environmental stimulus to stress response still remains to be elucidated (Toyoda and Inui 2016b).

3.2.6 σ^M, the Most Variable Sigma Factor in Corynebacteria

The *C. glutamicum sigM* gene encodes σ^M, which is the least understood *C. glutamicum* sigma factor. It was suggested that σ^M is involved in the regulation of oxidative and heat stress response in *C. glutamicum* (Nakunst et al. 2007).

C. glutamicum σ^M only shares 37% identity with the σ^M from *M. tuberculosis*, which is much lower than the degrees of identities of other *C. glutamicum* ECF σ factors with corresponding *M. tuberculosis* ECF σ factors (49–71%). The function of *C. glutamicum* σ^M may thus differ from that of *M. tuberculosis* σ^M. The supposed difference in the function of σ^M factors of corynebacteria from that of *M. tuberculosis* σ^M is further supported by the fact that the activity of *M. tuberculosis* σ^M is controlled by a specific anti-sigma factor (Sklar et al. 2010), whereas no similar anti-sigma factor was found in any corynebacterial species.

C. glutamicum σ^M shares a high level of identity (65–87%) with the σ^M from only three other species of the genus *Corynebacterium* (*C. callunae*, *C. crudilactis*, *C. efficiens*). The corynebacterial σ factors, designated σ^M, seem to form a heterogenous group of σ^M-like factors, since only a low level of mutual identity (<45%) was found, even between the individual corynebacterial species.

The *sigM* transcript was not detected by *C. glutamicum* transcriptome sequencing (Pfeifer-Sancar et al. 2013), which suggests that the gene is not significantly expressed under non-stressed conditions. The transcription of *C. glutamicum sigM* was induced by disulfide stress and after heat and cold shock. The *sigM* gene could be deleted, however the σ^M-deficient strain was more sensitive to heat, cold and disulfide stress. The influence of salt, alcohol or peroxide on growth was the same in the WT and Δ*sigM* strains (Nakunst et al. 2007).

The *sigM* gene was found to be expressed from a σ^H-dependent promoter according to qPCR (Nakunst et al. 2007). The *sigM* promoter was also proven to be σ^H-specific when tested by in vivo two-plasmid assay, however, it did not exhibit any activity with any σ factor in vitro (Dostálová, unpublished results). It is worth noting that very similar P*sigM* promoters of two other *Corynebacterium* species were found to be both σ^H- and σ^M-dependent (Dostálová, unpublished results). In fact, the *C. glutamicum* P*sigM* sequence does not fit the defined σ^H consensus sequence well (Pátek and Nešvera 2011).

The *sigM* deletion strain was used to establish the σ^M regulon by microarray transcriptome analysis (Nakunst et al. 2007). The promoters of the supposed directly σ^M-dependent genes were deduced from TSSs determined by 5′RACE. In this way 4 σ^M-dependent genes (*trxB*, *trxB1*, *trxC*, *sufR*) and their respective promoters were defined. All these genes are involved in the disulfide stress response. The consensus sequence GGGAAT—(17–18 nt)—GTTG is quite similar to the σ^H-consensus. In agreement with the consensus sequence similarity, the same promoters were found to be σ^H-dependent by in vitro transcription and in vivo two-plasmid assay (Dostálová et al. 2017).

The genes *trxC* and *sufR* were determined to be σ^H-dependent in the first σ^H study (Ehira et al. 2009). All four genes (*trxB*, *trxB1*, *trxC*, *sufR*) were also found to be σ^H-dependent in other study (Busche et al. 2012). In a hunt for other σ^M-specific promoters, we have found a few promoters which were both σ^M- and σ^H-dependent (Dostálová, unpublished results). It seems, therefore, that the potential σ^M regulon is another candidate for an overlap with the σ^H regulon. However, the gene group is not homogenous in terms of their functions and the role played by σ^M in the *C. glutamicum* cell is therefore still uncertain.

4 Regulatory Overlaps of Sigma Factors

Expression of the *sig* genes, activity of σ factors and the activation of the corresponding σ regulons in response to various stresses and environmental stimuli (Fig. 6) is orchestrated by many regulatory factors at the level of transcription, translation and proteolysis (Fang 2005). The participation of transcriptional regulators in the expression of *sig* genes, recognition of promoters by more than one holoenzyme (RNAP core + σ), overlaps of regulons, subsequent switching on of the

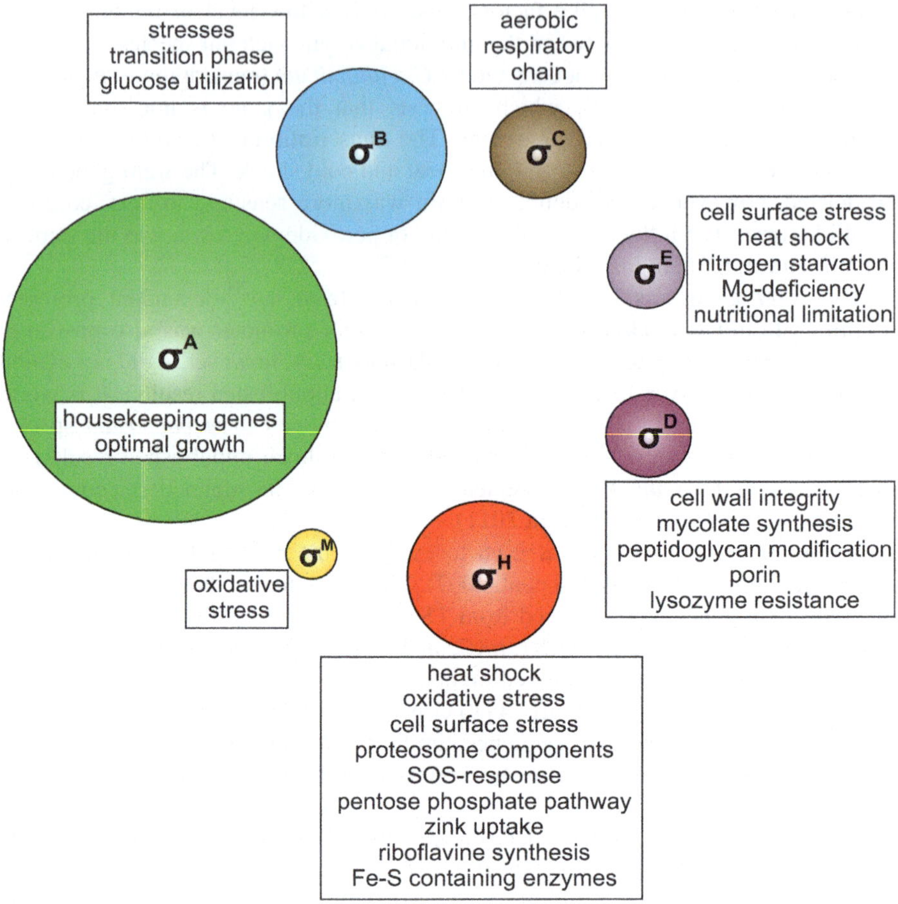

Fig. 6 Sigma regulons. The size of the circles approximately represents the number of genes within the regulon reported to date. Processes and stress responses regulated by the respective σ factors are boxed

sig genes by the action of another σ factor in a cascade, and the regulatory network controlled by σ factors are the features of the regulatory system which allow bacterial cells to adapt to any environmental conditions they encounter. At the same time, such entangled causal relationships make understanding the reasons for and consequences of individual regulatory steps more difficult.

Although the stress conditions may be different (e.g., oxidative stress, heat shock and cell envelope stress), the impairments of cell can be similar: damaged proteins. Such denatured proteins can be re-folded by chaperones or degraded by proteases. To ensure the production of these crucial stress response proteins under different conditions, the respective genes can be activated by different σ factors in response to specific conditions.

Sigma	−35 element	Spacer (nt)	extended −10 element
A	TTGNCA	16 - 18	TGN**TATAAT**NG
B	CGGCAA	16 - 18	TGNGN**TATAAT**GG
C	GGGAACT	15 - 16	CGAC
D	GTAAC	16 - 18	CGAT
E	N	N	N
H	GGAA	18 - 20	GTT
M	GGGAA	17 - 18	GTTG

Fig. 7 Consensus sequences of *C. glutamicum* of −35 and −10 promoter elements recognized by the individual σ factors. The spacer for σ^A- and σ^B-dependent promoters was counted between the −35 hexamer and −10 hexamer (in bold). *N* not yet described. Nucleotides within the consensus of ECF σ-specific promoters identical to the σ^H consensus are highlighted

Indeed, a number of promoter recognition overlaps between various stress σ factors were observed in *C. glutamicum*. The expression of *C. glutamicum* genes encoding some proteases (ClpB) and chaperones (DnaK and DnaJ2) was found to be induced by both σ^H and σ^E (Šilar et al. 2016). The same σ^H and σ^E overlap was observed in the transcription of *clgR* which encodes a heat-stress responsive regulator (Engels et al. 2005; Šilar et al. 2016). Expression of the *C. glutamicum sigB* gene that encodes the general stress response sigma factor σ^B is also alternatively driven by σ^H or σ^E (Dostálová et al. 2017). It is also possible that both σ^H and σ^M recognize the same *C. glutamicum* promoters (Dostálová et al. 2017; Nakunst et al. 2007; Šilar et al. 2016). However, this overlap has not yet been clearly documented.

Recently, a weaker transcription from several mainly σ^D-dependent promoters with σ^H was discovered (Dostálová et al. 2019). It seems that only a highly sensitive two-plasmid assay could detect this σ^H activity.

Figure 7 shows that all of the described consensus sequences of the −35 and −10 promoter elements recognized by different ECF σ factors in *C. glutamicum* are to some extent alike, especially in the −35 region. This feature seems to be a reason for observed overlap of different ECF σ factors in recognition of some *C. glutamicum* promoters described above.

Transcription of a gene from several promoters recognized by different σ factors is another mechanism that ensures optimal response of gene expression to stress conditions. As an example, additional σ^H-dependent promoters were found upstream of the σ^D-activated *C. glutamicum* genes *lppS* and *fadD2*. Thus, expression control of *lppS* (encoding L,D-transpeptidase involved in crosslink formation in peptidoglycans), and *fadD2* (encoding a long-chain fatty acyl-CoA ligase required for membrane lipid synthesis), mediated by both σ^D-and σ^H was documented (Dostálová et al. 2019). Since a third promoter which is most likely controlled by σ^A and/or σ^B was localized upstream of *lppS*, this gene may be co-regulated by as many as four σ factors. Such a redundancy of σ factors regulating a single gene may ensure the appropriate expression of a gene important for cell envelope synthesis under various growth phases and in response to various conditions threatening cell integrity.

Toxic compounds may cause complex stresses requiring the coordinated activity of a large number of genes and several σ factors. Such chemostress effects were analyzed with ferulic acid, vanillin and phenol at the *C. glutamicum* genome level. Vanillin stress activated several stress genes, including those controlled by σ^H (Chen et al. 2016). Ferulic acid stress induced 517 genes, including genes *sigB* and *sigE*, and consequently genes controlled by σ^B and σ^E (Chen et al. 2017). Depending on the concentration, phenol may cause an oxidative stress, non-specific cell permeability, or the dislodging of cell wall components (Denyer 1995). At the level of the cytoplasmic membrane, phenol may displace phospholipids. It is expected that such damage to the cell envelope, which is the first barrier maintaining bacterial cell integrity, would induce a complex response of cooperating stress σ factors. More than 200 genes were upregulated by phenol stress in *C. glutamicum* (Dostálová and Busche, unpublished results). Among them, more than 50 genes were identified to be controlled by σ^B, σ^C, σ^D, and σ^H or σ^E (Dostálová, unpublished results). Detailed analysis of this chemostress response may further elucidate the mobilization of complex defense mechanisms commanded by the combined efforts of different sigma factors.

Acknowledgements This work was supported by Grant 17-06991S from the Czech Science Foundation, Mobility Grant DAAD-18-11 from Czech Academy of Sciences (CAS) and Deutscher Akademischer. Austauschdienst and Institutional Research Project RVO61388971 from the Institute of Microbiology of the CAS.

References

Albersmeier A, Pfeifer-Sancar K, Rückert C, Kalinowski J (2017) Genome-wide determination of transcription start sites reveals new insights into promoter structures in the actinomycete *Corynebacterium glutamicum*. J Biotechnol 257:99–109

Barreiro C, Gonzalez-Lavado E, Pátek M, Martin JF (2004) Transcriptional analysis of the *groES-groEL1*, *groEL2*, and *dnaK* genes in *Corynebacterium glutamicum*: characterization of heat shock-induced promoters. J Bacteriol 186:4813–4817

Boldrin F, Mazzabo LC, Anoosheh S, Palu G, Gaudreau L, Manganelli R, Provvedi R (2019) Assessing the role of Rv1222 (RseA) as an anti-sigma factor of the *Mycobacterium tuberculosis* extracytoplasmic sigma factor SigE. Sci Rep 9:4513

Burr T, Mitchell J, Kolb A, Minchin S, Busby S (2000) DNA sequence elements located immediately upstream of the −10 hexamer in *Escherichia coli* promoters: a systematic study. Nucleic Acids Res 28:1864–1870

Busche T, Šilar R, Pičmanova M, Pátek M, Kalinowski J (2012) Transcriptional regulation of the operon encoding stress-responsive ECF sigma factor SigH and its anti-sigma factor RshA, and control of its regulatory network in *Corynebacterium glutamicum*. BMC Genomics 13:445

Chang A, Smollett KL, Gopaul KK, Chan BH, Davis EO (2012) *Mycobacterium tuberculosis* H37Rv *sigC* is expressed from two promoters but is not auto-regulatory. Tuberculosis 92:48–55

Chen C, Pan J, Yang X, Guo C, Ding W, Si M, Zhang Y, Shen X, Wang Y (2016) Global transcriptomic analysis of the response of *Corynebacterium glutamicum* to vanillin. PLoS One 11:e0164955

Chen C, Pan J, Yang X, Xiao H, Zhang Y, Si M, Shen X, Wang Y (2017) Global transcriptomic analysis of the response of *Corynebacterium glutamicum* to ferulic acid. Arch Microbiol 199:325–334

Denyer SP (1995) Mechanisms of action of antibacterial biocides. Int Biodeterior Biodegr 36:221–225

Dostálová H, Holátko J, Busche T, Rucká L, Rapoport A, Halada P, Nešvera J, Kalinowski J, Pátek M (2017) Assignment of sigma factors of RNA polymerase to promoters in *Corynebacterium glutamicum*. AMB Express 7:133

Dostálová H, Busche T, Holátko J, Rucká L, Štěpánek V, Barvík I, Nešvera J, Kalinowski J, Pátek M (2019) Overlap of promoter recognition specificity of stress response sigma factors SigD and SigH in *Corynebacterium glutamicum* ATCC 13032. Front Microbiol 9:3287

Ehira S, Shirai T, Teramoto H, Inui M, Yukawa H (2008) Group 2 sigma factor SigB of *Corynebacterium glutamicum* positively regulates glucose metabolism under conditions of oxygen deprivation. Appl Environ Microbiol 74:5146–5152

Ehira S, Teramoto H, Inui M, Yukawa H (2009) Regulation of *Corynebacterium glutamicum* heat shock response by the extracytoplasmic-function sigma factor SigH and transcriptional regulators HspR and HrcA. J Bacteriol 191:2964–2972

Engels S, Schweitzer JE, Ludwig C, Bott M, Schäffer S (2004) *clpC* and *clpP1P2* gene expression in *Corynebacterium glutamicum* is controlled by a regulatory network involving the transcriptional regulators ClgR and HspR as well as the ECF sigma factor σ^H. Mol Microbiol 52:285–302

Engels S, Ludwig C, Schweitzer JE, Mack C, Bott M, Schaffer S (2005) The transcriptional activator ClgR controls transcription of genes involved in proteolysis and DNA repair in *Corynebacterium glutamicum*. Mol Microbiol 57:576–591

Fang FC (2005) Sigma cascades in prokaryotic regulatory networks. Proc Natl Acad Sci USA 102:4933–4934

Gaballa A, Guariglia-Oropeza V, Durr F, Butcher BG, Chen AY, Chandrangsu P, Helmann JD (2018) Modulation of extracytoplasmic function (ECF) sigma factor promoter selectivity by spacer region sequence. Nucleic Acids Res 46:134–145

Gebhardt H, Meniche X, Tropis M, Krämer R, Daffé M, Morbach S (2007) The key role of the mycolic acid content in the functionality of the cell wall permeability barrier in *Corynebacterineae*. Microbiology 153:1424–1434

Gourse RL, Ross W, Gaal T (2000) UPs and downs in bacterial transcription initiation: the role of the alpha subunit of RNA polymerase in promoter recognition. Mol Microbiol 37:687–695

Gruber TM, Gross CA (2003) Multiple sigma subunits and the partitioning of bacterial transcription space. Annu Rev Microbiol 57:441–466

Halgasova N, Bukovska G, Ugorcakova J, Timko J, Kormanec J (2002) The *Brevibacterium flavum* sigma factor SigB has a role in the environmental stress response. FEMS Microbiol Lett 216:77–84

Helmann JD (2016) *Bacillus subtilis* extracytoplasmic function (ECF) sigma factors and defense of the cell envelope. Curr Opin Microbiol 30:122–132

Holátko J, Elišáková V, Prouza M, Sobotka M, Nešvera J, Pátek M (2008) Metabolic engineering of the L-valine biosynthesis pathway in *Corynebacterium glutamicum* using promoter activity modulation. J Biotechnol 139:203–210

Jakob K, Satorhelyi P, Lange C, Wendisch VF, Silakowski B, Scherer S, Neuhaus K (2007) Gene expression analysis of *Corynebacterium glutamicum* subjected to long-term lactic acid adaptation. J Bacteriol 189:5582–5590

Kim TH, Kim HJ, Park JS, Kim Y, Kim P, Lee HS (2005) Functional analysis of *sigH* expression in *Corynebacterium glutamicum*. Biochem Biophys Res Commun 331:1542–1547

Kim MS, Dufour YS, Yoo JS, Cho YB, Park JH, Nam GB, Kim HM, Lee KL, Donohue TJ, Roe JH (2012) Conservation of thiol-oxidative stress responses regulated by SigR orthologues in actinomycetes. Mol Microbiol 85:326–344

Kim KY, Park JK, Park S (2016) In *Streptomyces coelicolor* SigR, methionine at the −35 element interacting region 4 confers the −31′-adenine base selectivity. Biochem Biophys Res Commun 470:257–262

Larisch C, Nakunst D, Huser AT, Tauch A, Kalinowski J (2007) The alternative sigma factor SigB of *Corynebacterium glutamicum* modulates global gene expression during transition from exponential growth to stationary phase. BMC Genomics 8:4

Li L, Fang C, Zhuang N, Wang T, Zhang Y (2019) Structural basis for transcription initiation by bacterial ECF sigma factors. Nat Commun 10:1153

Lisser S, Margalit H (1993) Compilation of *E. coli* mRNA promoter sequences. Nucleic Acids Res 21:1507–1516

Nakunst D, Larisch C, Hüser AT, Tauch A, Pühler A, Kalinowski J (2007) The extracytoplasmic function-type sigma factor SigM of *Corynebacterium glutamicum* ATCC 13032 is involved in transcription of disulfide stress-related genes. J Bacteriol 189:4696–4707

Nesvera J, Pátek M, Hochmannová J, Abrhámová Z, Becvárová V, Jelínkova M, Vohradský J (1997) Plasmid pGA1 from *Corynebacterium glutamicum* codes for a gene product that positively influences plasmid copy number. J Bacteriol 179(5):1525–1532

Pacheco LG, Castro TL, Carvalho RD, Moraes PM, Dorella FA, Carvalho NB, Slade SE, Scrivens JH, Feelisch M, Meyer R, Miyoshi A, Oliveira SC, Dowson CG, Azevedo V (2012) A role for sigma factor σ^E in *Corynebacterium pseudotuberculosis* resistance to nitric oxide/peroxide stress. Front Microbiol 3:126

Paget MS (2015) Bacterial sigma factors and anti-sigma factors: structure, function and distribution. Biomol Ther 5:1245–1265

Paget MS, Molle V, Cohen G, Aharonowitz Y, Buttner MJ (2001) Defining the disulphide stress response in *Streptomyces coelicolor* A3(2): identification of the σ^R regulon. Mol Microbiol 42:1007–1020

Park SD, Youn JW, Kim YJ, Lee SM, Kim Y, Lee HS (2008) *Corynebacterium glutamicum* σ^E is involved in responses to cell surface stresses and its activity is controlled by the anti σ factor CseE. Microbiology 154:915–923

Pátek M, Nešvera J (2011) Sigma factors and promoters in *Corynebacterium glutamicum*. J Biotechnol 154:101–113

Pátek M, Eikmanns BJ, Pátek J, Sahm H (1996) Promoters from *Corynebacterium glutamicum*: cloning, molecular analysis and search for a consensus motif. Microbiology 142:1297–1309

Pfeifer-Sancar K, Mentz A, Rückert C, Kalinowski J (2013) Comprehensive analysis of the *Corynebacterium glutamicum* transcriptome using an improved RNAseq technique. BMC Genomics 14:888

Raman S, Song T, Puyang X, Bardarov S, Jacobs WR Jr, Husson RN (2001) The alternative sigma factor SigH regulates major components of oxidative and heat stress responses in *Mycobacterium tuberculosis*. J Bacteriol 183:6119–6125

Rodrigue S, Provvedi R, Jacques PE, Gaudreau L, Manganelli R (2006) The sigma factors of *Mycobacterium tuberculosis*. FEMS Microbiol Rev 30:926–941

Ruiz JC, D'Afonseca V, Silva A, Ali A, Pinto AC, Santos AR, Rocha AA, Lopes DO, Dorella FA, Pacheco LG, Costa MP, Turk MZ, Seyffert N, Moraes PM, Soares SC, Almeida SS, Castro TL, Abreu VA, Trost E, Baumbach J, Tauch A, Schneider MP, McCulloch J, Cerdeira LT, Ramos RT, Zerlotini A, Dominitini A, Resende DM, Coser EM, Oliveira LM, Pedrosa AL, Vieira CU, Guimaraes CT, Bartholomeu DC, Oliveira DM, Santos FR, Rabelo EM, Lobo FP, Franco GR, Costa AF, Castro IM, Dias SR, Ferro JA, Ortega JM, Paiva LV, Goulart LR, Almeida JF, Ferro MI, Carneiro NP, Falcao PR, Grynberg P, Teixeira SM, Brommonschenkel S, Oliveira SC, Meyer R, Moore RJ, Miyoshi A, Oliveira GC, Azevedo V (2011) Evidence for reductive genome evolution and lateral acquisition of virulence functions in two *Corynebacterium pseudotuberculosis* strains. PLoS One 6:e18551

Rytter JV, Helmark S, Chen J, Lezyk MJ, Solem C, Jensen PR (2014) Synthetic promoter libraries for *Corynebacterium glutamicum*. Appl Microbiol Biotechnol 98:2617–3623

Sachdeva P, Misra R, Tyagi AK, Singh Y (2010) The sigma factors of *Mycobacterium tuberculosis*: regulation of the regulators. FEBS J 277:605–626

Schröder J, Tauch A (2010) Transcriptional regulation of gene expression in *Corynebacterium glutamicum*: the role of global, master and local regulators in the modular and hierarchical gene regulatory network. FEMS Microbiol Rev 34:685–737

Šilar R, Holátko J, Rucká L, Rapoport A, Dostálová H, Kadeřabková P, Nešvera J, Pátek M (2016) Use of in vitro transcription system for analysis of *Corynebacterium glutamicum* promoters recognized by two sigma factors. Curr Microbiol 73:401–408

Silberbach M, Hüser A, Kalinowski J, Pühler A, Walter B, Krämer R, Burkovski A (2005) DNA microarray analysis of the nitrogen starvation response of *Corynebacterium glutamicum*. J Biotechnol 119:357–367

Sklar JG, Makinoshima H, Schneider JS, Glickman MS (2010) *M. tuberculosis* intramembrane protease Rip1 controls transcription through three anti-sigma factor substrates. Mol Microbiol 77:605–617

Song T, Dove SL, Lee KH, Husson RN (2003) RshA, an anti-sigma factor that regulates the activity of the mycobacterial stress response sigma factor SigH. Mol Microbiol 50:949–959

Song T, Song SE, Raman S, Anaya M, Husson RN (2008) Critical role of a single position in the −35 element for promoter recognition by *Mycobacterium tuberculosis* SigE and SigH. J Bacteriol 190:2227–2230

Taniguchi H, Wendisch VF (2015) Exploring the role of sigma factor gene expression on production by *Corynebacterium glutamicum*: sigma factor H and FMN as example. Front Microbiol 6:740

Taniguchi H, Busche T, Patschkowski T, Niehaus K, Pátek M, Kalinowski J, Wendisch VF (2017) Physiological roles of sigma factor SigD in *Corynebacterium glutamicum*. BMC Microbiol 17:158

Toyoda K, Inui M (2016a) Regulons of global transcription factors in *Corynebacterium glutamicum*. Appl Microbiol Biotechnol 100:45–60

Toyoda K, Inui M (2016b) The extracytoplasmic function sigma factor σ^C regulates expression of a branched quinol oxidation pathway in *Corynebacterium glutamicum*. Mol Microbiol 100:486–509

Toyoda K, Inui M (2018) Extracytoplasmic function sigma factor σ^D confers resistance to environmental stress by enhancing mycolate synthesis and modifying peptidoglycan structures in *Corynebacterium glutamicum*. Mol Microbiol 107:312–329

Toyoda K, Teramoto H, Yukawa H, Inui M (2015) Expanding the regulatory network governed by the extracytoplasmic function sigma factor σ^H in *Corynebacterium glutamicum*. J Bacteriol 197:483–496

Vašicová P, Pátek M, Nešvera J, Sahm H, Eikmanns B (1999) Analysis of the *Corynebacterium glutamicum dapA* promoter. J Bacteriol 181:6188–6191

Weber H, Polen T, Heuveling J, Wendisch VF, Hengge R (2005) Genome-wide analysis of the general stress response network in *Escherichia coli*: σ^S-dependent genes, promoters, and sigma factor selectivity. J Bacteriol 187:1591–1603

Wittchen M, Busche T, Gaspar AH, Lee JH, Ton-That H, Kalinowski J, Tauch A (2018) Transcriptome sequencing of the human pathogen *Corynebacterium diphtheriae* NCTC 13129 provides detailed insights into its transcriptional landscape and into DtxR-mediated transcriptional regulation. BMC Genomics 19:82

Zhang S, Liu D, Mao Z, Mao Y, Ma H, Chen T, Zhao X, Wang Z (2018) Model-based reconstruction of synthetic promoter library in *Corynebacterium glutamicum*. Biotechnol Lett 40:819–827

Global Transcriptional Regulators Involved in Carbon, Nitrogen, Phosphorus, and Sulfur Metabolisms in *Corynebacterium glutamicum*

Koichi Toyoda and Masayuki Inui

Contents

Abstract Bacteria are equipped with sophisticated metabolic control systems to adapt to variable nutrient conditions. While metabolic pathways are controlled at enzyme activity levels and at mRNA levels, scarcity in preferred nutrients causes transcriptional induction of genes for alternative nutrient source utilization. Transcriptional regulatory systems controlling metabolisms of carbon, nitrogen, phosphate, and sulfur in *Corynebacterium glutamicum* have been extensively studied for

K. Toyoda
Research Institute of Innovative Technology for the Earth (RITE), Kyoto, Japan
e-mail: toyo51@rite.or.jp

M. Inui (✉)
Research Institute of Innovative Technology for the Earth (RITE), Kyoto, Japan

Graduate School of Biological Sciences, Nara Institute of Science and Technology, Nara, Japan
e-mail: inui@rite.or.jp

M. Inui, K. Toyoda (eds.), *Corynebacterium glutamicum*, Microbiology Monographs 23, https://doi.org/10.1007/978-3-030-39267-3_5

the last two decades. The knowledge of the regulators, including regulon members, operator sequences, and effectors, has deepened our understanding of the *C. glutamicum* physiology and has led to develop synthetic biology tools for metabolic engineering, maximizing the *C. glutamicum* potential as a production host. In this chapter, we review the studies of the transcriptional regulators, especially focus on those with global regulatory roles in the primary metabolism in *C. glutamicum*.

1 Introduction

In bacterial genomes, uptake and metabolic systems for versatile nutrient sources, including carbon, nitrogen, phosphorus, and sulfur sources, are encoded to adapt to fluctuating conditions. Elaborate mechanisms for sensing the environmental and cellular nutrient status and those for switching gene expression in response to the status are also encoded to express the corresponding genes at the appropriate degree and timing and to avoid the expression of unnecessary genes. When multiple resources exist, selective and successive utilization can be achieved by global transcriptional regulatory mechanisms. Hierarchical transcriptional regulation mediated by global regulators and pathway-specific local regulators constitute a transcriptional regulatory network. Genome-wide analysis, including transcriptome using microarray and high-throughput sequencer, has revealed the network structure in some organisms. Since the genomic sequence of *Corynebacterium glutamicum* was reported in 2003 (Ikeda and Nakagawa 2003; Kalinowski et al. 2003), genes for the primary metabolic pathways and transcriptional regulators involved in their regulation have been intensively studied [reviewed in (Schröder and Tauch 2010; Nešvera and Pátek 2011; Pátek and Nešvera 2011; Bott and Brocker 2012; Neshat et al. 2014)]. These studies revealed that the transcriptional regulatory systems different from the paradigms established in model microorganisms, like *Escherichia coli* and *Bacillus subtilis*, are working in the suborder Corynebacterineae in the class Actinobacteria. For example, coordinated simultaneous carbon metabolism not subjected to the catabolite control, adenylylation-mediated protein interaction of a transcriptional regulator is involved in nitrogen control and coordinated transcriptional regulation of methionine and cysteine synthesis pathways. In this chapter, we review the current understanding of transcriptional regulators involved in carbon, nitrogen, phosphorus, and sulfur metabolism in *C. glutamicum*.

2 Carbon Catabolism Control

C. glutamicum is capable of utilizing a variety of carbon sources, including sugars, sugar alcohols, organic acids, alcohols, amino acids, and aromatic compounds. Studies identifying genes involved in the metabolism of these compounds were followed by those identifying transcriptional regulators controlling the genes. All of these carbon compounds are finally catabolized through the central carbon metabolic pathways consisting of glycolysis, the pentose phosphate pathway, the TCA cycle, and the glyoxylate shunt. The genes constituting these pathways are coordinately regulated by multiple transcriptional regulators.

2.1 Sugar Uptake

2.1.1 Phosphotransferase System-Dependent Sugar Uptake

The phosphoenolpyruvate (PEP): carbohydrate phosphotransferase system (PTS) is the major sugar uptake system in most bacteria (Postma et al. 1993). It consists of three components: two common cytoplasmic proteins, so-called general components, enzyme I (EI) and HPr, and sugar-specific enzyme IIs (EIIs). Phosphoryl group of PEP is transferred to EII via EI and HPr, then EIIs couple sugar transport with phosphorylation. *C. glutamicum* possesses EIIs specific to glucose, fructose, and sucrose, which are encoded by the *ptsG*, *ptsF*, and *ptsS* genes (cg1537, cg2120, and cg2925 in the type strain ATCC 13032, cgR_1425, cgR_1766, and cgR_2547 in strain R; hereafter the locus tag "cg----" indicates a gene in strain ATCC 13032, "cgR_----" indicates that in strain R), respectively, while the general components EI and HPr are encoded by the *ptsI* and *ptsH* genes (cg2117 or cgR_1763 and cg2121 or cgR_1767) (Kotrba et al. 2001; Parche et al. 2001). While *ptsG* and *ptsS* are dispersed in the genome, the other components are encoded in a cluster; an operon consisting of *fruR-pfkB1-ptsF-ptsH* is located upstream of *ptsI*, which is divergently transcribed. The genes, *fruR* (cg2118 or cgR_1764) and *pfkB1* (cg2119 or cgR_1765) encode a DeoR-type transcriptional regulator and fructose-1-phosphate kinase, respectively. Sugar-dependent transcriptional regulation of the *pts* genes has been first reported in 2008 (Tanaka et al. 2008a). Northern analysis demonstrated that fructose and sucrose induce all *pts* genes whereas glucose upregulates *ptsI*, *ptsH*, and *ptsG*, the last two of which are generally expressed even in the absence of PTS sugars. The analysis also showed that *ptsH* is transcribed not only as a part of the operon but from its own promoter.

FruR, encoded by the first gene of the *fruR-pfkB1-ptsF-ptsH* operon, is not involved in the sugar-dependent upregulation because, in the *fruR* deletion mutant, the fructose-dependent upregulation of the *fruR* operon and *ptsI* was enhanced (Tanaka et al. 2008a), which led to the idea that FruR attenuates the induction of expression of *ptsI*, *ptsH*, and its own operon in the presence of fructose. Although

FruR was obtained by DNA affinity purification using a DNA fragment encompassing the intergenic region between *ptsI* and *fruR* (Tanaka et al. 2008a), its binding motif was not determined.

Another DeoR-type transcriptional regulator, SugR, globally represses the *pts* genes. SugR is encoded by the *sugR* gene (cg2115 or cgR_1761), whose genomic locus is close to the *ptsI-ptsF* cluster, motivating its functional analysis (Engels and Wendisch 2007). Transcriptome analysis using microarray revealed that in the absence of PTS sugars, expression of all *pts* genes in the *sugR* deletion mutant was higher than in the wild type (Engels and Wendisch 2007; Gaigalat et al. 2007) (Fig. 1). The recombinant protein of SugR bound to the promoter region of these genes in vitro. The binding of SugR to the *ptsG* promoter region in vitro was inhibited by fructose 6-phosphate (Engels and Wendisch 2007). However, the binding to the promoter region of *ptsI* was not inhibited by fructose-6-phosphate, but by some sugar phosphates including fructose 1-phosphate and glucose 6-phosphate, each of which is yielded by fructose and glucose uptake via PtsF and PtsG, respectively, and fructose 1,6-bisphosphate (Gaigalat et al. 2007). Of the sugar phosphates tested, fructose-1-phosphate has the strongest inhibitory effect. This is consistent with the fact that the *pts* genes were highly induced in the presence of fructose compared with that in the presence of glucose (Tanaka et al. 2008a). The binding sites of SugR have been investigated by different groups, confining the region required for the SugR binding, but the consensus motif presented were different from each other (Engels and Wendisch 2007; Gaigalat et al. 2007; Tanaka et al. 2008b). Chromatin immunoprecipitation with microarray (ChIP-chip) analysis detected the SugR binding in vivo to the promoter regions of the *pts* genes (Engels et al. 2008). The alignment of the detected promoter regions determined the consensus binding motif as TCRRACA (R: A or G) (Engels et al. 2008).

Functionally redundant two GntR-type transcriptional regulators GntR1 and GntR2, the latter of which is strain-specific, have been first identified as transcriptional repressors of gluconate utilization genes, encoding gluconate permease, gluconate kinase, and 6-phosphogluconate dehydrogenase in strain ATCC 13032 (Frunzke et al. 2008). Transcriptome analysis of *gntR1/R2* (cg2783 or cgR_2434/cg1935) double deletion mutant using microarray revealed the downregulation of *ptsG* and *ptsS* as well as upregulation of the pentose phosphate pathway genes and the gluconate metabolic genes. In vitro DNA binding assay and in vivo promoter reporter assay confirmed that GntR1/R2 also function as transcriptional activators of the *pts* genes. The DNA binding activity of GntR1/R2 was inhibited by gluconate and glucono-δ-lactone (Frunzke et al. 2008). Consistent with this, the glucose uptake rate was reduced by gluconate. The coordinated regulation of gluconate utilization pathway and PTS sugar uptake likely represents the molecular mechanism of simultaneous utilization of different carbon sources of *C. glutamicum* rather than sequential utilization observed in other model organisms. A ChIP-chip analysis of GntR1 in *C. glutamicum* strain R, in which GntR2 is not encoded, not only confirmed the previous results but also expanded its regulon and identified

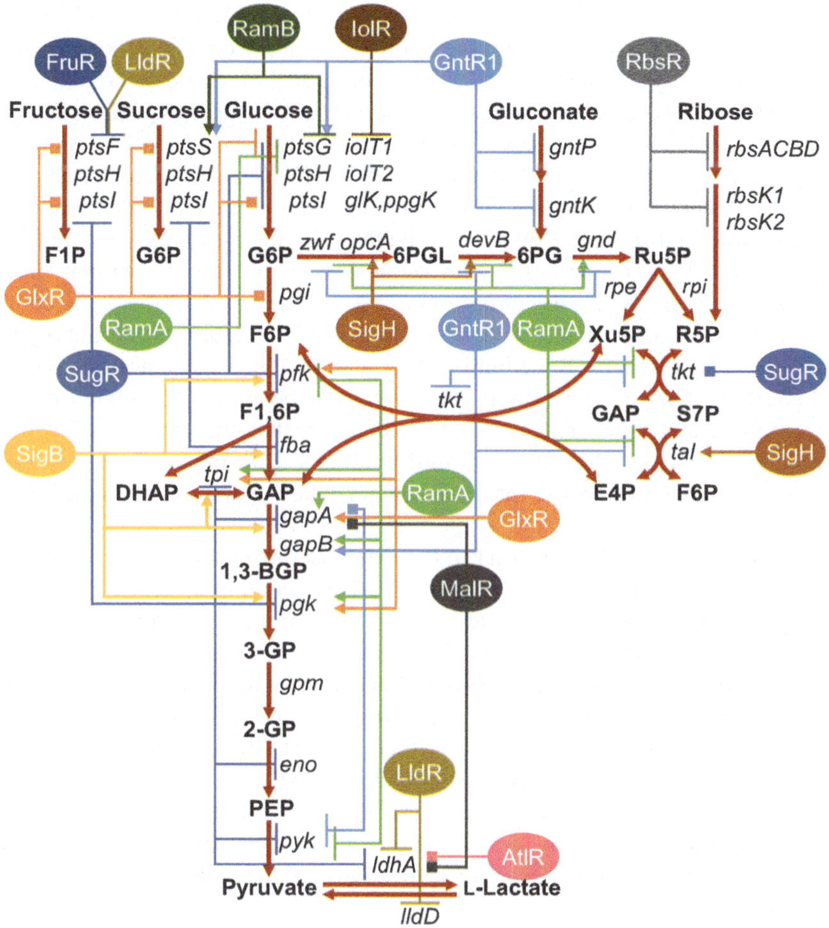

Fig. 1 Transcriptional regulators of genes involved in sugar uptake and metabolism. Metabolic pathways are indicated by thick red arrows. Genes constituting the metabolic pathways are indicated by italic letters. Sugars and metabolites are indicated by bold letters. Transcriptional regulators are indicated by ovals. Arrows from the ovals indicate transcriptional activation, whereas T-bars indicate repression. Lines ended with squares indicate the in vivo or in vitro binding of transcriptional regulators with unknown function. *G6P* glucose 6-phosphate, *F1P* fructose 1-phosphate, *6PGL* 6-phosphoglucono-1,5-lactone, *6PG* 6-phosphogluconate, *Ru5P* ribulose 5-phosphate, *Xu5P* xylulose 5-phosphate, *R5P* ribose 5-phosphate, *S7P* sedoheptulose 7-phosphate, *F6P* fructose 6-phosphate, *E4P* erythrose 4-phosphate, *F1,6BP* fructose 1,6-bisphosphate, *DHAP* dihydroxyacetone phosphate, *GAP* glyceraldehyde 3-phosphate, *1,3-BGP* glycerate 1,3-bisphosphate, *3-GP* glycerate 3-phosphate, *2-GP* glycerate 2-phosphate, *PEP* phosphoenolpyruvate

consensus binding motif more precisely, 5′-AWWGGTMRYACCWWT-3′ (Tanaka et al. 2014), unveiling the more global physiological function of GntR1 as described later.

The MerR-type transcriptional regulator RamB (cg0444 or cgR_0442) has been first identified as a transcriptional repressor of genes for acetate metabolism, including glyoxylate shunt genes, *aceA* (cg2560 or cgR_2212) and *aceB* (cg2559 or cgR_2211), and acetate activation genes, *pta* (cg3048 or cgR_2656) and *ackA* (cg3047 or cgR_2655) (Gerstmeir et al. 2004). This study identified the RamB binding motif, 5′-ARAACTTTGCAAA-3′ (R: A or G) and in silico search for the motif in the genome identified a putative binding site in the *ptsG* promoter region. Transcriptome analysis followed by in vitro DNA binding assay indicated that RamB directly activates *ptsG* and *ptsS* expression during growth in minimal medium containing glucose as a sole carbon source (Auchter et al. 2011a). In contrast, promoter reporter assay demonstrated that RamB acts as a transcriptional repressor of *ptsG* during growth in LB medium with or without glucose and acetate (Subhadra and Lee 2013). The putative binding site is located at position +72 with respect to the transcriptional start site (TSS) of *ptsG*, indicating that the repressor function is seemingly reasonable. The putative RamB binding site upstream of *ptsS* is centered at position −47 with respect to the TSS, which is likely consistent with the activator function.

RamA (cg2831 or cgR_2464), a LuxR-type transcriptional regulator, was isolated by DNA affinity purification using DNA fragments containing either the promoter region of the *pta-ackA* operon or the intergenic region of *aceA* and *aceB*, as RamB (Gerstmeir et al. 2004; Cramer and Eikmanns 2007). Thus, the names of these regulators RamA and RamB (regulators of acetate metabolism A and B) represent genes first identified to be regulated by themselves. Transcriptome analysis of the *ramA* deletion mutant and in vitro DNA binding assay demonstrated that *ptsG* and *ptsH* are directly repressed by RamA (Auchter et al. 2011a). There are seven and two putative binding sites (consecutive four to six G or C) of RamA in the *ptsG* and *ptsH* promoter regions, respectively, despite no direct evidence that RamA binds to all of these sites. A part of the sites is located downstream of the TSS, implying the repressor function of RamA.

GlxR (cg0350 or cgR_0377) is a homolog of *E. coli* cyclic AMP (cAMP) receptor protein (CRP). It has been first identified as a transcriptional repressor of the *aceB* gene encoding malate synthase in the glyoxylate shunt (Kim et al. 2004). In silico search for its binding motif, 5′-TGTGANNNNNNTCACA-3′, which is almost the same as that of *E. coli* CRP, and in vitro DNA binding assay confirmation detected 77 binding sites in the genome, identifying putative 150 genes belonging to the GlxR regulon (Kohl et al. 2008; Kohl and Tauch 2009). ChIP-chip and ChIP-seq (ChIP with high-throughput sequencing) analyses (Toyoda et al. 2011; Jungwirth et al. 2013) confirmed and extended the GlxR regulon by detecting GlxR binding to DNA in vivo. The GlxR binding sites upstream of *ptsG* and in the intergenic region between *ptsI* and the *fruR-pfkB1-ptsF* operon were detected by all the analyses whereas those upstream of *ptsS* was detected only by ChIP-chip (Toyoda et al. 2011). Although the function of GlxR in vivo at those binding sites has not been demonstrated, the position of the binding sites with respect to the TSS suggest a repressor function at these sites. Indeed, promoter reporter assay has recently shown that the *ptsG* promoter activity was upregulated in deletion mutants of *glxR* (cg0350 or cgR_0377) and *cyaB* (cg0375 or cgR_0397) (Subhadra and Lee 2013), the latter of which encodes adenylate cyclase producing cAMP. As the intracellular cAMP levels of the glucose-grown cells are higher than those of acetate-grown cells (Kim et al. 2004), GlxR may suppress *ptsG* expression in the presence of glucose where SugR-mediated repression is alleviated.

2.1.2 β-Glucoside Uptake and Utilization

Unlike the type strain ATCC 13032, strain R can utilize β-glucosides, including salicin and arbutin (Kotrba et al. 2003). Genes for uptake and utilization of the glucosides form a gene cluster, *bglF-bglA-bglG* (cgR_2729-2727), each gene of which encodes β-glucoside-specific enzyme II of PTS, phospho-β-glucosidase, and an antiterminator, respectively. The antiterminator is required for substrate-dependent expression of the operon. In the absence of a substrate, transcription of the operon terminates at a terminator present in the 5′-UTR (untranslated region) of mRNA. In the presence of a substrate, the antiterminator is postulated to bind to the ribonucleic antiterminator sequence in the terminator, thereby disrupting the terminator structure and allowing elongation of transcripts. *C. glutamicum* R carries another orthologous cluster consisting of *bglF2-bglA2-bglG2* (cgR_2610-2608) for β-glucosides utilization. Genetic analysis demonstrated that the components are functionally equivalent, but the antiterminators BglG and BglG2 are specific to cognate transcripts with minor cross-talk regulation (Tanaka et al. 2009). In contrast to other PTS components, SugR is not involved in the expression of the gene clusters (Tanaka et al. 2009). However, ChIP-chip analysis identified GlxR binding sites in the promoter region of the operons (unpublished data), suggesting its involvement in the regulation of the gene clusters.

In the type strain ATCC 13032 genome, the *bglF-bglA-bglG* cluster is present as pseudogenes and *bglA2* and *bglG2* in the *bglF2-bglA2-bglG2* cluster were lost (Kalinowski et al. 2003; Yukawa et al. 2007). This suggests that the clusters were present in the common ancestor of the two strains. This is supported by the fact that, at the end of 2018, NCBI BLAST search revealed that other than strain R, at least eight strains of *C. glutamicum*, whose genomic sequences are available, possess orthologous gene clusters for β-glucosides utilization.

2.1.3 Non-PTS Sugars

Non-PTS sugars are transported into the cytosol via other transporters but not PTS. The genes, *iolT1* (cg0226 or cgR_0261) and *iolT2* (cg3387 or cgR_2943), encoding the *myo*-inositol transporters IolT1 and IolT2, were identified as inositol-responsive genes by transcriptome analysis of cells grown on *myo*-inositol (Krings et al. 2006). These transporters belong to the major facilitator superfamily. Genetic analysis and sugar uptake study demonstrated that these transporters are functionally complementary inositol transporters (Krings et al. 2006). The gene cluster I and II comprising 13 and 8 genes, respectively, were upregulated in response to inositol. The *iolT1* gene is located about 10 kb downstream of the cluster I. Gene deletion experiment showed that the cluster II, containing *iolT2*, is dispensable for *myo*-inositol utilization. The *iolR* gene (cg0196 or cgR_0236), encoding a GntR-type transcriptional regulator, is located upstream of the *iol* operon in the gene cluster I and transcribed divergently. Transcriptome analysis of the *iolR* deletion mutant revealed that the *iol* operon (cg0197-cg207 or cgR_0237-cgR_0246) and *iolT1* (cg0223 or cgR_261) were highly upregulated compared with the wild type (Klaffl et al. 2013), demonstrating the transcriptional repressor function of IolR. The DNA-binding activity of IolR is probably controlled by inositol or its catabolite. The consensus binding site KGWCHTRACA (K: G or T, W: A or T, H: not G, R: A or G) was deduced using the MEME tool (Bailey and Elkan 1994) and confirmed by in vitro DNA binding assay (Klaffl et al. 2013). ChIP-chip analysis of GntR1 detected the GntR1 binding site in the *iolT1* promoter region, but deletion of *gntR1* had little effect on the *iolT1* expression (Tanaka et al. 2014). A factor directly involved in inositol-responsive induction of the gene cluster II has not been identified. IpsA, a LacI-type transcriptional regulator, activates expression of *ino1*, encoding *myo*-inositol phosphate synthase (cg3323 or cgR_2885), a key enzyme for endogenous inositol synthesis, in the absence of inositol (Baumgart et al. 2013). Transcriptome analysis of the *ipsA* (cg2910 or cgR_2534) deletion mutant showed that a part of genes (but not including *iolT2*) in the gene cluster II were upregulated in the mutant, suggesting repression of the gene cluster II by IpsA in the absence of inositol (Baumgart et al. 2013).

The in silico search detected the GlxR binding sites upstream of *iolT1* and *iolT2* (Kohl et al. 2008; Kohl and Tauch 2009). ChIP-chip and ChIP-seq analyses also confirmed the binding sites and additionally detected the binding site in the intergenic region between *iolR* and *iolC* (the first gene of the *iol* operon) (Toyoda et al. 2011; Jungwirth et al. 2013). These results suggest the GlxR involvement in the regulation of inositol uptake and catabolism, although whether it acts as a repressor or activator is not clear.

The IolT1 and IolT2 transporters are biotechnologically important owing to their broad substrate specificity for glucose (Ikeda et al. 2011; Lindner et al. 2011), fructose (Bäumchen et al. 2009), and xylose (Brüsseler et al. 2018), in addition to *myo*-inositol. Deletion of *iolR*, mutations in the *iolR* promoter, or *myo*-inositol supplementation, all of which induced *iolT1* and *iolT2*, rescued the growth defect of the deletion mutant of *ptsH*, encoding the PTS general component HPr, on glucose as a sole carbon source. Substitution of glucose uptake pathway via PTS with that via IolT1 was achieved by inactivation of PTS components and overexpression of IolT1 and glucokinase (Ikeda et al. 2011; Lindner et al. 2011). This enables cells to save PEP consumption during glucose uptake and use it for bioproduction of PEP-derived compounds, including amino acids and aromatic compounds, contributing to the improvement of production yield.

Another example of non-PTS sugar is maltose, a disaccharide composed of glucose. Uptake of maltose is mediated by an ABC transporter system, MusEFGK2I (Henrich et al. 2013), which is essential for maltose uptake and utilization. Deletion of *ramA* significantly decreased expression of the transporter-encoding gene cluster, *musK2* (cg2708 or cgR_2369), *musE* (cg2705 or cgR_2367), and *musFGI* (cg2704-2703-2701 or cgR_2365-2364-2363), indicating an activator role of RamA (Auchter et al. 2011a). In addition, as predicted by in silico analysis (Kohl et al. 2008; Kohl and Tauch 2009), the GlxR binding sites were detected in the upstream regions of *musK2* and *musE* by ChIP-chip and ChIP-seq analyses (Toyoda et al. 2011; Jungwirth et al. 2013). Because the binding sites are located downstream of the TSS of *musK2* and overlap with the −10 region of the *musE* promoter, GlxR probably represses the *mus* genes. Uptake of maltose enhances the expression of *ptsG*, encoding the glucose-specific EII of PTS. The upregulation of *ptsG* was abolished in the *ramA* mutant. As RamA is involved in upregulation of *mus* genes and *ptsG*, it is likely that RamA mediates the maltose effect on the *ptsG* upregulation (Henrich et al. 2013).

The wild-type *C. glutamicum* is incapable of utilizing the sugar alcohol mannitol. A deletion mutant of cgR_0187 (cg0146) is able to catabolize it (Peng et al. 2011). cgR_0187 encodes a DeoR-type transcriptional regulator, which has been first identified as the repressor SucR of the *sucCD* operon, encoding succinyl-CoA synthetase (Cho et al. 2010). Later, SucR was also shown to repress its own gene *sucR* and *adhA* (cg3107 or cgR_2695), encoding alcohol dehydrogenase (Auchter et al. 2011b). *mtlD*, encoding mannitol 2-dehydrogenase (MtlD), is encoded upstream of cgR_0187 (*sucR*) with *rbtT*, encoding a putative ribitol transporter. These genes are transcribed divergently from *sucR*. *xylB*, encoding xylulose kinase, is located downstream of cgR_0187 (*sucR*) in the same direction. These genes were upregulated in the cgR_0187 (*sucR*) deletion mutant and shown to be required for mannitol catabolism (Peng et al. 2011), demonstrating the repressor function of the regulator. Mannitol is oxidized to fructose, which is excreted once, then imported again by PtsF, yielding fructose-1-phosphate. Taken together, this gene cgR_0187 was renamed *mtlR* (Peng et al. 2011). However, mannitol did not induce the genes regulated by MtlR. Finally, arabitol was shown to be a genuine inducer of the regulator (Laslo et al. 2012), leading to the designation of the regulator AtlR. The adjacent genes, *mtlD*, *rbtT*, and *xylB*, were induced by arabitol supplementation, enabling the wild type *C. glutamicum* to grow on the carbon source. *rbtT* encodes a probable transporter for arabitol, not mannitol, and xylulokinase encoded by *xylB* is required for catabolism of xylulose generated by arabitol oxidation (Laslo et al. 2012). The ChIP-chip analysis detected the GlxR binding site overlapping the TSS of *rbtT* (Toyoda et al. 2011), indicating that GlxR also represses the *rbtT-mtlD* operon.

2.2 Glycolysis

Sugars transported by the uptake systems described above are metabolized in the glycolytic pathway. *C. glutamicum* uses the Embden–Meyerhof–Parnas (EMP) pathway comprising nine reactions for conversion of glucose-6-phosphate into pyruvate. For the glycolytic genes, their transcript organization and transition of their expression levels during the different growth phase on different carbon sources have been demonstrated (Han et al. 2007). The *gapA-pgk-tpi-ppc* gene cluster (cg1791-1790-1789-1787 or cgR_1636-1635-1634-1633) encodes the key enzyme glyceraldehyde-3-phosphate dehydrogenase, which is the first glycolytic enzyme that generates the reducing equivalent NADH, with phosphoglycerate kinase, triosephosphate isomerase, and phosphoenolpyruvate carboxylase, respectively. Of the glycolytic genes, this cluster was specifically and highly upregulated during the growth-arrested bioprocess, where cells were packed to a high density under oxygen deprivation (Inui et al. 2007). To clarify the molecular mechanism underlying the upregulation, transcriptional regulators interacting with the *gapA* promoter were isolated using DNA affinity purification. GlxR, GntR1, SugR, RamA, and MalR (cg3315 or cgR_2877), the former four of which have been shown to globally regulate gene expression, as described above, have been obtained (Toyoda et al. 2008, 2009a), illustrating the importance of the control of the cluster expression. As none of the regulators had been shown to regulate the glycolytic genes at that time, this study first indicated their global regulatory function. ChIP-chip and ChIP-seq analyses of GlxR and GntR1 detected the *gapA* promoter region, confirming their binding in vivo (Toyoda et al. 2011; Jungwirth et al. 2013; Tanaka et al. 2014).

SugR represses the *gapA* expression during growth on a gluconeogenic substrate like acetate or complex medium as it does the *pts* genes (Toyoda et al. 2008). ChIP-chip analysis detected the SugR binding at the promoter regions of *pfk* (cg1409 or cgR_1327, phosphofructokinase), *eno* (cg1111 or cgR_1071, enolase), *pyk* (cg2291 or cgR_1973, pyruvate kinase), and *fba* (cg3068 or cgR_2667, fructose-bisphosphate aldolase) of the glycolytic genes as well as those of *pts* genes (Engels et al. 2008). Although the binding of SugR to these promoter regions was confirmed by in vitro DNA binding assay, transcriptome analysis of the *sugR* deletion mutant revealed the significant upregulation of only *eno* owing to derepression. This suggests that complex regulation of the genes by multiple transcriptional regulators as observed for *gapA*. Even though not detected by ChIP-chip, another pyruvate kinase-encoding gene, *pyk2* (cg3218 or cgR_2811), was upregulated in the *sugR* mutant, but this is probably due to the derepression of its upstream gene, *ldhA* (cg3219 or cgR_2812), which is also repressed by SugR (Engels et al. 2008; Dietrich et al. 2009; Toyoda et al. 2009c).

RamA activates *gapA* expression by binding to three sites in the promoter with a different affinity (Toyoda et al. 2009a). In contrast, transcriptome analysis followed by in vitro DNA binding assay demonstrated that RamA negatively regulates *pfk* (6-phosphofructokinase) and *pyk* (pyruvate kinase) during growth on glucose (Auchter et al. 2011a). The transcriptome study showed that RamB is not involved in the regulation of the glycolytic genes.

GlxR also activates *gapA* expression (Toyoda et al. 2011) through its binding to the promoter region at position −244 with respect to the TSS. ChIP-chip analysis detected the GlxR binding sites in the promoter regions of the glycolytic genes *pgi*, *pfk*, *gapA*, *pgk*, and *eno* (Toyoda et al. 2011; our unpublished data). The putative binding sites were found at positions −65, −174, −244, −56, and −100 with respective to the TSSs of these genes, suggesting the activator role of GlxR. Indeed, promoter reporter assay demonstrated that GlxR activates the expression of *pfk* by binding to the putative site (Toyoda et al. 2011). The *gapA* operon yields four transcripts: *gapA*, *gapA-pgk-tpi*, *pgk-tpi*, and *pgk-tpi-ppc* (Schwinde et al. 1993). GlxR regulates the promoters upstream of *gapA* and *pgk*.

DNA affinity purification and ChIP-chip analysis detected the GntR1 binding to the *gapA* promoter (Toyoda et al. 2009a; Tanaka et al. 2014). The binding site is located farther upstream of the GlxR binding site. Based on these findings, GntR1 may be involved in activation of *gapA* along with other transcriptional regulators, although transcriptome analysis showed no discernible change in the *gapA* expression in the *gntR1* mutant. The ChIP-chip analysis also detected the GntR1 binding at the promoter regions of *pyk* and *gapB* (also known as *gapX*), encoding glyceraldehyde-3-phosphate dehydrogenase isozyme with a different cofactor preference (Omumasaba et al. 2004). Combined with the results of transcriptome analysis and promoter reporter assay, GntR1 directly activates *gapB* expression and represses the *pyk* expression in the absence of gluconate, the negative effector of GntR1 (Tanaka et al. 2014). Transcriptome analysis and in vitro DNA binding assay showed that *gapB* expression is also directly activated by RamA (Auchter et al. 2011a).

The group 2 sigma factor SigB globally activates the expression of the glycolytic genes, namely, *pfk*, *fba*, *gapA*, *pgk*, *tpi*, and *eno* (Ehira et al. 2008). This finding suggested the SigB-dependent promoter motif, distinct from the primary sigma factor SigA-dependent one. The promoter motifs specific for the sigma factors encoded in the *C. glutamicum* genome are described in detail in the Chap. 4.

2.3 Pentose Phosphate Pathway and Pentose Utilization

Pentoses, including ribose, xylose, and arabinose, are metabolized in the pentose phosphate pathway. The pathway generates not only NADPH for reducing equivalents, but also important intermediates, including riboflavin, phosphoribosyl pyrophosphate (PRPP), and erythrose-4-phosphate (E4P). PRPP is used for nucleic acids and histidine synthesis whereas E4P enters the shikimate pathway, through which chorismate, a precursor for aromatic amino acids, folate, and menaquinone, is synthesized. Although the wild-type *C. glutamicum* is incapable of utilizing xylose and arabinose, the metabolic engineering successfully developed strains that are able to utilize the pentoses simultaneously with glucose as carbon and energy source (Kawaguchi et al. 2006, 2008). Xylose and arabinose enter the pentose phosphate pathway as xylulose-5-phosphate.

Genes involved in the pentose phosphate pathway are: *tkt* (cg1774 or cgR_1624, transketolase), *tal* (cg1776 or cgR_1625, transaldolase), *zwf-opcA* (cg1778-1779 or cgR_1626-1627, glucose 6-phosphate dehydrogenase), *devB* (cg1780 or cgR_1628, 6-phosphogluconolactonase), all of which form an operon in this order, *gnd* (cg1643 or cgR_1513, 6-phosphogluconate dehydrogenase), *rpi* (cg2658 or cgR_2318, ribose 5-phosphate isomerase), and *rpe* (cg1801 or cgR_1646, ribulose 5-phosphate 3-epimerase), three of which are dispersed in the genome. As described above, GntR1 negatively regulates the *tkt* operon and *gnd* gene, as well as the gluconate utilization genes (Frunzke et al. 2008).

RamA represses the *tkt* operon (Auchter et al. 2011a), whereas it activates *gnd* (Tanaka et al. 2012). As the RamA binding site, upstream of the *gnd* gene, overlaps with one of the two binding sites of GntR1, RamA-mediated activation of the *gnd* gene is blocked in the presence of GntR1.

The binding of SugR to the upstream region of the *tkt* operon was detected by ChIP-chip analysis and further corroborated by the in vitro DNA binding assay, although deletion of *sugR* had little effect on the expression of the operon (Engels et al. 2008). ChIP-chip analysis also detected the binding of GlxR to the upstream region of the *tkt* operon (Toyoda et al. 2011), but the role of GlxR in the regulation of the operon has not been demonstrated.

ChIP-chip analysis of the alternative sigma factor SigH identified the SigH-dependent promoters upstream of *tal* in the *tkt-tal-zwf-opcA-devB* operon and upstream of *ribC* in the *ribGCAH* cluster (cg1800-1799-1798-1797 or cgR_1645-1644-1643-1642), which encodes riboflavin biosynthetic enzymes (Toyoda et al. 2015). SigH is responsible for heat and oxidative stress response and upregulates genes that are involved in protein quality control and oxidative stress protection as well as genes for enzymes carrying oxidative stress labile cofactors (Ehira et al. 2009; Busche et al. 2012; Toyoda et al. 2015). Deletion of *rshA*, encoding the cognate anti-sigma factor RshA, upregulated expression of *tal-zwf-opcA-devB* and *ribCAH*, enhancing the production of NADPH and riboflavin. NADPH is utilized by the thioredoxin system and dehydrogenases under the control of SigH. Riboflavin is the precursor of FMN and FAD, redox cofactors for flavoenzymes, which are also induced by SigH.

An ABC transporter essential for ribose uptake is encoded by the genes *rbsACBD*, which form the operon *rbsRACBD* (cg1410-1411-1412-1413-1414 or cgR_1328-1329-1330-1331-1332) with *rbsR*, encoding a LacI-type transcriptional regulator. Genetic analysis and in vitro DNA binding assay demonstrated that RbsR acts as a transcriptional repressor of its own operon in the absence of ribose (Nentwich et al. 2009). Ribokinases phosphorylating ribose to ribose 5-phosphate, which enters the pentose phosphate pathway, are encoded by *rbsK1* in the *uriR-rbsK1-uriT-uriH* operon (cg1547-1546-1545-1544 or cgR_1432-1431-1430-1429) and *rbsK2* (cg2554 or cgR_2206). RbsR binding sites were found upstream of the operon and the *rbsK2*. The *uri* operon is required for uridine uptake and metabolism. UriR, a LacI-type transcriptional regulator, represses its own operon (Brinkrolf et al. 2008). The *rbs* operon was no longer induced by ribose in the *rbsK1-rbsK2* double mutant, suggesting that an effector molecule affecting RbsR DNA binding activity is derived from ribose-5-phosphate, although ribose-5-phosphate itself does not affect the activity (Nentwich et al. 2009). The GlxR binding site with the unknown physiological role was identified by ChIP-chip at position −90 with respect to the TSS of *rbsR* (our unpublished data). In addition, the *rbs* operon, the *uri* operon, and *rbsK2* were significantly downregulated in the *gntR1* deletion mutant (Tanaka et al. 2014). As only a weak ChIP-chip peak was detected at the intragenic region of *uriT*, GntR1 may be indirectly involved in the regulation of these genes. Transcriptome analysis of *gntR1-gntR2* double deletion mutant of strain ATCC 13032 did not detect the downregulation of these genes (Frunzke et al. 2008). The difference in the results may be attributable to the basal medium used.

2.4 The Tricarboxylic Acid (TCA) Cycle and Glyoxylate Shunt

The TCA cycle generates reducing equivalents (in the form of NADH and menaquinol), which are used for ATP synthesis by oxidative phosphorylation. Meanwhile, the intermediates 2-oxoglutarate and oxaloacetate in the cycle are used as precursors of glutamate and aspartate family amino acids. The TCA cycle is essential for organic acid metabolism as well. For the acetate metabolism, the glyoxylate shunt bypasses decarboxylation reactions in the cycle. For the metabolism of propionate, the 2-methyl citrate cycle generates the TCA cycle intermediates for further metabolism (Plassmeier et al. 2012). Genes involved in these metabolic pathways are also controlled by the global regulators, like the genes for sugar uptake and the glycolytic pathway as described above.

Pyruvate generated by glycolysis is first transformed into acetyl-CoA by pyruvate dehydrogenase complex to enter the TCA cycle. Of the genes encoding the three subunits AceE, AceF (SucB), and Lpd, *aceE* (cg2466 or cgR_2120) is positively regulated by GlxR and RamB and *aceF* (cg2421 or cgR_2087) is activated by RamA (Blombach et al. 2009; Auchter et al. 2011a; Toyoda et al. 2011) (Fig. 2). The binding of SugR to the *lpd* (cg0441 or cgR_0439) promoter region was detected in vivo and in vitro, although *sugR* deletion had no apparent effect on the *lpd* expression (Engels et al. 2008). Lpd and AceF constitute 2-oxoglutarate dehydrogenase complex with OdhA in place of AceE (Hoffelder et al. 2010).

In silico analysis, in vitro DNA binding assay, ChIP-chip and ChIP-seq analyses detected the potential GlxR binding sites upstream of the TCA cycle genes, namely, *gltA* (cg0949 or cgR_0943, citrate synthase), *acn* (cg1737 or cgR_1598, aconitase), *sucC* (cg2837 or cgR_2469, succinyl-CoA synthetase), *sdhCD* (cg0445 or cgR_0443, succinate dehydrogenase), and *fum* (cg1145 or cgR_1099, fumarase) (Han et al. 2008; Kohl et al. 2008; Kohl and Tauch 2009; Toyoda et al. 2011; Jungwirth et al. 2013) (Fig. 2). Promoter reporter assay and *glxR* overexpression study showed that GlxR negatively regulates *gltA* and *sdhCDAB* (Bussmann et al. 2009; van Ooyen et al. 2011). The GlxR binding sites are found around the TSS of *sucC*, indicating the repressor function of GlxR. The GlxR binding sites are located upstream of the *fum* promoter, but the regulatory role of GlxR in the regulation of *fum* expression should be confirmed by in vivo experiment.

ChIP-chip analysis and promoter reporter assay demonstrated that GntR1 directly activates the expression of *icd* (cg0766 or cgR_0784), encoding isocitrate dehydrogenase (Tanaka et al. 2014). This is the first transcriptional regulator directly involved in the *icd* regulation. GntR1 also positively regulates *malE* (*maeB*)(cg3335 or cgR_2895), encoding malic enzyme for $NADP^+$-dependent malate decarboxylation (Gourdon et al. 2000) (Fig. 2). Isocitrate dehydrogenase from *C. glutamicum* is a monomer and catalyzes the oxidation of isocitrate with $NADP^+$, not NAD^+, as an electron acceptor (Eikmanns et al. 1995). Thus, GntR1 activates expression of genes encoding NADPH-generating enzymes. This activation is mitigated in the presence of gluconate, which also alleviates the GntR1-mediated repression of genes for the

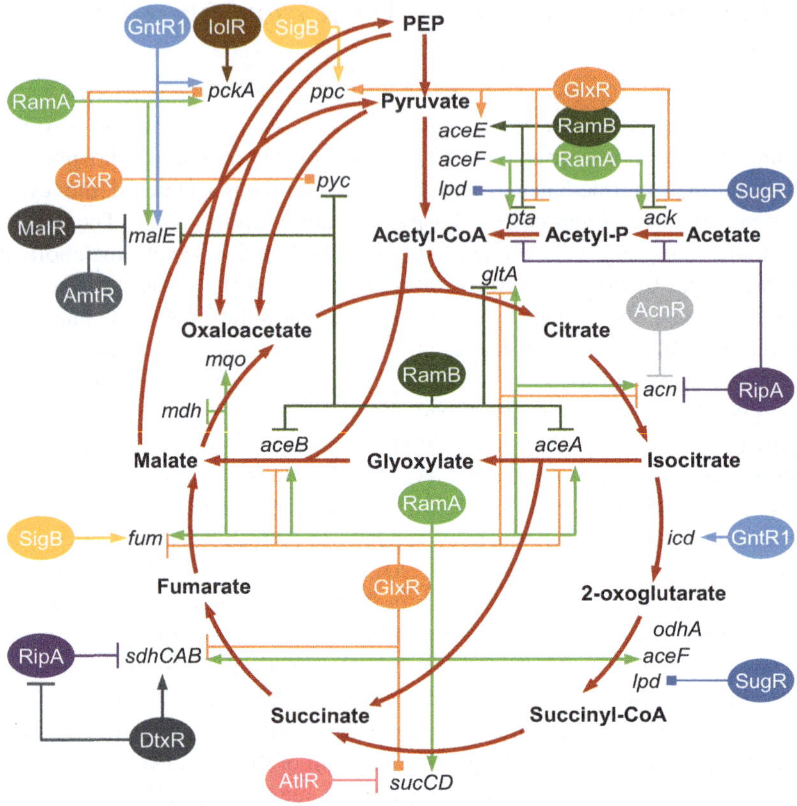

Fig. 2 Transcriptional regulators of genes involved in the TCA cycle, glyoxylate shunt, and acetate activation. Similar to Fig. 1, the metabolic pathways are indicated by thick red arrows, genes constituting the metabolic pathways are indicated by italic letters, and intermediate metabolites are indicated by bold letters. Transcriptional regulators are indicated by ovals. Arrows from the ovals indicate transcriptional activation, whereas T-bars indicate repression. Lines ended with squares indicate the in vivo binding of transcriptional regulators with unknown function. PEP, phosphoenolpyruvate; Acetyl-P, acetyl phosphate

pentose phosphate pathway, resulting in compensation for NADPH generation reactions (Tanaka et al. 2014).

Transcriptome analysis of the *ramA* deletion mutant followed by in vitro DNA binding assay indicated the involvement of RamA in the regulation of *gltA*, *acn*, *aceF*, *sucC*, *sdhCD*, *fum*, *mdh* (cg2613 or cgR_2262), and *mqo* (cg2192 or cgR_1830) (Auchter et al. 2011a). RamA activates the expression of these genes, except that of *mdh*, which is repressed by RamA (Fig. 2).

The two anaplerotic enzymes PEP carboxylase and pyruvate carboxylase are encoded by *ppc* and *pyc* (cg0791 or cgR_0809). Since *ppc* is transcribed as a part of the glycolytic genes operon *pgk-tpi-ppc*, it is positively regulated by GlxR (Toyoda et al. 2011; Jungwirth et al. 2013) and SigB (Ehira et al. 2008), as described above in the glycolysis part 2.2. Expression of *pyc* is negatively regulated by RamB, although its binding sites are over 150 bp upstream of the TSS (Auchter et al. 2011a). The binding sites of GlxR were located downstream of the TSS of *pyc*, suggesting the repressor role of GlxR (Kohl et al. 2008; Kohl and Tauch 2009; Toyoda et al. 2011). The in vitro DNA binding assay clearly confirmed RamA binding to the *pyc* promoter region and identified the putative RamA binding sites. However, *ramA* deletion had no apparent effect on the *pyc* expression during growth in glucose minimal medium and complex medium (Auchter et al. 2011a).

The *pckA* gene (cg3169 or cgR_2751), encoding PEP carboxykinase, which catalyzes the reverse reaction of PEP carboxylase for gluconeogenesis, is also under complex regulation. The binding of GlxR to the *pckA* promoter was detected in vivo and in vitro studies (Han et al. 2007; Toyoda et al. 2011). The position of the binding site is located at 30 bp upstream of the TSS, which suggests a repressor function of GlxR. RamA, GntR1/R2, and IolR were isolated by DNA affinity purification (Klaffl et al. 2013). Gene deletion analysis combined with in vitro DNA binding assay demonstrated that RamA acts as a repressor and GntR1/2 and IolR act as activators. Deletion of *iolR* upregulated gluconate permease and gluconate kinase genes as well as the *iol* gene cluster for inositol uptake and catabolism as described above and downregulated glyoxylate shunt genes and *pckA* (Klaffl et al. 2013). However, direct interaction was confirmed only for the *iol* gene cluster and *pckA*.

DNA affinity purification using the promoter of the *malE* (*maeB*) gene, encoding malic enzyme, isolated the MarR-type transcriptional regulator MalR, which is characterized as a transcriptional repressor (Krause et al. 2012). MalR has been first isolated using DNA affinity purification using the *gapA* promoter as a ligand (Toyoda et al. 2009a), although whether it acts as a repressor or activator was not demonstrated. Thus, the malic enzyme gene *malE* (*maeB*) is regulated by two repressors, MalR and RamB, and two activators, RamA and GntR1/R2.

Global transcriptional regulators, not described earlier in this chapter, are involved in the regulation of the TCA cycle enzymes containing iron as a cofactor, namely, aconitase and succinate dehydrogenase (Fig. 2). The global iron regulator DtxR positively regulates *sdhCDAB* (succinate dehydrogenase) under iron-rich conditions (Brune et al. 2006). Upon iron limitation, DtxR-dependent repression of *ripA* (cg1120 or cgR_1076), encoding an AraC-type regulator, is relieved, resulting in RipA-dependent repression of *sdhCDAB* and *acn* (Wennerhold et al. 2005). *acn* is also repressed by the TetR-type regulator AcnR, encoded by the immediately downstream gene *acnR* (cg1738 or cgR_1599) (Krug et al. 2005), although a sensing molecule of AcnR remains elusive.

Isocitrate lyase and malate synthase, encoded by the *aceA* and *aceB* genes, respectively, form the glyoxylate shunt. This pathway is required for the metabolism of simple carbon sources, such as acetate. Acetate is activated to acetyl-CoA by phosphotransacetylase and acetate kinase, which are encoded by the *pta-ackA* operon, then enters the TCA cycle. To prevent the loss of the acetate-derived two carbons, the glyoxylate shunt bypasses the decarboxylation reactions in the cycle. These four genes are specifically induced during growth on acetate when compared with on glucose (Wendisch et al. 1997; Hayashi et al. 2002; Muffler et al. 2002). Due to the evident expression alteration, the promoters of the genes have been utilized to isolate transcriptional regulators, GlxR (Kim et al. 2004), RamA (Cramer et al. 2006), and RamB (Gerstmeir et al. 2004), all of which were later identified as key global regulators in *C. glutamicum* as described above. GlxR and RamB were isolated as transcriptional repressors of *aceB*, whereas RamA was as a transcriptional activator.

2.5 L-Lactate Metabolism

NADH oxidation under oxygen deprivation depends on fermentation. *C. glutamicum* produces L-lactate as one of the fermentation products from pyruvate using L-lactate dehydrogenase, encoded by the *ldhA* gene (Inui et al. 2004) (Fig. 1). ChIP-chip analysis of the sugar-responsive regulator SugR identified its binding sites in the *ldhA* promoter region. Transcriptome analysis of SugR demonstrated that SugR acts as a transcriptional repressor in the absence of sugar (Engels et al. 2008). Thus, SugR coordinates sugar uptake and metabolism with fermentative pyruvate reduction. This was corroborated by DNA affinity purification identifying SugR as a protein binding to the *ldhA* promoter region, in vitro DNA binding assay, and genetic analysis (Dietrich et al. 2009; Toyoda et al. 2009c). The DNA affinity purification using the *ldhA* promoter region as a ligand also identified a GntR-type transcriptional regulator, LldR, which has been shown to repress the cg3226 (cgR2818)-*lldD* (cg3227 or cgR_2819) operon required for L-lactate utilization as a carbon source (Stansen et al. 2005; Georgi et al. 2008). DNA binding activity of LldR is inactivated by L-lactate, leading to derepression of the operon for uptake and oxidation of L-lactate (Georgi et al. 2008). *lldR* (cg3224 or cgR_2816) deletion had no apparent effect on *ldhA* expression. However, the finding that the *ldhA* promoter activity was decreased in the *ldhA* deletion mutant and that the promoter activity in the mutant was restored upon deletion of *lldR* indicates that L-lactate-mediated repression of *ldhA* by LldR is not relieved in the *ldhA* deletion mutant background in contrast to in the wild type (Toyoda et al. 2009b). The SugR-mediated repression is epistatic to the LldR-mediated one. This is consistent with the position of the binding sites for the regulators with respect to the TSS: around the −10 and −35 regions of the *ldhA* promoter for SugR and LldR, respectively (Toyoda et al. 2009b). Transcriptome analysis of the *lldR* mutant revealed that *ptsF*, encoding the fructose-specific EII of PTS, is also repressed by LldR, indicating its global function (Gao et al. 2008). Because it was isolated by DNA affinity purification using the *ldhA* promoter region as a ligand, AtlR, a repressor of the arabitol utilization operon, is likely involved in the expression of *ldhA* (Dietrich et al. 2009; Toyoda et al. 2009c), in addition to that of *sucCD* and *adhA*. As none of these genes (*sucCD*, *adhA*, and *ldhA*) were detected by transcriptome analysis of the *atlR* deletion mutant (Laslo et al. 2012), the role of AtlR in the regulation of these genes may be masked by other regulators. In addition,

MalR was also isolated by the DNA affinity purification study, although its role in the regulation of *ldhA* expression was unclear (Dietrich et al. 2009).

2.6 Carbon Catabolite Control

Carbon catabolite control (repression) enables the preferential consumption of one carbon source over the others, leading to a biphasic (diauxic) growth. CRP in *E. coli* and catabolite control protein A (CcpA) in *B. subtilis* represent transcriptional regulators and govern the catabolite repression. The intracellular cAMP levels in *E. coli* are lowered in the presence of glucose due to the abolishment of the activation of adenylate cyclase by glucose-specific EII of PTS (Reddy and Kamireddi 1998; Takahashi et al. 1998), deactivating catabolic genes for other carbon sources. The *B. subtilis* CcpA represses catabolic genes for various carbon sources by forming a protein complex with the PTS general component HPr, which is phosphorylated by HPr kinase in the presence of glucose (Deutscher et al. 1995; Jones et al. 1997).

In contrast to *E. coli* and *B. subtilis*, *C. glutamicum* co-metabolizes glucose with a variety of other carbon sources, including fructose, gluconate, lactate, pyruvate, acetate, propionate, vanillate, shikimate, and serine (Teramoto et al. 2009; Cocaign-Bousquet et al. 1993; Dominguez et al. 1997; Wendisch et al. 2000; Claes et al. 2002; Netzer et al. 2004; Merkens et al. 2005; Stansen et al. 2005; Frunzke et al. 2008). Ethanol and glutamate have been exceptionally shown to cause diauxic growth when provided in the mixture of glucose (Kronemeyer et al. 1995; Arndt and Eikmanns 2007; Kotrbova-Kozak et al. 2007). Although the *E. coli* CRP ortholog GlxR is encoded and involved in multiple carbon metabolic genes in *C. glutamicum* as described above, the intracellular cAMP levels of glucose-grown cells are higher than those of acetate-grown cells in contrast to *E. coli* (Kim et al. 2004). The intracellular cAMP levels of *C. glutamicum* are controlled by the adenylate cyclase CyaB and phosphodiesterase CpdA (Cha et al. 2010; Schulte et al. 2017b). In addition, HPr kinase, a component of the CcpA system, is not encoded in the *C. glutamicum* genome. Thus, *C. glutamicum* has been thought to use a molecular mechanism distinct from that functions in the model organisms.

GlxR is involved in the catabolite repression observed for these carbon sources utilization (Park et al. 2010; Subhadra and Lee 2013). The *gluABCD* operon (cg2136-2139 or cgR_1780-1783) encoding glutamate uptake transporter is upregulated by glutamate, but that upregulation is suppressed in the presence of glucose. The glucose-mediated suppression of the *gluA* promoter activity was alleviated in the *glxR* deletion mutant (Park et al. 2010). GlxR was shown to bind to the consensus site found in the *gluA* promoter region in vitro (Kohl et al. 2008). Ethanol is oxidized to acetate by alcohol dehydrogenase encoded by *adhA* and aldehyde dehydrogenase encoded by *ald* (cg3096 or cgR_2686). These genes are positively and negatively regulated by RamA and RamB, respectively (Arndt and Eikmanns 2007; Auchter et al. 2009). However, an additional regulator is suggested to be involved in the glucose-mediated repression of these genes. The activities of these ethanol catabolic enzymes in the *glxR* deletion mutant in the presence of glucose were higher than those in the wild type (Subhadra and Lee 2013). The two GlxR binding sites are located downstream of the TSS of the *adhA* and *ald* genes (Kohl et al. 2008; Kohl and Tauch 2009). Taken together, the CRP ortholog GlxR is involved in the catabolite repression observed during utilization of these carbon sources, which is different from *E. coli* CRP that functions as a transcriptional activator.

3 Global Nitrogen Control

Nitrogen is an essential component of almost all macromolecules, including nucleic acids, proteins, and cell wall. To supply an appropriate amount of nitrogen and to cope with environmental changes in nitrogen sources, bacteria use global nitrogen control systems optimized to their metabolisms and life cycles. Regarding *C. glutamicum*, which is used as a host for amino acids production, the nitrogen control system is important for not only homeostatic function but also for the industrial production of amino acids.

3.1 AmtR, a Global Nitrogen Regulator

As ammonium, the most favorable nitrogen source for *C. glutamicum*, is in equilibrium with ammonia, which diffuses into the cytosol via the cytoplasmic membrane, no specific protein is needed for uptake unless its concentration is limiting. When nitrogen source is scarce, various nitrogen uptake and assimilation genes are induced to utilize diverse nitrogen sources in the environment. A TetR-type transcriptional regulator, AmtR (encoded by cg0986 or cgR_0978), is responsible for the induction in *C. glutamicum*; it represses these genes under nitrogen-rich conditions. Since AmtR has been first identified as a transcriptional repressor of the ammonium transporter gene *amtA* under nitrogen surplus conditions (Jakoby et al. 2000), bioinformatic analysis and multi-omics analysis expanded its regulon to comprise over 30 genes for transport and assimilation of ammonium (*amtA*, *amtB*, *glnA*, *gltBD*; cg1785 or cgR_1631, cg2261 or cgR_1948, cg2429 or cgR_2093, cg0029-0030 or cgR_0263-0264), creatinine (*codA*, *crnT*; cg0104 or cgR_0096; cg0103 or cgR_0095) and urea uptake and metabolism (*urtABCDE*, *ureABCEFGD*, cg1061-1062-1064-1066 or cgR_1031-1035), signal transduction proteins (*glnD*, *glnK*, cg2260-2259 or cgR_1947-1946), *meso*-diaminopimelate synthesis (*dapD*, cg1256 or cgR_1189), malic enzyme (*malE*), and proteins with unknown function (Jakoby et al. 2000; Beckers et al. 2005; Buchinger et al. 2009). The *vanABK* operon (cg2616-2618 or cgR_2265-2267) encoding vanillate uptake transporter (VanK) and a metabolic enzyme (VanAB) is included as an indirect target due to the lack of the AmtR binding site (AmtR box) upstream of the operon. This operon is directly negatively regulated by VanR, which is encoded by the upstream gene *vanR* (cg2615 or cgR_2264) (Morabbi Heravi et al. 2015).

The AmtR box represented by the consensus motif, 5′–TTTCTATN6ATAGAWA–3′, with the strong conserved CTAT motif, is found upstream of the AmtR regulon. The similarity of the AmtR box to the consensus motif is likely correlated with the stringency of repression (Hasselt et al. 2011). Whereas transcription of ammonium uptake systems, encoded by *amtA* and *amtB*, as well as that of the *gltBD* operon encoding glutamate synthase are tightly regulated and fully repressed in nitrogen-rich medium, expression of *gdh* encoding glutamate dehydrogenase and *glnA* encoding glutamine synthetase is only loosely controlled. The strictly repressed genes possess multiple AmtR boxes with the conserved palindromic motif, CTATN6ATAG, while the AmtR boxes found upstream of the loosely regulated genes have an imperfect motif (Hasselt et al. 2011).

The importance of the conserved motif in the AmtR box for the interaction with AmtR was demonstrated by the structure of AmtR in free and DNA-bound forms (Palanca and Rubio 2016). AmtR exists as a homodimer with N-terminal DNA-binding (DNB) domain containing the helix-turn-helix (HTH) motif and a C-terminal ligand binding and dimerization (LBD) domain as other TetR-type transcriptional regulators. At the C-terminal region, there is an extra helix being a characteristic of corynebacterial AmtR proteins. Crystal structure of the DNA-bound form was obtained by mixing AmtR and self-complementary quasi-palindromic oligonucleotides containing the AmtR box. The recognition helices of the HTH motif of each subunit in the dimer were inserted into the two successive turns of the major groove of the double helix containing the AmtR box. The interacting residues were reasonable for the strongly conserved motif CTAT in the AmtR box. In addition, the N-terminal extension, which is rich in Arg and Pro and characteristic to corynebacterial AmtR, also interacted with DNA via the minor groove flanking to the major groove interacting with the helix of the HTH motif, which explains AT bases flanking the conserved CTAT motif. The mutant AmtR protein lacking this N-terminal 19-residues does not have the DNA binding ability in vitro. This extension may guide AmtR to the canonical box. The structure of the DNA-bound form revealed that the DNB domain of each subunit undergoes movement with the hinge at the junction between the two domains. This movement might be involved in triggering the dissociation of AmtR from DNA.

GlnK, a sole PII-type signal transduction protein in this organism, is involved in the derepression of the AmtR regulon expression (Nolden et al. 2001) (Fig. 3). The conserved Tyr residue 51 of GlnK is adenylylated by the adenylyltransferase GlnD during nitrogen starvation. The replacement of this Tyr residue with Phe, or the deletion of *glnK* or *glnD*, all of which result in permanent repression of the AmtR regulon (Nolden et al. 2001). These findings led to the assumption that direct interaction between GlnK and AmtR is required for release of AmtR from DNA. Indeed, adenylylated GlnK interacted with AmtR in vitro and in vivo and inhibited the binding of AmtR to DNA in vitro (Beckers et al. 2005). Deadenylylated GlnK is trapped by membrane via interaction with the ammonium transporter AmtB (Strösser et al. 2004), encoded in the operon *amtB-glnK-glnD*. GlnK is subjected

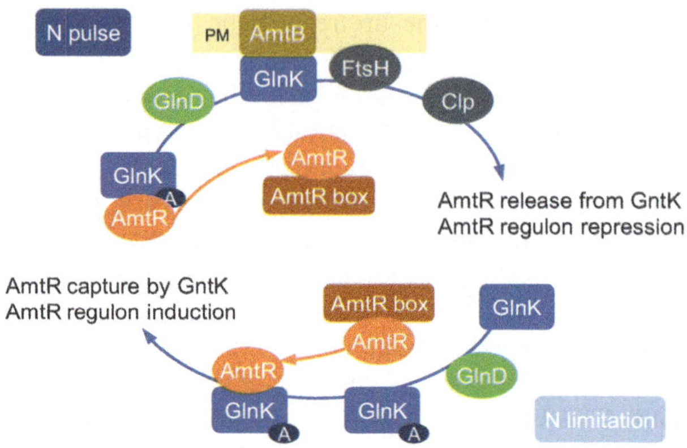

Fig. 3 Regulation of the AmtR regulon in response to nitrogen availability. The PII protein GlnK is adenylylated by the adenylyltransferase GlnD upon nitrogen limitation. Adenylation is indicated with small circles indicated with A. The adenylylated GlnK captures AmtR, resulting in derepression of the AmtR regulon. The adenylylated GlnK is deadenylylated by GlnD upon nitrogen supplementation. GlnK in the deadenylylated form is bound to the ammonium transporter AmtB and digested by FtsH and the Clp system. AmtR released from GlnK binds to the AmtR box, thereby repressing its regulon. PM, plasma membrane

to proteolysis by the membrane-bound proteinase FtsH, followed by the Clp system. Membrane localization is prerequisite for GlnK proteolysis because, in the absence of AmtB, GlnK is stable in the deadenylylated form, which is incapable of interacting with AmtR (Strösser et al. 2004). The structure of the AmtR-GlnK complex will provide a clue for understanding the molecular mechanism of AmtR dissociation from DNA. In this context, the crystal structure of AmtR suggested a loop possibly interacting with the GlnK T-loop, which is subjected to adenylylation (Palanca and Rubio 2016). Switching the interaction partner of the loop from the other subunit of AmtR to the T-loop of adenylylated GlnK may cause structural alteration resulting in dissociation of AmtR from the AmtR box.

3.2 *Nitrogen Sensing*

In *E. coli*, nitrogen limitation reduces the ratio of glutamine to 2-oxoglutarate, activating the uridylyltransferase GlnD to modify the PII proteins GlnB and GlnK. The uridylylated PII proteins induce positive autoregulation of the two-component NtrBC system, which directly and indirectly activates the expression of genes for nitrogen uptake and metabolism (reviewed in Merrick and Edwards 1995; Ninfa et al. 2000). In *B. subtilis*, cellular nitrogen availability is sensed by the glutamine synthetase GlnA, controlling the activity of two transcriptional regulators, TnrA and GlnR. Under nitrogen limitation, TnrA activates genes for ammonium transport and utilization of other nitrogen sources and represses those for glutamine and glutamate synthesis (Wray et al. 2001; Yoshida et al. 2003). Under nitrogen-rich conditions, GlnR represses *tnrA* and *glnA* encoding glutamine synthetase (Brown and Sonenshein 1996). The DNA binding activity of these regulators is controlled by the interaction with GlnA in the feedback-inhibited form (Wray et al. 2001; Fisher and Wray 2008, 2009; Schumacher et al. 2015). In contrast to the situations in these model organisms, glutamine is unlikely a sensing molecule in *C. glutamicum*, based on the finding that the AmtR regulon is derepressed in the glutamine-grown cells, in which glutamine concentrations are higher than those in ammonium-grown cells (Rehm et al. 2010). Intracellular ammonium levels, which are reduced upon nitrogen starvation (Nolden et al. 2001), are the possible indicators of the nitrogen status, although a sensor protein is still not identified.

3.3 Cross-Talk with Transcriptional Regulators for Other Metabolisms

The GlxR regulon contains the AmtR regulon including the *gltBD* operon (cgR_0263-0264), the *amtB-glnK-glnD* operon, the *urtABCDE* operon, *gdh*, the cg2181-cg2184 (cgR_1819-cgR_1822) operon, the *gluABCD* operon, and *glnA1*, thus connecting the carbon and nitrogen metabolisms (Kohl and Tauch 2009). Although its binding in vitro and in vivo was confirmed, its physiological role remains elusive, except for the repression of the *gluABCD* operon (Park et al. 2010). ChIP-chip analysis detected GntR1 binding to the intergenic region between *gdh* and *glxK*, although the *gdh* expression was not affected by *gntR1* deletion (Tanaka et al. 2014).

4 Phosphate Control

Another macromolecule essential for cell growth is phosphorus, which is used for nucleic acid synthesis and protein modification. Bacteria are able to utilize inorganic and organic phosphorus, including orthophosphate, pyrophosphate, sugar phosphates, nucleotides, and phospholipids, to adapt to fluctuating phosphorus concentrations in the environment. Molecular mechanism of adaptation to phosphate limitation, namely, upregulation of genes for transporters, extracellular enzymes, and metabolic enzymes for phosphonates and organophosphates utilization, is intensively studied in model organisms and known to be under the control of two-component systems consisting of a sensor kinase and a response regulator (Santos-Beneit 2015).

4.1 The Two-Component System PhoRS-Mediated Phosphate-Limitation Response

C. glutamicum utilizes a variety of phosphorus-containing compounds as a phosphorus source (Wendisch and Bott 2008). Transcriptional response to phosphate limitation in *C. glutamicum* was investigated by transcriptome analysis using microarray (Ishige et al. 2003). Phosphate limitation by transferring cells from phosphate sufficient (13 mM) medium to phosphate limiting (0.13 mM) medium induced the genes for phosphate uptake and metabolism as observed in other bacteria. These genes include the *pstSCAB* operon (cg2846-2843 or cgR_2478-2475, high-affinity ABC-type phosphate uptake system), *nucH* (cg2868 or cgR_2495, extracellular nuclease), *phoH1* (cg0085 or cgR_0084, ATPase with unknown function), the *ugpAEBC* operon (cg1568-1571 or cgR_1446-1449, ABC-type sn-glycerol 3-phosphate uptake system), *glpQ1* (cg3215 or cgR_2805, glycerophosphoryl diester phosphodiesterase), *phoC* (cg3393 or cgR_2949, putative phosphoesterase), the *pctABCD* operon (cg1652-1649 or cgR_1522-1519, ABC-type transporter for -phosphorus-containing compounds), and *ushA* (cg0397 or cgR_0412, putative 5-′-nucleotidase) (Rittmann et al. 2005). Genes encoding proteins unrelated to phosphate metabolism were also upregulated but their roles in adaptation to phosphate limitation are not known.

C. glutamicum intracellularly accumulates polyphosphate synthesized by a polyphosphate kinase encoded by cg3007 (cgR_2616) (Lindner et al. 2007). The polyphosphate is hydrolyzed by exopolyphosphatases encoded by *ppx1* (cg0488 or cgR_0480) and *ppx2* (cg1115 or cgR_1074) to release orthophosphate (Lindner et al. 2009). Although the polyphosphate can be utilized as a phosphorus source, expression of the genes involved in its synthesis and hydrolysis was not affected by phosphate limitation.

PhoRS, a two-component system comprising a pair of sensor kinase and response regulator encoded by the operon *phoR-phoS* (cg2888-2887 or cgR_2511-2510), was transiently induced in response to phosphate limitation. Screening of deletion mutants of the 13 two-component systems encoded in the *C. glutamicum* genome for growth under phosphate limitation confirmed PhoRS as a key system for induction of the genes in response to phosphate limitation (Kočan et al. 2006). Indeed, microarray analysis and primer extension analysis revealed that the upregulation of the phosphate limitation-responsive genes was abolished in the *phoRS* deletion mutant, except for the *pstSCAB* operon, which was still induced with lower induction levels than in the wild type. Thus, PhoRS positively regulates the expression of genes induced under phosphate limitation. In vitro analysis demonstrated that the sensor kinase PhoS, in the form lacking its transmembrane region, is capable of autophosphorylation and transferring the phosphoryl group to the response regulator PhoR (Schaaf and Bott 2007). In vitro DNA binding assay revealed that PhoR binds to the promoter regions of the genes that are induced under phosphate limitation. The affinity of PhoR to DNA was enhanced upon phosphorylation by PhoS. The affinity of the promoter regions of the PhoR regulon to PhoR are different; the *pstS* promoter has the highest affinity and the *glpQ1* has the lowest

(Schaaf and Bott 2007). In the course of the in vitro DNA binding assay, two additional genes were found to belong to the PhoR regulon: the *porB* gene (cg1109 or cgR_1069), encoding an anion-specific porin (Costa-Riu et al. 2003), and *pitA* (cg0545 or cgR_0533), encoding low-affinity phosphate uptake transporter (Schaaf and Bott 2007). Expression of *porB* is independent of phosphate concentration, but it is activated by PhoR. In contrast to the other PhoR regulon, *pitA* is negatively regulated by PhoR. Indeed, *pitA* expression decreased after phosphate limitation shift (Schaaf and Bott 2007). Successive deletion and mutation of the DNA fragments used in in vitro DNA binding assay confined the location of the PhoR binding site and identified important nucleotides in the site. Since an artificial 8-bp direct repeat with a 3-bp spacer, 5′-CCTGTGAANNNCCTGTGAA-3′, had the currently highest affinity to PhoR, the PhoR consensus binding motif is the direct repeat, although there is no identical sequence upstream of any of the PhoR regulon. Promoter reporter assay of the *phoR* promoter showed that, in contrast to the full-length promoter, the promoter fragment lacking the PhoR binding site lost induction of a reporter gene under phosphate limitation, demonstrating physiological importance of the PhoR binding site (Schaaf and Bott 2007). In the *pstS* promoter region, two PhoR binding sites were found, but only upstream one, more closely matched with the consensus motif, is required for the PhoR-dependent upregulation of *pstS* (Schaaf and Bott 2007).

4.2 Involvement of Other Transcriptional Regulators

Although the two-component system has served as a paradigm of the transcriptional response to phosphate limitation in bacteria, interaction with other transcriptional regulators involved in different metabolisms is important to coordinate whole metabolism (reviewed in Santos-Beneit 2015). Of the phosphate-responsive genes in *C. glutamicum*, the *pstSCAB* operon encoding ABC-type phosphate uptake system has been shown to be still induced upon phosphate limitation in the absence of PhoRS (Kočan et al. 2006), as described above. DNA affinity purification using the *pstS* promoter region as a ligand identified the global regulator GlxR (Panhorst et al. 2011). Bioinformatic analysis followed by in vitro DNA binding assay identified the GlxR binding site at position −133 to −117 with respect to the TSS. Mutational inactivation of the site decreased the *pstS* promoter activity after phosphate limitation, demonstrating the transcriptional activator function of GlxR. Overexpression of GlxR promotes the growth under phosphate-limiting conditions, supporting the GlxR function as an activator. The effect of the binding site mutations was larger in glucose-grown cells compared with the acetate-grown cells. This is likely because intracellular cAMP levels in glucose-grown cells were higher than in acetate-grown cells (Kim et al. 2004).

The *pstS* promoter lacking the cis elements (binding sites) of PhoR and GlxR is still induced upon phosphate limitation. These findings motivated the search for transcriptional regulators involved in the PhoRS-independent regulation. DNA affinity purification using the DNA fragment containing the *pstS* promoter region but lacking the binding sites for PhoR and GlxR identified RamB (Sorger-Herrmann et al. 2015). Two RamB binding sites having similarity to the consensus binding motif were identified around the position −80 and −4 with respect to the TSS and demonstrated to be bound by RamB in vitro. These sites are important for phosphate limitation induction of *pstS*. Thus, RamB functions as a transcriptional activator of the operon.

5 Sulfur Control

Sulfur is essential for the biosynthesis of biomolecules including cysteine, methionine, Coenzyme A, and iron-sulfur clusters. Bacteria utilize sulfate, sulfite, and sulfonate as sulfur sources. These compounds should be degraded and reduced to disulfide for the biosynthesis. Because the reduction of sulfate is a high energy-requiring process and the resulting disulfide is highly toxic, expression of genes encoding enzymes involved in the reduction and the biosynthesis should be coordinately regulated.

C. glutamicum utilizes organic and inorganic sulfur compounds as sulfur sources. Organic sulfur compounds that *C. glutamicum* utilizes are sulfonates like ethane sulfonate, 3-(N-morpholino)propane sulfonate (MOPS), and taurine, and sulfonate esters like busulfan, ethyl methanesulfonate, and propanesulfonate (Koch et al. 2005b). Combining genomic analysis based on the knowledge obtained from model organisms (Kertesz 2000) with genetic studies identified genes responsible for the utilization of these compounds (Koch et al. 2005b). The ABC transporter encoded by *ssuCBA* (cg1377-1379-1380 or cgR_1299-1301), which has a similarity to the reported sulfonate transporter SsuABC, is required for uptake of sulfonates and sulfonate esters, although the involvement of additional transporter is suggested by gene deletion analysis. Sulfonates and sulfonate esters are degraded to liberate sulfite for anabolism. Two functionally equivalent sulfonatases encoded by *ssuD1* (cg1376 or cgR_1298) and *ssuD2* (cg1156) decompose sulfonates, whereas monooxygenases encoded by *seuA*, *seuB*, and *seuC* (cg1151, 1152, and 1153) are also involved in the degradation of sulfonate esters in addition to the sulfonatases. These sulfonatases and monooxygenases are $FMNH_2$-dependent. $FMNH_2$ is provided by the reductase SsuI, which is essential for the utilization of all sulfonates and sulfonate esters. *ssuD1* and *ssuCBA* apparently form the operon *ssuD1CBA*. Although the gene cluster *seuABC* (cg1151-1153)-*ssuD2* (cg1156) is encoded in the genome of limited strains of *C. glutamicum*, including ATCC 13032, *ssuI* (cg1147 or cgR_1100), which is closely located to the cluster, is still present in the same locus of other strain genomes, likely supporting the ability of the strains not carrying the *seuABC-ssuD2* cluster to utilize the organic sulfur compounds.

Genomic analysis followed by genetic analysis also identified a gene cluster, *fpr2–cysIXHDNYZ* (cg3119-3112 or cgR_2704-2703-cgR_6148-cgR_2702-2698), for transporter and enzymes responsible for uptake and reduction of inorganic sulfur compounds, sulfate, and sulfite, to sulfide, which is utilized in the cysteine and methionine biosynthetic pathways (Rückert et al. 2005). Sulfate and sulfite are transported via CysZ. Sulfate is adenylated by CysDN, then reduced by CysH to sulfite. Sulfite, which is also generated from sulfonates and sulfonate esters degradation, is reduced to sulfide by CysI, for which the reducing equivalents are provided by the ferredoxin CysX and the ferredoxin reductases Fpr1 (cg3049 or cgR_2657) and Fpr2.

These sulfur compound utilization genes, together with genes constituting the biosynthetic pathway of sulfur-containing amino acids, cysteine and methionine (Rückert et al. 2003), are under the control of the global regulator McbR (Rey et al. 2005). McbR was first identified using DNA affinity purification as a transcriptional repressor of *metY* (cg0755 or cgR_0775) encoding *O*-acetyl-L-homoserine sulfhydrylase (Rey et al. 2003). The McbR regulon identified using transcriptome analysis of the *mcbR* (cg3253 or cgR_2850) deletion mutant and in vitro DNA binding assay comprises at least 45 genes, including the *ssu* and *seu* genes for sulfonates and sulfonate esters utilization, the *cys* cluster for sulfate uptake and reduction, and genes for cysteine and methionine biosynthesis. Bioinformatic analysis of the upstream region of the genes upregulated in the *mcbR* mutant identified the McbR binding motif, TAGAC-N6-GTCTA. This analysis also indicated the autoregulation of *mcbR*. The DNA binding activity of McbR is modulated by *S*-adenosylmethionine (SAM) and *S*-adenosylhomocysteine, which are a substrate and a product, respectively, of methyltransferases (Fig. 4) (Rey et al. 2005; Suda et al. 2008).

The *ssu* and *seu* genes are also transcriptionally regulated by the ROK (Repressor ORF kinase) family transcriptional regulator SsuR (Koch et al. 2005a) (Fig. 4), which belongs to the McbR regulon (Rey et al. 2005). The deletion of the *ssuR* gene (cg0012 or cgR_0010) lost the capability of sulfonate utilization as a sulfur source and abolished the expression of the *ssu* and *seu* genes, demonstrating that SsuR acts as a transcriptional activator. Competitive EMSA (electrophoretic mobility shift assay) using non-labeled oligonucleotides, followed by alignment analysis, revealed that the SsuR binding motif consisting of a T-, a GC- and an A-rich domain with a length of six, nine and six nucleotides, respectively. The motif was found around or upstream of the −35 region of the promoter of the target genes, consistent with the SsuR activator function. The genome-wide search for the motif revealed that no gene other than the *ssu* and *seu* genes is directly regulated by SsuR. Although the DNA binding activity of SsuR is inhibited by sulfate, APS, sulfite, and sulfide, it is not affected by the sulfonates including ethanesulfonate and methyl-ethanesulfonate, which are degraded by the Ssu and Seu systems. The induction of the *ssu* and *seu* genes during growth with sulfonates as a sole sulfur source was abolished in the presence of sulfate. Taken together, the absence of more preferred inorganic sulfur source is a signal for the SsuR regulon (Koch et al. 2005a).

The *fpr2–cysIXHDNYZ* gene cluster, required for sulfate uptake and reduction, is under the control of another ROK family transcriptional regulator CysR (Rückert et al. 2008), which also belongs to the McbR regulon. The deletion mutant of the *cysR* gene (cg0156 or cgR_0197) loses the ability to utilize sulfate and sulfonates as a sulfur source and has no expression of the *ssu* and *seu* genes as well as that of the

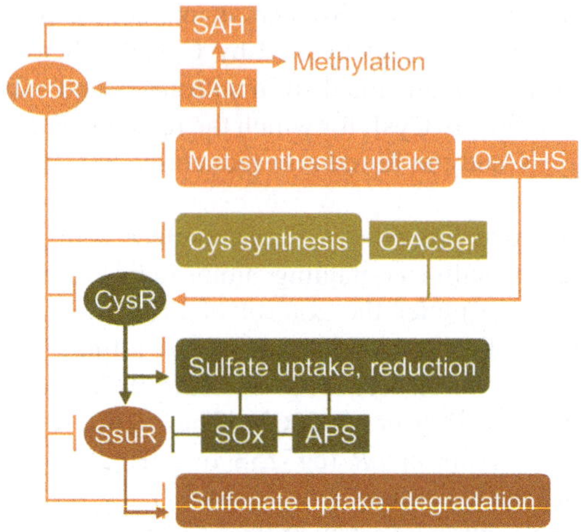

Fig. 4 The McbR regulon. Transcriptional regulators are indicated with ovals. Uptake systems and metabolic pathways under the control of the regulators are shown by rounded rectangles, whereas molecules and metabolites involved in the pathways are indicated with rectangles. Arrows and T-bars from the transcriptional regulators indicate transcriptional activation and repression, respectively. Those from the molecules and metabolites indicate activation and inactivation of the transcriptional regulators, except for arrows from SAM, indicating the methylation reaction catalyzed by *O*-methyltransferase (not indicated). SAH, *S*-adenosyl homocysteine; SAM, *S*-adenosyl methionine; O-AcHS, *O*-acetylhomoserine; O-AcSer, *O*-acetylserine; Sox, sulfate and sulfite; APS, adenosine-5′-phosphosulfate

cys gene cluster. Transcriptional analysis of the *cysR* deletion mutant revealed that CysR activates the *ssuR* gene encoding the transcriptional activator of the *ssu* and *seu* genes, indicating a hierarchical transcriptional regulation (Fig. 4). Comparison of the transcriptome profiles of the triple mutant of *mcbR*, *ssuR*, and *cysR*, and the mutant with constitutive expression of *cysR* distinguish the direct and indirect effect of the *cysR* deletion. While the *cys* gene cluster is directly activated by CysR, the *ssu* and *seu* genes are indirectly activated via activation of *ssuR* by CysR. In addition to the *cys* gene cluster, CysR positively regulates two transcript units, cg3372-3375 (cgR_2929-2932) and cg1514-4005 (cgR_6095-6094), encoding proteins with unknown function, and negatively regulates other two transcript units cg2810 (cgR_2451, dicarboxylate symporter family protein) and cg3138-3139 (cgR_2721-2722, proteins with unknown function), indicating its dual transcriptional regulatory function (Rückert et al. 2008). Alignment of the promoter regions of the genes activated by CysR revealed the CysR binding motif consisting of two 10-bp motifs forming an inverted repeat with a variable spacer of 6–8 bp. The 3′ region of the CysR binding site resembles that of the SsuR binding site. The DNA binding activity of CysR requires *O*-acetyl-L-serine or *O*-acetyl-L-homoserine, each of which serves as a sulfide acceptor for cysteine and methionine synthesis, respectively. Although sequence-specific binding in vitro was not confirmed, the importance of the binding site was confirmed by promoter reporter assay using DNA fragments with or without the binding site. Thus, CysR controls the expression of genes for sulfur uptake and reduction in response to the availability of sulfide acceptor to avoid an accumulation of toxic sulfide (Rückert et al. 2008). Since McbR also regulates the expression of the *ssuR* gene, the *ssu* and *seu* genes are only activated when additional sulfur is required and the concentration of inorganic sulfur sources is low. Because the genes with unknown function in the CysR regulon are co-regulated by McbR, proteins encoded are probably involved in the sulfur metabolic pathway. In contrast to the regulons of AmtR and PhoR, the involvement of other global transcriptional regulators has not been reported.

6 Concluding Remarks and Perspectives

The studies summarized in this chapter provide the overview of transcriptional regulatory systems for the metabolism of primary elements, i.e., carbon, nitrogen, sulfur, and phosphorus, in *C. glutamicum*. These studies demonstrated that the key global regulators were first identified as a transcriptional repressor or activator of a single gene in the metabolism. Such regulators have been isolated by either screening of the genomic library using reporter expression, which identified GlxR and AmtR, or by purification of proteins bound to the promoter region of the target gene, which identified RamA, RamB, IolR, AtlR, and McbR. Moreover, the genomic locus and orthologous regulators function also provided important clues to the identification of the global regulators, such as SugR, GntR1/2, and PhoRS. Accurate identification of the regulon members supports the understanding of the transcriptional regulators' functions in three contexts. Firstly, a common feature among the regulon genes regulated in the same way gives a hint to understand the function of the regulator. SugR has been first identified as a transcriptional repressor of one of the PTS components, but it was demonstrated to repress genes involved in sugar uptake and metabolism in response to sugar availability. Secondly, the alignment of the upstream regions of the regulon genes gives a clue to determine the binding motif of the regulator as a matrix. Finally, the establishment of the regulon list also helps identify effector molecules controlling the DNA binding activity of the regulators as described for the regulators involved in sulfur control. In this context, ChIP-chip or ChIP-seq analysis directly detects the binding of the transcriptional regulator to the chromosomal DNA in vivo, enabling identification of the regulon and the binding motif.

The knowledge of the global transcriptional regulators in *C. glutamicum* has been accumulated for the last decade. It has not only deepened the understanding of the *C. glutamicum* physiology at the molecular level but also has been recently applied to metabolic engineering. For example, deletion of *sugR* was applied to enhance sugar metabolic pathways (Bartek et al. 2010; Xu et al. 2016; Pérez-García et al. 2018) and that of *mcbR* has increased the productivity of methionine and SAM (Rey et al. 2003; Han et al. 2016). Moreover, the knowledge has been used to construct biosensors to detect changes in metabolites (Mustafi et al. 2012; Schulte et al. 2017a). The sensors coupled with fluorescence-activated cell sorting were used together with random mutagenesis to screen for beneficial mutations that improve productivity (Mahr et al. 2015; Zhang et al. 2018). There are over 50 transcriptional regulators, including response regulators of the two-component systems, that have not been characterized yet in *C. glutamicum*. Investigation of these regulators will provide further insights into the physiology and transcriptional regulatory network, and contribute to the development of synthetic biological tools.

References

Arndt A, Eikmanns BJ (2007) The alcohol dehydrogenase gene *adhA* in *Corynebacterium glutamicum* is subject to carbon catabolite repression. J Bacteriol 189:7408–7416

Auchter M, Arndt A, Eikmanns BJ (2009) Dual transcriptional control of the acetaldehyde dehydrogenase gene *ald* of *Corynebacterium glutamicum* by RamA and RamB. J Biotechnol 140:84–91

Auchter M et al (2011a) RamA and RamB are global transcriptional regulators in *Corynebacterium glutamicum* and control genes for enzymes of the central metabolism. J Biotechnol 154:126–139

Auchter M et al (2011b) Control of *adhA* and sucR expression by the SucR regulator in *Corynebacterium glutamicum*. J Biotechnol 152:77–86

Bailey TL, Elkan C (1994) Fitting a mixture model by expectation maximization to discover motifs in biopolymers. Proc Int Conf Intell Syst Mol Biol 2:28–36

Bartek T et al (2010) Studies on substrate utilisation in L: -valine-producing *Corynebacterium glutamicum* strains deficient in pyruvate dehydrogenase complex. Bioprocess Biosyst Eng 33:873–8883

Bäumchen C, Krings E, Bringer S, Eggeling L, Sahm H (2009) *Myo*-inositol facilitators IolT1 and IolT2 enhance D-mannitol formation from D-fructose in *Corynebacterium glutamicum*. FEMS Microbiol Lett 290:227–235

Baumgart M, Luder K, Grover S, Gätgens C, Besra GS, Frunzke J (2013) IpsA, a novel LacI-type regulator, is required for inositol-derived lipid formation in *Corynebacteria* and *Mycobacteria*. BMC Biol 11:122. https://doi.org/10.1186/1741-7007-11-122

Beckers G et al (2005) Regulation of AmtR-controlled gene expression in *Corynebacterium glutamicum*: mechanism and characterization of the AmtR regulon. Mol Microbiol 58:580–595

Blombach B, Cramer A, Eikmanns BJ, Schreiner M (2009) RamB is an activator of the pyruvate dehydrogenase complex subunit E1p gene in *Corynebacterium glutamicum*. J Mol Microbiol Biotechnol 16:236–239

Bott M, Brocker M (2012) Two-component signal transduction in *Corynebacterium glutamicum* and other corynebacteria: on the way towards stimuli and targets. Appl Microbiol Biotechnol 94:1131–1150. https://doi.org/10.1007/s00253-012-4060-x

Brinkrolf K et al (2008) The LacI/GalR family transcriptional regulator UriR negatively controls uridine utilization of Corynebacterium glutamicum by binding to catabolite-responsive element (cre)-like sequences. Microbiology 154:1068–1081

Brown SW, Sonenshein AL (1996) Autogenous regulation of the *Bacillus subtilis glnRA* operon. J Bacteriol 178:2450–2454

Brune I, Werner H, Hüser AT, Kalinowski J, Pühler A, Tauch A (2006) The DtxR protein acting as dual transcriptional regulator directs a global regulatory network involved in iron metabolism of *Corynebacterium glutamicum*. BMC Genomics 7:21

Brüsseler C, Radek A, Tenhaef N, Krumbach K, Noack S, Marienhagen J (2018) The *myo*-inositol/proton symporter IolT1 contributes to d-xylose uptake in *Corynebacterium glutamicum*. Bioresour Technol 249:953–961. https://doi.org/10.1016/j.biortech.2017.10.098

Buchinger S et al (2009) A combination of metabolome and transcriptome analyses reveals new targets of the *Corynebacterium glutamicum* nitrogen regulator AmtR. J Biotechnol 140:68–74

Busche T, Šilar R, Pičmanová M, Pátek M, Kalinowski J (2012) Transcriptional regulation of the operon encoding stress-responsive ECF sigma factor SigH and its anti-sigma factor RshA, and control of its regulatory network in *Corynebacterium glutamicum*. BMC Genomics 13:445. https://doi.org/10.1186/1471-2164-13-445

Bussmann M, Emer D, Hasenbein S, Degraf S, Eikmanns BJ, Bott M (2009) Transcriptional control of the succinate dehydrogenase operon *sdhCAB* of *Corynebacterium glutamicum* by the cAMP-dependent regulator GlxR and the LuxR-type regulator RamA. J Biotechnol 143:173–182

Cha PH et al (2010) Characterization of an adenylate cyclase gene (*cyaB*) deletion mutant of *Corynebacterium glutamicum* ATCC 13032. Appl Microbiol Biotechnol 85:1061–1068

Cho HY, Lee SG, Hyeon JE, Han SO (2010) Identification and characterization of a transcriptional regulator, SucR, that influences *sucCD* transcription in *Corynebacterium glutamicum*. Biochem Biophys Res Commun 401:300–305

Claes WA, Pühler A, Kalinowski J (2002) Identification of two *prpDBC* gene clusters in *Corynebacterium glutamicum* and their involvement in propionate degradation via the 2-methylcitrate cycle. J Bacteriol 184:2728–2739

Cocaign-Bousquet M, Monnet C, Lindley ND (1993) Batch kinetics of *Corynebacterium glutamicum* during growth on various carbon substrates: use of substrate mixtures to localise metabolic bottlenecks. Appl Microbiol Biotechnol 40:526–530

Costa-Riu N, Maier E, Burkovski A, Krämer R, Lottspeich F, Benz R (2003) Identification of an anion-specific channel in the cell wall of the Gram-positive bacterium *Corynebacterium glutamicum*. Mol Microbiol 50:1295–1308

Cramer A, Eikmanns BJ (2007) RamA, the transcriptional regulator of acetate metabolism in *Corynebacterium glutamicum*, is subject to negative autoregulation. J Mol Microbiol Biotechnol 12:51–59

Cramer A, Gerstmeir R, Schaffer S, Bott M, Eikmanns BJ (2006) Identification of RamA, a novel LuxR-type transcriptional regulator of genes involved in acetate metabolism of *Corynebacterium glutamicum*. J Bacteriol 188:2554–2567

Deutscher J, Küster E, Bergstedt U, Charrier V, Hillen W (1995) Protein kinase-dependent HPr/CcpA interaction links glycolytic activity to carbon catabolite repression in gram-positive bacteria. Mol Microbiol 15:1049–1053

Dietrich C, Nato A, Bost B, Le Maréchal P, Guyonvarch A (2009) Regulation of *ldh* expression during biotin-limited growth of *Corynebacterium glutamicum*. Microbiology 155:1360–1375

Dominguez H, Cocaign-Bousquet M, Lindley ND (1997) Simultaneous consumption of glucose and fructose from sugar mixtures during batch growth of *Corynebacterium glultamicum*. Appl Microbiol Biotechnol 47:600–603

Ehira S, Shirai T, Teramoto H, Inui M, Yukawa H (2008) Group 2 sigma factor SigB of *Corynebacterium glutamicum* positively regulates glucose metabolism under conditions of oxygen deprivation. Appl Environ Microbiol 74:5146–5152

Ehira S, Teramoto H, Inui M, Yukawa H (2009) Regulation of *Corynebacterium glutamicum* heat shock response by the extracytoplasmic-function sigma factor SigH and transcriptional regulators HspR and HrcA. J Bacteriol 191:2964–2972

Eikmanns BJ, Rittmann D, Sahm H (1995) Cloning, sequence analysis, expression, and inactivation of the *Corynebacterium glutamicum icd* gene encoding isocitrate dehydrogenase and biochemical characterization of the enzyme. J Bacteriol 177:774–782

Engels V, Wendisch VF (2007) The DeoR-Type regulator SugR represses expression of *ptsG* in *Corynebacterium glutamicum*. J Bacteriol 189:2955–2966. https://doi.org/10.1128/jb.01596-06

Engels V, Lindner SN, Wendisch VF (2008) The global repressor SugR controls expression of genes of glycolysis and of the L-lactate dehydrogenase LdhA in *Corynebacterium glutamicum*. J Bacteriol 190:8033–8044

Fisher SH, Wray LV Jr (2008) *Bacillus subtilis* glutamine synthetase regulates its own synthesis by acting as a chaperone to stabilize GlnR-DNA complexes. Proc Natl Acad Sci U S A 105:1014–1019. https://doi.org/10.1073/pnas.0709949105

Fisher SH, Wray LV Jr (2009) Novel trans-acting *Bacillus subtilis glnA* mutations that derepress *glnRA* expression. J Bacteriol 191:2485–2492. https://doi.org/10.1128/JB.01734-08

Frunzke J, Engels V, Hasenbein S, Gatgens C, Bott M (2008) Co-ordinated regulation of gluconate catabolism and glucose uptake in *Corynebacterium glutamicum* by two functionally equivalent transcriptional regulators, GntR1 and GntR2. Mol Microbiol 67:305–322

Gaigalat L et al (2007) The DeoR-type transcriptional regulator SugR acts as a repressor for genes encoding the phosphoenolpyruvate:sugar phosphotransferase system (PTS) in *Corynebacterium glutamicum*. BMC Mol Biol 8:104

Gao Y-G et al (2008) Structural and functional characterization of the LldR from *Corynebacterium glutamicum*: a transcriptional repressor involved in L-lactate and sugar utilization. Nucleic Acids Res 36:7110–7123. https://doi.org/10.1093/nar/gkn827

Georgi T, Engels V, Wendisch VF (2008) Regulation of L-lactate utilization by the FadR-type regulator LldR of *Corynebacterium glutamicum*. J Bacteriol 190:963–971

Gerstmeir R, Cramer A, Dangel P, Schaffer S, Eikmanns BJ (2004) RamB, a novel transcriptional regulator of genes involved in acetate metabolism of *Corynebacterium glutamicum*. J Bacteriol 186:2798–2809

Gourdon P, Baucher MF, Lindley ND, Guyonvarch A (2000) Cloning of the malic enzyme gene from *Corynebacterium glutamicum* and role of the enzyme in lactate metabolism. Appl Environ Microbiol 66:2981–2987

Han SO, Inui M, Yukawa H (2007) Expression of *Corynebacterium glutamicum* glycolytic genes varies with carbon source and growth phase. Microbiology 153:2190–2202

Han SO, Inui M, Yukawa H (2008) Effect of carbon source availability and growth phase on expression of *Corynebacterium glutamicum* genes involved in the tricarboxylic acid cycle and glyoxylate bypass. Microbiology 154:3073–3083

Han G, Hu X, Qin T, Li Y, Wang X (2016) Metabolic engineering of *Corynebacterium glutamicum* ATCC13032 to produce S-adenosyl-L-methionine. Enzym Microb Technol 83:14–21. https://doi.org/10.1016/j.enzmictec.2015.11.001

Hasselt K, Rankl S, Worsch S, Burkovski A (2011) Adaptation of AmtR-controlled gene expression by modulation of AmtR binding activity in *Corynebacterium glutamicum*. J Biotechnol 154:156–162. https://doi.org/10.1016/j.jbiotec.2010.09.930

Hayashi M et al (2002) Transcriptome analysis of acetate metabolism in *Corynebacterium glutamicum* using a newly developed metabolic array. Biosci Biotechnol Biochem 66:1337–1344

Henrich A, Kuhlmann N, Eck AW, Krämer R, Seibold GM (2013) Maltose uptake by the novel ABC transport system MusEFGK2I causes increased expression of *ptsG* in *Corynebacterium glutamicum*. J Bacteriol 195:2573–2584. https://doi.org/10.1128/JB.01629-12

Hoffelder M, Raasch K, van Ooyen J, Eggeling L (2010) The E2 domain of OdhA of *Corynebacterium glutamicum* has succinyltransferase activity dependent on lipoyl residues of the acetyltransferase AceF. J Bacteriol 192:5203–5211. https://doi.org/10.1128/JB.00597-10

Ikeda M, Nakagawa S (2003) The *Corynebacterium glutamicum* genome: features and impacts on biotechnological processes. Appl Microbiol Biotechnol 62:99–109

Ikeda M, Mizuno Y, Awane S, Hayashi M, Mitsuhashi S, Takeno S (2011) Identification and application of a different glucose uptake system that functions as an alternative to the phosphotransferase system in *Corynebacterium glutamicum*. Appl Microbiol Biotechnol 90:1443–1451. https://doi.org/10.1007/s00253-011-3210-x

Inui M, Murakami S, Okino S, Kawaguchi H, Vertès AA, Yukawa H (2004) Metabolic analysis of *Corynebacterium glutamicum* during lactate and succinate productions under oxygen deprivation conditions. J Mol Microbiol Biotechnol 7:182–196

Inui M et al (2007) Transcriptional profiling of *Corynebacterium glutamicum* metabolism during organic acid production under oxygen deprivation conditions. Microbiology 153:2491–2504. https://doi.org/10.1099/mic.0.2006/005587-0

Ishige T, Krause M, Bott M, Wendisch VF, Sahm H (2003) The phosphate starvation stimulon of *Corynebacterium glutamicum* determined by DNA microarray analyses. J Bacteriol 185:4519–4529

Jakoby M, Nolden L, Meier-Wagner J, Krämer R, Burkovski A (2000) AmtR, a global repressor in the nitrogen regulation system of *Corynebacterium glutamicum*. Mol Microbiol 37:964–977

Jones BE, Dossonnet V, Küster E, Hillen W, Deutscher J, Klevit RE (1997) Binding of the catabolite repressor protein CcpA to its DNA target is regulated by phosphorylation of its corepressor HPr. J Biol Chem 272:26530–26535. https://doi.org/10.1074/jbc.272.42.26530

Jungwirth B et al (2013) High-resolution detection of DNA binding sites of the global transcriptional regulator GlxR in *Corynebacterium glutamicum*. Microbiology 159:12–22. https://doi.org/10.1099/mic.0.062059-0

Kalinowski J et al (2003) The complete *Corynebacterium glutamicum* ATCC 13032 genome sequence and its impact on the production of L-aspartate-derived amino acids and vitamins. J Biotechnol 104:5–25

Kawaguchi H, Vèrtes AA, Okino S, Inui M, Yukawa H (2006) Engineering of a xylose metabolic pathway in Corynebacterium glutamicum. Appl Environ Microbiol 72:3418–3428

Kawaguchi H, Sasaki M, Vertès AA, Inui M, Yukawa H (2008) Engineering of an L: -arabinose metabolic pathway in Corynebacterium glutamicum. Appl Microbiol Biotechnol 77:1053–1062

Kertesz MA (2000) Riding the sulfur cycle–metabolism of sulfonates and sulfate esters in Gram-negative bacteria. FEMS Microbiol Rev 24:135–175. https://doi.org/10.1016/S0168-6445(99)00033-9

Kim HJ, Kim TH, Kim Y, Lee HS (2004) Identification and characterization of *glxR*, a gene involved in regulation of glyoxylate bypass in *Corynebacterium glutamicum*. J Bacteriol 186:3453–3460

Klaffl S, Brocker M, Kalinowski J, Eikmanns BJ, Bott M (2013) Complex regulation of the phosphoenolpyruvate carboxykinase gene *pck* and characterization of its GntR-type regulator IolR as a repressor of *myo*-inositol utilization genes in *Corynebacterium glutamicum*. J Bacteriol 195:4283–4296. https://doi.org/10.1128/JB.00265-13

Kočan M, Schaffer S, Ishige T, Sorger-Herrmann U, Wendisch VF, Bott M (2006) Two-component systems of *Corynebacterium glutamicum*: deletion analysis and involvement of the PhoS-PhoR system in the phosphate starvation response. J Bacteriol 188:724–732

Koch DJ et al (2005a) The transcriptional regulator SsuR activates expression of the *Corynebacterium glutamicum* sulphonate utilization genes in the absence of sulphate. Mol Microbiol 58:480–494

Koch DJ, Rückert C, Rey DA, Mix A, Pühler A, Kalinowski J (2005b) Role of the *ssu* and *seu* genes of *Corynebacterium glutamicum* ATCC 13032 in utilization of sulfonates and sulfonate esters as sulfur sources. Appl Environ Microbiol 71:6104–6114

Kohl TA, Tauch A (2009) The GlxR regulon of the amino acid producer *Corynebacterium glutamicum*: detection of the corynebacterial core regulon and integration into the transcriptional regulatory network model. J Biotechnol 143:239–246

Kohl TA, Baumbach J, Jungwirth B, Pühler A, Tauch A (2008) The GlxR regulon of the amino acid producer *Corynebacterium glutamicum*: *in silico* and *in vitro* detection of DNA binding sites of a global transcription regulator. J Biotechnol 135:340–350

Kotrba P, Inui M, Yukawa H (2001) Bacterial phosphotransferase system (PTS) in carbohydrate uptake and control of carbon metabolism. J Biosci Bioeng 92:502–517

Kotrba P, Inui M, Yukawa H (2003) A single V317A or V317M substitution in Enzyme II of a newly identified beta-glucoside phosphotransferase and utilization system of Corynebacterium glutamicum R extends its specificity towards cellobiose. Microbiology 149:1569–1580

Kotrbova-Kozak A, Kotrba P, Inui M, Sajdok J, Yukawa H (2007) Transcriptionally regulated *adhA* gene encodes alcohol dehydrogenase required for ethanol and *n*-propanol utilization in *Corynebacterium glutamicum* R. Appl Microbiol Biotechnol 76:1347–1356

Krause JP, Polen T, Youn JW, Emer D, Eikmanns BJ, Wendisch VF (2012) Regulation of the malic enzyme gene *malE* by the transcriptional regulator MalR in *Corynebacterium glutamicum*. J Biotechnol 159:204–215

Krings E et al (2006) Characterization of *myo*-inositol utilization by *Corynebacterium glutamicum*: the stimulon, identification of transporters, and influence on L-lysine formation. J Bacteriol 188:8054–8061

Kronemeyer W, Peekhaus N, Krämer R, Sahm H, Eggeling L (1995) Structure of the *gluABCD* cluster encoding the glutamate uptake system of *Corynebacterium glutamicum*. J Bacteriol 177:1152–1158

Krug A, Wendisch VF, Bott M (2005) Identification of AcnR, a TetR-type repressor of the aconitase gene *acn* in *Corynebacterium glutamicum*. J Biol Chem 280:585–595

Laslo T et al (2012) Arabitol metabolism of *Corynebacterium glutamicum* and its regulation by AtlR. J Bacteriol 194:941–955. https://doi.org/10.1128/JB.06064-11

Lindner SN, Vidaurre D, Willbold S, Schoberth SM, Wendisch VF (2007) NCgl2620 encodes a class II polyphosphate kinase in *Corynebacterium glutamicum*. Appl Environ Microbiol 73:5026–5033

Lindner SN, Knebel S, Wesseling H, Schoberth SM, Wendisch VF (2009) Exopolyphosphatases PPX1 and PPX2 from *Corynebacterium glutamicum*. Appl Environ Microbiol 75:3161–3170

Lindner SN, Seibold GM, Henrich A, Krämer R, Wendisch VF (2011) Phosphotransferase system-independent glucose utilization in *Corynebacterium glutamicum* by inositol permeases and glucokinases. Appl Environ Microbiol 77:3571–3581. https://doi.org/10.1128/AEM.02713-10

Mahr R, Gätgens C, Gätgens J, Polen T, Kalinowski J, Frunzke J (2015) Biosensor-driven adaptive laboratory evolution of l-valine production in *Corynebacterium glutamicum*. Metab Eng 32:184–194. https://doi.org/10.1016/j.ymben.2015.09.017

Merkens H, Beckers G, Wirtz A, Burkovski A (2005) Vanillate metabolism in *Corynebacterium glutamicum*. Curr Microbiol 51:59–65

Merrick MJ, Edwards RA (1995) Nitrogen control in bacteria. Microbiol Rev 59:604–622

Morabbi Heravi K, Lange J, Watzlawick H, Kalinowski J, Altenbuchner J (2015) Transcriptional regulation of the vanillate utilization genes (*vanABK* operon) of *Corynebacterium glutamicum* by VanR, a PadR-like repressor. J Bacteriol 197:959–972. https://doi.org/10.1128/JB.02431-14

Muffler A et al (2002) Genome-wide transcription profiling of *Corynebacterium glutamicum* after heat shock and during growth on acetate and glucose. J Biotechnol 98:255–268

Mustafi N, Grünberger A, Kohlheyer D, Bott M, Frunzke J (2012) The development and application of a single-cell biosensor for the detection of L-methionine and branched-chain amino acids. Metab Eng 14:449–457. https://doi.org/10.1016/j.ymben.2012.02.002

Nentwich SS et al (2009) Characterization of the LacI-type transcriptional repressor RbsR controlling ribose transport in *Corynebacterium glutamicum* ATCC 13032. Microbiology 155:150–164

Neshat A, Mentz A, Rückert C, Kalinowski J (2014) Transcriptome sequencing revealed the transcriptional organization at ribosome-mediated attenuation sites in *Corynebacterium glutamicum* and identified a novel attenuator involved in aromatic amino acid biosynthesis. J Biotechnol 190:55–63. https://doi.org/10.1016/j.jbiotec.2014.05.033

Nešvera J, Pátek M (2011) Tools for genetic manipulations in *Corynebacterium glutamicum* and their applications. Appl Microbiol Biotechnol 90:1641–1654

Netzer R, Peters-Wendisch P, Eggeling L, Sahm H (2004) Cometabolism of a nongrowth substrate: L-serine utilization by *Corynebacterium glutamicum*. Appl Environ Microbiol 70:7148–7155

Ninfa AJ, Jiang P, Atkinson MR, Peliska JA (2000) Integration of antagonistic signals in the regulation of nitrogen assimilation in *Escherichia coli*. Curr Top Cell Regul 36:31–75

Nolden L, Ngouoto-Nkili CE, Bendt AK, Krämer R, Burkovski A (2001) Sensing nitrogen limitation in *Corynebacterium glutamicum*: the role of *glnK* and *glnD*. Mol Microbiol 42:1281–1295

Omumasaba CA, Okai N, Inui M, Yukawa H (2004) *Corynebacterium glutamicum* glyceraldehyde-3-phosphate dehydrogenase isoforms with opposite, ATP-dependent regulation. J Mol Microbiol Biotechnol 8:91–103

Palanca C, Rubio V (2016) Structure of AmtR, the global nitrogen regulator of *Corynebacterium glutamicum*, in free and DNA-bound forms. FEBS J 283:1039–1059. https://doi.org/10.1111/febs.13643

Panhorst M, Sorger-Herrmann U, Wendisch VF (2011) The *pstSCAB* operon for phosphate uptake is regulated by the global regulator GlxR in *Corynebacterium glutamicum*. J Biotechnol 154:149–155. https://doi.org/10.1016/j.jbiotec.2010.07.015

Parche S, Burkovski A, Sprenger GA, Weil B, Krämer R, Titgemeyer F (2001) *Corynebacterium glutamicum*: a dissection of the PTS. J Mol Microbiol Biotechnol 3:423–428

Park SY, Moon MW, Subhadra B, Lee JK (2010) Functional characterization of the *glxR* deletion mutant of *Corynebacterium glutamicum* ATCC 13032: involvement of GlxR in acetate metabolism and carbon catabolite repression. FEMS Microbiol Lett 304:107–115

Pátek M, Nešvera J (2011) Sigma factors and promoters in *Corynebacterium glutamicum*. J Biotechnol 154:101–113

Peng X, Okai N, Vertès A, Inatomi K, Inui M, Yukawa H (2011) Characterization of the mannitol catabolic operon of *Corynebacterium glutamicum*. Appl Microbiol Biotechnol 91:1375–1387. https://doi.org/10.1007/s00253-011-3352-x

Pérez-García F, Jorge JMP, Dreyszas A, Risse JM, Wendisch VF (2018) Efficient production of the dicarboxylic acid glutarate by *Corynebacterium glutamicum* via a novel synthetic pathway. Front Microbiol 9:2589. https://doi.org/10.3389/fmicb.2018.02589

Plassmeier P, Persicke M, Pühler A, Sterthoff C, Rückert C, Kalinowski J (2012) Molecular characterization of PrpR, the transcriptional activator of propionate catabolism in *Corynebacterium glutamicum*. J Biotechnol 159:1–11

Postma PW, Lengeler JW, Jacobson GR (1993) Phosphoenolpyruvate:carbohydrate phosphotransferase systems of bacteria. Microbiol Mol Biol Rev 57:543–594

Reddy P, Kamireddi M (1998) Modulation of *Escherichia coli* adenylyl cyclase activity by catalytic-site mutants of protein IIA^{Glc} of the phosphoenolpyruvate:sugar phosphotransferase system. J Bacteriol 180:732–736

Rehm N et al (2010) L-Glutamine as a nitrogen source for *Corynebacterium glutamicum*: derepression of the AmtR regulon and implications for nitrogen sensing. Microbiology 156:3180–3193. https://doi.org/10.1099/mic.0.040667-0

Rey DA, Pühler A, Kalinowski J (2003) The putative transcriptional repressor McbR, member of the TetR-family, is involved in the regulation of the metabolic network directing the synthesis of sulfur containing amino acids in *Corynebacterium glutamicum*. J Biotechnol 103:51–65

Rey DA et al (2005) The McbR repressor modulated by the effector substance *S*-adenosylhomocysteine controls directly the transcription of a regulon involved in sulphur metabolism of *Corynebacterium glutamicum* ATCC 13032. Mol Microbiol 56:871–887

Rittmann D, Sorger-Herrmann U, Wendisch VF (2005) Phosphate starvation-inducible gene *ushA* encodes a 5' nucleotidase required for growth of *Corynebacterium glutamicum* on media with nucleotides as the phosphorus source. Appl Environ Microbiol 71:4339–4344

Rückert C, Pühler A, Kalinowski J (2003) Genome-wide analysis of the L-methionine biosynthetic pathway in *Corynebacterium glutamicum* by targeted gene deletion and homologous complementation. J Biotechnol 104:213–228

Rückert C et al (2005) Functional genomics and expression analysis of the *Corynebacterium glutamicum fpr2-cysIXHDNYZ* gene cluster involved in assimilatory sulphate reduction. BMC Genomics 6:121

Rückert C, Milse J, Albersmeier A, Koch DJ, Pühler A, Kalinowski J (2008) The dual transcriptional regulator CysR in *Corynebacterium glutamicum* ATCC 13032 controls a subset of genes of the McbR regulon in response to the availability of sulphide acceptor molecules. BMC Genomics 9:483

Santos-Beneit F (2015) The Pho regulon: a huge regulatory network in bacteria. Front Microbiol 6:402. https://doi.org/10.3389/fmicb.2015.00402

Schaaf S, Bott M (2007) Target genes and DNA-binding sites of the response regulator PhoR from *Corynebacterium glutamicum*. J Bacteriol 189:5002–5011

Schröder J, Tauch A (2010) Transcriptional regulation of gene expression in *Corynebacterium glutamicum*: the role of global, master and local regulators in the modular and hierarchical gene regulatory network. FEMS Microbiol Rev 34:685–737

Schulte J, Baumgart M, Bott M (2017a) Development of a single-cell GlxR-based cAMP biosensor for *Corynebacterium glutamicum*. J Biotechnol 258:33–40. https://doi.org/10.1016/j.jbiotec.2017.07.004

Schulte J, Baumgart M, Bott M (2017b) Identification of the cAMP phosphodiesterase CpdA as novel key player in cAMP-dependent regulation in *Corynebacterium glutamicum*. Mol Microbiol 103:534–552. https://doi.org/10.1111/mmi.13574

Schumacher MA, Chinnam NB, Cuthbert B, Tonthat NK, Whitfill T (2015) Structures of regulatory machinery reveal novel molecular mechanisms controlling *B. subtilis* nitrogen homeostasis. Genes Dev 29:451–464. https://doi.org/10.1101/gad.254714.114

Schwinde JW, Thum-Schmitz N, Eikmanns BJ, Sahm H (1993) Transcriptional analysis of the *gap-pgk-tpi-ppc* gene cluster of *Corynebacterium glutamicum*. J Bacteriol 175:3905–3908

Sorger-Herrmann U, Taniguchi H, Wendisch VF (2015) Regulation of the *pstSCAB* operon in *Corynebacterium glutamicum* by the regulator of acetate metabolism RamB. BMC Microbiol 15:113. https://doi.org/10.1186/s12866-015-0437-1

Stansen C, Uy D, Delaunay S, Eggeling L, Goergen JL, Wendisch VF (2005) Characterization of a *Corynebacterium glutamicum* lactate utilization operon induced during temperature-triggered glutamate production. Appl Environ Microbiol 71:5920–5928

Strösser J, Ludke A, Schaffer S, Kramer R, Burkovski A (2004) Regulation of GlnK activity: modification, membrane sequestration and proteolysis as regulatory principles in the network of nitrogen control in *Corynebacterium glutamicum*. Mol Microbiol 54:132–147

Subhadra B, Lee JK (2013) Elucidation of the regulation of ethanol catabolic genes and *ptsG* using a *glxR* and adenylate cyclase gene (*cyaB*) deletion mutants of *Corynebacterium glutamicum* ATCC 13032. J Microbiol Biotechnol 23:1683–1690

Suda M, Teramoto H, Imamiya T, Inui M, Yukawa H (2008) Transcriptional regulation of *Corynebacterium glutamicum* methionine biosynthesis genes in response to methionine supplementation under oxygen deprivation. Appl Microbiol Biotechnol 81:505–513

Takahashi H, Inada T, Postma P, Aiba H (1998) CRP down-regulates adenylate cyclase activity by reducing the level of phosphorylated IIA^{Glc}, the glucose-specific phosphotransferase protein, in *Escherichia coli*. Mol Gen Genet 259:317–326

Tanaka Y, Okai N, Teramoto H, Inui M, Yukawa H (2008a) Regulation of the expression of phosphoenolpyruvate: carbohydrate phosphotransferase system (PTS) genes in *Corynebacterium glutamicum* R. Microbiology 154:264–274

Tanaka Y, Teramoto H, Inui M, Yukawa H (2008b) Regulation of expression of general components of the phosphoenolpyruvate: carbohydrate phosphotransferase system (PTS) by the global regulator SugR in *Corynebacterium glutamicum*. Appl Microbiol Biotechnol 78:309–318

Tanaka Y, Teramoto H, Inui M, Yukawa H (2009) Identification of a second beta-glucoside phosphoenolpyruvate: carbohydrate phosphotransferase system in *Corynebacterium glutamicum* R. Microbiology 155:3652–3660

Tanaka Y, Ehira S, Teramoto H, Inui M, Yukawa H (2012) Coordinated regulation of *gnd*, which encodes 6-phosphogluconate dehydrogenase, by the two transcriptional regulators GntR1 and RamA in *Corynebacterium glutamicum*. J Bacteriol 194:6527–6536. https://doi.org/10.1128/JB.01635-12

Tanaka Y, Takemoto N, Ito T, Teramoto H, Yukawa H, Inui M (2014) Genome-wide analysis of the role of global transcriptional regulator GntR1 in *Corynebacterium glutamicum*. J Bacteriol 196:3249–3258. https://doi.org/10.1128/JB.01860-14

Teramoto H, Inui M, Yukawa H (2009) Regulation of expression of genes involved in quinate and shikimate utilization in *Corynebacterium glutamicum*. Appl Environ Microbiol 75:3461–3468

Toyoda K, Teramoto H, Inui M, Yukawa H (2008) Expression of the *gapA* gene encoding glyceraldehyde-3-phosphate dehydrogenase of *Corynebacterium glutamicum* is regulated by the global regulator SugR. Appl Microbiol Biotechnol 81:291–301

Toyoda K, Teramoto H, Inui M, Yukawa H (2009a) Involvement of the LuxR-type transcriptional regulator, RamA, in regulation of expression of the *gapA* gene encoding glyceraldehyde-3-phosphate dehydrogenase of *Corynebacterium glutamicum*. J Bacteriol 191:968–977

Toyoda K, Teramoto H, Inui M, Yukawa H (2009b) The *ldhA* gene, encoding fermentative L-lactate dehydrogenase of *Corynebacterium glutamicum*, is under the control of positive feedback regulation mediated by LldR. J Bacteriol 191:4251–4258. https://doi.org/10.1128/jb.00303-09

Toyoda K, Teramoto H, Inui M, Yukawa H (2009c) Molecular mechanism of SugR-mediated sugar-dependent expression of the *ldhA* gene encoding L-lactate dehydrogenase in *Corynebacterium glutamicum*. Appl Microbiol Biotechnol 83:315–327

Toyoda K, Teramoto H, Inui M, Yukawa H (2011) Genome-wide identification of *in vivo* binding sites of GlxR, a cyclic AMP receptor protein-type regulator in *Corynebacterium glutamicum*. J Bacteriol 193:4123–4133. https://doi.org/10.1128/JB.00384-11

Toyoda K, Teramoto H, Yukawa H, Inui M (2015) Expanding the regulatory network governed by the extracytoplasmic function sigma factor σ^H in *Corynebacterium glutamicum*. J Bacteriol 197:483–496. https://doi.org/10.1128/JB.02248-14

van Ooyen J, Emer D, Bussmann M, Bott M, Eikmanns BJ, Eggeling L (2011) Citrate synthase in *Corynebacterium glutamicum* is encoded by two *gltA* transcripts which are controlled by RamA, RamB, and GlxR. J Biotechnol 154:140–148

Wendisch VF, Bott M (2008) Phosphorus metabolism and its regulation. In: Burkovski A (ed) Corynebacteria: genomics and molecular biology. Caister Academic Press, pp 203–216

Wendisch VF, Spies M, Reinscheid DJ, Schnicke S, Sahm H, Eikmanns BJ (1997) Regulation of acetate metabolism in *Corynebacterium glutamicum*: transcriptional control of the isocitrate lyase and malate synthase genes. Arch Microbiol 168:262–269

Wendisch VF, de Graaf AA, Sahm H, Eikmanns BJ (2000) Quantitative determination of metabolic fluxes during coutilization of two carbon sources: comparative analyses with *Corynebacterium glutamicum* during growth on acetate and/or glucose. J Bacteriol 182:3088–3096

Wennerhold J, Krug A, Bott M (2005) The AraC-type regulator RipA represses aconitase and other iron proteins from *Corynebacterium* under iron limitation and is itself repressed by DtxR. J Biol Chem 280:40500–40508

Wray LV Jr, Zalieckas JM, Fisher SH (2001) *Bacillus subtilis* glutamine synthetase controls gene expression through a protein-protein interaction with transcription factor TnrA. Cell 107:427–435

Xu J, Zhang J, Liu D, Zhang W (2016) Increased glucose utilization and cell growth of *Corynebacterium glutamicum* by modifying the glucose-specific phosphotransferase system (PTS (Glc)) genes. Can J Microbiol 62:983–992. https://doi.org/10.1139/cjm-2016-0027

Yoshida K, Yamaguchi H, Kinehara M, Ohki YH, Nakaura Y, Fujita Y (2003) Identification of additional TnrA-regulated genes of *Bacillus subtilis* associated with a TnrA box. Mol Microbiol 49:157–165

Yukawa H et al (2007) Comparative analysis of the *Corynebacterium glutamicum* group and complete genome sequence of strain R. Microbiology 153:1042–1058

Zhang X, Xu G, Shi J, Xu Z (2018) Integration of ARTP mutagenesis with biosensor-mediated high-throughput screening to improve L-serine yield in *Corynebacterium glutamicum*. Appl Microbiol Biotechnol 102:5939–5951. https://doi.org/10.1007/s00253-018-9025-2

Post-Translational Modifications in *Corynebacterium glutamicum*

Saori Kosono

Contents

Abstract Newly synthesized proteins are subject to several post-translational modifications (PTMs). These PTMs can create diverse proteins from a single gene and are important for the function of certain proteins in a given situation. Recent advances in mass spectrometry (MS)-based proteomics have enabled the global detection of PTMs of bacterial proteins. In *Corynebacterium* and other actinobacteria, common PTMs (e.g., phosphorylation and acetylation) as well as uncommon PTMs (e.g., pupylation, mycoloylation, and mycothiolation) have been detected. In this chapter, the features of six representative PTMs found in *Corynebacterium glutamicum*, phosphorylation, acylation, pupylation, mycoloylation, mycothiolation, and glycosylation, and their roles in protein regulation are described.

S. Kosono (✉)
Biotechnology Research Center, The University of Tokyo, Tokyo, Japan
e-mail: uskos@mail.ecc.u-tokyo.ac.jp

M. Inui, K. Toyoda (eds.), *Corynebacterium glutamicum*, Microbiology Monographs 23, https://doi.org/10.1007/978-3-030-39267-3_6

1 Phosphorylation

Phosphorylation is a ubiquitous post-translational modification (PTM) of proteins that occurs in most organisms. Phosphorylation can directly control the activity of target proteins by functioning as a molecular switch, it can also control the localization of target proteins by inducing conformational changes in localization signals and regulate protein-protein interactions through phosphorylation of domains recognizing phosphorylated residues (e.g., Forkhead-associated (FHA) domain) (Humphrey et al. 2015; Almawi et al. 2017). Protein phosphorylation is reversible, and it functions in signal transduction. The major bacterial kinases include two-component systems, Hanks-type serine/threonine kinases, bacterial tyrosine (BY) kinases, and sugar phosphotransferase systems (PTSs) (Mijakovic and Macek 2011). In the *Corynebacterium glutamicum* genome, genes encoding putative low molecular weight phosphotyrosine protein phosphatases (LMW-PTPs, Pfam 01451) are present (Grangeasse et al. 2007), but tyrosine phosphorylation has not been experimentally demonstrated. An early global analysis of phosphoproteins in *C. glutamicum* using two-dimensional gel electrophoresis identified 41 phosphorylated proteins, most of which are enzymes in central metabolic pathways (Bendt et al. 2003). The biological impact of phosphorylation on metabolism seems to be interesting, but it remains to be elucidated.

1.1 Histidine and Aspartate Phosphorylation by Two-Component Signal Transduction Systems

Bacterial two-component signal transduction systems (TCS) usually consist of a membrane-bound sensory histidine kinase (HK) and a response regulator (RR) that typically functions as a transcriptional regulator. The HK responds to a certain stimulus by autophosphorylating a conserved histidine residue in its HisKA domain (Pfam 00512). This phosphoryl group is subsequently transferred to an aspartate residue in the receiver domain (Pfam 00072) of the cognate RR. In many RRs, phosphorylation elicits homodimer or homomultimer formation, which is thought to promote DNA binding and transcriptional activation (Gao and Stock 2010; Zschiedrich et al. 2016). The genome of the *C. glutamicum* type strain ATCC 13032 encodes 13 HKs and 13 RRs, which are present as 13 paired sets (Bott and Brocker 2012). A systematic deletion study revealed that all TCS genes, except SenX3 (cg0483/NCgl0391)-RegX3 (cg0484/NCgl0392), could be deleted, thus the SenX3-RegX3 pair is considered to be essential for growth (Kocan et al. 2006). To date, the roles of seven TCS have been elucidated, which are briefly described below.

The CitA-CitB System

The sensor kinase CitA (cg0089/NCgl0067) and its cognate RR CitB (cg0090/NCgl0068) are involved in the control of citrate utilization. *C. glutamicum* is able to grow aerobically on citrate as the sole carbon and energy source, whereas a mutant lacking the *citAB* genes was not (Brocker et al. 2009). Interestingly, in the presence of glucose and citrate, *C. glutamicum* consumes both substrates simultaneously, suggesting that the growth of the bacterium on citrate is not subject to carbon catabolite repression (Brocker et al. 2009). The CitA-CitB system positively regulates the genes encoding two citrate transporters, *citH* (encoding a citrate-Mg^{2+} (Ca^{2+})/H^+ symporter) and *tctCBA* (encoding a tripartite tricarboxylate transporter), in response to citrate (Brocker et al. 2009). The *citH* gene is located immediately upstream of *citAB* and is transcribed in the opposite direction. Recombinant CitB protein purified from *E. coli* was shown to bind to the promoter regions of *citH* and *tctABC*, although no putative CitB binding site, i.e., a highly conserved sequence motif, was identified (Brocker et al. 2009).

The MtrB-MtrA System

This system is composed of the sensor kinase MtrB (cg0864/NCgl0722) and its cognate RR MtrA (cg0862/NCgl0721) and is highly conserved in actinobacteria. Although MtrA was shown to be essential in *Mycobacterium tuberculosis*, *C. glutamicum* Δ*mtrA* and Δ*mtrAB* mutants are viable but show severe morphological phenotypes and sensitivity to cell wall-targeting compounds, such as penicillin, vancomycin, and ethambutol (Möker et al. 2004). In vitro phosphorylation of MtrA led to its dimerization and enhanced its DNA binding affinity, suggesting that MtrA is activated by phosphorylation (Brocker et al. 2011). MtrA acts as both an activator and a repressor, depending on the location of the binding sites. MtrA positively regulates *betP* and *proP*, two genes encoding uptake carriers of compatible solutes (glycine betaine, proline, and ectoine) and negatively regulates *mepA*, *mepB*, and *nlpC*, which encode putative cell wall peptidases (Brocker and Bott 2006; Brocker et al. 2011). Therefore, it was proposed that the MtrAB system is involved in cell wall homeostasis and osmoprotection (Möker et al. 2004, 2007).

The PhoS-PhoR System

This system is composed of the HK PhoS (cg2887/NCgl2517) and its cognate RR PhoR (cg2888/NCgl2518) and was identified in a screen for growth under phosphate (Pi) limitation. A Δ*phoRS* mutant showed a growth defect under Pi limitation, but not under Pi excess (Kocan et al. 2006). The *phoRS* genes are rapidly and transiently induced in response to Pi limitation (Ishige et al. 2003). The sensor kinase PhoS was shown to have constitutive autokinase activity and rapidly phosphorylate the RR PhoR (Schaaf and Bott 2007). Phosphorylated PhoR binds to its own promoter and positively autoregulates the *phoRS* genes; it also activates phosphate starvation-inducible genes, including *pstSCAB* (encoding a high affinity phosphate uptake system), *phoC* (encoding a cell wall-associated phosphatase), *ugpAEBC* (encoding an ABC transporter for glycerol-3-phosphate), and *glpQ* (encoding a glycerophosphoryl diester phosphodiesterase), and represses *pitA* (encoding a low-affinity phosphate uptake system). It is likely that basal levels of phosphorylated PhoR are sufficient to induce the high-affinity *pstSCAB* and *phoRS* promoters, whereas elevated levels of phosphorylated PhoR are required to induce low-affinity promoters such as *glpQ* (Schaaf and Bott 2007).

The CopS-CopR System

The CopS (cg3284/NCgl2862)-CopR (cg3285/NCgl2863) system is activated by elevated copper levels and plays a role in copper homeostasis. A Δ*copSR* mutant showed increased susceptibility to copper ions (Schelder et al. 2011). There is a single binding site for the RR CopR in the intergenic region between *copR* and *cg3286* (*NCgl2864*), and CopR activates the expression of the divergent gene clusters encoding a putative ATP-dependent copper exporter (CopB) and a putative multicopper oxidase (CopO) that may convert Cu^{+} to less toxic Cu^{2+}. In vitro phosphorylation of CopR was shown to increase its DNA binding affinity (Schelder et al. 2011).

The HrrS-HrrA and ChrS-ChrA Systems are Involved in Heme Homeostasis and Adaptation

Both the HrrS-HrrA and ChrS-ChrA systems are involved in the control of heme homeostasis, but they play different roles. HrrSA (cg3248-cg3247/NCgl2835-NCgl2834) plays a crucial role in the utilization of heme as an alternative iron source, whereas ChrSA (cg2201-cg2200/NCgl1935-NCgl1934) is required for tolerance to high (toxic) levels of heme. Mutants of *hrrSA* and *chrSA* show growth defects under different conditions with respect to heme; *hrrSA* mutants show growth defects when hemin is available as a sole iron source, and *chrSA* mutants show growth defects when hemin is present in addition to free iron (Hentschel et al. 2014). The RR HrrA activates the heme oxygenase gene *hmuO* as well as genes encoding the heme-containing components of the cytochrome bc_1-aa_3 supercomplex and represses genes involved in heme biosynthesis (Frunzke et al. 2011). In contrast, the RR ChrA directly activates the divergently transcribed *hrtBA* operon, which encodes a putative ABC transporter conferring resistance to heme toxicity (Heyer et al. 2012). The DNA binding affinities of both HrrA and ChrA were shown to be increased by phosphorylation (Frunzke et al. 2011; Heyer et al. 2012). HrrA was also shown to bind to the intergenic region between *hrtBA* and *chrSA* and negatively regulate both operons. Thus, these two TCS are functionally interrelated and manage heme homeostasis and adaptation, as was shown in *C. diphtheriae* (Bibb et al. 2007; Burgos and Schmitt 2016).

There is crosstalk between the HrrSA and ChrSA systems. ChrS and HrrS possess dual kinase and phosphatase activities. HrrS and ChrS can phosphorylate both HrrA and ChrA, but their phosphatase activities are specific for their cognate RR in vitro (Hentschel et al. 2014). Cross-phosphorylation of ChrA by the non-cognate HrrS kinase contributes to the transient activation of the *hrtBA*-encoded detoxification system in response to external heme, and the strong phosphatase activity of ChrS is important to quickly turn off *hrtBA* expression upon stimulus decline (Keppel et al. 2019).

The EsrISR Three-Component System

A pair of the HK EsrS and its cognate RR EsrR (envelope stress response, formerly CgtSR7, cg0707-0709/NCgl0585-0586) is distinguished from other TCS systems in that they constitute a three-component system with an additional integral membrane protein EsrI (cg0706/NCgl0584). The EsrISR three-component system is highly conserved in the phylum Actinobacteria and is involved in the cell envelope stress response (Kleine et al. 2017). The *esrI* gene is divergently transcribed in relation to the *esrSR* locus. Both the HK EsrS and EsrI possess an N-terminal phage shock protein C (PspC) domain (Pfam 04024), which has been implicated in sensing a signal that triggers the phage shock response (Flores-Kim and Darwin 2016). The RR EsrR activates several genes, including *esrSR*, *esrI*, *rsmP* (a conserved cytoskeletal protein with a PspA domain), *cg3322-3320* (*NCgl2893-2981*, encoding an ABC transporter involved in bacitracin resistance), *clpB* (an ATP-dependent protease), and *dnaK* (a heat shock protein) (Kleine et al. 2017). The expression of *esrSR*, *esrI*, and the direct target genes of EsrR was induced in response to cell wall-targeting antibiotics, such as bacitracin and vancomycin. A Δ*esrSR* mutant was highly susceptible to bacitracin, likely due to a lack of *cg3322-3320* induction (Kleine et al. 2017). Deletion of *esrI* upregulated the expression of EsrR target genes without an inducing stimulus (Kleine et al. 2017). This evidence suggests that EsrI acts as an inhibitor of the EsrSR TCS, similar to LiaF in the LiaFSR system from *Bacillus subtilis* (Jordan et al. 2006). It was proposed that the interaction of EsrI with EsrS maintains the EsrSR system in the "off" state under non-stressed conditions and that their dissociation, induced by a signal sensed by the PspC domain-containing EsrI and/or EsrS, converts the system to the "on" state, although direct interaction between EsrS and EsrI remains to be shown (Kleine et al. 2017).

1.2 Protein Phosphorylation of Serine and Threonine Residues

Since the first characterization of Pkn1 as a serine/threonine protein kinase in *Myxococcus xanthus* (Muñoz-Dorado et al. 1991), Hanks-type serine/threonine kinases have been shown to control many physiological processes across all domains of life (Pereira et al. 2011; Stancik et al. 2018). Hanks-type kinase-encoding genes are enriched in the genomes of actinobacteria (Perez et al. 2008). *C. glutamicum* possesses four Hanks-type kinases (PknA, PknB, PknG, and PknL) and a single phosphoserine/phosphothreonine phosphatase (Ppp) (Fig. 1). PknA (cg0059/NCgl0041), PknB (cg0057/NCgl0040), PknL (cg2388/NCgl2095), and Ppp (cg0062/NCgl0044) contain a single transmembrane helix, indicating that they are integral membrane proteins, whereas PknG (cg3046/NCgl2655) does not and is predicted to be a soluble kinase. PknB and PknL possess several PASTA domains (*p*enicillin binding protein *a*nd *s*erine/*t*hreonine kinase *a*ssociated domain, Pfam 03793) in their C-terminal extracytoplasmic regions that are involved in binding to peptidoglycan components (Yeats et al. 2002), suggesting their possible role in cell wall biosynthesis. The kinase PknG has an extended C-terminal domain, consisting of tetratricopeptide repeats (Pfam 16918), that is involved in protein-protein interaction (Blatch and Lässle 1999). The kinase domains of PknA, PknB, and PknL were shown to possess autokinase (autophosphorylation-dependent protein kinase) activity (Fiuza et al. 2008b; Schultz et al. 2009). There are conflicting reports on PknG autophosphorylation: Fiuza et al. (2008b) reported that both full-length and the kinase

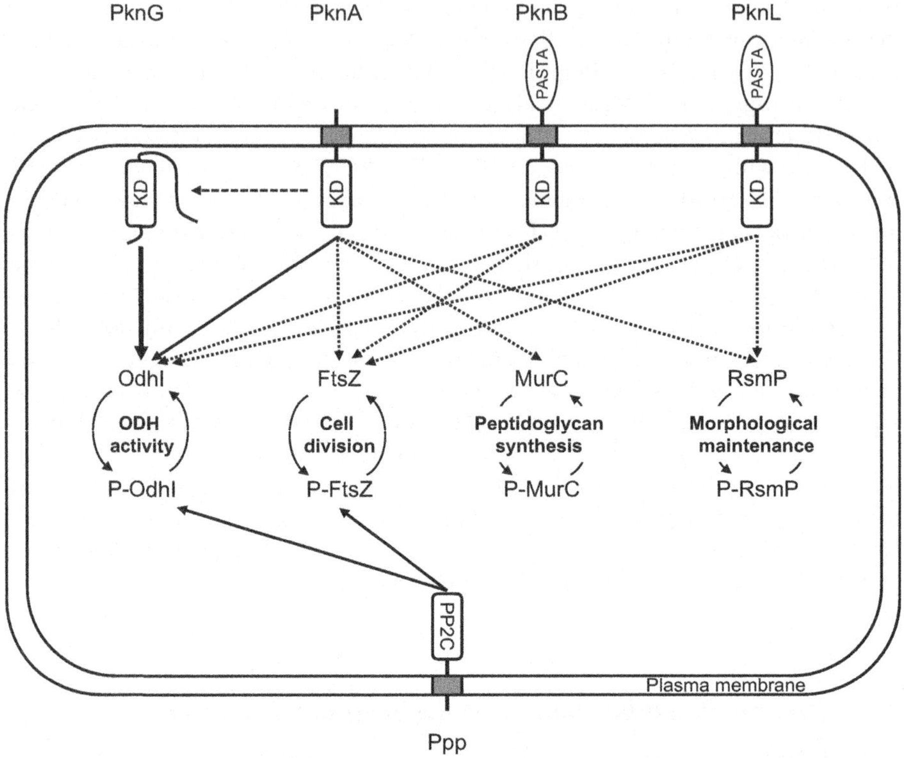

Fig. 1 Schematic of serine/threonine phosphorylation-dependent regulation in *Corynebacterium glutamicum*. The Hanks-type serine/threonine kinases PknA, PknB, PknL, and PknG and the phosphoserine/phosphothreonine phosphatase Ppp are encoded by genes in the type strain ATCC 13032 genome. *KD* kinase domain, *PP2C* phosphatase domain, *PASTA* penicillin binding protein and serine/threonine kinase associated domain. PknG does not have a transmembrane helix (indicated by a gray box) and is predicted to be a soluble kinase. OdhI, FtsZ, MurC, and RsmP are targeted by Ser/Thr kinases and the Ppp phosphatase as indicated by the solid lines (shown in vivo and in vitro) and dotted lines (shown in vitro). Hyphenated capital P (P-) indicates a phosphorylated protein

domain of recombinant PknG did not exhibit autophosphorylation activity and suggested that PknG requires PknA-dependent phosphorylation for its activation. In contrast, Schultz et al. (2009) reported that PknG purified from a *pknA*-deleted *C. glutamicum* exhibited autophosphorylation activity and was capable of phosphorylating OdhI, strongly suggesting that PknG is an authentic autokinase.

1.2.1 OdhI, a Regulator of 2-Oxoglutarate Dehydrogenase Complex, and Its Impact on L-Glutamate Overproduction

Both Δ*pknG* and Δ*ppp* mutants showed severe growth defects when glutamine was present as a sole carbon and nitrogen source, suggesting that they are involved in glutamine utilization and nitrogen metabolism (Niebisch et al. 2006; Schultz et al. 2009). Mutants of the other three kinase genes (*pknA*, *pknB*, and *pknL*) do not provide the growth phenotype on glutamine. PknG and Ppp respectively function as a kinase and a phosphatase for the 15-kDa acidic protein OdhI (cg1630/NCgl1385) (Fig. 1). OdhI has a forkhead-associated domain (FHA, Pfam 00498) that recognizes phosphorylated threonine. OdhI is primarily phosphorylated by PknG but also by PknA, PknB, and PknL in vitro. The N-terminal Thr14 and Thr15 residues of OdhI were shown to be phosphorylated in vivo (Niebisch et al. 2006; Schultz et al. 2009), and a phospho-ablative variant of OdhI (T14A) showed a phenotype similar to that of the Δ*pknG* mutant, suggesting that Thr14 is a critical phosphorylation site. OdhA (cg1280/NCgl1084), which is the E1o subunit of the 2-oxoglutarate dehydrogenase (ODH) complex, was identified as an OdhI-interacting protein by affinity purification. Unphosphorylated OdhI binds to OdhA to inhibit ODH activity. Upon phosphorylation, OdhI is folded via interaction of phosphorylated Thr14 and/or Thr15 with the FHA domain and it no longer inhibits ODH activity. Thus, the phosphorylation of OdhI functions as a molecular switch regulating ODH activity (Niebisch et al. 2006).

The carbon flux distribution at the 2-oxoglutarate node is important for glutamate overproduction, and the attenuation of ODH activity during glutamate overproduction is well documented (Kawahara et al. 1997; Shimizu et al. 2003; Shirai et al. 2005). Phosphorylation-dependent regulation of OdhI affects glutamate overproduction in *C. glutamicum*. An *odhI* deletion severely decreased glutamate production (Schultz et al. 2007) and overexpression of *odhI* caused glutamate production without any triggers (Kim et al. 2010; Kim et al. 2011). A Δ*pknG* mutant showed increased glutamate production under strong biotin limitation (1 μg/L) and in the presence of ethambutol, but not in the presence of Tween 40, and decreased glutamate production under weak biotin limitation (2.5 μg/L) (Schultz et al. 2007). The positive effect of the *pknG* deletion on glutamate production is likely due to an increased level of unphosphorylated OdhI that inhibits ODH activity, but it may be not true of all conditions. It is speculated that the other kinases (PknA, PknB, and PknL) are responsible for OdhI phosphorylation and/or Ppp phosphatase activity against OdhI varies depending on the induction conditions.

1.2.2 Serine and Threonine Phosphorylation Targets Proteins Involved in Peptidoglycan Biosynthesis, Cell Division, and Cell Morphology

Systematic deletions of the PknA, PknB, PknG, and PknL kinases and the Ppp phosphatase were conducted, and many of the deletion mutants showed defects in morphology and growth. It was reported that neither *pknA* nor *pknB* could be deleted in the *C. glutamicum* ATCC 13869 background, suggesting that both genes are essential for growth (Fiuza et al. 2008b). Whereas in another study using the *C. glutamicum* ATCC 13032 strain, *pknA* or *pknB* single mutants were viable, but a double *pknA*-*pknB* mutant could not be obtained, suggesting that either *pknA* or *pknB* is required for growth (Schultz et al. 2009). Single mutants of *pknG* and *pknL* had growth curves similar to that of the wild-type strain but showed an elongated cell morphology. A Δ*ppp* mutant showed both a growth defect and abnormal cell morphology when grown in complex medium (BHI with 4% glucose) (Schultz et al. 2009). The genome organization also suggests the possible involvement of serine/threonine phosphorylation in cell division and cell morphology: *ppp*, *pknA*, and *pknB* are clustered with genes encoding two small proteins with FHA domains, cg0064/NCgl0046 and cg0063/NCgl0045, as well as genes encoding a putative FtsW/RodA/SpoE-family cell cycle protein [cg0061/NCgl0043], PbpA (penicillin-binding protein A) [cg0060/NCgl0042], and a putative septation inhibitor protein [cg0055/NCgl0039]).

Three other proteins, MurC, FtsZ, and RsmP, were also shown to be targeted by serine/threonine phosphorylation (Fig. 1). MurC (cg2368/NCgl2077) is involved in cell wall peptidoglycan synthesis and catalyzes the addition of the first L-alanine residue to the nucleotide precursor UDP-MurNAc. MurC was shown to be phosphorylated in vivo in a phosphoproteome study of *C. glutamicum* (Bendt et al. 2003). MurC was shown to be phosphorylated by PknA in vitro, and phosphorylation of MurC at Thr120, Thr133, and Thr362 is critical for inhibiting enzymatic activity. Thr133 is located close to the 129-GKT-131 motif that is involved in ATP binding, and its phosphorylation likely affects ATP-dependent MurC activity (Fiuza et al. 2008a).

FtsZ is involved in septum formation and plays an essential role in bacterial cell division. FtsZ (cg2366/NCgl2075) was shown to be differentially phosphorylated at several serine and threonine residues by PknA, PknB, and PknL in vitro and was shown to be a substrate of Ppp in vivo (Schultz et al. 2009). Although the exact role of FtsZ phosphorylation in *C. glutamicum* remains to be elucidated, it might affect GTPase activity, resulting in decreased polymerization activity as described in *M. tuberculosis* (Thakur and Chakraborti 2006).

RsmP (rod-shaped morphology protein, cg3264/NCgl2848) is a cytoskeletal protein required for maintenance of the rod-shaped morphology of *C. glutamicum*. RsmP forms filamentous structures in vitro. RsmP is preferentially phosphorylated by the PknA kinase and to a lesser extent by PknL. A phosphomimetic variant of RsmP (with substitutions of the Ser6, Thr168, and Thr211 phosphorylation sites by aspartate) lost the filamentous structure and accumulated at the cell pole. In contrast, a phospho-ablative RsmP (with substitutions of the phosphorylation sites by alanine) retained the filamentous structure, similar to the wild-type RsmP, suggesting that phosphorylation of RsmP negatively regulates filament formation of the protein and is required for directing cell growth at the cell poles (Fiuza et al. 2010).

1.3 The PTS System

The phosphoenolpyruvate (PEP)-dependent sugar phosphotransferase system (PTS) catalyzes the transport of sugar substrates across the cell membrane, concomitant with substrate phosphorylation (Deutscher et al. 2006, 2014). The PTS comprises a phosphorylation cascade on histidine residues and in some cases, on cysteine residues. The source of phosphate in this system is PEP. In general, enzyme I (EI) receives the phosphoryl group from PEP and transfers it to a histidine residue in the phosphoryl carrier protein HPr. Another component, the membrane-bound sugar-specific permease (enzyme II or EII) consists of two hydrophilic cytoplasmic domains (EIIA and EIIB) and a hydrophobic integral membrane domain (EIIC); the enzyme transfers the phosphoryl group from Hpr to the EIIA domain and then to the EIIB domain and finally to the sugar substrate after EIIC-dependent translocation of the substrate across the membrane. EII enzymes exist either as a single multidomain protein or as separate proteins. The type strain ATCC 13032 has genes encoding the general EI phosphotransferase (PtsI, cg2117/NCgl1858), Hpr (cg2121/NCgl1862), and four enzyme II (EII) permeases that are specific for glucose (cg1537/NCgl1305), fructose (cg2120/NCgl1861), sucrose (cg2925/NCgl2553), and an unknown substrate (putatively ribitol) (cg3366-3365/NCgl2934-2933). The putative phosphorylation sites in the above PTS proteins are well conserved. The PTS system of *C. glutamicum* has been described in detail elsewhere (Moon et al. 2006).

2 Acetylation and Succinylation

Protein acetylation is another ubiquitous PTM. This modification mostly targets the ε-amino group of lysine residues (*N*ε-acetylation) but also modifies the hydroxyl group of serine and threonine (*O*-acetylation) in some cases (Mukherjee et al. 2006; Birhanu et al. 2017). Since the first report of the bacterial acetyl proteome (acetylome) in *E. coli* (Yu et al. 2008; Zhang et al. 2009), many acetylated proteins have been discovered in bacterial genera, including *Corynebacterium* (Mizuno et al. 2016), *Mycobacterium* (Liu et al. 2014; Xie et al. 2015; Singhal et al. 2015; Guo et al. 2016; Birhanu et al. 2017), and *Streptomyces* (Liao et al. 2014; Ishigaki et al. 2017). Furthermore, *N*ε-lysine acyl-modifications, including propionylation (Chen et al. 2007), butyrylation (Chen et al. 2007), crotonylation (Tan et al. 2011), malonylation (Du et al. 2011; Peng et al. 2011), succinylation (Zhang et al. 2010), glutarylation (Tan et al. 2014), 2-hydroxyisobutyrylation (Dai et al. 2014), and β-hydroxybutyrylation (Xie et al. 2016) have been newly discovered. Lysine succinylation utilizing succinyl-CoA as an acylating agent is a secondary major acyl-modification in eukaryotes and bacteria (Weinert et al. 2013b). In this section, protein acetylation and acylations refer to *N*ε-lysine acetylation and acyl-modifications, respectively.

There are two known mechanisms of protein acetylation (Hentchel and Escalante-Semerena 2015; Carabetta and Cristea 2017; VanDrisse and Escalante-Semerena 2019) (Fig. 2a). One is enzymatic acetylation catalyzed by lysine acetyltransferases (KATs) utilizing acetyl-CoA as an acetyl-group donor. Although several families of KATs, including GNAT, MYST, and CBP/p300, have been identified in eukaryotes (Sadoul et al. 2011), only KATs belonging to the family of Gcn5-related *N*-acetyltransferases (GNAT, Pfam 00583) are found in bacteria. The genome of *C. glutamicum* ATCC 13032 encodes 18 proteins with the GNAT motif; however, no

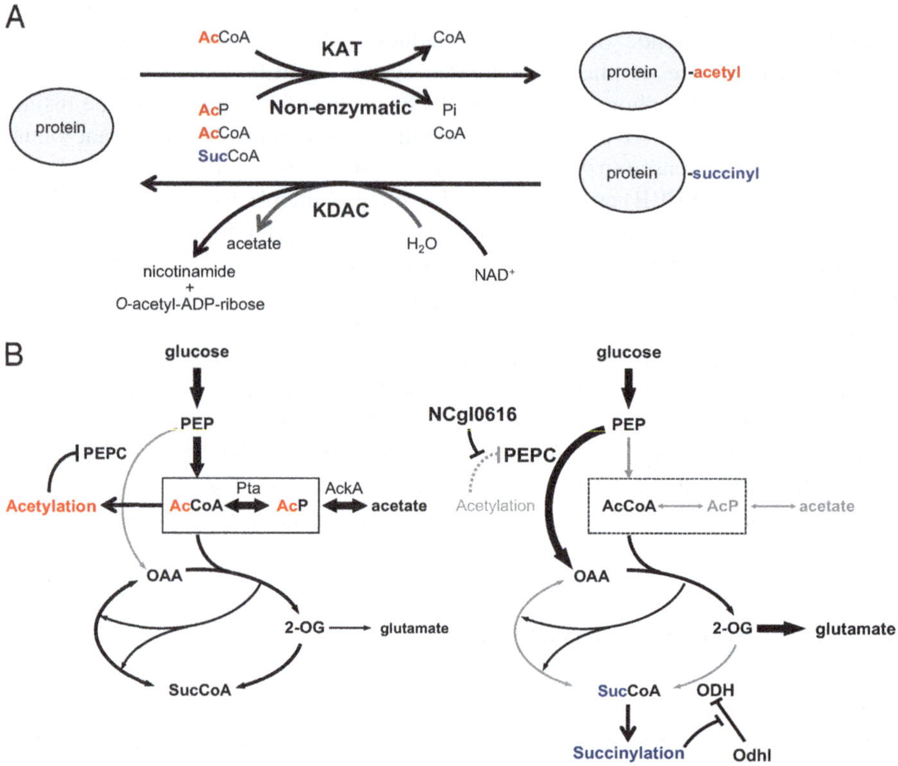

Fig. 2 (**a**) Schematic of protein lysine acetylation and succinylation in *C. glutamicum.* Acetylation occurs via two mechanisms, lysine acetyltransferase (KAT)-dependent and non-enzymatic mechanisms. Acetyl-CoA (AcCoA) and acetyl phosphate (AcP) are utilized as acetylating agents. Succinylation utilizing succinyl-CoA (SucCoA) likely occurs non-enzymatically. The acetylation and succinylation of some proteins are reversed by lysine deacetylase (KDAC). Sirtuin-type KDACs utilize NAD^+ as a co-substrate and generate nicotinamide and *O*-acetyl-ADP-ribose as by-products. Hydrolase-type KDACs produce acetate as a by-product. (**b**) Changes in protein acetylation and succinylation in response to glutamate overproduction. Under non-glutamate-producing conditions (left), carbon flux from glucose toward acetate production via the Pta-AckA pathway is high, and acetylating agents (indicated by a box) are supplied to induce protein acetylation. Acetylation of PEPC at Lys653 inhibits its enzymatic activity. Under glutamate-producing conditions (right), PEPC flux is increased to supply oxaloacetate and flux of the Pta-AckA pathway is reduced, resulting in depletion of acetylating agents and reduced protein acetylation. The NCgl0616 deacetylase activates PEPC to maintain flux despite its decreased expression. Depletion of acetylating agents may enhance protein succinylation utilizing succinyl-CoA, which is highly reactive. Succinylation of OdhI at Lys132 inhibits its binding to the E1o (OdhA) subunit of ODH. *PEP* phosphoenolpyruvate, *OAA* oxaloacetate, *2-OG* 2-oxoglutarate, *PEPC* phosphoenolpyruvate carboxylase, *Pta* phosphotransacetylase, *AckA* acetate kinase, *ODH* 2-oxoglutarate dehydrogenase

KAT has been reported. The other mechanism is non-enzymatic acetylation utilizing acetyl-CoA or acetyl phosphate (acetyl-P) as the acetylating agent. Acetyl-P dependent non-enzymatic acetylation is common in many bacteria. Some acetylated lysine residues can be deacetylated by lysine deacetylases (KDACs). There are two families of KDACs, one is classified as sirtuin (Pfam 02146), which requires NAD^+ as a co-substrate, the other is a Zn^{2+}-dependent hydrolase (Pfam 00850). *C. glutamicum* ATCC 13032 possesses two sirtuin-type KDAC homologues, cg0108/NCgl0078 and cg0745/NCgl0616.

Some KATs and KDACs target other lysine acylations in addition to acetylation (Sabari et al. 2016). The sirtuin-type deacetylase CobB in *E. coli* was reported to have both deacetylase and desuccinylase activities (Colak et al. 2013). Protein succinylation is considered to occur non-enzymatically in the mitochondria and in many bacteria (Wagner and Payne 2013; Wagner et al. 2017).

2.1 Altered Protein Acetylation and Succinylation Under L-Glutamate Overproduction

Mizuno et al. (2016) found that protein acetylation and succinylation were altered in response to conditions inducing L-glutamate overproduction in *C. glutamicum*: protein acetylation was increased with cultivation time under non-producing conditions; this increase was reduced by the addition of Tween 40, which induced glutamate overproduction and increased succinylation. Alteration of acetylation and succinylation was also observed in response to other triggers, such as biotin limitation and penicillin addition (Mizuno et al. 2016).

MS-based shotgun proteomic analysis of *C. glutamicum* ATCC 13032 identified 604 acetylated and 288 succinylated proteins when grown under Tween 40-dependent glutamate producing and non-producing conditions, which were categorized as involved in metabolism, translation, and other biological functions (Mizuno et al. 2016). Many enzymes in central carbon metabolic pathways were acetylated and succinylated, and their acetylation and succinylation status were different under glutamate-producing and non-producing conditions, with increased succinylation under glutamate-producing conditions and frequent acetylation under non-producing conditions. It has been reported that metabolic fluxes from glucose toward glutamate synthesis are increased under glutamate overproduction (Shirai et al. 2007), even though the expression levels of the corresponding enzymes are not increased (Kataoka et al. 2006). It is interesting to consider the possibility that post-translational regulation of metabolic enzymes through protein acetylation and succinylation have a role in the metabolic flux change.

2.2 Regulation of PEP Carboxylase by Reversible Acetylation

Anaplerotic enzymes that supply oxaloacetate are important for efficient production of various amino acids, including L-glutamate, and other compounds. *C. glutamicum* possesses two anaplerotic enzymes, PEP carboxylase (PEPC, cg1787/NCgl1523) and pyruvate carboxylase (PC, cg0791/NCgl0659). In the ATCC 13032 (Kyowa Hakko) strain, deletion of PEPC severely impaired glutamate production, whereas deletion of PC did not, indicating that PEPC provides the major anaplerotic function under Tween 40-dependent glutamate-producing conditions in the strain (Nagano-Shoji et al. 2017). Meanwhile, Peters-Wendisch et al. (2001) reported that PC has a major role in Tween 60-induced glutamate production of the ATCC 13032 strain. The reason for this discrepancy is not known, but it is speculated that different anaplerotic enzymes are used in different conditions and strains. PEPC was shown to be subjected to acetylation and succinylation (Mizuno et al. 2016). Substitution of Lys653 by acetylation mimic glutamine (K653Q) severely impaired glutamate production and in vitro PEPC activity compared to the levels of the wild type enzyme, whereas the acetylation-ablative (K653R) PEPC retained both. Comparison of the phenotypes of the K653Q and K653R variants of PEPC suggested that acetylation of Lys653 negatively affects activity. An enzyme assay using an acetyllysine-incorporated PEPC protein at position 653 confirmed that acetylation of Lys653 is necessary and sufficient to inhibit its activity (Nagano-Shoji et al. 2017). The sirtuin-type KDAC homologue NCgl0616 could deacetylate Lys653-acetylated PEPC in vitro (Nagano-Shoji et al. 2017). These results indicate that PEPC is regulated by reversible acetylation of Lys653.

PEPC protein levels in lysates were reproducibly lower under glutamate-producing conditions than under non-producing conditions, but PEPC activity was retained, suggesting that PEPC is probably post-translationally regulated. The apparent specific activity of PEPC in the lysates was estimated to be 1.7-times higher under glutamate-producing conditions than under non-producing conditions, which was blocked by the acetylation-ablative K653R substitution or double deletion of *NCgl0078 and NCgl0616* that encode KDACs. These observations suggest the biological significance of PEPC regulation by reversible acetylation in glutamate production: NCgl0616-dependent deacetylation likely contributes to the activation of PEPC and maintains PEPC flux despite of decreasing protein levels under glutamate-producing conditions (Nagano-Shoji et al. 2017).

2.3 Regulation of OdhI by Succinylation

Proteomic analysis identified Lys52 and Lys132 in the FHA domain of OdhI as acetylation and succinylation sites (Mizuno et al. 2016). In a structural model of OdhA in complex with the unphosphorylated form of OdhI (Raasch et al. 2014), Lys132 of OdhI is predicted to electrostatically interact with Asp523 of OdhA, and succinylation of Lys132 was predicted to cause electric repulsion and inhibit the interaction between OdhA and OdhI. As expected, an OdhI variant protein with a succinylation mimic at Lys132 (K132E) showed highly impaired OdhA-binding ability and ODH activity-inhibiting ability in vitro when compared to the function of a succinylation-ablative K132R variant and the wild-type proteins. In vitro succinylation of OdhI also impaired its ODH inhibitory ability. However, this effect of in vitro succinylation was not observed with the OdhI K132R variant. These results suggest that succinylation of Lys132 inhibits OdhI-dependent ODH inhibition (Fig. 2b). A *C. glutamicum* strain producing a K132E OdhI variant protein showed impaired glutamate production likely due to retained ODH activity (Komine-Abe et al. 2017). Similarly, a Lys141 to glutamate substitution (K141E) in *M. tuberculosis* GarA, which corresponds to the K132E substitution in OdhI, abolished binding to α-ketoglutarate dehydrogenase, glutamate dehydrogenase, and glutamate synthase as well as its inhibitory effects on the target enzymes (Nott et al. 2009).

As described above in Sect. 1.2.1, phosphorylation status is a critical determinant of OdhI function. PknG-mediated phosphorylation of OdhI led to a loss of ODH inhibition activity, which was not affected by succinylation, suggesting that phosphorylation is a major modification controlling OdhI, while succinylation may have a minor tuning role (Komine-Abe et al. 2017).

2.4 Molecular Rationale for Altered Protein Acetylation and Succinylation Upon Glutamate Production

The molecular basis for the altered acetylation and succinylation upon glutamate production has been addressed. Deletion of the *pta* gene (cg3048/NCgl2657) that encodes a phosphotransacetylase, which blocks the conversion of acetyl-CoA to acetyl-P, was shown to decrease total acetylation levels under non-producing conditions, suggesting that acetyl-P dependent non-enzymatic acetylation is functional in *C. glutamicum* (Mizuno et al. 2016). Meanwhile, deletion of *ackA* (cg3047/Cgl2656) that encodes an acetate kinase, which blocks the conversion of acetyl-P to acetate, did not increase the acetylation levels under either glutamate-producing or non-producing conditions, which is different from what was observed in *E. coli* (Weinert et al. 2013a; Kuhn et al. 2014) and *B. subtilis* (Kosono et al. 2015). Double deletion of the two KDAC homologues (NCgl0078 and NCgl0616) in *C. glutamicum* also did not increase acetylation levels (Mizuno et al. 2016). Interestingly, deletion of PEPC, but not PC, enhanced acetylation levels under glutamate-producing conditions (Nagano-Shoji et al. 2017). These observations suggest that carbon flux in the Pta-Ack pathway is reduced and PEPC flux is abundant under glutamate-producing conditions, which were reversed by PEPC deletion. This provides an answer to why total acetylation levels are reduced under glutamate-producing conditions; it is likely due to depletion of the acetylating agents (acetyl-CoA and acetyl-P) supplied from the Pta-Ack pathway (Fig. 2b). This result also supports that PEPC, rather than PC, is responsible for a large portion of carbon flux under glutamate-producing conditions in the Kyowa Hakko strain.

As for the supply of succinyl-CoA as a succinylating agent, two possible routes, one from 2-oxoglutarate via ODH and another from isocitrate via succinate through isocitrate lyase (cg2560/NCgl2248) and succinyl-CoA synthetase have been proposed. The former route is not likely since decreased ODH activity under glutamate-producing conditions has been well documented. Deletion of NCgl2248 had no apparent effect on total succinylation levels under glutamate-producing conditions (Mizuno et al. 2016). It is possible that succinyl-CoA, which is a highly reactive acyl-CoA species (Wagner et al. 2017), more easily reacts with target lysine sites under conditions where acetylating agents are depleted. It seems that the altered protein acetylation and succinylation observed in glutamate-producing conditions may reflect the metabolic state in the cell.

3 Pupylation

Protein pupylation, which is a modification of lysine residues with a prokaryotic ubiquitin-like protein (Pup), was first discovered in *Mycobacterium* (Pearce et al. 2008; Burns et al. 2009) and is conserved among genera in the phylum Actinobacteria, including *Corynebacterium*, *Nocardia*, *Rhodococcus*, and *Streptomyces* (Darwin 2009; Striebel et al. 2014; Delley et al. 2017). Pup is attached to a target protein via an isopeptide bond between the C-terminal glutamate of Pup and a lysine residue in the target protein, which is catalyzed by the Pup ligase PafA (proteasome accessory factor A). Pupylation is reversed by a single depupylation enzyme, Dop (deamidase of Pup) (Fig. 3). PafA and Dop are homologous proteins and catalyze the ATP-dependent ligation of the terminal γ-carboxylate of glutamate to the ε-amine of lysine, similar to γ-glutamyl-cysteine synthetase and glutamine synthetase (EC 6.3.1.19). Several Pup variants possess the C-terminal glutamine residue, and Dop also catalyzes the deamination of the C-terminal glutamine (Q) to glutamate (E) to activate Pup (Fig. 3). In mycobacteria and other actinobacteria, pupylation functions as a recruitment tag for proteasomal degradation. Pupylated substrates are unfolded by a ring-shaped complex consisting of an AAA (ATPases associated with diverse cellular activities) family ATPase (referred to as Mpa in mycobacteria and ARC in other actinobacteria) and are then translocated to the 20S proteasome chamber, which is composed of PrcA and PrcB subunits. This Pup-proteasome system is functionally related to the eukaryotic ubiquitin-proteasome system, although different players are involved in the two systems and their origins appear to be independent.

In the *C. glutamicum* genome, genes encoding PafA (cg1688/NCgl1437), Pup (cg1689/NCgl1438), Dop (cg1690/NCgl1439), and ARC (cg1691/NCgl1440) are located in a cluster; however, genes encoding the proteasome components PrcA and PrcB are not present (Fig. 3a). Proteome-wide analysis of pupylated proteins (the pupylome) in the type strain ATCC 13032 identified 66 pupylation sites on 55 proteins, the majority of which are involved in metabolism and translation (Küberl et al. 2014). To investigate the physiological function of protein pupylation, the phenotype of a Δ*pup* mutant was screened under various carbon and stress conditions and was found to have a severe growth defect under iron limitation (Küberl et al. 2016). Mutants of Δ*arc* and to a lesser extent Δ*dop* also showed growth defects under iron limitation. A pupylation-disabled variant (Pup-E64A) did not rescue the growth defect of the Δ*pup* mutant. The iron-storage protein ferritin (Ftn, cg2782/NCgl2439) was included in the *C. glutamicum* pupylome. Lysine-78, which is

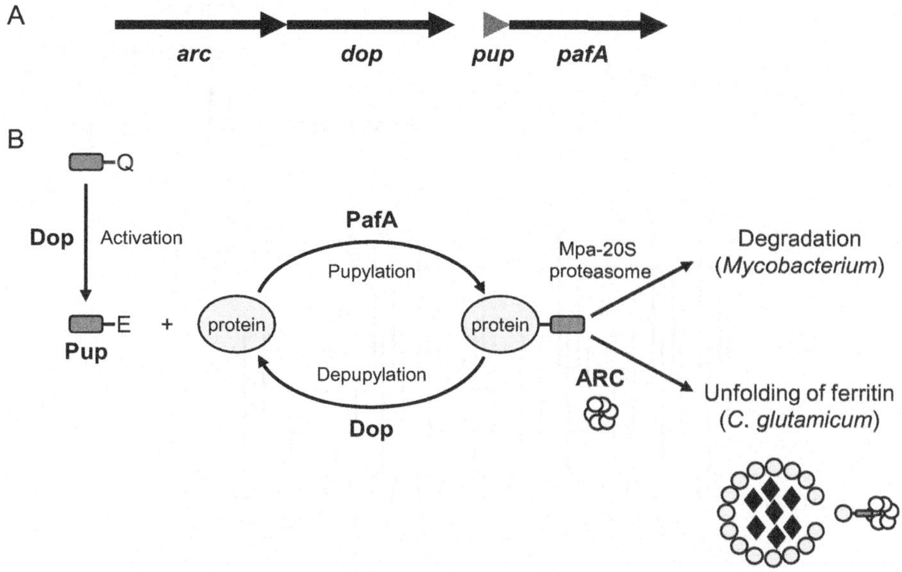

Fig. 3 Protein pupylation in *C. glutamicum* and other actinobacteria. (**a**) Gene organization of the pupylation system in *C. glutamicum*. (**b**) PafA catalyzes ATP-dependent ligation of the C-terminal glutamate (E) form of Pup to lysine residues of target proteins, which is reversed by Dop. Dop also catalyzes deamidation of the C-terminal glutamine (Q) of Pup for its activation. In *Mycobacteria*, pupylated proteins are targeted to the Mpa-20S proteasome for degradation. In *C. glutamicum*, which lacks a proteasome, it was proposed that pupylation triggers the disassembly of multimeric ferritin by ARC to release the stored iron, which has a crucial role in iron homeostasis

located on the outer surface of Ftn, was identified as the only pupylation site in vivo and was shown to be pupylated in vitro. A mutant strain expressing a pupylation-ablative Ftn-K78A variant showed the same phenotype as the Δ*pup* mutant: a growth defect under iron limitation and normal growth under iron replete conditions. These observations indicate that protein pupylation of Ftn plays a critical role in iron homeostasis in *C. glutamicum* (Fig. 3b). Pupylation may trigger the disassembly of multimeric ferritin by ARC to release stored iron without degradation (Küberl et al. 2016). This work gives an important suggestion for the physiological function of pupylation in bacteria that lack a proteasome.

4 Mycoloylation

Protein lipidation is a PTM found in all domains of life (Jiang et al. 2017). Of several kinds of protein lipidation, *O*-fatty acylation is a very rare modification compared to *S*- and *N*-fatty acylation and is found in the Corynebacteriales. Protein *O*-mycoloylation was first discovered in *C. glutamicum* by Huc et al. (2010). Genera of the order Corynebacteriales, including *Corynebacterium*, *Mycobacterium*, *Nocardia*, and *Rhodococcus*, contain α-alkyl and β-hydroxylated long-chain fatty

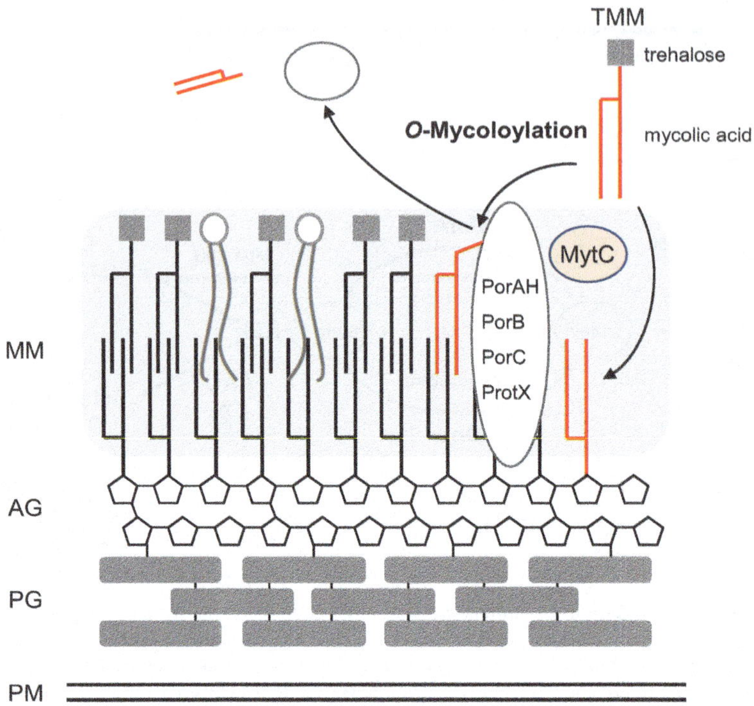

Fig. 4 Mycoloylation in *C. glutamicum*. The mycolyltransferase MytC/Cmt1 (Cg0413/NCgl0336) catalyzes the transfer of a mycolate moiety (shown in red) from trehalose monomycolate (TMM) to the porins PorA, PorH, PorB, and PorC (*O*-mycoloylation), which functions as an anchor in the mycomembrane. Extracellularly released forms of these porins are free of mycoloylation. *MM* mycomembrane, *AG* arabinogalactan, *PG* peptidoglycan, *PM* plasma membrane

acids, called mycolic acids, in their cell envelope. These mycolic acids organize in a bilayer structure to form an outer membrane, called the mycomembrane. The mycolic acids in the inner leaflet of the mycomembrane are covalently linked to arabinogalactan, which in turn is covalently attached to peptidoglycan. Mycolic acids can also be attached to the serine residues of proteins through an *O*-ester linkage to provide protein *O*-mycoloylation. Mycoloyltransferases catalyze the transfer of a mycolic acid moiety from a common precursor, trehalose monomycolate (TMM), to a hydroxyl group of the substrate polypeptide and arabinogalactan to generate *O*-mycoloylated proteins and arabinogalactan mycolate, respectively (Lanéelle et al. 2013) (Fig. 4).

In *C. glutamicum*, protein *O*-mycoloylation has been detected on the small porins PorA and PorH (cg3008 and cg3009, respectively; the coding sequences are upstream of NCgl2620, but NCgl numbers have not been assigned), which form a cation-selective channel; PorB (cg1109/NCgl0933) and PorC (cg1108/NCgl0932), which are anion-selective channels; and ProtX (cg2875/NCgl number not assigned) (Huc et al. 2010; Issa et al. 2017; Carel et al. 2017), a protein of unknown function.

These proteins are localized to the mycomembrane and secreted. *O*-Mycoloylation of PorA is required for pore-forming activity, whereas mycoloylation of PorH is dispensable (Huc et al. 2010; Rath et al. 2011). MytC (also called Cmt1, cg0413/NCgl0336), which belongs to the α/β hydrolase superfamily, was identified as a mycoloyltransferase for PorA, PorH, PorB, PorC, and ProtX (Huc et al. 2013; Carel et al. 2017). MytC is also located in the mycomembrane, and MytC-dependent *O*-mycoloylation is required for retention of PorH, PorB, and PorC in the mycomembrane, but not PorA, which contains predominantly hydrophobic residues (Issa et al. 2017; Carel et al. 2017). *O*-Mycoloylation plays an essential role in targeting proteins to the mycomembrane and functions as a stable membrane anchor. PorB, PorC, PorH, and a trace of PorA are detected in the extracellular medium, and the secreted version of the proteins are no longer acylated. This observation implies the reversibility of *O*-mycoloylation, although the enzymes that reverse *O*-mycoloylation remain to be determined (Carel et al. 2017).

Two short conserved motifs, [F/L]SS and VLSG, have been proposed as *O*-mycoloylation motifs. These motifs are found in cell envelope-associated proteins, such as mycoloyltransferases and the peptidoglycan hydrolases in *C. glutamicum*, suggesting their association with the mycomembrane (Carel et al. 2017).

5 Mycothiolation

Protein *S*-thiolation is a widespread redox-modification. This modification occurs on the cysteine residues of proteins and occurs via the addition of non-protein low-molecular weight (LMW) thiols such as glutathione, bacillithiol, mycothiol, and coenzyme A. Under oxidative or hypochlorite (HOCl) stress, redox-sensitive protein thiols are susceptible to thiol oxidation, including *S*-thiolation with LMW thiols, the formation of inter- or intramolecular protein disulfide bonds, and irreversible oxidation to cysteine sulfinic or sulfonic acids. Reversible *S*-thiolation with LMW thiols can protect protein thiols from undesirable thiol oxidation and control the activity of proteins in response to redox states (Loi et al. 2015; Imber et al. 2019).

Although all eukaryotes and most Gram-negative bacteria, including *E. coli*, utilize glutathione (GSH) as a major LMW thiol to cope with oxidative stress and redox regulation of proteins, Gram-positive actinobacteria, including corynebacteria, mycobacteria, and streptomycetes, utilize mycothiol (MSH) as their major LMW thiol (Newton et al. 2008). MSH is composed of acetyl cysteine, glucosamine, and *myo*-inositol and has a molecular mass of 486 Da (Fig. 5). Under oxidative or HOCl stress conditions, MSH forms disulfides with the protein thiols of cysteine residues, which is called *S*-mycothiolation (Fig. 5). Protein *S*-mycothiolation is reversed by the thiol-disulfide-reducing Mrx1/MSH/Mtr system to regenerate protein function. Mycoredoxin-1 (Mrx1, cg0964/NCgl0808) catalyzes the reduction of *S*-mycothiolated proteins coupled with the generation of Mrx1-SSM, which is recycled by MSH and the NADPH-dependent MSSM reductase Mtr (cg2194/NCgl1928). MSH is also used as a cofactor for many redox enzymes,

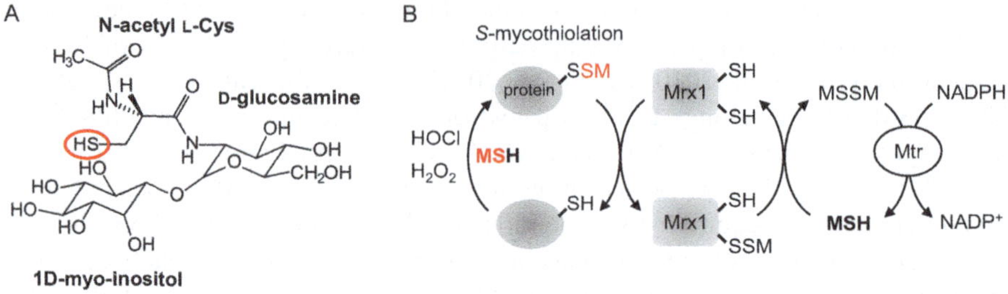

Fig. 5 Mycothiolation. (**a**) Mycothiol (1-D-myo-inositol-2-(N-acetyl-L-cysteinyl)amido-2-deoxy-α-D-glucopyranoside) is a major LMW thiol in *C. glutamicum* and other actinobacteria. (**b**) In *S*-mycothiolation, mycothiol (MSH) forms disulfide bonds with the thiols of cysteine residues. Protein *S*-mycothiolation is reversed by the Mrx1-MSH-Mtr electron pathway, at the expense of NADPH. *Mrx1* mycoredoxin-1; *Mtr* MSH disulfide reductase

including a putative MSH peroxidase (Mpx), MSH-dependent formaldehyde dehydrogenase (AdhE/FadH), MSH-dependent arsenate reductases (ArsC1 and ArsC2), and MSH-dependent maleylpyruvate isomerase (cg3349/NCgl2918) in *C. glutamicum* (Loi et al. 2015; Imber et al. 2019).

In *C. glutamicum*, 25 proteins with *S*-mycothiolation were identified under NaOCl stress conditions by shotgun proteomics, which included metabolic enzymes, translation-related proteins, and antioxidant enzymes (Chi et al. 2014). The thiol peroxidase Tpx (cg1236/NCgl1041) and the MSH peroxidase Mpx (cg2867/NCgl2502) are *S*-mycothiolated at Cys residues in their active sites. *S*-Mycothiolation of Tpx inhibits its peroxidase activity in vitro, which is restored by reduction via the Mrx1/MSH/Mtr electron pathway. The antioxidative activity of Mpx toward H_2O_2 and alkyl hydroperoxides was shown to be restored by both the Mrx1/MSH/Mtr and thioredoxin/thioredoxin reductase (Trx/TrxR) pathways (Si et al. 2015). The glycogen phosphorylase MalP (cg1479/NCgl1255) is one of most frequently *S*-mycothiolated proteins under NaOCl stress conditions, and NaOCl stress leads to reduced glycogen degradation, suggesting that *S*-mycothiolation of MalP may inhibit glycogen degradation (Chi et al. 2014). Tuf (elongation factor Tu), GuaB1 and GuaB2 (inosine-5′-mono phosphate dehydrogenases), SerA (phosphoglycerate dehydrogenase), and MetE (methionine synthase) are conserved targets for *S*-thiolation across Gram-positive bacteria (Chi et al. 2014).

6 Glycosylation

Both *N*-linked (attached to the amide group of asparagine) and *O*-linked (attached to the hydroxyl group of serine and threonine) protein glycosylation are widespread in bacteria (Nothaft and Szymanski 2010; Eichler and Koomey 2017). In *C. glutamicum*, Rpf2 (cg1037/NCgl0872) was first identified as a glycosylated protein, and the Rpf2 glycosylation was shown to contain mannose and galactose as components (Hartmann et al. 2004). Rpf (resuscitation promoting factor) homologs are widely distributed in the actinobacteria, including corynebacteria, mycobacteria, and streptomycetes (Mukamolova et al. 1998). *C. glutamicum* has two *rpf* genes in the genome, *rpf1* [cg0936/NCgl0785] and *rpf2*, and a double deletion mutant showed an elongated lag phase and impaired growth when grown in fresh medium after a long storage period (on an LBG plate at 4 °C for 25 days) (Hartmann et al. 2004). An extracellular subproteome analysis using 2D-PAGE and subsequent Western blotting revealed several glycosylated proteins in *C. glutamicum* ATCC 13032, and three of them were identified by MALDI-TOF-MS: Rpf2, a putative L,D-transpeptidase LppS (cg2720/NCgl2388), and a hypothetical secreted protein (cg1859/NCgl1588) (Mahne et al. 2006). Deletion of the *pmt* gene (*cg1014/NCgl0971*), which encodes a putative homolog of the eukaryotic protein *O*-mannosyltransferase, caused a complete loss of glycosylation, indicating that Pmt is a major glycosylation enzyme in *C. glutamicum* (Mahne et al. 2006). Both *pmt* and *rpf2* genes are well conserved in the genomes of actinobacteria. However, direct evidence for the role of Rpf2 glycosylation in growth stimulation after storage has yet to be described.

7 Outlook

An increasing number of systematic genome-wide studies have revealed that protein activity levels and phenotypes are not always correlated with gene expression, suggesting the importance of post-translational regulation. This is true for microbial metabolism, which is coordinated by complex and multifaceted regulatory networks: changing gene expression as well as the modulation of protein activity via PTMs and the allosteric binding of small metabolic molecules (Chubukov et al. 2013, 2014). To understand the exact role and impact of PTMs on protein activity and metabolism will be increasingly important for the utilization of *C. glutamicum* as a host for the production of useful compounds.

References

Almawi AW, Matthews LA, Guarné A (2017) FHA domains: phosphopeptide binding and beyond. Prog Biophys Mol Biol 127:105–110

Bendt AK, Burkovski A, Schaffer S et al (2003) Towards a phosphoproteome map of *Corynebacterium glutamicum*. Proteomics 3:1637–1646

Bibb LA, Kunkle CA, Schmitt MP (2007) The ChrA-ChrS and HrrA-HrrS signal transduction systems are required for activation of the *hmuO* promoter and repression of the *hemA* promoter in *Corynebacterium diphtheriae*. Infect Immun 75:2421–2431

Birhanu AG, Yimer SA, Holm-Hansen C et al (2017) *Nε*- and *O*-Acetylation in *Mycobacterium tuberculosis* lineage 7 and lineage 4 strains: proteins involved in bioenergetics, virulence, and antimicrobial resistance are acetylated. J Proteome Res 16:4045–4059

Blatch GL, Lässle M (1999) The tetratricopeptide repeat: a structural motif mediating protein-protein interactions. BioEssays 21:932–939

Bott M, Brocker M (2012) Two-component signal transduction in *Corynebacterium glutamicum* and other corynebacteria: on the way towards stimuli and targets. Appl Microbiol Biotechnol 94:1131–1150

Brocker M, Bott M (2006) Evidence for activator and repressor functions of the response regulator MtrA from *Corynebacterium glutamicum*. FEMS Microbiol Lett 264:205–212

Brocker M, Schaffer S, Mack C, Bott M (2009) Citrate utilization by *Corynebacterium glutamicum* is controlled by the CitAB two-component system through positive regulation of the citrate transport genes *citH* and *tctCBA*. J Bacteriol 191:3869–3880

Brocker M, Mack C, Bott M (2011) Target genes, consensus binding site, and role of phosphorylation for the response regulator MtrA of *Corynebacterium glutamicum*. J Bacteriol 193:1237–1249

Burgos JM, Schmitt MP (2016) The ChrSA and HrrSA two-component systems are required for transcriptional regulation of the *hemA* promoter in *Corynebacterium diphtheriae*. J Bacteriol 198:2419–2430

Burns KE, Liu W-T, Boshoff HIM et al (2009) Proteasomal protein degradation in Mycobacteria is dependent upon a prokaryotic ubiquitin-like protein. J Biol Chem 284:3069–3075

Carabetta VJ, Cristea IM (2017) Regulation, function, and detection of protein acetylation in bacteria. J Bacteriol 199:e00107–e00117

Carel C, Marcoux J, Réat V et al (2017) Identification of specific posttranslational *O*-mycoloylations mediating protein targeting to the mycomembrane. Proc Natl Acad Sci U S A 114:4231–4236

Chen Y, Sprung R, Tang Y et al (2007) Lysine propionylation and butyrylation are novel post-translational modifications in histones. Mol Cell Proteomics 6:812–819

Chi BK, Busche T, Van Laer K et al (2014) Protein *S*-mycothiolation functions as redox-switch and thiol protection mechanism in *Corynebacterium glutamicum* under hypochlorite stress. Antioxid Redox Signal 20:589–605

Chubukov V, Uhr M, Le Chat L et al (2013) Transcriptional regulation is insufficient to explain substrate-induced flux changes in *Bacillus subtilis*. Mol Syst Biol 9:709

Chubukov V, Gerosa L, Kochanowski K, Sauer U (2014) Coordination of microbial metabolism. Nat Rev Microbiol 12:327–340

Colak G, Xie Z, Zhu AY et al (2013) Identification of lysine succinylation substrates and the succinylation regulatory enzyme CobB in *Escherichia coli*. Mol Cell Proteomics 12:3509–3520

Dai L, Peng C, Montellier E et al (2014) Lysine 2-hydroxyisobutyrylation is a widely distributed active histone mark. Nat Chem Biol 10:365–370

Darwin KH (2009) Prokaryotic ubiquitin-like protein (pup), proteasomes and pathogenesis. Nat Rev Microbiol 7:485–491

Delley CL, Müller AU, Ziemski M, Weber-Ban E (2017) Prokaryotic ubiquitin-like protein and its ligase/deligase enyzmes. J Mol Biol 429:3486–3499

Deutscher J, Francke C, Postma PW (2006) How phosphotransferase system-related protein phosphorylation regulates carbohydrate metabolism in bacteria. Microbiol Mol Biol Rev 70:939–1031

Deutscher J, Aké FMD, Derkaoui M et al (2014) The bacterial phosphoenolpyruvate:carbohydrate phosphotransferase system: regulation by protein phosphorylation and phosphorylation-dependent protein-protein interactions. Microbiol Mol Biol Rev 78:231–256

Du J, Zhou Y, Su X et al (2011) Sirt5 is a NAD-dependent protein lysine demalonylase and desuccinylase. Science 334:806–809

Eichler J, Koomey M (2017) Sweet new roles for protein glycosylation in prokaryotes. Trends Microbiol 25:662–672

Fiuza M, Canova MJ, Patin D et al (2008a) The MurC ligase essential for peptidoglycan biosynthesis is regulated by the serine/threonine protein kinase PknA in *Corynebacterium glutamicum*. J Biol Chem 283:36553–36563

Fiuza M, Canova MJ, Zanella-Cléon I et al (2008b) From the characterization of the four serine/threonine protein kinases (PknA/B/G/L) of *Corynebacterium glutamicum* toward the role of PknA and PknB in cell division. J Biol Chem 283:18099–18112

Fiuza M, Letek M, Leiba J et al (2010) Phosphorylation of a novel cytoskeletal protein (RsmP) regulates rod-shaped morphology in *Corynebacterium glutamicum*. J Biol Chem 285:29387–29397

Flores-Kim J, Darwin AJ (2016) The phage shock protein response. Annu Rev Microbiol 70:83–101

Frunzke J, Gätgens C, Brocker M, Bott M (2011) Control of heme homeostasis in *Corynebacterium glutamicum* by the two-component system HrrSA. J Bacteriol 193:1212–1221

Gao R, Stock AM (2010) Molecular strategies for phosphorylation-mediated regulation of response regulator activity. Curr Opin Microbiol 13:160–167

Grangeasse C, Cozzone A, Deutccher J, Mijakovic I (2007) Tyrosine phosphorylation: an emerging regulatory device of bacterial physiology. Trends Biochem Sci 32:86–94

Guo J, Wang C, Han Y et al (2016) Identification of lysine acetylation in *Mycobacterium abscessus* using LC–MS/MS after immunoprecipitation. J Proteome Res 15:2567–2578

Hartmann M, Barsch A, Niehaus K et al (2004) The glycosylated cell surface protein Rpf2, containing a resuscitation-promoting factor motif, is involved in intercellular communication of *Corynebacterium glutamicum*. Arch Microbiol 182:299–312

Hentchel KL, Escalante-Semerena JC (2015) Acylation of biomolecules in prokaryotes: a widespread strategy for the control of biological function and metabolic stress. Microbiol Mol Biol Rev 79:321–346

Hentschel E, Mack C, Gätgens C et al (2014) Phosphatase activity of the histidine kinases ensures pathway specificity of the ChrSA and HrrSA two-component systems in *Corynebacterium glutamicum*. Mol Microbiol 92:1326–1342

Heyer A, Gatgens C, Hentschel E et al (2012) The two-component system ChrSA is crucial for haem tolerance and interferes with HrrSA in haem-dependent gene regulation in *Corynebacterium glutamicum*. Microbiology 158:3020–3031

Huc E, Meniche X, Benz R et al (2010) *O*-Mycoloylated proteins from *Corynebacterium*. J Biol Chem 285:21908–21912

Huc E, de Sousa-D'Auria C, de la Sierra-Gallay IL et al (2013) Identification of a mycoloyl transferase selectively involved in *O*-acylation of polypeptides in *Corynebacteriales*. J Bacteriol 195:4121–4128

Humphrey SJ, James DE, Mann M (2015) Protein phosphorylation: a major switch mechanism for metabolic regulation. Trends Endocrinol Metab 26:676–687

Imber M, Pietrzyk-Brzezinska AJ, Antelmann H (2019) Redox regulation by reversible protein *S*-thiolation in gram-positive bacteria. Redox Biol 20:130–145

Ishigaki Y, Akanuma G, Yoshida M et al (2017) Protein acetylation involved in streptomycin biosynthesis in *Streptomyces griseus*. J Proteome 155:63–72

Ishige T, Krause M, Bott M et al (2003) The phosphate starvation stimulon of *Corynebacterium glutamicum* determined by DNA microarray analyses. J Bacteriol 185:4519–4529

Issa H, Huc-Claustre E, Reddad T et al (2017) Click-chemistry approach to study mycoloylated proteins: evidence for PorB and PorC porins mycoloylation in *Corynebacterium glutamicum*. PLoS One 12:e0171955

Jiang H, Zhang X, Chen X et al (2017) Protein lipidation: occurrence, mechanisms, biological functions, and enabling technologies. Chem Rev 118:919–988

Jordan S, Junker A, Helmann JD, Mascher T (2006) Regulation of LiaRS-dependent gene expression in *Bacillus subtilis*: identification of inhibitor proteins, regulator binding sites, and target genes of a conserved cell envelope stress-sensing two-component system. J Bacteriol 188:5153–5166

Kataoka M, Hashimoto KI, Yoshida M et al (2006) Gene expression of *Corynebacterium glutamicum* in response to the conditions inducing glutamate overproduction. Lett Appl Microbiol 42:471–476

Kawahara Y, Takahashi-Fuke K, Shimizu E et al (1997) Relationship between the glutamate production and the activity of 2-oxoglutarate dehydrogenase in *Brevibacterium lactofermentum*. Biosci Biotechnol Biochem 61:1109–1112

Keppel M, Piepenbreier H, Gätgens C et al (2019) Toxic but tasty – temporal dynamics and network architecture of heme-responsive two-component signaling in *Corynebacterium glutamicum*. Mol Microbiol 111:1367–1381

Kim J, Fukuda H, Hirasawa T et al (2010) Requirement of de novo synthesis of the OdhI protein in penicillin-induced glutamate production by *Corynebacterium glutamicum*. Appl Microbiol Biotechnol 86:911–920

Kim J, Hirasawa T, Saito M et al (2011) Investigation of phosphorylation status of OdhI protein during penicillin- and tween 40-triggered glutamate overproduction by *Corynebacterium glutamicum*. Appl Microbiol Biotechnol 91:143–151

Kleine B, Chattopadhyay A, Polen T et al (2017) The three-component system EsrISR regulates a cell envelope stress response in *Corynebacterium glutamicum*. Mol Microbiol 106:719–741

Kocan M, Schaffer S, Ishige T et al (2006) Two-component systems of *Corynebacterium glutamicum*: deletion analysis and involvement of the PhoS-PhoR system in the phosphate starvation response. J Bacteriol 188:724–732

Komine-Abe A, Nagano-Shoji M, Kubo S et al (2017) Effect of lysine succinylation on the regulation of 2-oxoglutarate dehydrogenase inhibitor, OdhI, involved in glutamate production in *Corynebacterium glutamicum*. Biosci Biotechnol Biochem 81:2130–2138

Kosono S, Tamura M, Suzuki S et al (2015) Changes in the acetylome and succinylome of *Bacillus subtilis* in response to carbon source. PLoS One 10:e0131169

Küberl A, Fränzel B, Eggeling L et al (2014) Pupylated proteins in *Corynebacterium glutamicum* revealed by MudPIT analysis. Proteomics 14:1531–1542

Küberl A, Polen T, Bott M (2016) The pupylation machinery is involved in iron homeostasis by targeting the iron storage protein ferritin. Proc Natl Acad Sci U S A 113:4806–4811

Kuhn ML, Zemaitaitis B, Hu LI et al (2014) Structural, kinetic and proteomic characterization of acetyl phosphate-dependent bacterial protein acetylation. PLoS One 9:e94816

Lanéelle M-A, Tropis M, Daffé M (2013) Current knowledge on mycolic acids in *Corynebacterium glutamicum* and their relevance for biotechnological processes. Appl Microbiol Biotechnol 97:9923–9930

Liao G, Xie L, Li X et al (2014) Unexpected extensive lysine acetylation in the trump-card antibiotic producer *Streptomyces roseosporus* revealed by proteome-wide profiling. J Proteome 106:260–269

Liu F, Yang M, Wang X et al (2014) Acetylome analysis reveals diverse functions of lysine acetylation in *Mycobacterium tuberculosis*. Mol Cell Proteomics 13:3352–3366

Loi VV, Rossius M, Antelmann H (2015) Redox regulation by reversible protein *S*-thiolation in bacteria. Front Microbiol 6:1748–1722

Mahne M, Tauch A, Pühler A, Kalinowski JR (2006) The *Corynebacterium glutamicum* gene *pmt* encoding a glycosyltransferase related to eukaryotic protein-*O*-mannosyltransferases is essential for glycosylation of the resuscitation promoting factor (Rpf2) and other secreted proteins. FEMS Microbiol Lett 259:226–233

Mijakovic I, Macek B (2011) Impact of phosphoproteomics on studies of bacterial physiology. FEMS Microbiol Rev 36:877–892

Mizuno Y, Nagano-Shoji M, Kubo S et al (2016) Altered acetylation and succinylation profiles in *Corynebacterium glutamicumin* response to conditions inducing glutamate overproduction. Microbiology 5:152–173

Möker N, Brocker M, Schaffer S et al (2004) Deletion of the genes encoding the MtrA-MtrB two-component system of *Corynebacterium glutamicum* has a strong influence on cell morphology, antibiotics susceptibility and expression of genes involved in osmoprotection. Mol Microbiol 54:420–438

Möker N, Reihlen P, Krämer R, Morbach S (2007) Osmosensing properties of the histidine protein kinase MtrB from *Corynebacterium glutamicum*. J Biol Chem 282:27666–27677

Moon M-W, Park S-Y, Choi S-K, Lee J-K (2006) The phosphotransferase system of *Corynebacterium glutamicum*: features of sugar transport and carbon regulation. J Mol Microbiol Biotechnol 12:43–50
Mukamolova GV, Kaprelyants AS, Young DI, Young M (1998) A bacterial cytokine. Proc Natl Acad Sci U S A 95:8916–8921
Mukherjee S, Keitany G, Li Y et al (2006) *Yersinia* YopJ acetylates and inhibits kinase activation by blocking phosphorylation. Science 312:1208–1211
Muñoz-Dorado J, Inouye S, Inouye M (1991) A gene encoding a protein serine/threonine kinase is required for normal development of *M. xanthus*, a gram-negative bacterium. Cell 67:995–1006
Nagano-Shoji M, Hamamoto Y, Mizuno Y et al (2017) Characterization of lysine acetylation of a phosphoenolpyruvate carboxylase involved in glutamate overproduction in *Corynebacterium glutamicum*. Mol Microbiol 104:677–689
Newton GL, Buchmeier N, Fahey RC (2008) Biosynthesis and functions of mycothiol, the unique protective thiol of Actinobacteria. Microbiol Mol Biol Rev 72:471–494
Niebisch A, Kabus A, Schultz C et al (2006) Corynebacterial protein kinase G controls 2-oxoglutarate dehydrogenase activity via the phosphorylation status of the OdhI protein. J Biol Chem 281:12300–12307
Nothaft H, Szymanski CM (2010) Protein glycosylation in bacteria: sweeter than ever. Nat Rev Microbiol 8:765–778
Nott TJ, Kelly G, Stach L et al (2009) An intramolecular switch regulates phosphoindependent FHA domain interactions in *Mycobacterium tuberculosis*. Sci Signal 2:ra12
Pearce MJ, Mintseris J, Ferreyra J et al (2008) Ubiquitin-like protein involved in the proteasome pathway of *Mycobacterium tuberculosis*. Science 322:1104–1107
Peng C, Lu Z, Xie Z et al (2011) The first identification of lysine malonylation substrates and its regulatory enzyme. Mol Cell Proteomics 10:M111.012658
Pereira SFF, Goss L, Dworkin J (2011) Eukaryote-like serine/threonine kinases and phosphatases in bacteria. Microbiol Mol Biol Rev 75:192–212
Perez J, Castaneda-Garcia A, Jenke-Kodama H, Muller R, Munoz-Dorado J (2008) Eukaryotic-like protein kinases in the prokaryotes and the myxobacterial kinome. Proc Natl Acad Sci 105(41):15950–15955
Peters-Wendisch PG, Schiel B, Wendisch VF et al (2001) Pyruvate carboxylase is a major bottleneck for glutamate and lysine production by *Corynebacterium glutamicum*. J Mol Microbiol Biotechnol 3:295–300
Raasch K, Bocola M, Labahn J et al (2014) Interaction of 2-oxoglutarate dehydrogenase OdhA with its inhibitor OdhI in *Corynebacterium glutamicum*: mutants and a model. J Biotechnol 191:99–105
Rath P, Demange P, Saurel O et al (2011) Functional expression of the PorAH channel from *Corynebacterium glutamicum* in cell-free expression systems. J Biol Chem 286:32525–32532
Sabari BR, Zhang D, Allis CD, Zhao Y (2016) Metabolic regulation of gene expression through histone acylations. Nat Rev Mol Cell Biol 18:90–101
Sadoul K, Wang J, Diagouraga B, Khochbin S (2011) The tale of protein lysine acetylation in the cytoplasm. J Biomed Biotechnol 2011:970382
Schaaf S, Bott M (2007) Target genes and DNA-binding sites of the response regulator PhoR from *Corynebacterium glutamicum*. J Bacteriol 189:5002–5011
Schelder S, Zaade D, Litsanov B et al (2011) The two-component signal transduction system CopRS of *Corynebacterium glutamicum* is required for adaptation to copper-excess stress. PLoS One 6:e22143–e22113
Schultz C, Niebisch A, Gebel L, Bott M (2007) Glutamate production by *Corynebacterium glutamicum*: dependence on the oxoglutarate dehydrogenase inhibitor protein OdhI and protein kinase PknG. Appl Microbiol Biotechnol 76:691–700
Schultz C, Niebisch A, Schwaiger A et al (2009) Genetic and biochemical analysis of the serine/threonine protein kinases PknA, PknB, PknG and PknL of *Corynebacterium glutamicum*: evidence for non-essentiality and for phosphorylation of OdhI and FtsZ by multiple kinases. Mol Microbiol 74:724–741

Shimizu H, Tanaka H, Nakato A et al (2003) Effects of the changes in enzyme activities on metabolic flux redistribution around the 2-oxoglutarate branch in glutamate production by *Corynebacterium glutamicum*. Bioprocess Biosyst Eng 25:291–298
Shirai T, Nakato A, Izutani N et al (2005) Comparative study of flux redistribution of metabolic pathway in glutamate production by two coryneform bacteria. Metab Eng 7:59–69
Shirai T, Fujimura K, Furusawa C et al (2007) Study on roles of anaplerotic pathways in glutamate overproduction of *Corynebacterium glutamicum* by metabolic flux analysis. Microb Cell Factories 6:19
Si M, Xu Y, Wang T et al (2015) Functional characterization of a mycothiol peroxidase in *Corynebacterium glutamicum* that uses both mycoredoxin and thioredoxin reducing systems in the response to oxidative stress. Biochem J 469:45–57
Singhal A, Arora G, Virmani R et al (2015) Systematic analysis of mycobacterial acylation reveals first example of acylation-mediated regulation of enzyme activity of a bacterial phosphatase. J Biol Chem 290:26218–26234
Stancik IA, Šestak MS, Ji B et al (2018) Serine/threonine protein kinases from bacteria, archaea and eukarya share a common evolutionary origin deeply rooted in the tree of life. J Mol Biol 430:27–32
Striebel F, Imkamp F, Özcelik D, Weber-Ban E (2014) Pupylation as a signal for proteasomal degradation in bacteria. BBA - Mol Cell Res 1843:103–113
Tan M, Luo H, Lee S et al (2011) Identification of 67 histone marks and histone lysine crotonylation as a new type of histone modification. Cell 146:1016–1028
Tan M, Peng C, Anderson KA et al (2014) Lysine glutarylation is a protein posttranslational modification regulated by SIRT5. Cell Metab 19:605–617
Thakur M, Chakraborti PK (2006) GTPase activity of mycobacterial FtsZ is impaired due to its transphosphorylation by the eukaryotic-type Ser/Thr kinase, PknA. J Biol Chem 281:40107–40113
Van Drisse CM, Escalante-Semerena JC (2019) Protein acetylation in bacteria. Annu Rev Microbiol 73:5.1–5.22
Wagner GR, Payne RM (2013) Widespread and enzyme-independent *N*ε-acetylation and *N*-ε-succinylation of proteins in the chemical conditions of the mitochondrial matrix. J Biol Chem 288:29036–29045
Wagner GR, Bhatt DP, O'Connell TM et al (2017) A class of reactive acyl-CoA species reveals the non-enzymatic origins of protein acylation. Cell Metab 25:823–837
Weinert BT, Iesmantavicius V, Wagner SA et al (2013a) Acetyl-phosphate is a critical determinant of lysine acetylation in *E. coli*. Mol Cell 51:265–272
Weinert BT, Schölz C, Wagner SA et al (2013b) Lysine succinylation is a frequently occurring modification in prokaryotes and eukaryotes and extensively overlaps with acetylation. Cell Rep 4:842–851
Xie L, Wang X, Zeng J et al (2015) Proteome-wide lysine acetylation profiling of the human pathogen *Mycobacterium tuberculosis*. Int J Biochem Cell Biol 59:193–202
Xie Z, Di Z, Chung D et al (2016) Metabolic regulation of gene expression by histone lysine β-hydroxybutyrylation. Mol Cell 62:194–206
Yeats C, Finn RD, Bateman A (2002) The PASTA domain: a β-lactam-binding domain. Trends Biol Sci 27(9):438–440
Yu BJ, Kim JA, Moon JH et al (2008) The diversity of lysine-acetylated proteins in *Escherichia coli*. J Microbiol Biotechnol 18:1529–1536
Zhang J, Sprung R, Pei J et al (2009) Lysine acetylation is a highly abundant and evolutionarily conserved modification in *Escherichia coli*. Mol Cell Proteomics 8:215–225
Zhang Z, Tan M, Xie Z et al (2010) Identification of lysine succinylation as a new post-translational modification. Nat Chem Biol 7:58–63
Zschiedrich CP, Keidel V, Szurmant H (2016) Molecular mechanisms of two-component signal transduction. J Mol Biol 428:3752–3775

Part III
Amino Acids

Recent Advances in Amino Acid Production

Masato Ikeda and Seiki Takeno

Contents

Abstract The annual world production of amino acids is currently estimated at more than seven million tons and is expected to reach ten million tons by 2022. This giant market has been underpinned largely by amino acid fermentation technologies in which *Corynebacterium glutamicum* has played a leading role. Various genetic engineering tools and global analysis techniques for this bacterium have been developed and successfully applied with a great impact on the amino acid industry. In particular, systems biology for this bacterium is almost fully capable of predicting targets to be engineered and metabolic states that will yield maximum production, thus allowing "systems metabolic engineering" and development of industrially competitive production strains. Additionally, whole genomes of classically derived industrial producers have been analyzed by "reverse engineering" to identify

M. Ikeda (✉) · S. Takeno
Department of Agricultural and Life Sciences, Faculty of Agriculture, Shinshu University, Nagano, Japan
e-mail: m_ikeda@shinshu-u.ac.jp

M. Inui, K. Toyoda (eds.), *Corynebacterium glutamicum*, Microbiology Monographs 23, https://doi.org/10.1007/978-3-030-39267-3_7

important genetic traits, enabling the establishment of new industrial processes and the creation of genetically defined producers from scratch. This "genome breeding" strategy was first developed using *C. glutamicum* as a model and currently yields producers that are more efficient than classical ones. These advances in strain development technology have almost achieved the optimization of entire cellular systems as cell factories for amino acid production, as demonstrated by their ability to produce glutamate and lysine at concentrations now exceeding 150 g/L with estimated production yields towards sugar at almost 70%. This chapter describes advances in the production of amino acids by *C. glutamicum* and presents the latest details of the technology and strategies used for molecular strain improvement.

1 Introduction

Amino acids have a wide variety of characteristics in terms of nutritional value, taste, medicinal action, and chemical properties, and thus have many potential uses as food additives, feed supplements, pharmaceuticals, cosmetics, polymer materials, and agricultural chemicals. As each new use is developed, demand for that type of amino acid grows rapidly and is followed by the development of mass production technology for that amino acid. The annual world production of amino acids has increased year by year (Fig. 1), from 0.7 million tons in 1985 to 1.7 million tons in 1996, 3.7 million tons in 2006, and 7.0 million tons in 2016, and is expected to reach ten million tons by 2022, growing at a CAGR (Compound Annual Growth Rate) of 5.6% from 2015 to 2022 (Ikeda 2003; Hermann 2003; Ajinomoto 2007, 2016; Sanchez et al. 2018). According to a recent market research report (Research and Markets 2018) and other relevant publications (Ajinomoto 2016; Sanchez et al. 2018), the global market for amino acids is estimated to be approximately US$13 billion in 2016 and is growing at an annual rate of 7%. Figure 2 shows the estimated global markets for amino acids of different applications in 2016. The feed amino acids, namely, lysine, methionine, threonine, and tryptophan, have the largest share of the market, generating US$7.0 billion. The second largest share (US$5.3 billion) belongs to food additives, which are comprised mainly of the flavor-enhancer monosodium glutamate and the amino acids aspartate and phenylalanine, both used as ingredients in the peptide sweetener aspartyl phenylalanyl methyl ester (Aspartame).

Most L-amino acids are manufactured through microbial processes, mainly through fermentation. *Corynebacterium glutamicum*, which plays a principal role in amino acid fermentation, is therefore highly important, as demonstrated by the increasing number of relevant research papers (Fig. 3). Figure 3 also shows the main topics in amino acid fermentation and strain development technology during the decades since such research began. Amino acid fermentation was developed primarily in Japan and has extended across East Asia and into Europe, North America, and South America. Today amino acid fermentation is a global industry. It should be

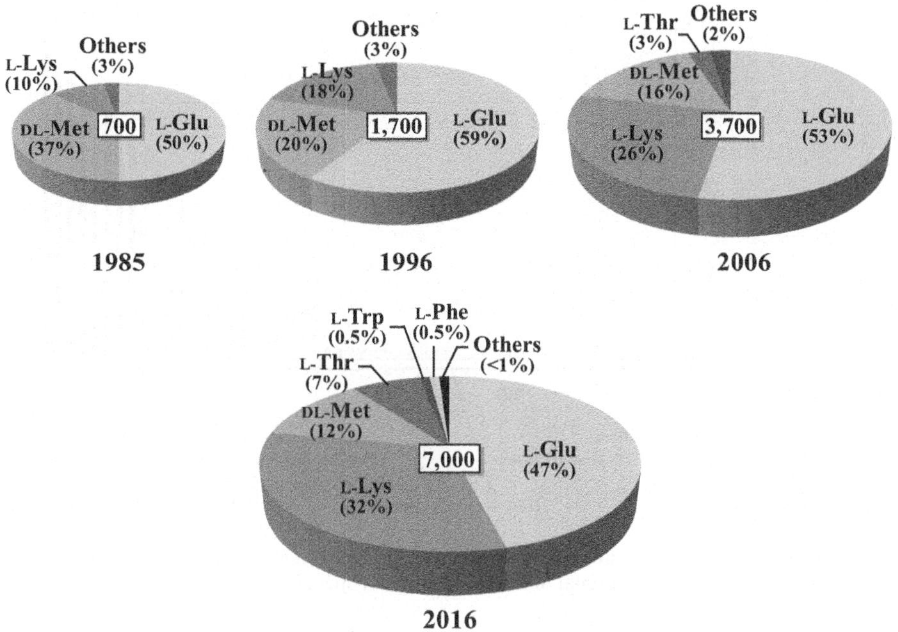

Fig. 1 Changes in world annual production quantities of amino acids. The *numbers in the squares* indicate the estimated amounts of amino acids produced (1000 metric tons)

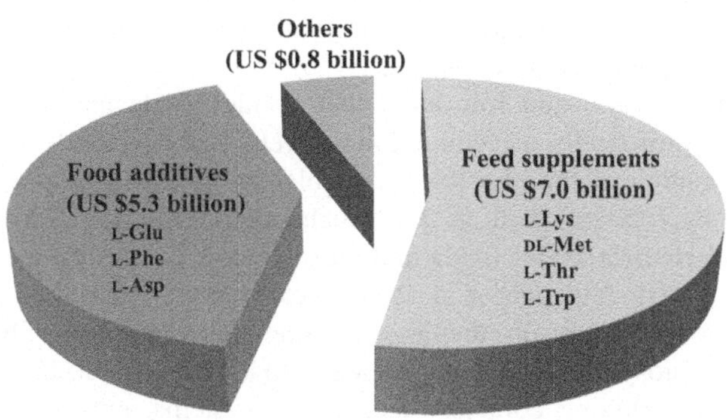

Fig. 2 Estimated 2016 global markets for amino acids segmented by applications

noted that China has achieved a prominent presence in this field, as reflected by the drastic increase in the number of research papers relevant to amino acid production (Ma et al. 2017; Zhang et al. 2017b; Cheng et al. 2018).

In general, commercially potent producers have been developed by the stepwise accumulation of beneficial genetic and phenotypic characteristics in one background through classical mutagenesis and/or recombinant DNA technology. Such improvements involve strains capable not only of producing amino acids at higher yields but

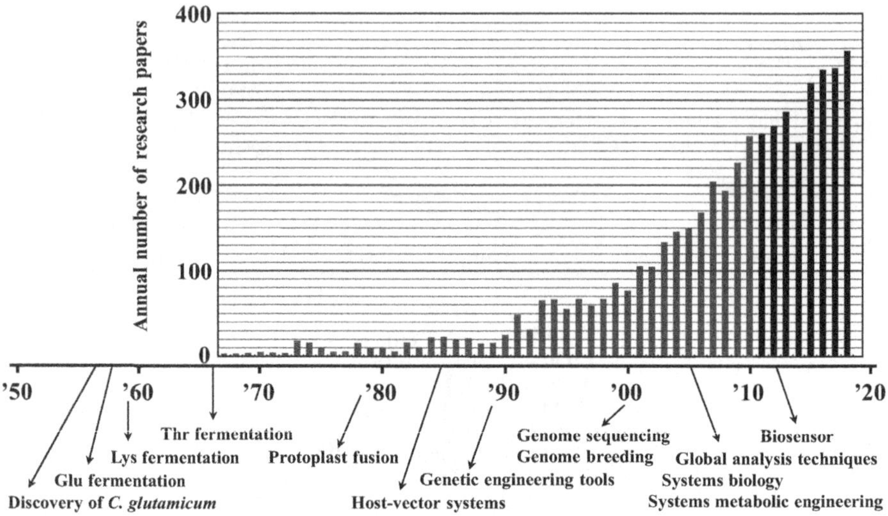

Fig. 3 History of amino acid fermentation and strain development technology in *Corynebacterium glutamicum*, together with the annual number of research papers relevant to this microbe

also of producing lower quantities of by-products, as the removal of by-products dominates the costs of downstream processing (Ikeda 2003; Marienhagen and Eggeling 2008; Feng et al. 2018). The current production yields towards sugar (w/w %) can be estimated as follows: lysine hydrochloride, 60–70; glutamate, 60–70; arginine, 40–50; isoleucine, 20–30; valine, 35–40; leucine, 25–30; tryptophan, 20–30; and phenylalanine, 30–35.

Since the year 2000, genomic and other "omics" data have accumulated for *C. glutamicum*, profoundly affecting strain development methods and providing a global understanding of the physiology, regulatory networks, and unknown functions of this microbe as well as the mechanisms underlying hyperproduction (Wittmann and Heinzle 2002; Ikeda and Nakagawa 2003; Kalinowski et al. 2003; Strelkov et al. 2004; Yukawa et al. 2007; Ikeda 2017; Yokota and Ikeda 2017; Becker et al. 2018). As a result, the targets of metabolic engineering have expanded beyond the core biosynthetic pathways leading to amino acids of interest into entire cellular systems including cofactor-regeneration systems, uptake and export systems, energy metabolism, global regulation, and stress responses. Such global and systematic metabolic engineering has repeatedly led to successful yield improvements for amino acid production by *C. glutamicum* (Eggeling and Bott 2005; Wendisch 2007; Burkovski 2008; Mitsuhashi 2014; Yokota and Ikeda 2017; Becker et al. 2018). In addition, the product spectrum of *C. glutamicum* has also been expanded, and metabolic engineering has been applied to the production of amino acids that formerly could not be produced effectively from glucose, such as serine, methionine, and cysteine. The present chapter describes the technologies and strategies that have been used in strain development in recent years, then reports the latest findings on rational metabolic engineering of *C. glutamicum* to develop efficient amino acid producers. Representative work using other bacteria, such as *Escherichia coli* and *Pantoea ananatis*, is also included for reference.

2 Recent Technologies for Strain Development

The focus in strain development technologies after the year 2000 has been directed to the development of new methodologies and tools employing genomic information, multi-omics data, bioinformatics, systems and synthetic biology, high-throughput single cell screening, and so on. These efforts have led to the development of several powerful new approaches, such as "systems metabolic engineering", "genome breeding", and "biosensor-driven single cell screening", which has rejuvenated strain development for amino acid production. Such new approaches for strain development in *C. glutamicum* are highlighted here.

2.1 *From Genome to Producers*

Advances in microbial genomics have dramatically transformed our approaches to strain development. Their largest benefit to this field is obviously the availability of high-throughput DNA sequencing, which has made it feasible to decode the genomes of classical industrial producers and thereby to identify important genetic traits that distinguish them from their wild-type ancestors. As a result, the conventional style of selecting improved strains by their phenotypes, formerly the standard practice in the industry, is rapidly being replaced by a new method called "genome breeding" (Fig. 4), where desirable genotypes are systematically assembled in a wild-type genome (Ohnishi et al. 2002; Ikeda et al. 2006; Lee et al. 2012; Kim et al. 2013; Wu et al. 2015; Ma et al. 2018). The strains reconstructed from scratch through genome breeding can be more robust, give higher fermentation yields in less time, and resist stressful conditions better than classical industrial producers (Ohnishi et al. 2003). Meanwhile, microbial genomics allows *in silico* reconstruction of the whole metabolic map of a relevant microorganism. In fact, the genome information that is now available for a diverse variety of microorganisms has already revealed the numerous metabolic pathways that sustain their lives. This is now opening the way for the creation of a new methodology centered around redesigning a particular metabolic pathway in a desired microorganism on the model of one of the other metabolic pathways for which information is now publicly available. One example of this practice is the recent re-engineering of 'Glutamic acid bacteria' on the model of the unique redox metabolism seen in 'Tooth decay bacteria' (Takeno et al. 2010, 2016). Here, the two approaches, genome breeding and metabolic redesign, are described to show how we can take advantage of genome information to enable more efficient amino acid production.

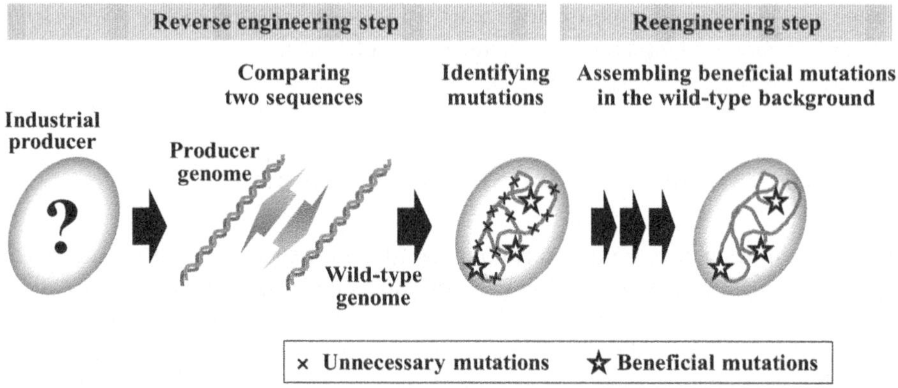

Fig. 4 The "genome breeding" methodology for the creation of defined mutants that carry only beneficial mutations. This methodology starts with decoding the genomes of classical industrial producers to identify the important genetic traits that distinguish them from their wild-type ancestors (the *Reverse engineering step*) and progresses to systematically assembling the beneficial genetic properties in a single wild-type background (the *Reengineering step*)

2.1.1 Genome Breeding

Because classical strain breeding is based on random mutation and selection, we cannot eliminate the possibility that this method will introduce detrimental or unnecessary mutations into a genome. Genome breeding methodology, however, can overcome this limitation. In this program, biotechnologically useful mutations identified through the genome analysis of classical mutants are systematically introduced into the wild-type genome in a pinpointed manner (Ikeda et al. 2005), thus allowing the creation of a defined mutant that carries only useful mutations (Figs. 4 and 5). As an example, one industrial lysine producer that had undergone years of mutagenesis and screening was found to have more than 1000 mutations accumulated in its genome (Ikeda 2017). Among these, only six mutations were identified as positive mutations for lysine production; two (*hom59* and *lysC311*) that are located in the terminal pathway to lysine (Ohnishi et al. 2002), three (*pyc458, gnd361,* and *mqo224*) that are involved in central metabolism (Ohnishi et al. 2002, 2005; Mitsuhashi et al. 2006), and one (*leuC456*) that causes global induction of the amino acid-biosynthetic genes and thereby further increases production (Hayashi et al. 2006a). The assembly of these six useful mutations in a robust wild-type strain of *C. glutamicum* (Fig. 5a) was shown to substantially improve producer performance, resulting in a final titer of 100 g/L after 30 h of 5-L jar fermentor cultivation at a suboptimal temperature of 40 °C (Ohnishi et al. 2003; Ikeda et al. 2006).

The usefulness of the genome breeding approach has been also demonstrated in the production of arginine and citrulline (Ikeda et al. 2009, 2010a). In this case, the assembly of three positive mutations (*argB26*, Δ*argR*, and *argB31*) derived from three different lines of classical producers in a single wild-type background (Fig. 5b) has led to the new strain RBid, characterized by dramatically increased productivity

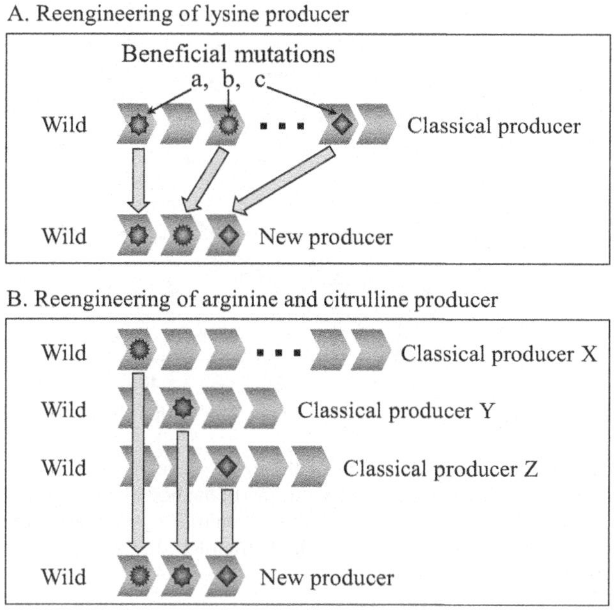

Fig. 5 Reengineering of a lysine producer (**a**) and an arginine and citrulline producer (**b**). In the case of the new lysine producer, six beneficial mutations identified from the genome of a classical lysine producer were assembled in a single wild-type background (Ikeda et al. 2006), while the new arginine and citrulline producer was created by assembly of three positive mutations derived from three different lines of classical producers (Ikeda et al. 2009)

of arginine and citrulline compared with the best classical producer, A-27, even at the suboptimal temperature of 38 °C (Fig. 6). In this approach, not only identification of beneficial mutations but also screening for the specific host that will give the best performance is an important consideration because the wild-type background can have a significant impact on the ultimate outcome (Ohnishi and Ikeda 2006; Ikeda et al. 2009). The host strain into which the three mutations were incorporated was ATCC 13032, which was identified through screening from among several *C. glutamicum* wild-type strains as the strain with the highest potential for industrial arginine/citrulline production at elevated temperatures (Ikeda et al. 2009). If another wild-type strain had been used as the host, the result would have been unsatisfactory.

2.1.2 Metabolic Redesign

Progress in genomics has made it possible to construct an organism's entire metabolic map *in silico*. Diverse metabolic pathways of approximately 3500 bacterial species have already been constructed and are available at the Kyoto Encyclopedia of Genes and Genomes (http://www.kegg.jp/kegg-bin/get_htext?htext=br08601_map00010.keg&hier=5) (Kanehisa et al. 2014). Most well-known bacteria possess the complete pentose phosphate pathway which enables them to generate NADPH as a reducing power. In some unique microorganisms, however, the database has revealed a defective pentose phosphate pathway. One such microorganism is *Streptococcus mutans*, a tooth decay bacterium reportedly harboring an unusual glycolytic pathway that can generate NADPH at the step of glyceraldehyde 3-phosphate dehydrogenase

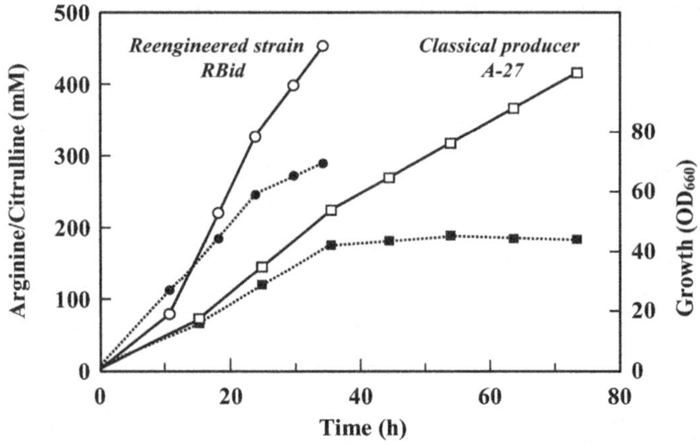

Fig. 6 Fermentation kinetics of the newly developed strain RBid at 38 °C in 5-L jar fermentor cultivation. For comparison, the profiles of the best classical producer A-27 when cultured under its optimal 30 °C conditions are shown as controls. *Open circles* arginine and citrulline production by strain RBid, *closed circles* growth of strain RBid, *open squares* arginine and citrulline production by strain A-27, *closed squares* growth of strain A-27

(Fig. 7). Based on these findings, an attempt has been made to recreate the *S. mutans*-type NADPH-generating glycolytic pathway in *C. glutamicum* (Takeno et al. 2010, 2016). In this study, endogenous NAD-dependent glyceraldehyde 3-phosphate dehydrogenase (GapA) of *C. glutamicum* was replaced with nonphosphorylating NADP-dependent glyceraldehyde 3-phosphate dehydrogenase (GapN) of *S. mutans* (Fig. 8). Unfortunately, the resulting strain (RE2) exhibited severely retarded growth, probably because the engineering attempt had favored the generation of reducing power while theoretically restricting ATP generation. The strategy for solving this problem was to use GapA together with GapN in the early growth phase where more ATP is required for growth, and thereafter to shift the combination-type glycolytic pathway to one that depends only on GapN for production in the subsequent growth phase (Fig. 8). To achieve this, the gene for GapA was expressed under the *myo*-inositol-inducible promoter of *iolT1* encoding a *myo*-inositol transporter. In strain RE2A^{iol} which was thus engineered, a well-balanced use of GapA and GapN has led to both improved growth and high-level lysine production. Moreover, it has been demonstrated that blockade of the oxidative pentose phosphate pathway through a defect in glucose 6-phosphate dehydrogenase did not significantly affect lysine production in the engineered strain (Fig. 8), while a drastic decrease in lysine production was observed for the reference strain (Takeno et al. 2016). This study was the first to demonstrate efficient lysine production independent of the oxidative pentose phosphate pathway.

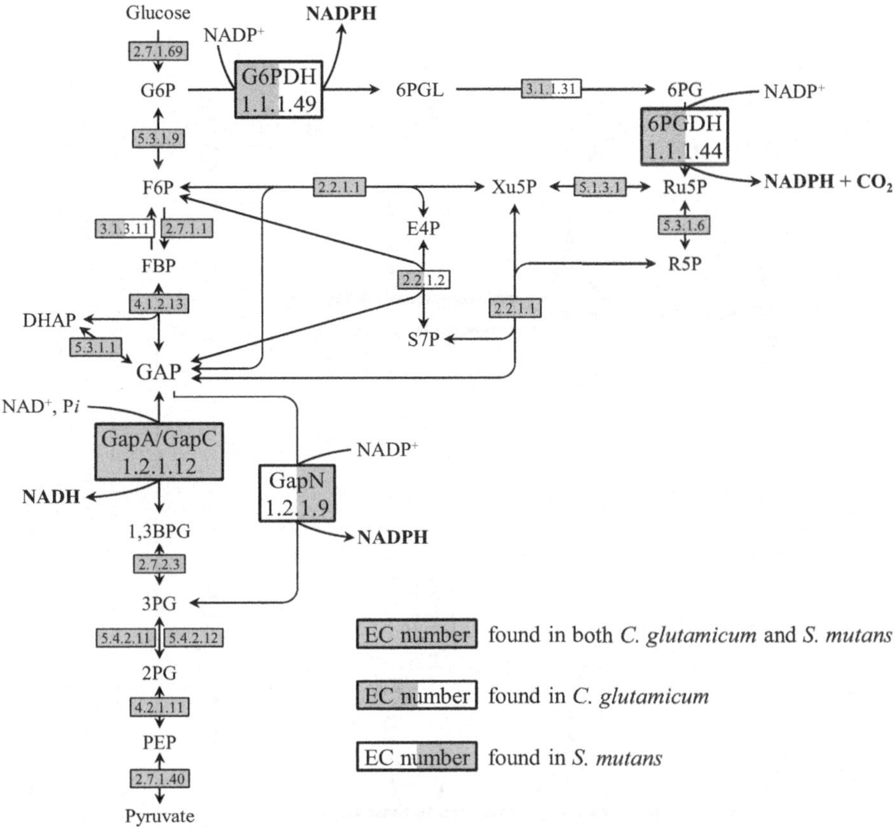

Fig. 7 Glycolysis and the pentose phosphate pathway in *C. glutamicum* ATCC 13032 and *S. mutans* UA159. The metabolic pathways are extracted from the KEGG pathway database (Kanehisa et al. 2014). Each enzyme is presented as an EC number. *GapA* NAD-dependent glyceraldehyde 3-phosphate dehydrogenase from *C. glutamicum*, *GapC* NAD-dependent glyceraldehyde 3-phosphate dehydrogenase from *S. mutans*, *GapN* nonphosphorylating NADP-dependent glyceraldehyde 3-phosphate dehydrogenase

2.2 Systems Metabolic Engineering

The cumulative body of knowledge on cellular metabolism and physiological properties of amino acid-producing microorganisms was combined with "omics" technologies and computational methods, including metabolic flux profiling and *in silico* modeling, to facilitate metabolic engineering in a systematic and global manner (Dai and Nielsen 2015; Hirasawa and Shimizu 2016; Ma et al. 2017; Lee and Wendisch 2017; Zhang et al. 2017b; Becker et al. 2018). This "systems metabolic engineering" approach is particularly useful for predicting a combination of genetic modifications that would lead to the theoretically best flux scenario for

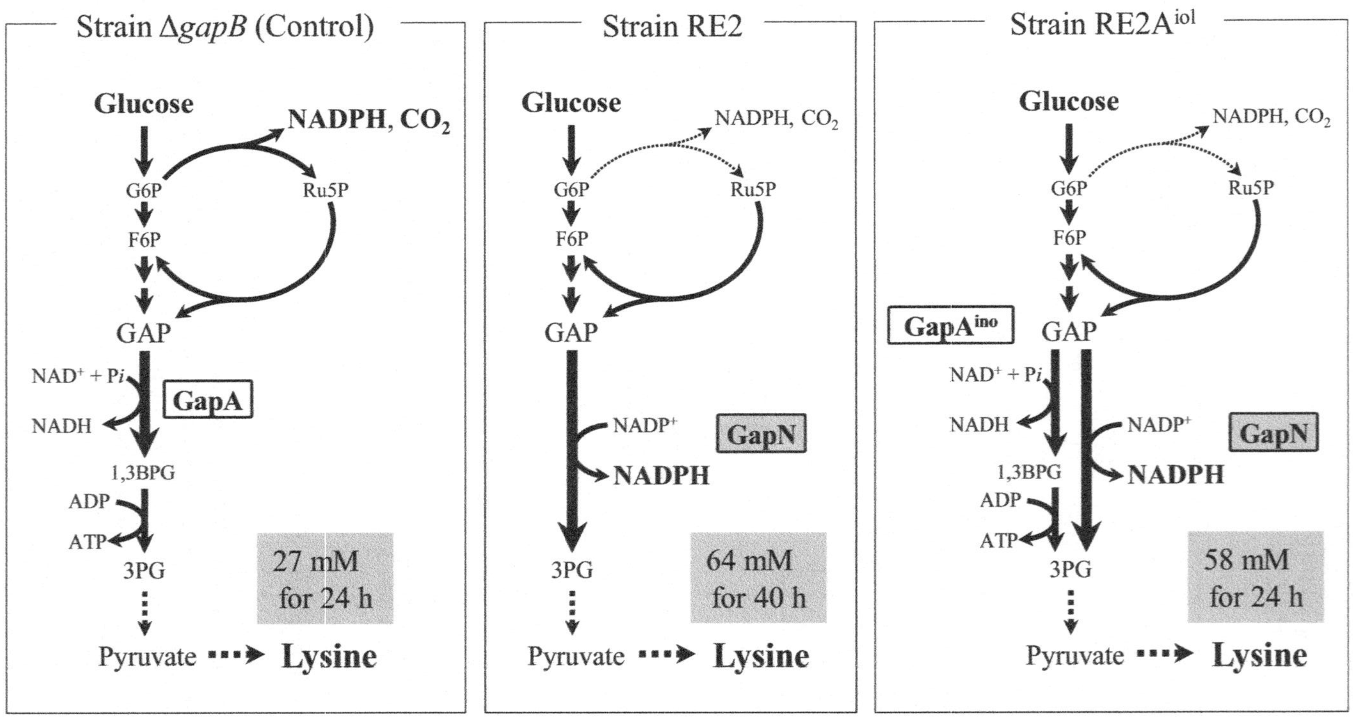

Fig. 8 Engineering of *C. glutamicum* with the *S. mutans*-type NADPH-generating glycolytic pathway for lysine production. The control strain *C. glutamicum* Δ*gapB* has an NADH-generating glycolytic pathway (*left*). Replacement of endogenous GapA with *S. mutans* GapN created an NADPH-generating glycolytic pathway, leading to increased lysine production but retarded fermentation (*middle*). Further engineering to supplementarily express GapA only in the early growth phase resulted in efficient lysine production independent of the oxidative pentose phosphate pathway (*right*)

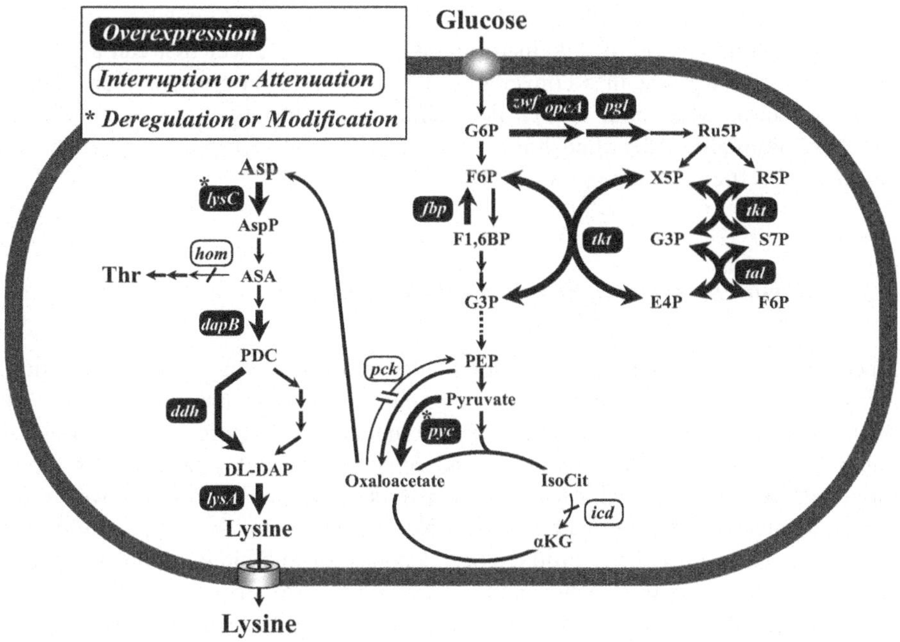

Fig. 9 Schematic diagram of genetic modifications to a wild-type genome leading to the lysine hyper-producer *C. glutamicum* LYS-12

amino acid production. Several applications of this have recently led to the successful creation of efficient amino acid producers in both *C. glutamicum* and *E. coli*.

Becker et al. (2011) used flux model analysis with a genome-scale metabolic model to predict the target steps to be modified for optimum lysine production by *C. glutamicum*. Ultimately, this analysis identified twelve stages of modifications to a wild-type genome resulting in the lysine hyper-producer LYH-12 (Fig. 9), which can achieve a final titer of 120 g/L with a conversion yield of 55% on glucose after 30 h of 5-L jar fermentor cultivation at 30 °C. Among the twelve modifications were six (introduction of the *lysC311* and *hom59* mutations, duplication of the *ddh* and *lysA* genes, and overexpression of the *lysC* and *dapB* genes under a strong promoter) that cause increased flux through the lysine biosynthetic pathway, three (introduction of the *pyc458* mutation, overexpression of the *pyc* gene under a strong promoter, and deletion of the *pck* gene) that cause increased flux towards oxaloacetate through anaplerotic carboxylation, two (overexpression of the *fbp* gene and the *zwf-opcA-tkt-tal* operon under strong promoters) that cause increased flux through the pentose phosphate pathway for NADPH supply, and one (replacement of the start codon ATG by the rare GTG in the *icd* gene) that causes reduced flux through the TCA cycle and thereby increases the availability of oxaloacetate.

Lee et al. (2007) and Park et al. (2007a) reported the strategies for systems metabolic engineering of *E. coli* for the production of threonine and valine, respectively. For threonine production, the target genes to be engineered were identified through transcriptome profiling and *in silico* flux response analysis, ultimately resulting in construction of a defined threonine hyper-producer that can achieve a final titer of 82.4 g/L with a conversion yield of 39.3% on glucose after 50 h of fed-batch culture (Lee et al. 2007). For valine production, likewise, *in silico* gene knockout simulation identified three target genes to be disrupted, leading to the design of an efficient valine producer (Park et al. 2007a).

2.3 Biosensor-Driven Single Cell Screening

Since rational strain improvement generally depends on known genetic information, relying on this approach alone can cause researchers to miss unknowns, which are often difficult to predict. For this reason, the classical approach consisting of multiple rounds of random mutation and screening is still significant for strain improvement. The classical approach also offers opportunities to find novel information applicable to the rational approach. Yet screening almost always requires the cultivation and subsequent productivity analysis of individual mutants, which requires considerable time, labor, and cost. Biosensor-driven single cell screening is a technique that allows researchers to overcome such disadvantages and thereby accelerate strain improvement. The major technique that has been applied to *C. glutamicum* is based on the principle that transcriptional regulators (TRs) activate expression of their target gene fused to a reporter gene *eyfp* (encoding enhanced yellow fluorescent protein, eYFP) in response to intracellular concentration of a specific metabolite. This setup enables translation of the intracellular metabolite concentration into a fluorescent output and the high-throughput screening of single cells via fluorescence-activated cell sorting (FACS). The representative achievements have been performed using the homologous TRs: LysG and Lrp (Fig. 10).

In response to cytosolic concentrations of basic amino acids such as lysine, arginine, or histidine, LysG activates expression of the *lysE* gene encoding an exporter for lysine and arginine (Bellmann et al. 2001). A plasmid containing the *eyfp* gene under the control of the promoter of the *lysE* gene enables the translation of the intracellular lysine concentration into a fluorescent signal (Binder et al. 2012). A mutant library consisting of 7×10^6 cells obtained from wild-type *C. glutamicum* ATCC 13032 carrying this plasmid was subjected to FACS analysis, which resulted in the isolation of 185 lysine-producing mutants. It should be noted that increased intracellular concentration of lysine correlates with its increased concentration in the culture supernatant. Whole-genome sequencing of the mutants revealed novel mutations in the *murE* gene encoding UDP-*N*-acetylmuramoyl-L-alanyl-D-glutamate-2,6-diaminopimelate ligase that utilizes D, L-diaminopimelate as a substrate in the lysine synthetic pathway. One of these mutations, mutation *murE*^{G81E}, was introduced into the genome of the defined lysine producer *C. glutamicum* DM1933 (Blombach et al. 2009b) where it increased lysine production by approximately 1.3-fold on glucose (Binder et al. 2012). The LysG-based biosensor has also been utilized to isolate less-feedback variants of key enzymes in lysine, arginine, and histidine biosynthesis

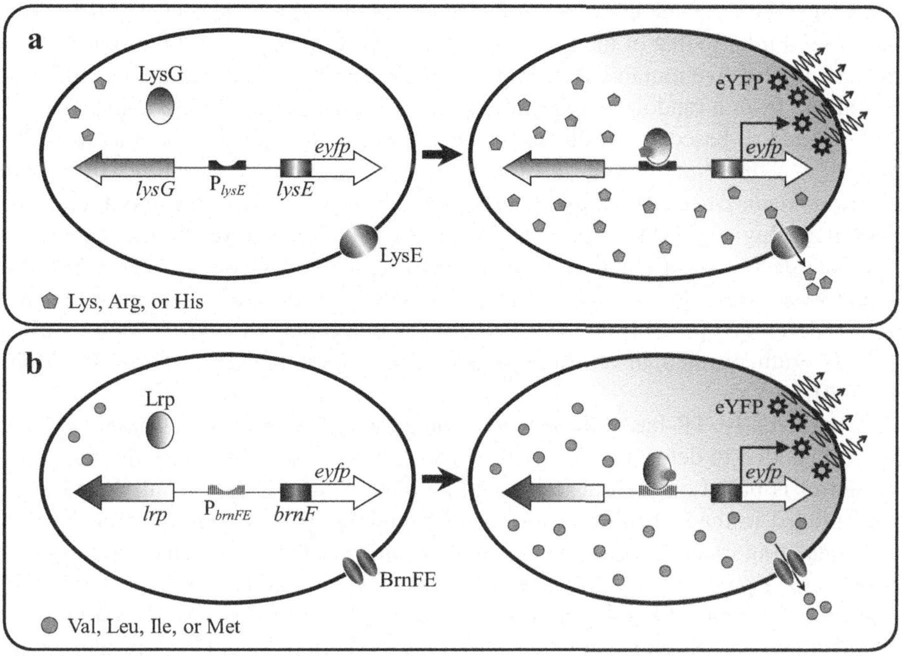

Fig. 10 Biosensors using transcriptional regulators LysG (**a**) and Lrp (**b**). The biosensors are based on the principle that transcriptional regulators activate the target gene fused to a reporter gene *eyfp* in response to intracellular concentration of each effector metabolite. Eventually, increased intracellular concentration of the metabolite is reflected as increased fluorescence, which enables screening of single cells by fluorescence-activated cell sorting (FACS). (**a**) LysG activates expression of the *lysE* gene in the presence of increased levels of the effectors lysine, arginine, or histidine. (**b**) Lrp activates expression of the *brnFE* operon in the presence of increased levels of valine, leucine, isoleucine, or methionine

(Schendzielorz et al. 2014). This technique was performed on a mutant library consisting of approximately 2.2×10^6 cells of the *C. glutamicum* Δ*argR* strain carrying a plasmid containing a randomly mutated *argB* gene encoding *N*-acetyl-L-glutamate kinase. Analysis of 96 selected positive cells revealed the presence of ArgB variants. The *argB* gene, which was modified based on information on these variants, encoded a feedback-resistant enzyme to inhibition by arginine and allowed the Δ*argR* strain to produce 35 mM arginine on glucose (Schendzielorz et al. 2014). A similar approach has identified feedback-resistant variants of aspartate kinase (LysC) and ATP phosphoribosyl transferase (HisG), which are key enzymes in lysine and histidine biosynthesis, respectively, and revealed that these mutations lead to the production of the respective amino acids by the wild-type strain (Schendzielorz et al. 2014).

Another representative example of biosensor-driven single cell screening is based on the transcriptional regulator Lrp. Lrp activates expression of the *brnFE* operon encoding the export system for valine, leucine, isoleucine, and methionine in response to intracellular concentration of these amino acids (Lange et al. 2012). By a principle similar to that of the LysG-based biosensor, the Lrp-based biosensor led to the isolation of five mutants that produced different amounts of valine, leucine, and isoleucine from a random mutant library of wild-type ATCC 13032 (Mustafi et al. 2012). The Lrp-based approach has also been utilized for adaptive laboratory evolution for valine production. An evolved consortium originating from the *C. glutamicum* Δ*aceE* strain and having gone through five iterative evolution steps exhibited a higher growth rate, up to 25% increased valine production, and three- to four-fold decreased production of the by-product alanine. Genome sequencing of a single isolate revealed a loss-of-function mutation in the *ureD* gene encoding the urease accessory protein, and introduction of this mutation into the non-evolved Δ*aceE* strain resulted in an increase in valine production up to two-fold (Mahr et al. 2015).

Very recently, TR-based biosensors coupled with FACS for *C. glutamicum* have been expanded to detection of shikimic acid, serine, and cAMP. The shikimic acid biosensor is based on the transcriptional regulator ShiR (Kubota et al. 2015). This was utilized in the screening of a ribosome binding site (RBS) library for the *tktA* gene encoding transketolase, an enzyme that catalyzes the formation of erythrose-4-phosphate, a precursor of shikimic acid, and in the identification of a useful RBS sequence for improved shikimic acid production (Liu et al. 2018a). In contrast, the serine biosensor employs the transcriptional regulator NCgl0581 (Binder et al. 2012). This system was used to screen the random mutant library of a serine-producing *C. glutamicum* strain that was obtained through a combination of random mutagenesis and rational engineering, which resulted in the isolation of a mutant with 1.3-fold higher serine production than its parental strain (Zhang et al. 2018e). The mutant produced 35 g/L of serine with a conversion yield of 35% on sucrose after 120 h of batch culture (Zhang et al. 2018e). Although all of the TRs described above bind with the corresponding effectors and thereafter can activate expression of the reporter gene, the cAMP sensor is composed of the transcriptional regulator GlxR and a promoter repressed by cAMP-bound GlxR. This biosensor has been used successfully to separate cells with different cAMP levels (Schulte et al. 2017).

3 Current Status of Amino Acid Production

Recently, various genetic engineering tools and global analysis techniques for *C. glutamicum* as well as high-throughput genomic analysis technologies have been successfully applied and have contributed both to the understanding of the molecular mechanisms underlying high-level production and to the development of more advanced production strains of this microbe. At the same time, the new approaches to strain development such as genome breeding and systems metabolic engineering have allowed the creation of nearly optimal genetically defined and industrially competitive producers from scratch. Here, the current status of production of various amino acids by *C. glutamicum* is highlighted, with a special focus on the amino acids whose production methods have been significantly advanced in the 2000s. These include glutamate, lysine, arginine, citrulline, ornithine, the branched-chain amino acids valine, leucine, and isoleucine, and the sulfur-containing amino acids methionine, *S*-adenosyl-methionine, and cysteine. Tryptophan is also included because the scale of its market has been growing rapidly. Production technology aimed at other industrially important amino acids such as serine, alanine, and threonine has been omitted because it was discussed in the first edition of this book and/or other publications (Ikeda 2003, Willis et al. 2005; Sprenger 2007; Rieping and Hermann 2007; Dong et al. 2011; Ikeda and Takeno 2013; Yokota and Ikeda 2017).

3.1 Glutamate

Since the discovery of *C. glutamicum* as a producer of the food flavoring monosodium glutamate, commercial production of glutamate has been conducted using this microbe exclusively. The industrial glutamate titer is assumed to exceed 150 g/L (Sanchez et al. 2018) with an estimated production yield towards sugar of almost 70%. The global demand for monosodium glutamate amounted to over three million tons in 2016 (Ajinomoto 2016; Sanchez et al. 2018) and is expected to surpass four million tons by 2023 (Global market insights 2016). Glutamate production by *C. glutamicum* is induced by biotin limitation or by treatment with certain fatty acid ester surfactants or with β-lactam antibiotics such as penicillin. Although induction treatment is the core technology involved in industrial glutamate production processes, the molecular basis of the induction of glutamate secretion was long unknown. In recent years, however, a valuable insight into the secretion mechanism has been gained with the identification of the NCgl1221 gene product as a glutamate exporter (Nakamura et al. 2007). An intriguing finding is that only a specific point mutation in the NCgl1221gene resulted in glutamate secretion without any induction treatments. It has also been shown that amplification of the wild-type NCgl1221gene increases glutamate secretion while its disruption substantially abolishes secretion accompanied by an increase in the intracellular glutamate pool under the induction conditions mentioned above. The gene in question encodes the YggB protein, which was originally described as a putative mechanosensitive channel (Nottebrock et al. 2003). Later electrophysiological studies using an *E. coli* or *Bacillus subtilis* strain devoid of mechanosensitive channels indicated that the NCgl1221 gene product actually possesses the activity of a mechanosensitive channel (Börngen et al. 2010; Hashimoto et al. 2010). It has also been shown that glutamate excretion through the channel was mediated by passive diffusion (Hashimoto et al. 2012), while carrier-mediated glutamate secretion by *C. glutamicum* was shown to be energy-dependent (Gutmann et al. 1992). Based on this possible function as a mechanosensitive channel, the following mechanism has been proposed: the induction conditions, such as biotin limitation and penicillin treatment, alter membrane tension by inhibiting lipid or peptidoglycan synthesis.

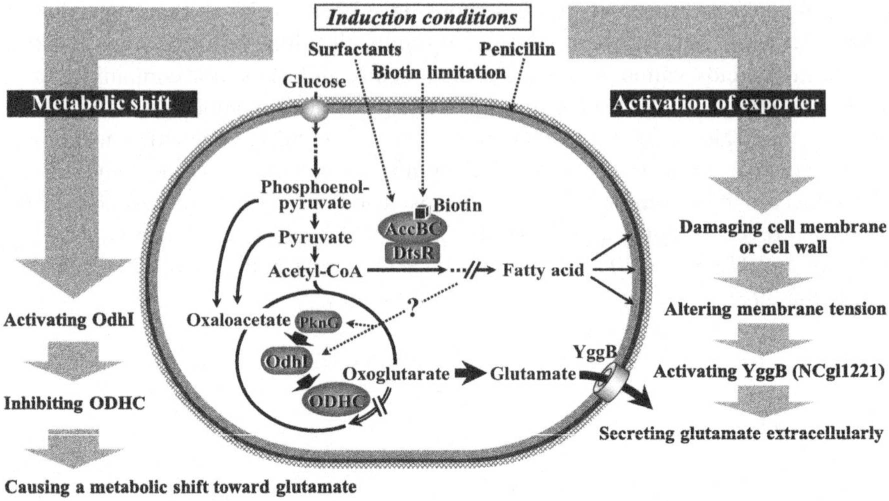

Fig. 11 Possible mechanisms triggering glutamate overproduction under the induction conditions in *C. glutamicum*. In this model, proteins AccBC and DtsR form the biotin-dependent acetyl-CoA carboxylase complex required for fatty acid biosynthesis; this biotin-enzyme complex is thought to be the primary target of biotin limitation and surfactant addition

This triggers conformational changes in the NCgl1221 gene product, which in turn enables the protein to export glutamate (Fig. 11).

Very recently, a different type of mechanosensitive channel was identified as a second glutamate exporter in some limited strains of *C. glutamicum* (Wang et al. 2018b). Although the channel MscCG2 shares only 23% identity with the NCgl1221 gene product, both channels have an important common feature: they need to be activated for glutamate production by the alteration of membrane tension, which is triggered by biotin limitation or penicillin treatment. This observation raises the question of how the new model may be congruent with the accepted notion that a decrease in the activity of the 2-oxoglutarate dehydrogenase complex (ODHC) is crucial for glutamate production (Shingu and Terui 1971; Kawahara et al. 1997; Kimura 2003; Asakura et al. 2007; Kim et al. 2009a, b). Although the new model seems to explain the basics of the mechanism underlying the induction of glutamate secretion, it is probably not sufficient to explain the entire process of glutamate production by *C. glutamicum*. Recently, a possible connection at a molecular level has been uncovered between ODHC activity and glutamate production (Fig. 11). A novel 15 kDa protein known as OdhI was identified as a regulator of ODHC (Niebisch et al. 2006). The unphosphorylated form of OdhI binds to the OdhA protein, one of the subunits of ODHC, and inhibits ODHC activity. This inhibition can be prevented by the PknG-catalyzed phosphorylation of OdhI. A phosphoserine/threonine protein phosphatase responsible for dephosphorylation of OdhI has also been identified (Schultz et al. 2007). Interestingly, disruption of the *odhI* gene was shown to abolish glutamate production even under the induction conditions (Schultz et al. 2007), suggesting a close relationship between the regulator protein and the reduction of ODHC activity that occurs during glutamate production. It is also worth noting that proteome analyses have revealed a significant increase in the OdhI protein upon penicillin treatment, which has become a conventional industrial method of triggering glutamate production (Kim et al. 2009a, b). These findings have confirmed the existence of a connection between ODHC activity and glutamate production, but it should be noted that an ODHC-activity-reducing metabolic change alone is not sufficient to induce glutamate production (Kim et al. 2009a, b).

Taken together, the evidence to date suggests a link between the induction treatments, such as biotin limitation and penicillin treatment, and glutamate production. In our proposed mechanism, the induction treatments enhance the synthesis of the regulator protein OdhI in its unphosphorylated form and thereby inhibit ODHC activity. This causes a metabolic shift at the branch point of 2-oxoglutarate, which channels carbon toward glutamate. Intracellularly-accumulated glutamate is then secreted into the medium via the NCgl1221 gene product YggB which has been activated in response to altered membrane tension (Fig. 11). Questions for the future include why and how the OdhI protein is overexpressed in response to the induction treatment and what conditions are required for the phosphorylation and dephosphorylation of OdhI.

Acetylome and succinylome analyses of glutamate-producing *C. glutamicum* have suggested that protein acetylation and succinylation are involved in glutamate production through the post-translational control of key enzymes such as phosphoenolpyruvate carboxylase (PPC) and ODHC (Mizuno et al. 2016). It has been shown that acetylation of PPC at lysine 653 caused decreased enzymatic activity, resulting in reduced glutamate production (Nagano-Shoji et al. 2017). On the contrary, deacetylation of the lysine residue has been suggested to improve glutamate production through activation of PPC because the increase in PPC activity during glutamate production was canceled by the defect of the deacetylases (Nagano-Shoji et al. 2017).

Continuous efforts have been made not only to understand glutamate production but also to improve the process. In addition to the general approaches, in which metabolic fluxes are directed into glutamate (Kimura 2003; Sato et al. 2008; Sawada et al. 2010), an innovative metabolic design allowing an increased maximum theoretical yield has recently been reported (Chinen et al. 2007). Glutamate biosynthesis from glucose in *C. glutamicum* is inevitably associated with the release of CO_2 in the pyruvate dehydrogenase reaction, but the creation of a novel metabolic route by installing the phosphoketolase pathway of *Bifidobacterium animalis* allowed the CO_2-releasing pyruvate dehydrogenase reaction to be bypassed via acetyl phosphate, and thereby led to increased glutamate production coupled with the suppression of CO_2 emission. On the other hand, expression of the *Vitreoscilla* hemoglobin gene *vgb* under a *tac* promoter in a wild-type *C. glutamicum* strain has been shown to increase glutamate production in both shake-flask and fermentor cultivations (Liu et al. 2008), probably due to the enhancement of respiration by the hemoglobin (Webster 1987; Kallio et al. 1994; Zhang et al. 2007). Very recently, a Chinese group has demonstrated efficient glutamate production in the biotin-excessive corn stover hydrolysate (lignocellulose biomass). To achieve this, an industrial strain of *C. glutamicum* was improved by two genetic modifications, that is, (1) truncating C-terminal amino acid residue of the mechanosensitive channel NCgl1221, leading to activation of glutamate secretion without any induction treatments, and (2) engineering *odhA* ribosome-binding site, leading to decreased ODHC activity. The engineered strain XW6 achieved a final titer of 65.2 g/L with a yield of 63% after 48 h of 3-L jar fermentor cultivation using corn stover hydrolysate as the feedstock (Wen and Bao 2019).

Aside from *C. glutamicum*, Ajinomoto isolated the gram-negative acid-tolerant bacterium *Pantoea ananatis* AJ13355 to generate a glutamate producer (Hara et al. 2012). This producer was shown to allow glutamate fermentation to be conducted under acidic conditions (pH 3–5) where the solubility of glutamate is low, leading to a new process called "glutamate crystallization fermentation" (Izui et al. 2006, Usuda et al. 2017). This new type of fermentation is considered to reduce the amounts of alkali (e.g., ammonia) and acid (e.g., sulfuric acid or hydrochloric acid) during the fermentation and subsequent purification steps, respectively, and also the amounts of by-product salts such as ammonium sulfate. This means that the new process is expected to decrease not only production costs but also the burden on the environment, thus indicating its potential as a sustainable production process.

3.2 Lysine

Lysine, an essential amino acid for animals, has significant commercial value as a feed additive to promote the growth of animals including swine and poultry, and thus is the second-ranking amino acid after glutamate in terms of worldwide annual production. Lysine is also used as a fish feed additive because it is generally the first limiting essential amino acid in many protein sources used in fish feeds (Hua 2013). The scale of the lysine market has been estimated at approximately 2.3 million tons in 2016 (Ajinomoto 2016) and is still growing at annual rates of around 10% (Ikeda 2017). As the scale of production has increased, lysine prices per kilogram have dropped to around US$1.4 (Ajinomoto 2016), fluctuating between US$1.2 and 2.5 over the past decade (Ajinomoto 2016; Eggeling and Bott 2015; Ikeda 2017), depending largely on competition from natural lysine sources such as soybean meal and sardine. The main suppliers are CJ CheilJedang (South Korea), Global Bio-Chem Technology Group (China), Ajinomoto (Japan), Archer Daniels Midland (USA), and Evonik Industries (Germany), among others (Eggeling and Bott 2015). Major commercial plants are located in the respective corn belts in China, North America, Brazil, Indonesia, and Russia. Because of the growing market for lysine, exhaustive studies have been undertaken in an attempt to engineer the metabolism of *C. glutamicum* for lysine production (Ikeda 2017). These studies have resulted in several effective strategies for rational strain improvement, including engineering of terminal pathways (Shiio and Miyajima 1969; Sano and Shiio 1971; Kase and Nakayama 1974), central metabolism (Petersen et al. 2001; Peters-

Wendisch et al. 2001; Riedel et al. 2001; Shiio et al. 1984; Becker et al. 2009; Chen et al. 2014; Blombach et al. 2007b; Radmacher and Eggeling 2007; van Ooyen et al. 2012; Mitsuhashi et al. 2006), NADPH-regeneration systems (Marx et al. 2003; Becker et al. 2005, 2007; Ohnishi et al. 2005; Kiefer et al. 2004; Moon et al. 2005; Takeno et al. 2010, 2016; Komati Reddy et al. 2015; Bommareddy et al. 2014; Kabus et al. 2007a; Xu et al. 2014c, 2018b, c), export systems (Burkovski and Krämer 2002; Vrljić et al. 1996), glucose uptake systems (Ikeda et al. 2010b, 2011, 2015; Lindner et al. 2011a, b; Ikeda 2012), energy metabolism (Bott and Niebisch 2003; Kabus et al. 2007b), and global regulation (Burkovski 2008; Brockmann-Gretza and Kalinowski 2006; Krömer et al. 2004, 2008; Hayashi et al. 2006a, b). In addition, recent genome-based and systems-level approaches such as genome breeding (Ohnishi et al. 2002; Ikeda et al. 2006) and systems metabolic engineering (Becker et al. 2011) have led to lysine producers with superior production performance in terms of yield, titer, and productivity, as described earlier in this chapter.

A genome-scale model of the *C. glutamicum* metabolic network has been constructed based on the annotated genome, available literature, and various "omic" data (Kjeldsen and Nielsen 2009). The constructed metabolic model consists of 446 reactions and 411 metabolites; the predicted metabolic fluxes during lysine production and growth under various conditions are highly consistent with experimental values. The ability to predict the metabolic state associated with maximum production yield can be used to guide strain engineering. This strategy has been proven through the rational design of high lysine-producing strains of *C. glutamicum* (Krömer et al. 2004; Becker et al. 2005; Wendisch et al. 2006).

Recently, Chinese groups have dramatically improved lysine productive performance of *C. glutamicum* by rational metabolic engineering (Xu et al. 2013, 2014a, b, c, 2018a, b, c). Seventeen stages of modification to a wild-type genome, aimed at increasing the carbon flow into the lysine-biosynthetic pathway, have led to the lysine hyper-producer Lys5-8, which achieved a final lysine titer of 130.82 g/L (163.52 g/L as lysine hydrochloride) with a conversion yield of 47.06% on glucose after 48 h of 7-L jar fermentor cultivation (Xu et al. 2014b). Very recently, different lines of lysine hyper-producers have been rationally constructed from a classical lysine producer. These include strains JL-6 Δ*dapB::Ec-dapB*$^{C115G,\ G116C}$ (Xu et al. 2018b) and JL-69$P_{tac\text{-}M}$gdh (Xu et al. 2018a). The typical strategy, used to create the former strain, was to lower dependency on NADPH during lysine biosynthesis. For this purpose, the NADPH-dependent DapB gene was replaced with an *E. coli* gene encoding a mutant type of NADH-dependent DapB to switch its nucleotide-cofactor specificities from NADPH to NADH, leading to an increase in final lysine titer from 82.6 to 117.3 g/L, an increase in conversion yield on sugar from 35 to 44%, and an increase in productivity from 2.07 to 2.93 g/L/h in 5-L jar fermentor cultivation. The latter strain, in contrast, was created by optimizing the carbon flux through the TCA cycle to balance cell growth and availability of oxaloacetate and glutamate for lysine biosynthesis from the viewpoint of maximizing precursor supply. Rational engineering of the phosphoenolpyruvate-pyruvate-oxaloacetate node and the TCA cycle, as well as the suitable feeding of biotin, has resulted in a final lysine titer of 181.5 g/L with a conversion yield of 64.6% on sugar after 48 h (productivity of 3.78 g/L/h) in

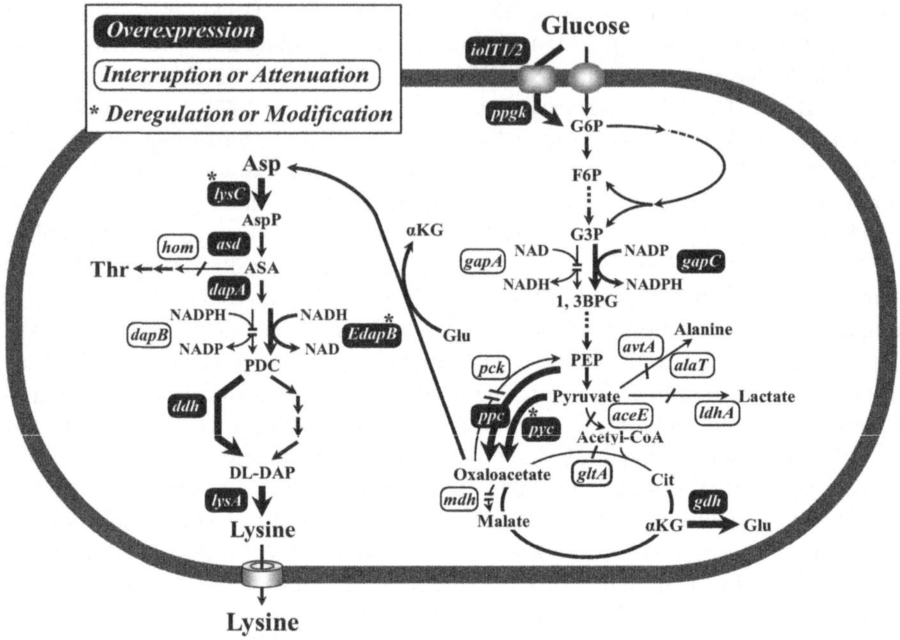

Fig. 12 Outline of metabolic engineering disclosed by the Chinese groups for lysine production

5-L jar fermentor cultivation. On the other hand, a classically derived lysine-producing mutant of *C. glutamicum* was improved by engineering glucose uptake systems. Adequate expression of the bypasses for glucose uptake (IolT1, IolT2, and PpgK), in addition to the native phosphoenolpyruvate-dependent sugar phosphotransferase system (PTS^{Glc}), in the strain has resulted in the lysine hyper-producer ZL-92, which achieved a final lysine titer of 201.6 g/L with a conversion yield of 65% on glucose after 40 h (productivity of 5.04 g/L/h) in 5-L jar fermentor cultivation (Xu et al. 2019). An outline of the metabolic engineering strategies by which these lysine hyper-producers were generated is schematically shown in Fig. 12.

The main feedstocks for lysine production by *C. glutamicum* are sugars from agricultural crops, such as cane molasses, beet molasses, and starch hydrolysates (glucose) from corn and cassava, but it is becoming necessary to engineer the use of alternative raw materials, in particular, materials that do not compete with human food or energy sources. To reduce the environmental impact of lysine production, *C. glutamicum* strains have been constructed that can utilize whey, which contains lactose and galactose (Barrett et al. 2004); glycerol, the main by-product of biodiesel production (Rittmann et al. 2008); lignocellulose, which contains the pentoses xylose and arabinose (Kawaguchi et al. 2006, 2008); and rice straw hydrolysate (Meiswinkel et al. 2013), though there are still technical challenges related to upstream raw material processing and carbon use efficiency. In addition to these attempts to employ nonedible second-generation renewables, there is an increasing interest in the development of a lysine production process using mannitol, a major constituent of marine microalgae (seaweed), as a third-generation renewable resource that might be more efficiently and sustainably supplied from ocean farms (Hoffmann et al. 2018). The sugar alcohol mannitol can be metabolized into lysine by an engineered *C. glutamicum* strain that expressed the NADP-dependent glyceraldehyde 3-phosphate dehydrogenase (GapN) of *S. mutans* so as to couple the glycolysis to NADPH formation, resulting in a lysine yield of 0.24 mol/mol and a productivity of 1.3 mmol/g/h.

In addition to strain engineering, continuous improvement of the process and the development of a comprehensive methodology for assessing the process (Anaya-Reza and Lopez-Arenas 2017) have resulted not only in fermentation processes with increased product yields and reduced loads on downstream processing, but also in the commercialization of various product forms for novel intended uses (Hirao et al. 1989; Ikeda 2003, 2017; Kelle et al. 2005; Kobayashi et al. 2011). Meanwhile, anaerobic production of lysine through a *C. glutamicum* process remains a great challenge, though several attempts, including the operation of a nitrate respiration system (Nishimura et al. 2007; Takeno et al. 2007) and the use of an anode such as ferricyanide as the extracellular electron carrier (Xafenias et al. 2017; Vassilev et al. 2018), have indicated that it may be possible to turn an aerobic production process into an anaerobic process.

3.3 *Arginine, Citrulline, and Ornithine*

Arginine, a semi-essential amino acid, has lately attracted considerable attention for being a precursor to nitric oxide (NO), a key component of endothelial-derived relaxing factor (Appleton 2002). Citrulline and ornithine, precursors of arginine biosynthesis as well as intermediates in the urea cycle, are also important for human health since they are sources of endogenous arginine in the body (Hayashi et al. 2005, 2006c; Curis et al. 2007; Ochiai et al. 2012; Mori et al. 2015). As the economic values of these amino acids have increased, considerable attention has been given to improving our understanding of their metabolism in microbes (Utagawa 2004; Glansdorff and Xu 2007; Lee et al. 2010; Petri et al. 2013; Huang et al. 2015, 2016; Lubitz et al. 2016). In parallel, increasing efforts have been directed to the development of more efficient production strains by using recent technologies for strain development. For example, the genome breeding approach has been successfully applied to develop an arginine and citrulline producer from a *C. glutamicum* wild-type strain, as described earlier in this chapter (Ikeda et al. 2009). The reengineered strain was constructed by assembling just three mutations (*argB26*, Δ*argR*, and *argB31*) derived from three different lines of classical producers, resulting in a final titer of over 80 g/L (as a sum of arginine and citrulline) in 30 h of 5-L jar fermentor cultivation at a suboptimal temperature of 38 °C (Fig. 5b and 6).

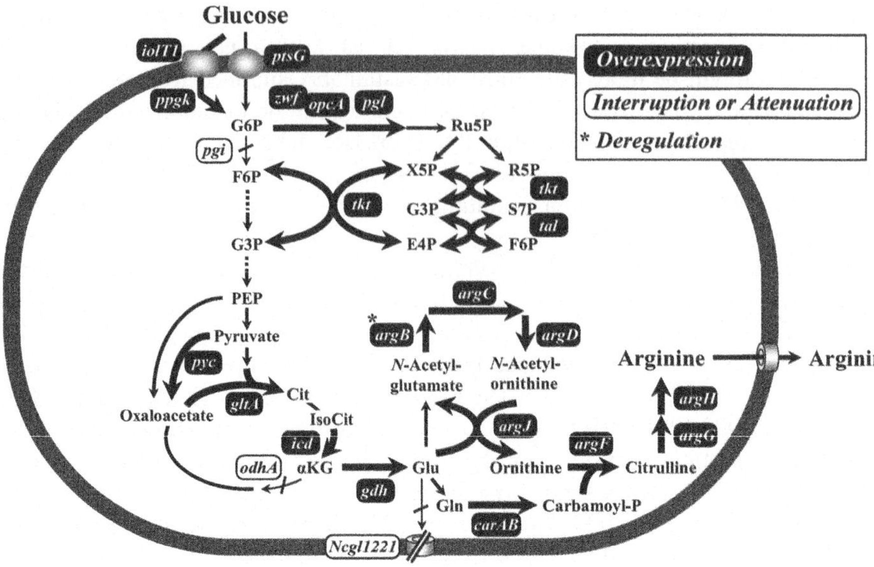

Fig. 13 Different approaches of metabolic engineering applied to the production of arginine

Recently, a classically derived arginine-producing mutant of *C. glutamicum* was improved through stepwise metabolic engineering, including deregulation of arginine biosynthesis, increased NADPH availability, disruption of the glutamate exporter, and adequate modifications of the terminal pathway (Fig. 13). This systematic approach has ultimately led to the arginine hyper-producer AR6, which achieved a final titer of 92.5 g/L with a conversion yield of 40% on glucose plus sucrose after 72 h of 5-L jar fermentor cultivation (Park et al. 2014). It is notable that citrulline by-production was not observed in the fermentation.

Likewise, a Chinese group has employed a similar metabolic engineering strategy for the development of the *C. glutamicum* arginine hyper-producer Cc6 (Fig. 13), which achieved a final titer of 87.3 g/L with a yield of 43.1% on glucose after 72 h of 5-L jar fermentor cultivation (Man et al. 2016b). They have also shown that modifications leading to decreased H_2O_2 synthesis and increased NADH and ATP levels contribute to improved arginine production (Man et al. 2016a).

Meanwhile, a German group took a different approach to developing producers of arginine, citrulline, and ornithine (Jensen et al. 2015). They first constructed the *C. glutamicum* ornithine-producing strain ORN2 (Δ*argF*, Δ*argR*, Δ*argG*) as a platform strain for subsequent development. Additional modifications, including deregulation of ArgB, overexpression of *gdh*, attenuation of *pgi*, and duplication of *argCJB*, resulted in the ornithine hyper-producer ORN6 with a yield of 52% on glucose. This ornithine hyper-producer was then converted into strains capable of producing citrulline (a yield of 41%) and arginine (a yield of 30%) by plasmid-mediated overexpression of *argFB* and *argGFB*, respectively. It is worth noting that the two specific mutations (A49V, M54 V) used for deregulation of ArgB are the same as those originally identified from the genomes of classical arginine producers during the genome breeding process mentioned earlier (Ikeda et al. 2009, 2010a).

Metabolic engineering of *C. glutamicum* has also led to industrially potent ornithine producers. For example, a Korean group has generated an ornithine producer by removal of competing pathways (Δ*argF*, Δ*proB*), deregulation of ornithine biosynthesis (Δ*argR*), overexpression of the terminal pathway with the use of plasmid pSY223 carrying *argCJBD*, and increased NADPH availability (pgi^{GTG}, zwf^{ATG}, P_{tkt}::P_{sod}). The resulting strain YW06 (pSY223) produced 51.5 g/L of ornithine with a yield of 24% on glucose after 40 h of 6.6-L jar fermentor cultivation (Kim et al. 2015).

Chinese groups have also developed *C. glutamicum* ornithine producers through systematic metabolic engineering (Jiang et al. 2013a, b; Zhang et al. 2017a, 2018a, b, c). One of their strategies was to convert an industrial glutamate producer into an ornithine producer by adequately attenuating *argF*, followed by combining the common strategies including deregulation of ornithine biosynthesis, increased NADPH availability, disruption of the glutamate exporter, removal of competing pathways, and adequate modifications of the terminal pathway. The resulting strain SO16 produced 32.3 g/L of ornithine with a yield of 39.5% on glucose in shake flask cultivation (Zhang et al. 2018b). It is notable that overexpression of *lysE* contributed to increased ornithine production (Zhang et al. 2017a). Considering that the lysine and arginine exporter LysE was shown not to accept ornithine and citrulline as substrates (Bellmann et al. 2001), the positive effect of LysE on ornithine production remains elusive.

3.4 Tryptophan

Tryptophan is one of the essential amino acids required in the diet of humans and other mammals such as pigs and poultry. Since tryptophan is particularly scarce in cereal grains, this amino acid is of considerable value for animal nutrition. Furthermore, tryptophan is known to improve the sleep state and mood as it is a precursor of serotonin which acts as a neurotransmitter in the nervous system (Bender 1985). Due to these nutritional and medicinal benefits, the amino acid has various application fields including food additives, pharmaceuticals, and feed supplements. Accordingly, the scale of the tryptophan market has expanded from about 500 tons in the year 2000 to approximately 33,000 tons in 2016 (Ajinomoto 2016). As the scale of production has increased, tryptophan prices per kilogram have dropped to around US$8 (Ajinomoto 2016).

Biosynthesis of 1 mol of tryptophan from glucose requires 1 mol each of erythrose 4-phosphate (E4P) and phosphoenolpyruvate (PEP) as starting precursors and consumes an additional 1 mol each of PEP, glutamine, phosphoribosyl-5-pyrophosphate, and serine on its biosynthetic pathways (Umbarger 1978). Therefore, a balanced supply of the precursors is required for efficient production of the

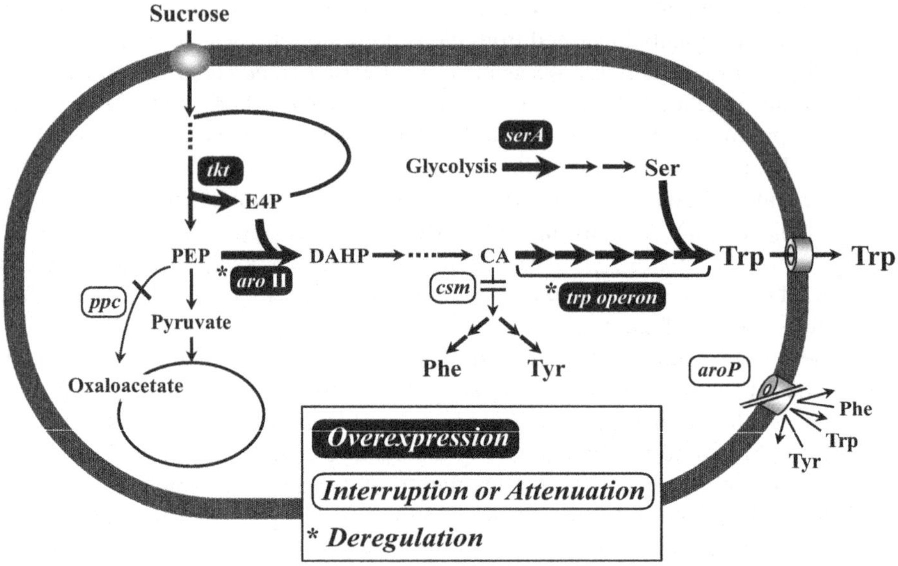

Fig. 14 Schematic diagram of a genetically engineered tryptophan-producing *C. glutamicum* strain

amino acid. Toward this goal, a sophisticated strategy has been applied to the pathway engineering of a classically derived tryptophan-producing *C. glutamicum* strain (Fig. 14) (Katsumata and Ikeda 1993; Ikeda et al. 1994). This strategy consisted of amplification of the first enzyme 3-deoxy-D-*arabino*-heptulosonate 7-phosphate synthase in the common pathway diverging from central metabolism to increase carbon flow down that pathway, followed by sequential removal of bottlenecks discerned by the accumulation of intermediates, resulting in a 61% increase in tryptophan production and a final yield of approximately 50 g/l. The remarkable improvement involves not only systematic genetic modifications to efficiently channel carbon towards tryptophan via plasmid-mediated amplification of all together eight genes of the pathways leading to tryptophan and serine, that is, *aro* II, *trpEGDCBA*, and *serA* (Fig. 14), but also construction of a plasmid stabilization system based on the presence of the *serA* gene on the plasmid and the gene's absence from the chromosome. Further modifications in the central metabolism to increase the availability of PEP and E4P through decreased PPC activity and increased transketolase activity, respectively, have ultimately resulted in a final titer of 58 g/l with a conversion yield of 23.2% on sucrose after 80 h of 2-L jar fermentor cultivation with no need for antibiotics (Katsumata and Kino 1989; Ikeda and Katsumata 1999). It is notable that more than half of the product crystallized in the medium.

In addition to pathway engineering, the impact of transport engineering on tryptophan production has been demonstrated in *C. glutamicum* (Ikeda and Katsumata 1994, 1995). A modification leading to a decreased rate of tryptophan uptake in a tryptophan-producing mutant resulted in increased production, while accelerated tryptophan uptake drastically decreased production. Considering this, the increased capacity of tryptophan efflux can be also a promising strategy for further increased production, although the excretion process of tryptophan in *C. glutamicum* remains to be elucidated.

Looking beyond *C. glutamicum*, the tryptophan exporter YddG has been reported in *E. coli* (Doroshenko et al. 2007; Airich et al. 2010). The YddG protein has been shown to mediate export of not only tryptophan but also phenylalanine and tyrosine and to enhance the production of the aromatic amino acids when overexpressed in *E. coli*. Following this, a Chinese group has improved a classically derived *E. coli* tryptophan producer by stepwise modifications including prevention of tryptophan uptake (Δ*mtr*), overexpression of *yddG*, and decreased acetate by-production (Δ*pta*), resulting in a final tryptophan titer of 48.68 g/l with a conversion yield of 21.87% on glucose after 38 h of 30-L fermentor cultivation (Wang et al. 2013).

Recently, starting from a wild-type background with respect to tryptophan biosynthesis, a genetically defined tryptophan-producing *E. coli* strain was developed through a rational metabolic engineering process that included interruption of tryptophan degradation (Δ*tnaA*), disruption of tryptophan importer (Δ*mtr*, Δ*tnaB*), and stepwise modifications of the terminal pathways leading to both tryptophan and serine (Chen and Zeng 2017). The resulting strain S028 produced 40.3 g/L of tryptophan with a yield of 15% on glucose after 61 h of fed-batch fermentation. Intracellular metabolite analysis of this strain suggested that availability of glutamine and export of tryptophan were likely to limit tryptophan production.

More recently, the same group has reported that tryptophan biosynthesis in some microorganisms, including *E. coli*, *Aspergillus niger*, and *Saccharomyces cerevisiae*, is regulated through anthranilate-associated feed-forward regulation at the indole-3-glycerol phosphate synthase step (TrpC), in addition to already-known regulations such as repression, attenuation, and feedback inhibition (Chen et al. 2018). Based on the findings, the anthranilate-activated TrpC enzyme from *A. niger* was expressed in the tryptophan producer S028, leading to an increase in a conversion yield on glucose from 15 to 18% in fed-batch cultivation.

3.5 Branched-Chain Amino Acids

The branched-chain amino acids valine, leucine, and isoleucine are all essential for human and animal nutrition, and all have increasing uses in various fields including pharmaceuticals, cosmetics, agricultural chemicals, dietary supplements, and feed additives. Currently, their most popular use is as a supplement for athletes to promote strength; this use is based on the nutraceutical effect of these amino acids on skeletal muscles (Shimomura et al. 2006). The intermediates for these amino acids can also be used for the production of biofuels (Atsumi et al. 2008). In *C. glutamicum*, all three of these amino acids share common uptake and export systems (Ebbighausen et al. 1989; Kennerknecht et al. 2002) as well as common substrates and enzymes for their biosynthesis, and thus are closely related in their metabolic fate.

Since the year 2000, rational metabolic engineering has been applied to the production of the branched-chain amino acids by *C. glutamicum* many times, with a special emphasis on valine production (Pátek 2007; Park and Lee 2010; Wang et al. 2018a). The strategies used to improve production of valine include (1) eliminating bottlenecks in the terminal pathway, either by conferring isoleucine auxotrophy which allows the attenuation control of the *ilvBNC* operon to be circumvented (Radmacher et al. 2002), by deregulating the key regulatory enzyme acetohydroxyacid synthase (Elisáková et al. 2005), or by overexpressing the gene set responsible for valine biosynthesis (Radmacher et al. 2002; Blombach et al. 2007a; Bartek et al. 2010); (2) increasing the availability of precursor pyruvate, either by blocking pantothenate synthesis (Radmacher et al. 2002; Bartek et al. 2008), by inactivating pyruvate dehydrogenase, pyruvate carboxylase, and pyruvate:quinine oxidoreductase (Blombach et al. 2007a, 2008, 2009a), or by introducing an H^+-ATPase defect which contributes to the enhanced glycolysis and thus to the increased supply of pyruvate (Li et al. 2007; Wada et al. 2008); and (3) increasing NADPH supply by inactivating phosphoglucose isomerase (Blombach et al. 2008; Bartek et al. 2010). These modifications have mostly been achieved through plasmid-mediated amplification and/or deletion of the targeted genes, possibly leading to perturbations of the natural homeostatic mechanisms of the cell. To alleviate such side-effects on cell physiology, the desired metabolic engineering has been achieved through purposeful mutagenesis of promoters of the chromosomal genes involved in the valine biosynthesis pathway and in competing pathways (Holátko et al. 2009). The resulting plasmid-free valine producer was auxotrophic to pantothenate and bradytrophic to isoleucine, carried a feedback-resistant acetohydroxy acid synthase, and expressed the genes *ilvD* and *ilvE* from strong mutant promoters. This new type strain with all mutations constructed within the chromosome has been shown to produce 136 mM (15.9 g/L) valine from 4% glucose after 48 h of flask cultivation.

In addition to such conventional aerobic processes, a different bioprocess has been reported for valine production (Hasegawa et al. 2012, 2013). In this alternative process, engineering of the redox balance in combination with the use of growth-arrested packed cells has allowed *C. glutamicum* to produce valine at high yields under anaerobic conditions. Theoretically, the biosynthesis of 1 mol of valine from 1 mol of glucose generates 2 mol of NADH during glycolysis but requires 2 mol of NADPH at the IlvC and IlvE steps, thus causing a redox imbalance during anaerobic valine production (Fig. 15). This problem was overcome by switching the cofactor requirement of IlvC from NADPH to NADH and introducing NAD-specific exogenous leucine dehydrogenase instead of NADPH-specific IlvE (Fig. 15). Further modifications intended to reconcile redox balance with high-yield valine production and low by-product formation have ultimately resulted in a final titer of 1280 mM (150 g/l) with a conversion yield of 88% (mol/mol) or 57.2% (w/w) on glucose after 24 h under growth-arrested anaerobic conditions.

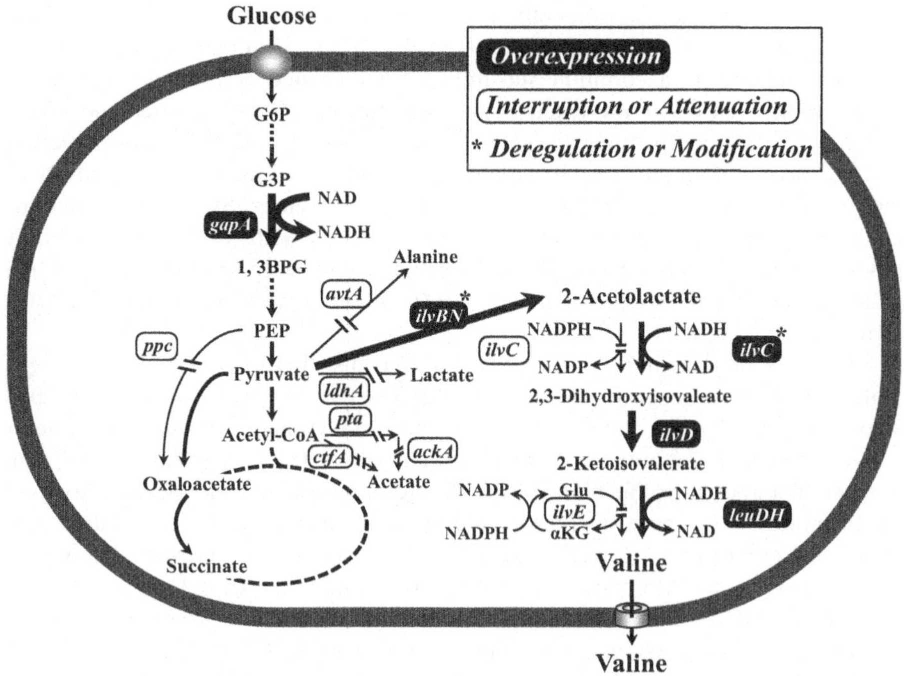

Fig. 15 Schematic diagram of a genetically engineered *C. glutamicum* strain for valine production under oxygen deprivation conditions

Recently, a Chinese group has developed valine producers starting from a different *C. glutamicum* wild-type strain, ATCC 13869, through systematic metabolic engineering involving deletion of the three genes *aceE*, *alaT*, and *ilvA* and overexpression of the six genes *ilvB*, *ilvN*, *ilvC*, lrp_1, *brnF*, and *brnE*, resulting in a final titer of 51.2 g/l with a conversion yield of 30.8% on glucose after 96 h of 5-L fermentor cultivation (Chen et al. 2015). The same group has also found, through transcriptomic and proteomic analysis of a classically derived *C. glutamicum* valine-producing mutant, that the up-regulation of the genes responsible for ribosome elongation factors and ribosomal proteins is involved in valine production (Zhang et al. 2018d).

On the other hand, evolutionary approaches such as biosensor-driven screening or applications of genetic suppression have revealed non-intuitive beneficial mutations for valine production. These include a knockout mutation in the *ureD* gene involved in the degradation of urea to carbon dioxide and ammonium and an *icd* knockdown mutation allowing a metabolic shift from the TCA cycle to the glyoxylate shunt (Mahr et al. 2015; Schwentner et al. 2018). Recent developments in valine production by metabolically engineered *C. glutamicum* strains have been summarized in a previous review (Wang et al. 2018a).

For leucine production, rationally designed *C. glutamicum* strains have also been reported by a German group (Vogt et al. 2014). Leucine biosynthesis has 2-ketoisovalerate as a common substrate for valine and additionally requires a supply of acetyl-CoA in the leucine-specific pathway. Therefore, researchers have sought to expand the pathways leading to the common substrate 2-ketoisovalerate, and the targets of engineering include citrate synthase for increased availability of the precursor acetyl-CoA and the leucine-specific enzymes. Systematic metabolic engineering starting from the wild-type strain ATCC 13032 has led to an industrially potent, genetically defined, and plasmid-free leucine producer capable of a final titer of 181 mM (23.7 g/L) with a yield of 21.8% (w/w) on glucose after 72 h of fed-batch fermentation.

Metabolic engineering of *C. glutamicum* has also led to isoleucine producers. Isoleucine is synthesized from threonine through five enzymatic steps. In the first step, catalysed by the key enzyme threonine dehydratase (IlvA), threonine is converted into 2-ketobutyrate, which is then converted into isoleucine by four enzymes common to valine biosynthesis. Therefore, metabolic engineering involves modifications that cause increased supply of threonine and deregulation and overexpression of IlvA, as well as modifications affecting the enzymes common to valine biosynthesis (Morbach et al. 1995, 1996; Yin et al. 2012; Vogt et al. 2015; Dong et al. 2016). Modifications leading to overexpression of the global regulator Lrp and the branched-chain amino acid exporter BrnFE (Xie et al. 2012; Yin et al. 2013) and increased availability of NADPH (Shi et al. 2013; Ma et al. 2016) are also useful for efficient production of isoleucine.

Recently, a Chinese group has reported that ribosome elongation factor G and recycling factor, both of which are relevant to protein synthesis, contribute to increased isoleucine production when overexpressed (Zhao et al. 2015). Plasmid-mediated amplification of the corresponding genes *fusA* and *frr*, together with the isoleucine-biosynthetic genes *ilvA*, *ilvB*, and *ilvN* and the NAD kinase gene *ppnk*, in a classical isoleucine producer of *C. glutamicum* has resulted in a final titer of 28.5 g/L with a yield of 13.9% on glucose after 72 h of fed-batch fermentation.

The branched-chain amino acids are mainly used for pharmaceutical purposes and are required to have the highest degree of purity. From this perspective, attempts have been made to minimize the by-production of other amino acids to a level at which supplementary purification of the desired amino acid is not necessary. For example, the by-production of alanine occurs during valine production but can be overcome by deletion of the alanine aminotransferase gene *alaT* in a *C. glutamicum* valine producer, thereby facilitating cost-effective downstream processing (Marienhagen and Eggeling 2008). On the other hand, production of leucine is often accompanied by accumulation of valine since *C. glutamicum* predominantly uses the single transaminase IlvE for the synthesis of the branched-chain amino acids from the respective keto acids, thus causing co-production of the amino acids. This problem has been overcome by using different types of aminotransferases (Feng et al. 2018). Overexpression of endogenous AspB or heterologous *E. coli* TyrB instead of native IlvE in an isoleucine-auxotrophic leucine producer of *C. glutamicum* has led to leucine production with almost no by-production of valine.

3.6 Methionine

Methionine, another essential amino acid for animals, has a great deal of commercial value as a feed additive. Moreover, methionine is important as a precursor of *S*-adenosylmethionine. For this purpose, methionine is produced exclusively by chemical synthesis in D, L-forms, as this amino acid is considered to have a similar effect on animal nutrition in both L- and D, L-forms. Nowadays, however, there is an increasing interest in the development of environmentally friendly fermentation methods using renewable feedstocks to produce methionine.

In the hope of discovering a method for the rational construction of a methionine producer, methionine biosynthesis and its regulation are being studied in *C. glutamicum*. This microbe possesses both transsulfuration and direct sulfhydrylation pathways, in contrast to *E. coli* and most other microorganisms, which utilize only one of these two pathways (Lee and Hwang 2003; Hwang et al. 2007). Two regulatory genes in *C. glutamicum* have been identified as relevant to methionine biosynthesis: *mcbR* (cg3253) and NCgl2640. Inactivation of either in wild-type *C. glutamicum* results in increased methionine production (Mampel et al. 2005; Rey et al. 2003, 2005). The common strategy underlying the metabolic engineering is to redirect carbon from the lysine pathway into the methionine pathway, and all the following achievements have in principle been performed using *C. glutamicum* lysine producers as platforms. For example, the introduction of feedback-resistant *lysC* and *hom* genes and deletion of the *thrB* gene resulted in 2.9 g/L of methionine together with 23.8 g/L of lysine (Park et al. 2007b). Overexpression of the homologous *metX* and *metY* genes in another lysine-producing *C. glutamicum* strain was reported in a patent by Möckel et al. (2002) as resulting in a final titer of 16 g/L of methionine. Deletion of the *mcbR* gene and overexpression of the *brnFE* genes encoding an exporter for methionine (Trötschel et al. 2005) led to 6.3 g/L of methionine production after 64 h in fed-batch fermentation (Qin et al. 2015).

Recently, Li et al. (2016) have developed a *C. glutamicum* methionine producer through the combination of rational metabolic engineering and random mutagenesis. The procedure and its effects on methionine production are summarized as follows (Fig. 16). The first step was to abolish the reuptake of methionine, followed by random mutagenesis. Deletion of the *metD* locus comprising genes *metQNI* for the methionine uptake system (Trötschel et al. 2008) and subsequent random mutagenesis conferred methionine production of 2.54 g/L on wild-type *C. glutamicum* ATCC 13032 after 72 h in fed-batch fermentation. Although the mutations obtained by the random mutagenesis remain undefined, they are associated with increased expression of the genes mainly involved in methionine biosynthesis, including *hom*, *metX*, *metY*, *metB*, *aecD*, *metE*, *metH*, and *metK*. The second step was to block or weaken competitive branch pathways. Deletion of the *thrB* gene and replacement of the start codon ATG by the rare GTG in the *dapA* gene ($dapA_{A1G}$) resulted in methionine production of 2.99 g/L. The third step was to enhance the precursor supply for methionine biosynthesis. Introduction of the feedback-resistant *lysC* gene ($lysC^{T311I}$)

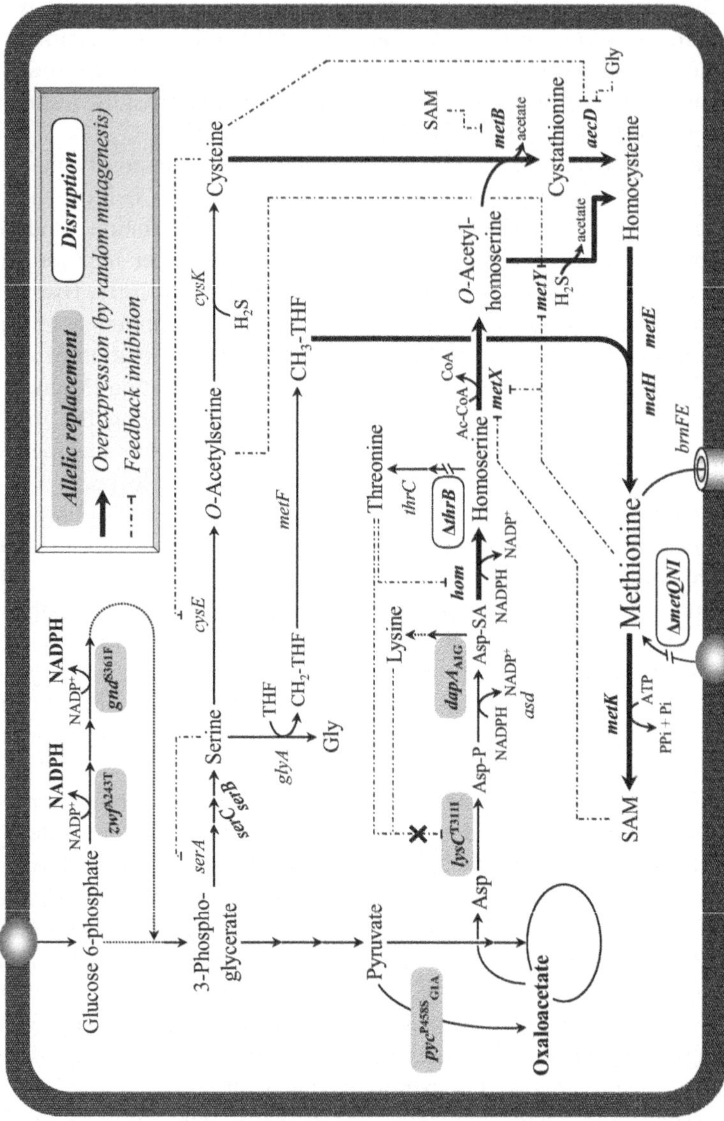

Fig. 16 Schematic diagram of a genetically engineered methionine-producing *C. glutamicum* strain

(Ohnishi et al. 2002) and a version of the *pyc* gene that had undergone a C-to-T exchange at nucleotide position 1327 and a GTG-to-ATG exchange at the start codon ($pyc^{P458S}{}_{G1A}$) resulted in methionine production of 5.89 g/L. The pyc^{P458S} gene is responsible for an increased supply of oxaloacetate (Ohnishi et al. 2002). The fourth step was to improve the NADPH supply. Introduction of mutant alleles of genes *zwf* (zwf^{A243T}) and *gnd* (gnd^{S361F}), both of which are responsible for an increased supply of NADPH (Ohnishi et al. 2005; Becker et al. 2007), achieved methionine production of 6.85 g/L after 72 h in fed-batch fermentation, which corresponds to a conversion yield of 8% (mol/mol) on glucose.

In *E. coli*, on the other hand, an attempt at systematic metabolic engineering resulted in a strain that produces methionine at an industrially useful level (Figge et al. 2009). The key to success here was achieving a balanced supply of three important precursors for methionine biosynthesis: *O*-succinylhomoserine, cysteine, and the C_1 carbon methyl-tetrahydrofolate (CH_3-THF). An imbalanced supply of these precursors causes the formation of undesired by-products such as homolanthionine and isoleucine through the involvement of certain methionine-biosynthetic enzymes themselves. The engineered *E. coli* strain has achieved a yield of 19.9% after 50 h in fed-batch fermentation without the formation of any detectable undesirable by-products. Based on this yield, the methionine titer is estimated at more than 35 g/L. The procedure and impact of this metabolic engineering project can be found in the first edition of this book. Very recently, in an engineered *E. coli* strain based on the similar concept of balancing the supply of the three precursors, deletion of the *metI* gene involved in methionine uptake (Merlin et al. 2002) and overexpression of the *yjeH* gene encoding an exporter of methionine (Liu et al. 2015) resulted in methionine production of 17 g/L after 48 h in fed-batch fermentation (Huang et al. 2018). The fact that the final titer in this report is lower than that reported by Figge et al. (2009) appears to be attributable to insufficient optimization of the expression of the manipulated genes or the culture conditions.

Though some progress has been made toward creating improved methionine producers, methionine yields still remain low compared with those attained for other amino acids. Metabolic pathway analysis has been used to evaluate the theoretical maximum yields of methionine production on the substrates glucose, sulfate, and ammonia in *C. glutamicum* and *E. coli* (Krömer et al. 2006). The theoretical yield (mol-C methionine per mol-C glucose) of *C. glutamicum* was 0.49, while that of *E. coli* was somewhat higher at 0.52. This analysis also showed that introduction of the *E. coli* glycine cleavage system into *C. glutamicum* as an additional C_1 source and the replacement of sulfate with thiosulfate or sulfide, thereby avoiding the need for reduction of oxidized sulfur, would increase the theoretical maximal methionine yields in *C. glutamicum* to 0.57 and 0.68, respectively. Furthermore, when methanethiol (also known as methylmercaptan) is used as a combined source for a C_1 carbon and sulfur in *C. glutamicum*, the theoretical yield is estimated to reach its highest potential value at 0.91 (Krömer et al. 2006).

Recently, the potential utilization of methanethiol and its dimeric form dimethyldisulfide as both the C_1 source and the sulfur source has been experimentally verified in *C. glutamicum* (Bolten et al. 2010). Isotope experiments have

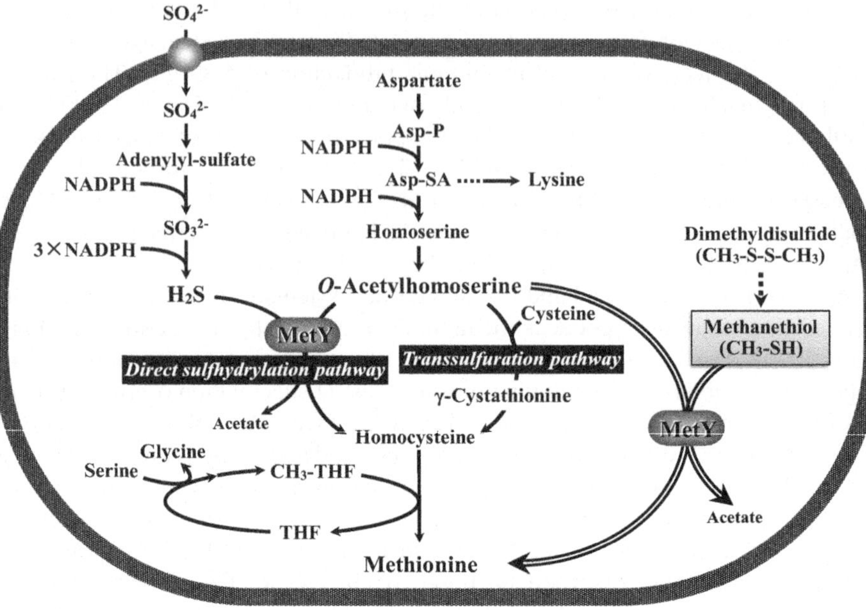

Fig. 17 Proposed pathway for assimilation in *C. glutamicum* of methanethiol and dimethyldisulfide into methionine in addition to two known pathways of transsulfuration and direct sulfhydrylation

revealed that the S-CH_3 group is entirely added to *O*-acetylhomoserine, directly yielding methionine (Fig. 17). This reaction has been shown to be catalyzed by MetY, creating a shortcut for methionine biosynthesis. The problem in this case would be the toxicity of these sulfur compounds to cells. A delivery system using a beaded macroporous polystyrene resin has been suggested as a potential way of alleviating the toxic effects (Bolten et al. 2010).

3.7 S-Adenosyl-Methionine

S-Adenosylmethionine (SAM) is a sulfonium compound recognized as a primary methyl donor for reactions catalyzed by methyltransferases. SAM is also required for a variety of reactions as a source of methylene groups, amino groups, ribosyl groups, aminoalkyl groups, and 5′-deoxyadenosyl radicals (Fontecave et al. 2004). In addition, SAM is also available worldwide as a drug and in the United States as a nutritional supplement (Lu and Mato 2012).

Recently, metabolic engineering for SAM production has been conducted using *C. glutamicum*. To achieve the following outcomes, SAM is accumulated in engineered cells; each SAM titer (g/L or mg/L) is thus provided as an intracellular concentration. The wild-type strain *C. glutamicum* ATCC 13032 has been reported to accumulate 0.32 g/L of SAM by the overexpression of the homologous *metK* gene encoding a methionine adenosyltransferase after 24 h of 72 h fed-batch culture with feeding methionine at a final concentration of 6 g/L (Han et al. 2016b). Since ATP as well as methionine is required for the MetK reaction, metabolic engineering aiming to enhance ATP supply by increasing oxygen availability for the respiratory chain has been performed using *Vitreoscilla* hemoglobin encoded by the *vgb* gene (Han et al. 2015). Overexpression of the *metK* and *vgb* genes in the isoleucine-producing strain *C. glutamicum* IWJ001 resulted in co-production of 0.67 g/L of SAM and 13.9 g/L of isoleucine after 72 h of fed-batch culture with feeding methionine (6 g/L), which quantities are 37-fold higher and 1.3-fold lower, respectively, than those produced by the control strain IWJ001.

Direct fermentation of SAM from glucose has also been attempted using an engineered *C. glutamicum* strain (Han et al. 2016a). This strategy is aimed at enhancing methionine biosynthesis and blocking competing branch pathways. The procedure and SAM production are summarized as follows (Fig. 18). First, the *mcbR* and *thrB* genes were deleted in wild type *C. glutamicum* ATCC 13032. This resulted in SAM accumulation of 33.2 mg/L (2.58 mg/g of dry cell weight) after 36 h on glucose in 50-mL batch culture. Second, in order to prevent homolanthionine accumulation and redirect the metabolic flux toward the direct sulfhydrylation pathway by *O*-acetylhomoserine sulfhydrylase encoded by the *metY* gene, the *metB* gene encoding cystathionine-γ-synthase was deleted. This resulted in SAM accumulation of 74.3 mg/L (6.12 mg/g of dry cell weight) after 36 h. Third, to improve assimilation of sulfur for methionine production using the sulfhydrylation pathway, NCgl2640, another regulatory gene for the *metY* gene, was deleted, which resulted in SAM accumulation of 95.4 mg/L (7.18 mg/g of dry cell weight) after 36 h. Finally, the *metK* and *vgb* genes were co-overexpressed, leading to SAM accumulation of 196.7 mg/L (12.15 mg/g of dry cell weight) after 48 h.

3.8 Cysteine

Cysteine, the other sulfur-containing amino acid, is nonessential but has a crucial function in metabolism as a precursor of sulfur-containing compounds such as methionine, thiamine, biotin, lipoic acid, and coenzyme A. In addition to its biological significance, cysteine is important commercially because of its various applications in the pharmaceutical, cosmetic, food, and livestock industries. As there is currently no efficient method of producing cysteine through fermentation, its production has depended on other methods including microbial conversion from DL-2-amino-Δ^2thiazoline 4-carboxylic acid (Sano et al. 1977) and extraction from natural protein-rich resources such as hair and keratin.

As with methionine, it has been difficult to engineer strains that produce high yields of cysteine, though this amino acid is synthesized in *C. glutamicum* from serine via *O*-acetyl-serine in only two steps (Haitani et al. 2006). Typical strategies

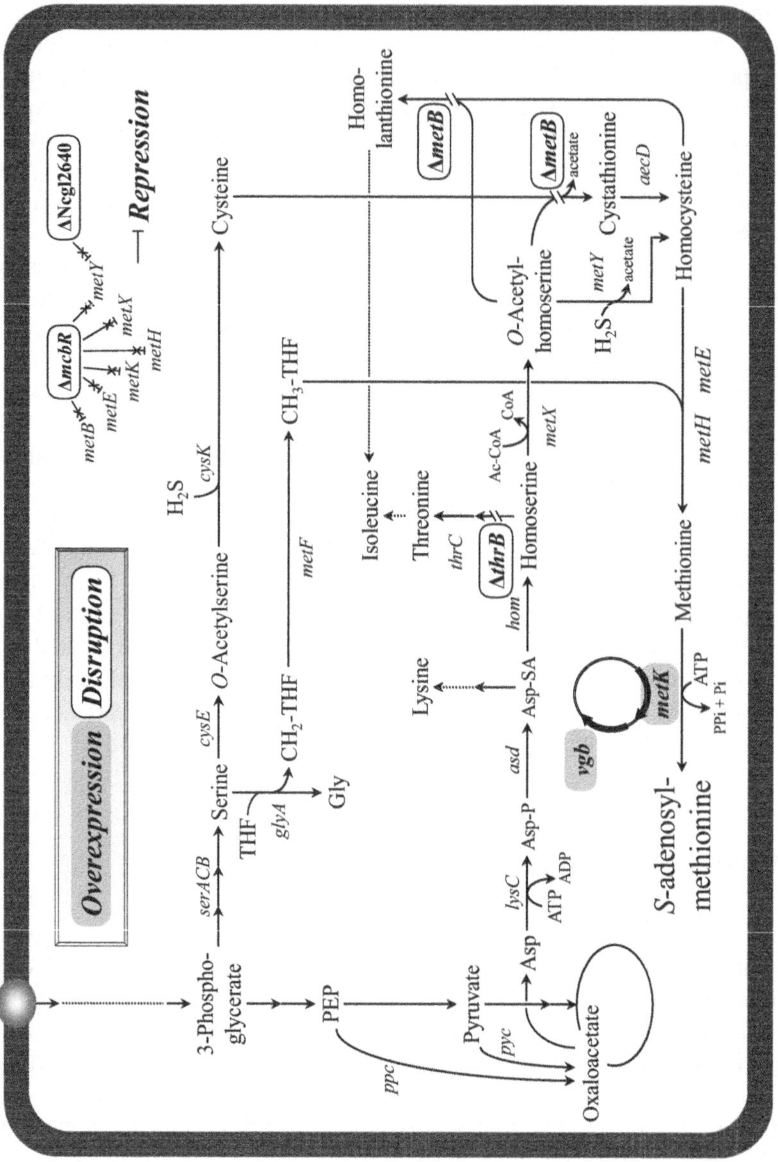

Fig. 18 Schematic diagram of a genetically engineered *S*-adenosylmethionine-producing *C. glutamicum* strain. Information regarding feedback regulation is omitted here since it was already provided in Fig. 16

include deregulation of the key regulatory enzyme serine *O*-acetyltransferase, deletion of the cysteine desulfhydrase gene that catalyzes cysteine degradation to pyruvate, and overexpression of cysteine exporters; these methods have been shown to increase cysteine production in both *E. coli* and *C. glutamicum*, though the final titers were below 2 g/L (Wada et al. 2002; Wada and Takagi 2006). It has been suggested that the combination of these strategies and the improvement of other factors including an increased supply of the precursor serine and a decreased reuptake of the product would lead to further improvement (Wada and Takagi 2006).

In *C. glutamicum*, the transcriptional regulator CysR, whose gene is repressed by McbR, activates the expression of the genes involved in assimilatory sulfate reduction and of the other regulatory gene *ssuR* (Rückert et al. 2008). Subsequently, SsuR activates the expression of the genes involved in sulfonate utilization (Koch et al. 2005). Therefore, overexpression of the *cysR* gene enables the simultaneous expression of many genes associated with the assimilatory reduction of sulfur source. Overexpression of the *cysE*, *cysK*, and *cysR* genes leads to intracellular accumulation of approximately 60 mg/L of cysteine on glucose after 15 h of 50-mL batch culture (Joo et al. 2017). In a more advanced study using *C. glutamicum*, deletion of *aecD* and *sdaA*, two genes that are involved in cysteine and serine degradation, respectively, along with reduction in the expression level of the *glyA* gene and overexpression of the feedback-insensitive *cysE* gene, the *cysK* gene, the *E. coli bcr* gene involved in cysteine export, and the feedback-insensitive *serA* gene together with the *serCB* genes, allowed the wild-type strain to produce 950 mg/L of cysteine with a yield of 2.73% on glucose after 36 h batch cultivation in a medium containing 6 g/L of thiosulfate (Wei et al. 2018).

Recently, in *E. coli*, a novel CysM (*O*-acetylserine sulfhydrylase B)-independent thiosulfate assimilation pathway was identified. This novel route, specifically a thiosulfate sulfurtransferase (GlpE)-mediated bypass from thiosulfate to sulfite, was evaluated for cysteine production (Kawano et al. 2017). Whereas overexpression of the feedback-insensitive *serA* and *cysE* genes together with the native *ydeD* gene involved in efflux of cysteine allowed a wild-type *E. coli* strain to produce approximately 1.0 g/L of cysteine on glycerol after 72 h in an experiment where sulfate and thiosulfate were added as sulfur sources, the addition of *glpE* overexpression achieved cysteine production of 1.5 g/L under the same conditions (Kawano et al. 2017). Very recently, it has been reported in *E. coli* that overexpression of the feedback-insensitive *cysE* gene, the *ydeD* gene, and the feedback-insensitive *serA* gene along with the native *serCB* genes, as well as deletion of the *sdaA* and *tnaA* genes, which are responsible for degradation of serine and cysteine, respectively, resulted in cysteine production of 5.1 g/L after 32 h on glucose in fed-batch culture (Liu et al. 2018b).

In recent years, *Pantoea ananatis* and closely related species have received attention in the fermentation industry for their potential to overproduce a wide variety of useful chemicals (Takumi et al. 2017). Metabolic engineering for cysteine production has been performed using *P. ananatis* strain SC17 (Fig. 19). This was initiated by the following manipulations (Takumi and Nonaka 2014). First, to deregulate the cysteine biosynthetic pathway, a mutant allele of the *E. coli cysE*

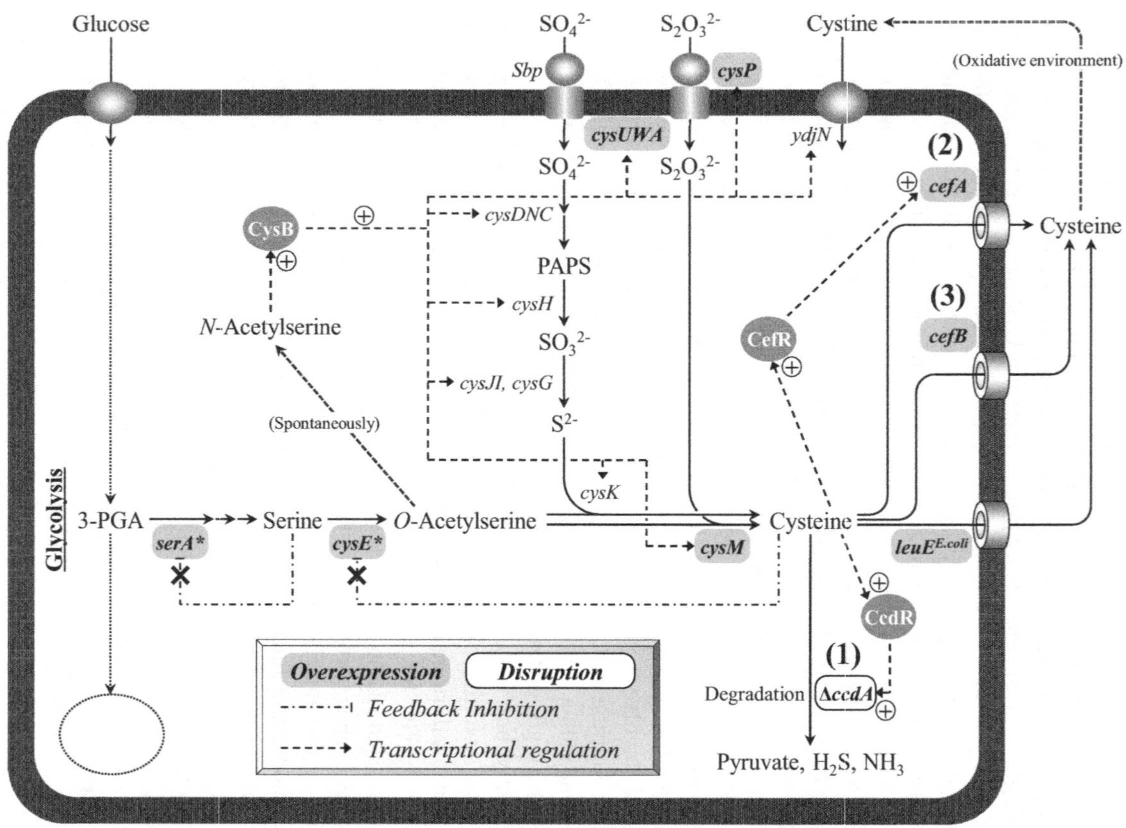

Fig. 19 Schematic diagram of a genetically engineered cysteine-producing *P. ananatis* strain. *Asterisked genes* encode enzymes resistant to feedback inhibition. Deletion of the *ccdA* gene (*1*), overexpression of the *cefA* gene (2), or overexpression of the *cefB* gene (*3*) was conducted in strain AG4854 where CysM activity was enhanced as described in the text

gene that encodes a feedback-resistant serine *O*-acetyltransferase was overexpressed. Simultaneously, to enhance the efflux of cysteine, the *E. coli leuE* encoding LeuE, which was originally identified as a leucine efflux pump (Kutukova et al. 2005) but which also exhibits cysteine efflux activity (Takumi and Nonaka 2014), was overexpressed. Second, to enhance thiosulfate uptake, the intrinsic *cysPUWA* genes encoding the thiosulfate transporter CysPUWA, a bottleneck in the intracellular sulfur supply pathway (Sirko et al. 1990), were overexpressed. Finally, to deregulate the biosynthetic pathway of serine that is the precursor of cysteine, a mutant allele of the homologous *serA* gene that encodes a feedback-resistant 3-phosphoglycerate dehydrogenase was overexpressed. Although the cysteine productivity of each intermediate strain derived during the above process has not been disclosed, the assembly of all these modifications to strain SC17 resulted in strain AG4854, which is capable of cysteine production of 1.3 g/L on glucose after 16 h cultivation with an estimated conversion yield of 3.1% on glucose (Takumi et al. 2017). While engineering of the transcriptional regulator CysB, a master regulator that induces most genes involved in sulfur assimilation and cysteine metabolism, is assumed to be beneficial to cysteine production as it simultaneously enhances the metabolism of sulfur and cysteine, it has a disadvantage in that it causes the expression of the *ydjN* gene encoding the cystine uptake transporter, resulting in reuptake of cysteine in the form of cystine, which is formed from cysteine in an oxidative environment (Nonaka 2018).

Based on the finding that CysM is the bottleneck of cysteine production, continued research has demonstrated that conferring moderately and excessively enhanced CysM activity on strain AG4854 (which created strains AG6181 and AG6184, respectively) resulted in yields of 3.5% and 0.4%, respectively. The failure of strain AG6184 to produce cysteine was accompanied by increased activity of cysteine-inducible CcdA (cysteine desulfhydrase), the only major cysteine degradation enzyme in *P. ananatis* (Takumi and Nonaka 2016), which is presumably induced by the elevated intracellular cysteine levels resulting from the increased activity of CysM. Disruption of the *ccdA* gene improved the yields of strains AG6181 and AG6184 to 3.9% and 1.6%, respectively, but had negative effects on growth and glucose consumption, which were especially severe in strain AG6184, presumably due to the toxicity of intracellularly accumulated cysteine. An effort to elicit only positive effects from CysM overexpression on cysteine production was made using the cysteine efflux pumps CefA (Takumi and Nonaka 2016) and CefB (Takumi et al. 2017). Yet without the implementation of fine-tuned expression, overexpression of the *cefA* gene in AG6184 led to cysteine production of approximately 0.4 g/L with an estimated yield of approximately 1.0% after 28 h cultivation, whereas overexpression of the *cefB* gene achieved cysteine production of approximately 2.2 g/L with an estimated yield of approximately 5.5% after 22 h cultivation (Takumi et al. 2017). A combination of *ccdA* deletion and *cefA* and/or *cefB* overexpression was not conducted in this report. Although not all details of this experiment were given, metabolic engineering of *P. ananatis* has achieved cysteine production of close to 5.0 g/L (Nonaka 2018).

4 Conclusion and Future Prospects

As already mentioned, the global amino acid market per year has expanded to more than seven million tons and over US$13 billion in response to increased demand for amino acids. This market growth is expected to continue due to the ongoing increases in the nutritional values of amino acids and the growth of the numerous fields that use them. The giant market has been underpinned largely by recent advances in amino acid fermentation technologies, especially strain development technology.

In the history of amino acid fermentation, the determination of the complete genome sequence of *C. glutamicum* was obviously an important milestone. The subsequent rapid progress in genomics, various "omics" technologies, and systems biology for this bacterium have dramatically transformed our approaches to strain development. For example, in-silico modeling and simulation approaches are now used routinely to help identify new targets for further engineering and strain improvement. The power of such systems-level approaches will surely increase as modeling is combined with the ever-accumulating "omics" data.

It should be noted, however, that not all purely rational approaches from scratch have necessarily resulted in commercially potent production strains, probably due to the existence of unknown mechanisms affecting industrially important properties, such as hyperproduction and high adaptability to large-scale processes. This means that there is a great deal more to learn from the genomes of classical strains. Since the dawn of the genomic era, new possibilities have emerged, including analysis of producer's genomes, leading to the genome breeding approach, and systems metabolic engineering, leading to tailor-made cell factories with designed properties (Becker and Wittmann 2015; Ikeda 2017). The next-generation strains are expected to be created through the synergy of these approaches and through integration of the knowledge accumulated over decades of industrial strain development with emerging technologies such as biosensor-driven single cell screening, in-silico modeling, and carbon flux simulation.

At the same time, the amino acid industry is beginning to consider sustainable and environmentally-friendly manufacturing systems in response to the continuing crisis of global warming. From this standpoint, the industry is expected to develop strains enabling the use of feedstocks that are renewable and that do not compete with human food or energy sources. The development of innovative technologies enabling reduction in effluents and wastes generated during fermentation and purification processes is also expected. These remain important themes for future engineering.

References

Airich LG, Tsyrenzhapova IS, Vorontsova OV, Feofanov AV, Doroshenko VG, Mashko SV (2010) Membrane topology analysis of the *Escherichia coli* aromatic amino acid efflux protein YddG. J Mol Microbiol Biotechnol 19:189–197

Ajinomoto (2007) Fact sheets: amino acids business and feed-use amino acids business. http://www.ajinomoto.com/ar/i_r/pdf/fact/Aminoacids-Oct2007.pdf, http://www.ajinomoto.com/ar/i_r/pdf/fact/Feed-useAA-Oct2007.pdf

Ajinomoto (2016) FY2015 market and other information. https://www.ajinomoto.com/en/ir/event/presentation/main/09/teaserItems1/00/linkList/02/link/FY15_Data_E.pdf

Anaya-Reza O, Lopez-Arenas T (2017) Comprehensive assessment of the L-lysine production process from fermentation of sugarcane molasses. Bioprocess Biosyst Eng 40:1033–1048

Appleton J (2002) Arginine: clinical potential of a semi-essential amino acid. Altern Med Rev 7:512–522

Asakura Y, Kimura E, Usuda Y, Kawahara Y, Matsui K, Osumi T, Nakamatsu T (2007) Altered metabolic flux due to deletion of *odhA* causes L-glutamate overproduction in *Corynebacterium glutamicum*. Appl Environ Microbiol 73:1308–1319

Atsumi S, Hanai T, Liao JC (2008) Non-fermentative pathways for synthesis of branched-chain higher alcohols as biofuels. Nature 451:86–89

Barrett E, Stanton C, Zelder O, Fitzgerald G, Ross RP (2004) Heterologous expression of lactose- and galactose-utilizing pathways from lactic acid bacteria in *Corynebacterium glutamicum* for production of lysine in whey. Appl Environ Microbiol 70:2861–2866

Bartek T, Makus P, Klein B, Lang S, Oldiges M (2008) Influence of L-isoleucine and pantothenate auxotrophy for L-valine formation in *Corynebacterium glutamicum* revisited by metabolome analyses. Bioprocess Biosyst Eng 31:217–225

Bartek T, Blombach B, Zönnchen E, Makus P, Lang S, Eikmanns BJ, Oldiges M (2010) Importance of NADPH supply for improved L-valine formation in *Corynebacterium glutamicum*. Biotechnol Prog 26:361–371

Becker J, Wittmann C (2015) Advanced biotechnology: metabolically engineered cells for the bio-based production of chemicals and fuels, materials, and health-care products. Angew Chem Int Ed Engl 54:3328–3350

Becker J, Klopprogge C, Zelder O, Heinzle E, Wittmann C (2005) Amplified expression of fructose 1,6-bisphosphatase in *Corynebacterium glutamicum* increases in vivo flux through the pentose phosphate pathway and lysine production on different carbon sources. Appl Environ Microbiol 71:8587–8596

Becker J, Klopprogge C, Herold A, Zelder O, Bolten CJ, Wittmann C (2007) Metabolic flux engineering of L-lysine production in *Corynebacterium glutamicum* - over expression and modification of G6P dehydrogenase. J Biotechnol 132:99–109

Becker J, Klopprogge C, Schröder H, Wittmann C (2009) Metabolic engineering of the tricarboxylic acid cycle for improved lysine production by *Corynebacterium glutamicum*. Appl Environ Microbiol 75:7866–7869

Becker J, Zelder O, Häfner S, Schröder H, Wittmann C (2011) From zero to hero: design-based systems metabolic engineering of *Corynebacterium glutamicum* for L-lysine production. Metab Eng 13:159–168

Becker J, Gießelmann G, Hoffmann SL, Wittmann C (2018) *Corynebacterium glutamicum* for sustainable bioproduction: from metabolic physiology to systems metabolic engineering. Adv Biochem Eng Biotechnol 162:217–263

Bellmann A, Vrljić M, Pátek M, Sahm H, Krämer R, Eggeling L (2001) Expression control and specificity of the basic amino acid exporter LysE of *Corynebacterium glutamicum*. Microbiology 147:1765–1774

Bender DA (1985) Amino acid metabolism, 2nd edn. Wiley, New York

Binder S, Schendzielorz G, Stäbler N, Krumbach K, Hoffmann K, Bott M, Eggeling L (2012) A high-throughput approach to identify genomic variants of bacterial metabolite producers at the single-cell level. Genome Biol 13:R40

Blombach B, Schreiner ME, Holátko J, Bartek T, Oldiges M, Eikmanns BJ (2007a) L-valine production with pyruvate dehydrogenase complex-deficient *Corynebacterium glutamicum*. Appl Environ Microbiol 73:2079–2084

Blombach B, Schreiner ME, Moch M, Oldiges M, Eikmanns BJ (2007b) Effect of pyruvate dehydrogenase complex deficiency on L-lysine production with *Corynebacterium glutamicum*. Appl Microbiol Biotechnol 76:615–623

Blombach B, Schreiner ME, Bartek T, Oldiges M, Eikmanns BJ (2008) *Corynebacterium glutamicum* tailored for high-yield L-valine production. Appl Microbiol Biotechnol 79:471–479

Blombach B, Arndt A, Auchter M, Eikmanns BJ (2009a) L-valine production during growth of pyruvate dehydrogenase complex-deficient *Corynebacterium glutamicum* in the presence of ethanol or by inactivation of the transcriptional regulator SugR. Appl Environ Microbiol 75:1197–1200

Blombach B, Hans S, Bathe B, Eikmanns BJ (2009b) Acetohydroxyacid synthase, a novel target for improvement of L-lysine production by *Corynebacterium glutamicum*. Appl Environ Microbiol 75:419–427

Bolten CJ, Schröder H, Dickschat J, Wittmann C (2010) Towards methionine overproduction in *Corynebacterium glutamicum*: methanethiol and dimethyldisulfide as reduced sulfur sources. J Microbiol Biotechnol 20:1196–1203

Bommareddy RR, Chen Z, Rappert S, Zeng AP (2014) A de novo NADPH generation pathway for improving lysine production of *Corynebacterium glutamicum* by rational design of the coenzyme specificity of glyceraldehyde 3-phosphate dehydrogenase. Metab Eng 25:30–37

Börngen K, Battle AR, Möker N, Morbach S, Marin K, Martinac B, Krämer R (2010) The properties and contribution of the *Corynebacterium glutamicum* MscS variant to fine-tuning of osmotic adaptation. Biochim Biophys Acta 1798:2141–2149

Bott M, Niebisch A (2003) The respiratory chain of *Corynebacterium glutamicum*. J Biotechnol 104:129–153

Brockmann-Gretza O, Kalinowski J (2006) Global gene expression during stringent response in *Corynebacterium glutamicum* in presence and absence of the *rel* gene encoding (p)ppGpp synthase. BMC Genomics 7:230

Burkovski A (2008) Corynebacteria: genomics and molecular biology. Caister Academic, Norfolk

Burkovski A, Krämer R (2002) Bacterial amino acid transport proteins: occurrence, functions, and significance for biotechnological applications. Appl Microbiol Biotechnol 58:265–274

Chen L, Zeng AP (2017) Rational design and metabolic analysis of *Escherichia coli* for effective production of L-tryptophan at high concentration. Appl Microbiol Biotechnol 101:559–568

Chen Z, Bommareddy RR, Frank D, Rappert S, Zeng AP (2014) Deregulation of feedback inhibition of phosphoenolpyruvate carboxylase for improved lysine production in *Corynebacterium glutamicum*. Appl Environ Microbiol 80:1388–1393

Chen C, Li Y, Hu J, Dong X, Wang X (2015) Metabolic engineering of *Corynebacterium glutamicum* ATCC13869 for L-valine production. Metab Eng 29:66–75

Chen L, Chen M, Ma C, Zeng AP (2018) Discovery of feed-forward regulation in L-tryptophan biosynthesis and its use in metabolic engineering of *E. coli* for efficient tryptophan bioproduction. Metab Eng 47:434–444

Cheng J, Chen P, Song A, Wang D, Wang Q (2018) Expanding lysine industry: industrial biomanufacturing of lysine and its derivatives. J Ind Microbiol Biotechnol 45:719–734

Chinen A, Kozlov YI, Hara Y, Izui H, Yasueda H (2007) Innovative metabolic pathway design for efficient L-glutamate production by suppressing CO_2 emission. J Biosci Bioeng 103:262–269

Curis E, Crenn P, Cynober L (2007) Citrulline and the gut. Curr Opin Clin Nutr Metab Care 10:620–626

Dai Z, Nielsen J (2015) Advancing metabolic engineering through systems biology of industrial microorganisms. Curr Opin Biotechnol 36:8–15

Dong X, Quinn PJ, Wang X (2011) Metabolic engineering of *Escherichia coli* and *Corynebacterium glutamicum* for the production of L-threonine. Biotechnol Adv 29:11–23

Dong X, Zhao Y, Zhao J, Wang X (2016) Characterization of aspartate kinase and homoserine dehydrogenase from *Corynebacterium glutamicum* IWJ001 and systematic investigation of L-isoleucine biosynthesis. J Ind Microbiol Biotechnol 43:873–885

Doroshenko V, Airich L, Vitushkina M, Kolokolova A, Livshits V, Mashko S (2007) YddG from *Escherichia coli* promotes export of aromatic amino acids. FEMS Microbiol Lett 275:312–318
Ebbighausen H, Weil B, Krämer R (1989) Transport of branched-chain amino acids in *Corynebacterium glutamicum*. Arch Microbiol 151:238–244
Eggeling L, Bott M (2005) Handbook of *Corynebacterium glutamicum*. CRC, Boca Raton
Eggeling L, Bott M (2015) A giant market and a powerful metabolism: L-lysine provided by *Corynebacterium glutamicum*. Appl Microbiol Biotechnol 99:3387–3394
Elisáková V, Pátek M, Holátko J, Nesvera J, Leyval D, Goergen JL, Delaunay S (2005) Feedback-resistant acetohydroxy acid synthase increases valine production in *Corynebacterium glutamicum*. Appl Environ Microbiol 71:207–213
Feng LY, Xu JZ, Zhang WG (2018) Improved L-leucine production in *Corynebacterium glutamicum* by optimizing the aminotransferases. Molecules 23:2102
Figge R, Soucaille P, Barbier G, Bestel-Corre G, Boisart C, Chateau M (2009) Increasing methionine yield. International Patent Application WO 2009/043803 A2
Fontecave M, Atta M, Mulliez E (2004) *S*-adenosylmethionine: nothing goes to waste. Trends Biochem Sci 29:243–249
Glansdorff N, Xu Y (2007) Microbial arginine biosynthesis: pathway, regulation and industrial production. In: Wendisch VF (ed) Microbiology monographs, amino acid biosynthesis - pathways, regulation and metabolic engineering. Springer, Berlin, pp 219–257
Global Market Insights (2016) Glutamic acid and monosodium glutamate (MSG) market size, potential, industry outlook, regional analysis, application development, competitive landscape & forecast, 2016–2023. https://www.gminsights.com/industry-analysis/glutamic-acid-and-monosodium-glutamate-msg-market
Gutmann M, Hoischen C, Krämer R (1992) Carrier-mediated glutamate secretion by *Corynebacterium glutamicum* under biotin limitation. Biochim Biophys Acta 1112:115–123
Haitani Y, Awano N, Yamazaki M, Wada M, Nakamori S, Takagi H (2006) Functional analysis of L-serine *O*-acetyltransferase from *Corynebacterium glutamicum*. FEMS Microbiol Lett 255:156–163
Han G, Hu X, Wang X (2015) Co-production of *S*-adenosyl-L-methionine and L-isoleucine in *Corynebacterium glutamicum*. Enzym Microb Technol 78:27–33
Han G, Hu X, Qin T, Li Y, Wang X (2016a) Metabolic engineering of *Corynebacterium glutamicum* ATCC 13032 to produce *S*-adenosyl-L-methionine. Enzym Microb Technol 83:14–21
Han G, Hu X, Wang X (2016b) Overexpression of methionine adenosyltransferase in *Corynebacterium glutamicum* for production of *S*-adenosyl-L-methionine. Biotechnol Appl Biochem 63:679–689
Hara Y, Kadotani N, Izui H, Katashkina JI, Kuvaeva TM, Andreeva IG, Golubeva LI, Malko DB, Makeev VJ, Mashko SV, Kozlov YI (2012) The complete genome sequence of *Pantoea ananatis* AJ13355, an organism with great biotechnological potential. Appl Microbiol Biotechnol 93:331–341
Hasegawa S, Uematsu K, Natsuma Y, Suda M, Hiraga K, Jojima T, Inui M, Yukawa H (2012) Improvement of the redox balance increases L-valine production by *Corynebacterium glutamicum* under oxygen deprivation conditions. Appl Environ Microbiol 78:865–875
Hasegawa S, Suda M, Uematsu K, Natsuma Y, Hiraga K, Jojima T, Inui M, Yukawa H (2013) Engineering of *Corynebacterium glutamicum* for high-yield L-valine production under oxygen deprivation conditions. Appl Environ Microbiol 79:1250–1257
Hashimoto K, Nakamura K, Kuroda T, Yabe I, Nakamatsu T, Kawasaki H (2010) The protein encoded by NCgl1221 in *Corynebacterium glutamicum* functions as a mechanosensitive channel. Biosci Biotechnol Biochem 74:2546–2549
Hashimoto K, Murata J, Konishi T, Yabe I, Nakamatsu T, Kawasaki H (2012) Glutamate is excreted across the cytoplasmic membrane through the NCgl1221 channel of *Corynebacterium glutamicum* by passive diffusion. Biosci Biotechnol Biochem 76:1422–1424

Hayashi T, Juliet PA, Matsui-Hirai H, Miyazaki A, Fukatsu A, Funami J, Iguchi A, Ignarro LJ (2005) L-Citrulline and L-arginine supplementation retards the progression of high-cholesterol-diet-induced atherosclerosis in rabbits. Proc Natl Acad Sci U S A 102:13681–13686

Hayashi M, Mizoguchi H, Ohnishi J, Mitsuhashi S, Yonetani Y, Hashimoto S, Ikeda M (2006a) A *leuC* mutation leading to increased L-lysine production and *rel*-independent global expression changes in *Corynebacterium glutamicum*. Appl Microbiol Biotechnol 72:783–789

Hayashi M, Ohnishi J, Mitsuhashi S, Yonetani Y, Hashimoto S, Ikeda M (2006b) Transcriptome analysis reveals global expression changes in an industrial L-lysine producer of *Corynebacterium glutamicum*. Biosci Biotechnol Biochem 70:546–550

Hayashi T, Matsui-Hirai H, Miyazaki-Akita A, Fukatsu A, Funami J, Ding QF, Kamalanathan S, Hattori Y, Ignarro LJ, Iguchi A (2006c) Endothelial cellular senescence is inhibited by nitric oxide: implications in atherosclerosis associated with menopause and diabetes. Proc Natl Acad Sci U S A 103:17018–17023

Hermann T (2003) Industrial production of amino acids by coryneform bacteria. J Biotechnol 104:155–172

Hirao T, Nakano T, Azuma T, Sugimoto M, Nakanishi T (1989) L-lysine production in continuous culture of an L-lysine hyper- producing mutant of *Corynebacterium glutamicum*. Appl Microbiol Biotechnol 32:269–273

Hirasawa T, Shimizu H (2016) Recent advances in amino acid production by microbial cells. Curr Opin Biotechnol 42:133–146

Hoffmann SL, Jungmann L, Schiefelbein S, Peyriga L, Cahoreau E, Portais JC, Becker J, Wittmann C (2018) Lysine production from the sugar alcohol mannitol: design of the cell factory *Corynebacterium glutamicum* SEA-3 through integrated analysis and engineering of metabolic pathway fluxes. Metab Eng 47:475–487

Holátko J, Elisáková V, Prouza M, Sobotka M, Nesvera J, Pátek M (2009) Metabolic engineering of the L-valine biosynthesis pathway in *Corynebacterium glutamicum* using promoter activity modulation. J Biotechnol 139:203–210

Hua K (2013) Investigating the appropriate mode of expressing lysine requirement of fish through non-linear mixed model analysis and multilevel analysis. Br J Nutr 109:1013–1021

Huang Y, Zhang H, Tian H, Li C, Han S, Lin Y, Zheng S (2015) Mutational analysis to identify the residues essential for the inhibition of N-acetyl glutamate kinase of *Corynebacterium glutamicum*. Appl Microbiol Biotechnol 99:7527–7537

Huang Y, Li C, Zhang H, Liang S, Han S, Lin Y, Yang X, Zheng S (2016) Monomeric *Corynebacterium glutamicum* N-acetyl glutamate kinase maintains sensitivity to L-arginine but has a lower intrinsic catalytic activity. Appl Microbiol Biotechnol 100:1789–1798

Huang JF, Shen ZY, Mao QL, Zhang XM, Zhang B, Wu JS, Liu ZQ, Zheng YG (2018) Systematic analysis of bottlenecks in a multibranched and multilevel regulated pathway: the molecular fundamentals of L-methionine biosynthesis in *Escherichia coli*. ACS Synth Biol 7:2577–2589

Hwang BJ, Park SD, Kim Y, Kim P, Lee HS (2007) Biochemical analysis on the parallel pathways of methionine biosynthesis in *Corynebacterium glutamicum*. J Microbiol Biotechnol 17:1010–1017

Ikeda M (2003) Amino acid production processes. In: Faurie R, Thommel J (eds) Advances in biochemical engineering/biotechnology, Microbial production of L-amino acids, vol 79. Springer, Berlin, pp 1–35

Ikeda M (2012) Sugar transport systems in *Corynebacterium glutamicum*: features and applications to strain development. Appl Microbiol Biotechnol 96:1191–1200

Ikeda M (2017) Lysine fermentation: history and genome breeding. In: Yokota A, Ikeda M (eds) Advances in biochemical engineering/biotechnology, Amino acid fermentation, vol 159. Springer, Japan, pp 73–102

Ikeda M, Katsumata R (1994) Transport of aromatic amino acids and its influence on overproduction of the amino acids in *Corynebacterium glutamicum*. J Ferment Bioeng 78:420–425

Ikeda M, Katsumata R (1995) Tryptophan production by transport mutants of *Corynebacterium glutamicum*. Biosci Biotechnol Biochem 59:1600–1602

Ikeda M, Katsumata R (1999) Hyperproduction of tryptophan by *Corynebacterium glutamicum* with the modified pentose phosphate pathway. Appl Environ Microbiol 65:2497–2502

Ikeda M, Nakagawa S (2003) The *Corynebacterium glutamicum* genome: features and impacts on biotechnological process. Appl Microbiol Biotechnol 62:99–109

Ikeda M, Takeno S (2013) Amino acid production by *Corynebacterium glutamicum*. In: Yukawa H, Inui M (eds) Microbiology monographs, vol 23. Springer, Berlin, pp 107–147

Ikeda M, Nakanishi K, Kino K, Katsumata R (1994) Fermentative production of tryptophan by a stable recombinant strain of *Corynebacterium glutamicum* with a modified serine-biosynthetic pathway. Biosci Biotech Biochem 58:674–678

Ikeda M, Ohnishi J, Mitsuhashi S (2005) Genome breeding of an amino acid-producing *Corynebacterium glutamicum* mutant. In: Barredo JLS (ed) Microbial processes and products. Humana Press, Totowa, pp 179–189

Ikeda M, Ohnishi J, Hayashi M, Mitsuhashi S (2006) A genome-based approach to create a minimally mutated *Corynebacterium glutamicum* strain for efficient L-lysine production. J Ind Microbiol Biotechnol 33:610–615

Ikeda M, Mitsuhashi S, Tanaka K, Hayashi M (2009) Reengineering of a *Corynebacterium glutamicum* L-arginine and L-citrulline producer. Appl Environ Microbiol 75:1635–1641

Ikeda M, Nakano T, Mitsuhashi S, Hayashi M, Tanaka K (2010a) Process for producing L-arginine, L-ornithine or L-citrulline. US Patent 7741081B2

Ikeda M, Takeno S, Mizuno Y, Mitsuhashi S (2010b) Process for producing useful substance. International Patent Application WO 2010/024267 A1

Ikeda M, Mizuno Y, Awane S, Hayashi M, Mitsuhashi S, Takeno S (2011) Identification and application of a different glucose uptake system that functions as an alternative to the phosphotransferase system in *Corynebacterium glutamicum*. Appl Microbiol Biotechnol 90:1443–1451

Ikeda M, Noguchi N, Ohshita M, Senoo A, Mitsuhashi S, Takeno S (2015) A third glucose uptake bypass in *Corynebacterium glutamicum* ATCC 31833. Appl Microbiol Biotechnol 99:2741–2750

Izui H, Moriya M, Hirano S, Hara Y, Ito H, Matsui K (2006) Method for producing L-glutamic acid by fermentation accompanied by precipitation. US Patent 7015010B1

Jensen JV, Eberhardt D, Wendisch VF (2015) Modular pathway engineering of *Corynebacterium glutamicum* for production of the glutamate-derived compounds ornithine, proline, putrescine, citrulline, and arginine. J Biotechnol 214:85–94

Jiang LY, Chen SG, Zhang YY, Liu JZ (2013a) Metabolic evolution of *Corynebacterium glutamicum* for increased production of L-ornithine. BMC Biotechnol 13:47

Jiang LY, Zhang YY, Li Z, Liu JZ (2013b) Metabolic engineering of *Corynebacterium glutamicum* for increasing the production of L-ornithine by increasing NADPH availability. J Ind Microbiol Biotechnol 40:1143–1151

Joo YC, Hyeon JE, Han SO (2017) Metabolic design of *Corynebacterium glutamicum* for production of L-cysteine with consideration of sulfur-supplemented animal feed. J Agric Food Chem 65:4698–4707

Kabus A, Georgi T, Wendisch VF, Bott M (2007a) Expression of the *Escherichia coli pntAB* genes encoding a membrane-bound transhydrogenase in *Corynebacterium glutamicum* improves L-lysine formation. Appl Microbiol Biotechnol 75:47–53

Kabus A, Niebisch A, Bott M (2007b) Role of cytochrome *bd* oxidase from *Corynebacterium glutamicum* in growth and lysine production. Appl Environ Microbiol 73:861–868

Kalinowski J, Bathe B, Bartels D, Bischoff N, Bott M, Burkovski A, Dusch N, Eggeling L, Eikmanns BJ, Gaigalat L, Goesmann A, Hartmann M, Huthmacher K, Krämer R, Linke B, McHardy AC, Meyer F, Möckel B, Pfefferle W, Pühler A, Rey DA, Rückert C, Rupp O, Sahm H, Wendisch VF, Wiegräbe I, Tauch A (2003) The complete *Corynebacterium glutamicum* ATCC 13032 genome sequence and its impact on the production of L-aspartate-derived amino acids and vitamins. J Biotechnol 104:5–25

Kallio PT, Kim DJ, Tsai PS, Bailey JE (1994) Intracellular expression of *Vitreoscilla* hemoglobin alters *Escherichia coli* energy metabolism under oxygen-limited conditions. Eur J Biochem 219:201–208

Kanehisa M, Goto S, Sato Y, Kawashima M, Furumichi M, Tanabe M (2014) Data, information, knowledge and principle: back to metabolism in KEGG. Nucleic Acids Res 42:D199–D205

Kase H, Nakayama K (1974) Mechanism of L-threonine and L-lysine production by analog-resistant mutants of *Corynebacterium glutamicum*. Agr Biol Chem 38:993–1000

Katsumata R, Ikeda M (1993) Hyperproduction of tryptophan in *Corynebacterium glutamicum* by pathway engineering. Bio/Technology 11:921–925

Katsumata R, Kino K (1989) Process for producing amino acids by fermentation. Japan Patent 01,317,395 A (P2,578,488)

Kawaguchi H, Vertès AA, Okino S, Inui M, Yukawa H (2006) Engineering of a xylose metabolic pathway in *Corynebacterium glutamicum*. Appl Environ Microbiol 72:3418–3428

Kawaguchi H, Sasaki M, Vertès AA, Inui M, Yukawa H (2008) Engineering of an L-arabinose metabolic pathway in *Corynebacterium glutamicum*. Appl Microbiol Biotechnol 77:1053–1062

Kawahara Y, Takahashi-Fuke K, Shimizu E, Nakamatsu T, Nakamori S (1997) Relationship between the glutamate production and the activity of 2-oxoglutarate dehydrogenase in *Brevibacterium lactofermentum*. Biosci Biotechnol Biochem 61:1109–1112

Kawano Y, Onishi F, Shiroyama M, Miura M, Tanaka N, Oshiro S, Nonaka G, Nakanishi T, Ohtsu I (2017) Improved fermentative L-cysteine overproduction by enhancing a newly identified thiosulfate assimilation pathway in *Escherichia coli*. Appl Microbiol Biotechnol 101:6879–6889

Kelle R, Hermann T, Bathe B (2005) L-Lysine production. In: Eggeling L, Bott M (eds) Handbook of *Corynebacterium glutamicum*. CRC, Boca Raton, pp 465–488

Kennerknecht N, Sahm H, Yen MR, Patek M, Saier MH Jr, Eggeling L (2002) Export of L-isoleucine from *Corynebacterium glutamicum*: a two-gene-encoded member of a new translocator family. J Bacteriol 184:3947–3956

Kiefer P, Heinzle E, Zelder O, Wittmann C (2004) Comparative metabolic flux analysis of lysine-producing *Corynebacterium glutamicum* cultured on glucose or fructose. Appl Environ Microbiol 70:229–239

Kim J, Fukuda H, Hirasawa T, Nagahisa K, Nagai K, Wachi M, Shimizu H (2009a) Requirement of de novo synthesis of the OdhI protein in penicillin-induced glutamate production by *Corynebacterium glutamicum*. Appl Microbiol Biotechnol 86(3):911–920. https://doi.org/10.1007/s00253-009-2360-6

Kim J, Hirasawa T, Sato Y, Nagahisa K, Furusawa C, Shimizu H (2009b) Effect of *odhA* overexpression and *odhA* antisense RNA expression on Tween-40-triggered glutamate production by *Corynebacterium glutamicum*. Appl Microbiol Biotechnol 81:1097–1106

Kim HI, Nam JY, Cho JY, Lee CS, Park YJ (2013) Next-generation sequencing-based transcriptome analysis of L-lysine-producing *Corynebacterium glutamicum* ATCC 21300 strain. J Microbiol 51:877–880

Kim SY, Lee J, Lee SY (2015) Metabolic engineering of *Corynebacterium glutamicum* for the production of L-ornithine. Biotechnol Bioeng 112:416–421

Kimura E (2003) Metabolic engineering of glutamate production. In: Faurie R, Thommel J (eds) Advances in biochemical engineering/biotechnology, Microbial production of L-amino acids, vol 79. Springer, Berlin, pp 37–58

Kjeldsen KR, Nielsen J (2009) In silico genome-scale reconstruction and validation of the *Corynebacterium glutamicum* metabolic network. Biotechnol Bioeng 102:583–597

Kobayashi M, Itoyama T, Mitani Y, Usui N (2011) Method for producing basic amino acid. European Patent 1182261 B1

Koch DJ, Rückert C, Albersmeier A, Hüser AT, Tauch A, Pühler A, Kalinowski J (2005) The transcriptional regulator SsuR activates expression of the *Corynebacterium glutamicum* sulphonate utilization genes in the absence of sulphate. Mol Microbiol 58:480–494

Komati Reddy G, Lindner SN, Wendisch VF (2015) Metabolic engineering of an ATP-neutral Embden-Meyerhof-Parnas pathway in *Corynebacterium glutamicum*: growth restoration by an adaptive point mutation in NADH dehydrogenase. Appl Environ Microbiol 81:1996–2005

Krömer JO, Sorgenfrei O, Klopprogge K, Heinzle E, Wittmann C (2004) In-depth profiling of lysine-producing *Corynebacterium glutamicum* by combined analysis of the transcriptome, metabolome, and fluxome. J Bacteriol 186:1769–1784

Krömer JO, Wittmann C, Schröder H, Heinzle E (2006) Metabolic pathway analysis for rational design of L-methionine production by *Escherichia coli* and *Corynebacterium glutamicum*. Metab Eng 8:353–369

Krömer JO, Bolten CJ, Heinzle E, Schröder H, Wittmann C (2008) Physiological response of *Corynebacterium glutamicum* to oxidative stress induced by deletion of the transcriptional repressor McbR. Microbiology 154:3917–3930

Kubota T, Tanaka Y, Takemoto N, Hiraga K, Yukawa H, Inui M (2015) Identification and expression analysis of a gene encoding a shikimate transporter of *Corynebacterium glutamicum*. Microbiology 161:254–263

Kutukova EA, Livshits VA, Altman IP, Ptisyn LR, Zyiatdinov MH, Tokmakova IL, Zakataeva NP (2005) The *yeaS* (*leuE*) gene of *Escherichia coli* encodes an exporter of leucine, and the Lrp protein regulates its expression. FEBS Lett 579:4629–4634

Lange C, Mustafi N, Frunzke J, Kennerknecht N, Wessel M, Bott M, Wendisch VF (2012) Lrp of *Corynebacterium glutamicum* controls expression of the *brnFE* operon encoding the export system for L-methionine and branched-chain amino acids. J Biotechnol 158:231–241

Lee HS, Hwang BJ (2003) Methionine biosynthesis and its regulation in *Corynebacterium glutamicum*: parallel pathways of transsulfuration and direct sulfhydrylation. Appl Microbiol Biotechnol 62:459–467

Lee JH, Wendisch VF (2017) Production of amino acids - genetic and metabolic engineering approaches. Bioresour Technol 245:1575–1587

Lee KH, Park JH, Kim TY, Kim KU, Lee SY (2007) Systems metabolic engineering of *Escherichia coli* for L-threonine production. Mol Syst Biol 3:149

Lee SY, Shin HS, Park JS, Kim YH, Min J (2010) Proline reduces the binding of transcriptional regulator ArgR to upstream of *argB* in *Corynebacterium glutamicum*. Appl Microbiol Biotechnol 86:235–242

Lee CS, Nam JY, Son ES, Kwon OC, Han W, Cho JY, Park YJ (2012) Next-generation sequencing-based genome-wide mutation analysis of L-lysine-producing *Corynebacterium glutamicum* ATCC 21300 strain. J Microbiol 50:860–863

Li L, Wada M, Yokota A (2007) A comparative proteomic approach to understand the adaptations of an H^+-ATPase-defective mutant of *Corynebacterium glutamicum* ATCC14067 to energy deficiencies. Proteomics 7:3348–3357

Li Y, Cong H, Liu B, Song J, Sun X, Zhang J, Yang Q (2016) Metabolic engineering of *Corynebacterium glutamicum* for methionine production by removing feedback inhibition and increasing NAPDH level. Antonie Van Leeuwenhoek 109:1185–1197

Lindner SN, Seibold GM, Henrich A, Krämer R, Wendisch VF (2011a) Phosphotransferase system-independent glucose utilization in *Corynebacterium glutamicum* by inositol permeases and glucokinases. Appl Environ Microbiol 77:3571–3581

Lindner SN, Seibold GM, Krämer R, Wendisch VF (2011b) Impact of a new glucose utilization pathway in amino acid-producing *Corynebacterium glutamicum*. Bioeng Bugs 2:291–295

Liu Q, Zhang J, Wei XX, Ouyang SP, Wu Q, Chen GQ (2008) Microbial production of L-glutamate and L -glutamine by recombinant *Corynebacterium glutamicum* harboring *Vitreoscilla* hemoglobin gene *vgb*. Appl Microbiol Biotechnol 77:1297–1304

Liu Q, Liang Y, Zhang Y, Shang X, Liu S, Wen J, Wen T (2015) YjeH is a novel exporter of L-methionine and branched-chain amino acids in *Escherichia coli*. Appl Environ Microbiol 81:7753–7766

Liu C, Zhang B, Liu YM, Yang KQ, Liu SJ (2018a) New intracellular shikimic acid biosensor for monitoring shikimate synthesis in *Corynebacterium glutamicum*. ACS Synth Biol 7:591–601

Liu H, Fang G, Wu H, Li Z, Ye Q (2018b) L-cysteine production in *Escherichia coli* based on rational metabolic engineering and modular strategy. Biotechnol J 13:e1700695

Lu SC, Mato JM (2012) *S*-adenosylmethionine in liver health, injury, and cancer. Physiol Rev 92:1515–1542

Lubitz D, Jorge JM, Pérez-García F, Taniguchi H, Wendisch VF (2016) Roles of export genes cgmA and lysE for the production of L-arginine and L-citrulline by *Corynebacterium glutamicum*. Appl Microbiol Biotechnol 100:8465–8474

Ma W, Wang J, Li Y, Hu X, Shi F, Wang X (2016) Enhancing pentose phosphate pathway in *Corynebacterium glutamicum* to improve L-isoleucine production. Biotechnol Appl Biochem 63:877–885

Ma Q, Zhang Q, Xu Q, Zhang C, Li Y, Fan X, Xie X, Chen N (2017) Systems metabolic engineering strategies for the production of amino acids. Synth Syst Biotechnol 2:87–96

Ma Y, Chen Q, Cui Y, Du L, Shi T, Xu Q, Ma Q, Xie X, Chen N (2018) Comparative genomic and genetic functional analysis of industrial L-leucine- and L-valine-producing *Corynebacterium glutamicum* strains. J Microbiol Biotechnol 28:1916–1927

Mahr R, Gätgens C, Gätgens J, Polen T, Kalinowski J, Frunzke J (2015) Biosensor-driven adaptive laboratory evolution of l-valine production in *Corynebacterium glutamicum*. Metab Eng 32:184–194

Mampel J, Schröder H, Haefner S, Sauer U (2005) Single-gene knockout of a novel regulatory element confers ethionine resistance and elevates methionine production in *Corynebacterium glutamicum*. Appl Microbiol Biotechnol 68:228–236

Man Z, Rao Z, Xu M, Guo J, Yang T, Zhang X, Xu Z (2016a) Improvement of the intracellular environment for enhancing l-arginine production of *Corynebacterium glutamicum* by inactivation of H_2O_2-forming flavin reductases and optimization of ATP supply. Metab Eng 38:310–321

Man Z, Xu M, Rao Z, Guo J, Yang T, Zhang X, Xu Z (2016b) Systems pathway engineering of *Corynebacterium crenatum* for improved L-arginine production. Sci Rep 6:28629

Marienhagen J, Eggeling L (2008) Metabolic function of *Corynebacterium glutamicum* aminotransferases AlaT and AvtA and impact on L-valine production. Appl Environ Microbiol 74:7457–7462

Marx A, Hans S, Mockel B, Bathe B, de Graaf AA (2003) Metabolic phenotype of phosphoglucose isomerase mutants of *Corynebacterium glutamicum*. J Biotechnol 104:185–197

Meiswinkel TM, Gopinath V, Lindner SN, Nampoothiri KM, Wendisch VF (2013) Accelerated pentose utilization by *Corynebacterium glutamicum* for accelerated production of lysine, glutamate, ornithine and putrescine. Microb Biotechnol 6:131–140

Merlin C, Gardiner G, Durand S, Masters M (2002) The *Escherichia coli metD* locus encodes an ABC transporter which includes Abc (MetN), YaeE (MetI), and YaeC (MetQ). J Bacteriol 184:5513–5517

Mitsuhashi S (2014) Current topics in the biotechnological production of essential amino acids, functional amino acids, and dipeptides. Curr Opin Biotechnol 26:38–44

Mitsuhashi S, Hayashi M, Ohnishi J, Ikeda M (2006) Disruption of malate:quinone oxidoreductase increases L-lysine production by *Corynebacterium glutamicum*. Biosci Biotechnol Biochem 70:2803–2806

Mizuno Y, Nagano-Shoji M, Kubo S, Kawamura Y, Yoshida A, Kawasaki H, Nishiyama M, Yoshida M, Kosono S (2016) Altered acetylation and succinylation profiles in *Corynebacterium glutamicum* in response to conditions inducing glutamate overproduction. Microbiology 5:152–173

Möckel B, Pfefferle W, Huthmacher K, Rückert C, Kalinowski J, Pühler A, Binder M, Greissinger D, Thierbach G (2002) Nucleotide sequences which code for the *metY* gene. International Patent Application WO 02/18613

Moon MW, Kim HJ, Oh TK, Shin CS, Lee JS, Kim SJ, Lee JK (2005) Analyses of enzyme II gene mutants for sugar transport and heterologous expression of fructokinase gene in *Corynebacterium glutamicum* ATCC 13032. FEMS Microbiol Lett 244:259–266

Morbach S, Sahm H, Eggeling L (1995) Use of feedback-resistant threonine dehydratases of *Corynebacterium glutamicum* to increase carbon flux towards L-isoleucine. Appl Environ Microbiol 61:4315–4320

Morbach S, Sahm H, Eggeling L (1996) L-isoleucine production with *Corynebacterium glutamicum*: further flux increase and limitation of export. App Environ Microbiol 62:4345–4351

Mori A, Morita M, Morishita K, Sakamoto K, Nakahara T, Ishii K (2015) L-Citrulline dilates rat retinal arterioles via nitric oxide- and prostaglandin-dependent pathways in vivo. J Pharmacol Sci 127:419–423

Mustafi N, Grünberger A, Kohlheyer D, Bott M, Frunzke J (2012) The development and application of a single-cell biosensor for the detection of L-methionine and branched-chain amino acids. Metab Eng 14:449–457

Nagano-Shoji M, Hamamoto Y, Mizuno Y, Yamada A, Kikuchi M, Shirouzu M, Umehara T, Yoshida M, Nishiyama M, Kosono S (2017) Characterization of lysine acetylation of a phosphoenolpyruvate carboxylase involved in glutamate overproduction in *Corynebacterium glutamicum*. Mol Microbiol 104:677–689

Nakamura J, Hirano S, Ito H, Wachi M (2007) Mutations of the *Corynebacterium glutamicum* NCgl1221 gene, encoding a mechanosensitive channel homolog, induce L-glutamic acid production. Appl Environ Microbiol 73:4491–4498

Niebisch A, Kabus A, Schultz C, Weil B, Bott M (2006) Corynebacterial protein kinase G controls 2-oxoglutarate dehydrogenase activity via the phosphorylation status of the OdhI protein. J Biol Chem 281:12300–12307

Nishimura T, Vertès AA, Shinoda Y, Inui M, Yukawa H (2007) Anaerobic growth of *Corynebacterium glutamicum* using nitrate as a terminal electron acceptor. Appl Microbiol Biotechnol 75:889–897

Nonaka G (2018) Bacterial metabolism and fermentative production of cysteine. Bioscience and Industry 76:110–104

Nottebrock D, Meyer U, Krämer R, Morbach S (2003) Molecular and biochemical characterization of mechanosensitive channels in *Corynebacterium glutamicum*. FEMS Microbiol Lett 218:305–309

Ochiai M, Hayashi T, Morita M, Ina K, Maeda M, Watanabe F, Morishita K (2012) Short-term effects of L-citrulline supplementation on arterial stiffness in middle-aged men. Int J Cardiol 155:257–261

Ohnishi J, Ikeda M (2006) Comparisons of potentials for L-lysine production among different *Corynebacterium glutamicum* strains. Biosci Biotechnol Biochem 70:1017–1020

Ohnishi J, Mitsuhashi S, Hayashi M, Ando S, Yokoi H, Ochiai K, Ikeda M (2002) A novel methodology employing *Corynebacterium glutamicum* genome information to generate a new L-lysine-producing mutant. Appl Microbiol Biotechnol 58:217–223

Ohnishi J, Hayashi M, Mitsuhashi S, Ikeda M (2003) Efficient 40°C fermentation of L-lysine by a new *Corynebacterium glutamicum* mutant developed by genome breeding. Appl Microbiol Biotechnol 62:69–75

Ohnishi J, Katahira R, Mitsuhashi S, Kakita S, Ikeda M (2005) A novel *gnd* mutation leading to increased L-lysine production in *Corynebacterium glutamicum*. FEMS Microbiol Lett 242:265–274

Park JH, Lee SY (2010) Fermentative production of branched chain amino acids: a focus on metabolic engineering. Appl Microbiol Biotechnol 85:491–506

Park JH, Lee KH, Kim TY, Lee SY (2007a) Metabolic engineering of *Escherichia coli* for the production of L-valine based on transcriptome analysis and *in silico* gene knockout simulation. Proc Natl Acad Sci U S A 104:7797–7802

Park SD, Lee JY, Sim SY, Kim Y, Lee HS (2007b) Characteristics of methionine production by an engineered *Corynebacterium glutamicum* strain. Metab Eng 9:327–336

Park SH, Kim HU, Kim TY, Park JS, Kim SS, Lee SY (2014) Metabolic engineering of *Corynebacterium glutamicum* for L-arginine production. Nat Commun 5:4618

Pátek M (2007) Branched-chain amino acids. In: Wendisch VF (ed) Microbiology monographs, amino acid biosynthesis - pathways, regulation and metabolic engineering. Springer, Berlin, pp 129–162

Petersen S, Mack C, de Graaf AA, Riedel C, Eikmanns BJ, Sahm H (2001) Metabolic consequences of altered phosphoenolpyruvate carboxykinase activity in *Corynebacterium glutamicum* reveal anaplerotic mechanisms in vivo. Metab Eng 3:344–361

Peters-Wendisch PG, Schiel B, Wendisch VF, Katsoulidis E, Möckel B, Sahm H, Eikmanns BJ (2001) Pyruvate carboxylase is a major bottleneck for glutamate and lysine production by *Corynebacterium glutamicum*. J Mol Microbiol Biotechnol 3:295–300

Petri K, Walter F, Persicke M, Rückert C, Kalinowski J (2013) A novel type of N-acetylglutamate synthase is involved in the first step of arginine biosynthesis in *Corynebacterium glutamicum*. BMC Genomics 14:713

Qin T, Hu X, Hu J, Wang X (2015) Metabolic engineering of *Corynebacterium glutamicum* strain ATCC 13032 to produce L-methionine. Biotechnol Appl Biochem 62:563–573

Radmacher E, Eggeling L (2007) The three tricarboxylate synthase activities of *Corynebacterium glutamicum* and increase of L-lysine synthesis. Appl Microbiol Biotechnol 76:587–589

Radmacher E, Vaitsikova A, Burger U, Krumbach K, Sahm H, Eggeling L (2002) Linking central metabolism with increased pathway flux: L-valine accumulation by *Corynebacterium glutamicum*. Appl Environ Microbiol 68:2246–2250

Research and Markets (2018) Amino acids market: global industry trends, share, size, growth, opportunity and forecast 2018–2023. https://www.researchandmarkets.com/reports/4514538/amino-acids-market-global-industry-trends

Rey DA, Pühler A, Kalinowski J (2003) The putative transcriptional regulator McbR, member of the TetR-family, is involved in the regulation of the metabolic network directing the synthesis of sulfur containing amino acids in *Corynebacterium glutamicum*. J Biotechnol 103:51–65

Rey DA, Nentwich SS, Koch DJ, Rückert C, Pühler A, Tauch A, Kalinowski J (2005) The McbR repressor modulated by the effector substance *S*-adenosylmethionine controls directly the transcription of a regulon involved in Sulphur metabolism of *Corynebacterium glutamicum* ATCC 13032. Mol Microbiol 56:871–887

Riedel C, Rittmann D, Dangel P, Möckel B, Sahm H, Eikmanns BJ (2001) Characterization, expression, and inactivation of the phosphoenolpyruvate carboxykinase gene from *Corynebacterium glutamicum* and significance of the enzyme for growth and amino acid production. J Mol Microbiol Biotechnol 3:573–583

Rieping M, Hermann T (2007) L-threonine. In: Wendisch VF (ed) Microbiology monographs, amino acid biosynthesis - pathways, regulation and metabolic engineering. Springer, Berlin, pp 71–92

Rittmann D, Lindner SN, Wendisch VF (2008) Engineering of a glycerol utilization pathway for amino acid production by *Corynebacterium glutamicum*. Appl Environ Microbiol 74:6216–6222

Rückert C, Milse J, Albersmeier A, Koch DJ, Pühler A, Kalinowski J (2008) The dual transcriptional regulator CysR in *Corynebacterium glutamicum* ATCC 13032 controls a subset of genes of the McbR regulon in response to the availability of sulphide acceptor molecules. BMC Genomics 9:483

Sanchez S, Rodríguez-Sanoja R, Ramos A, Demain AL (2018) Our microbes not only produce antibiotics, they also overproduce amino acids. J Antibiot 71:26–36. 2018

Sano K, Shiio I (1971) Microbial production of L-lysine. IV. Selection of lysine-producing mutants from *Brevibacterium flavum* by detecting threonine sensitivity or halo-forming method. J Gen Appl Microbiol 17:97–113

Sano K, Yokozeki K, Tamura F, Yasuda N, Noda I, Mitsugi K (1977) Microbial conversion of DL-2-amino-Δ^2thiazoline-4-carboxylic acid to L-cysteine and L-cystine: screening of microorganisms and identification of products. Appl Environ Microbiol 34:806–810

Sato H, Orishimo K, Shirai T, Hirasawa T, Nagahisa K, Shimizu H, Wachi M (2008) Distinct roles of two anaplerotic pathways in glutamate production induced by biotin limitation in *Corynebacterium glutamicum*. J Biosci Bioeng 106:51–58

Sawada K, Zen-In S, Wada M, Yokota A (2010) Metabolic changes in a pyruvate kinase gene deletion mutant of *Corynebacterium glutamicum* ATCC 13032. Metab Eng 12:401–407

Schendzielorz G, Dippong M, Grünberger A, Kohlheyer D, Yoshida A, Binder S, Nishiyama C, Nishiyama M, Bott M, Eggeling L (2014) Taking control over control: use of product sensing in single cells to remove flux control at key enzymes in biosynthesis pathways. ACS Synth Biol 3:21–29

Schulte J, Baumgart M, Bott M (2017) Development of a single-cell GlxR-based cAMP biosensor for *Corynebacterium glutamicum*. J Biotechnol 258:33–40

Schultz C, Niebisch A, Gebel L, Bott M (2007) Glutamate production by *Corynebacterium glutamicum*: dependence on the oxoglutarate dehydrogenase inhibitor protein OdhI and protein kinase PknG. Appl Microbiol Biotechnol 76:691–700

Schwentner A, Feith A, Münch E, Busche T, Rückert C, Kalinowski J, Takors R, Blombach B (2018) Metabolic engineering to guide evolution - creating a novel mode for L-valine production with *Corynebacterium glutamicum*. Metab Eng 47:31–41

Shi F, Li K, Huan X, Wang X (2013) Expression of NAD(H) kinase and glucose-6-phosphate dehydrogenase improve NADPH supply and L-isoleucine biosynthesis in *Corynebacterium glutamicum* ssp. *lactofermentum*. Appl Biochem Biotechnol 171:504–521

Shiio I, Miyajima R (1969) Concerted inhibition and its reversal by end products of aspartate kinase in *Brevibacterium flavum*. J Biochem 65:849–859

Shiio I, Toride Y, Sugimoto S (1984) Production of lysine by pyruvate dehydrogenase mutants of *Brevibacterium flavum*. Agric Biol Chem 48:3091–3098

Shimomura Y, Yamamoto Y, Bajotto G, Sato J, Murakami T, Shimomura N, Kobayashi H, Mawatari K (2006) Nutraceutical effects of branched-chain amino acids on skeletal muscle. J Nutr 136:529S–532S

Shingu H, Terui G (1971) Studies on process of glutamic acid fermentation at the enzyme level. Part I. on the change of α-ketoglutaric acid dehydrogenase in the course of culture. J Ferment Technol 49:400–405

Sirko A, Hryniewicz M, Hulanicka D, Böck A (1990) Sulfate and thiosulfate transport in *Escherichia coli* K-12: nucleotide sequence and expression of the *cysTWAM* gene cluster. J Bacteriol 172:3351–3357

Sprenger GA (2007) Aromatic amino acids. In: Wendisch VF (ed) Microbiology monographs, Amino acid biosynthesis - pathways, regulation and metabolic engineering. Springer, Berlin, pp 93–127

Strelkov S, von Elstermann M, Schomburg D (2004) Comprehensive analysis of metabolites in *Corynebacterium glutamicum* by gas chromatography/mass spectrometry. Biol Chem 385:853–861

Takeno S, Ohnishi J, Komatsu T, Masaki T, Sen K, Ikeda M (2007) Anaerobic growth and potential for amino acid production by nitrate respiration in *Corynebacterium glutamicum*. Appl Microbiol Biotechnol 75:1173–1182

Takeno S, Murata R, Kobayashi R, Mitsuhashi S, Ikeda M (2010) Engineering of *Corynebacterium glutamicum* with an NADPH-generating glycolytic pathway for L-lysine production. Appl Environ Microbiol 76:7154–7160

Takeno S, Hori K, Ohtani S, Mimura A, Mitsuhashi S, Ikeda M (2016) L-lysine production independent of the oxidative pentose phosphate pathway by *Corynebacterium glutamicum* with the *Streptococcus mutans gapN* gene. Metab Eng 37:1–10

Takumi K, Nonaka G (2014) An L-amino acid-producing bacterium and a method for producing an L-amino acid. European Patent EP 2218729 B1

Takumi K, Nonaka G (2016) Bacterial cysteine-inducible cysteine resistance systems. J Bacteriol 198:1384–1392

Takumi K, Ziyatdinov MK, Samsonov V, Nonaka G (2017) Fermentative production of cysteine by *Pantoea ananatis*. Appl Environ Microbiol 83:e02502–e02516

Trötschel C, Deutenberg D, Bathe B, Burkovski A, Krämer R (2005) Characterization of methionine export in *Corynebacterium glutamicum*. J Bacteriol 187:3786–3794

Trötschel C, Follmann M, Nettekoven JA, Mohrbach T, Forrest LR, Burkovski A, Marin K, Krämer R (2008) Methionine uptake in *Corynebacterium glutamicum* by MetQNI and by MetPS, a novel methionine and alanine importer of the NSS neurotransmitter transporter family. Biochemistry 47:12698–12709

Umbarger HE (1978) Amino acid biosynthesis and its regulation. Annu Rev Biochem 47:533–606

Usuda Y, Hara Y, Kojima H (2017) Toward sustainable amino acid production. In: Yokota A, Ikeda M (eds) Advances in biochemical engineering/biotechnology, Amino acid fermentation, vol 159. Springer, Tokyo, pp 289–304

Utagawa T (2004) Arginine metabolism: enzymology, nutrition, and clinical significance. J Nutr 134:2854S–2857S

van Ooyen J, Noack S, Bott M, Reth A, Eggeling L (2012) Improved L-lysine production with *Corynebacterium glutamicum* and systemic insight into citrate synthase flux and activity. Biotechnol Bioeng 109:2070–2081

Vassilev I, Gießelmann G, Schwechheimer SK, Wittmann C, Virdis B, Krömer JO (2018) Anodic electro-fermentation: anaerobic production of L-lysine by recombinant *Corynebacterium glutamicum*. Biotechnol Bioeng 115:1499–1508

Vogt M, Haas S, Klaffl S, Polen T, Eggeling L, van Ooyen J, Bott M (2014) Pushing product formation to its limit: metabolic engineering of *Corynebacterium glutamicum* for L-leucine overproduction. Metab Eng 22:40–52

Vogt M, Krumbach K, Bang WG, van Ooyen J, Noack S, Klein B, Bott M, Eggeling L (2015) The contest for precursors: channelling L-isoleucine synthesis in *Corynebacterium glutamicum* without byproduct formation. Appl Microbiol Biotechnol 99:791–800

Vrljić M, Sahm H, Eggeling L (1996) A new type of transporter with a new type of cellular function: L-lysine export from *Corynebacterium glutamicum*. Mol Microbiol 22:815–826

Wada M, Takagi H (2006) Metabolic pathways and biotechnological production of L-cysteine. Appl Microbiol Biotechnol 73:48–54

Wada M, Awano N, Haisa K, Takagi H, Nakamori S (2002) Purification, characterization and identification of cysteine desulfhydrase of *Corynebacterium glutamicum*, and its relationship to cysteine production. FEMS Microbiol Lett 217:103–107

Wada M, Hijikata N, Aoki R, Takesue N, Yokota A (2008) Enhanced valine production in *Corynebacterium glutamicum* with defective H^+-ATPase and C-terminal truncated acetohydroxyacid synthase. Biosci Biotechnol Biochem 72:2959–2965

Wang J, Cheng LK, Wang J, Liu Q, Shen T, Chen N (2013) Genetic engineering of *Escherichia coli* to enhance production of L-tryptophan. Appl Microbiol Biotechnol 97:7587–7596

Wang X, Zhang H, Quinn PJ (2018a) Production of L-valine from metabolically engineered *Corynebacterium glutamicum*. Appl Microbiol Biotechnol 102:4319–4330

Wang Y, Cao G, Xu D, Fan L, Wu X, Ni X, Zhao S, Zheng P, Sun J, Ma Y (2018b) A novel *Corynebacterium glutamicum* L-glutamate exporter. Appl Environ Microbiol 84(6):e02691–e02617

Webster DA (1987) Structure and function of bacterial hemoglobin and related proteins. In: Eichhorn GC, Marzilli LG (eds) Advances in inorganic chemistry. Elsevier, New York, pp 245–265

Wei L, Wang H, Xu N, Zhou W, Ju J, Liu J, Ma Y (2018) Metabolic engineering of *Corynebacterium glutamicum* for L-cysteine production. Appl Microbiol Biotechnol 103(3):1325–1338. https://doi.org/10.1007/s00253-018-9547-7

Wen J, Bao J (2019) Engineering *Corynebacterium glutamicum* triggers glutamic acid accumulation in biotin-rich corn Stover hydrolysate. Biotechnol Biofuels 12:86

Wendisch VF (2007) Microbiology monographs, amino acid biosynthesis - pathways, regulation and metabolic engineering. Springer, Berlin

Wendisch VF, Bott M, Kalinowski J, Oldiges M, Wiechert W (2006) Emerging *Corynebacterium glutamicum* systems biology. J Biotechnol 124:74–92

Willis LB, Lessard PA, Sinskey AJ (2005) Synthesis of L-threonine and branched-chain amino acids. In: Eggeling L, Bott M (eds) Handbook of *Corynebacterium glutamicum*. CRC, Boca Raton, pp 511–531

Wittmann C, Heinzle E (2002) Genealogy profiling through strain improvement by using metabolic network analysis: metabolic flux genealogy of several generations of lysine-producing *Corynebacteria*. Appl Environ Microbiol 68:5843–5859

Wu Y, Li P, Zheng P, Zhou W, Chen N, Sun J (2015) Complete genome sequence of *Corynebacterium glutamicum* B253, a Chinese lysine-producing strain. J Biotechnol 207:10–11

Xafenias N, Kmezik C, Mapelli V (2017) Enhancement of anaerobic lysine production in *Corynebacterium glutamicum* electrofermentations. Bioelectrochemistry 117:40–47

Xie X, Xu L, Shi J, Xu Q, Chen N (2012) Effect of transport proteins on L-isoleucine production with the L-isoleucine-producing strain *Corynebacterium glutamicum* YILW. J Ind Microbiol Biotechnol 39:1549–1556

Xu J, Zhang J, Guo Y, Zai Y, Zhang W (2013) Improvement of cell growth and L-lysine production by genetically modified *Corynebacterium glutamicum* during growth on molasses. J Ind Microbiol Biotechnol 40:1423–1432

Xu J, Han M, Zhang J, Guo Y, Qian H, Zhang W (2014a) Improvement of L-lysine production combines with minimization of by-products synthesis in *Corynebacterium glutamicum*. J Chem Technol Biotechnol 89:1924–1933

Xu J, Han M, Zhang J, Guo Y, Zhang W (2014b) Metabolic engineering *Corynebacterium glutamicum* for the L-lysine production by increasing the flux into L-lysine biosynthetic pathway. Amino Acids 46:2165–2175

Xu JZ, Zhang JL, Guo YF, Jia QD, Zhang WG (2014c) Heterologous expression of *Escherichia coli* fructose-1,6-bisphosphatase in *Corynebacterium glutamicum* and evaluating the effect on cell growth and L-lysine production. Prep Biochem Biotechnol 44:493–509

Xu JZ, Wu ZH, Gao SJ, Zhang W (2018a) Rational modification of tricarboxylic acid cycle for improving L-lysine production in *Corynebacterium glutamicum*. Microb Cell Factories 17:105

Xu JZ, Yang HK, Liu LM, Wang YY, Zhang WG (2018b) Rational modification of *Corynebacterium glutamicum* dihydrodipicolinate reductase to switch the nucleotide-cofactor specificity for increasing L-lysine production. Biotechnol Bioeng 115:1764–1777

Xu JZ, Yang HK, Zhang WG (2018c) NADPH metabolism: a survey of its theoretical characteristics and manipulation strategies in amino acid biosynthesis. Crit Rev Biotechnol 38:1061–1076

Xu JZ, Yu HB, Han M, Liu LM, Zhang WG (2019) Metabolic engineering of glucose uptake systems in *Corynebacterium glutamicum* for improving the efficiency of L-lysine production. J Ind Microbiol Biotechnol 46(7):937–949. https://doi.org/10.1007/s10295-019-02170-w

Yin L, Hu X, Xu D, Ning J, Chen J, Wang X (2012) Co-expression of feedback-resistant threonine dehydratase and acetohydroxy acid synthase increase L-isoleucine production in *Corynebacterium glutamicum*. Metab Eng 14:542–550

Yin L, Shi F, Hu X, Chen C, Wang X (2013) Increasing L-isoleucine production in *Corynebacterium glutamicum* by overexpressing global regulator Lrp and two-component export system BrnFE. J Appl Microbiol 114:1369–1377

Yokota A, Ikeda M (2017) Amino acid fermentation. In: Advances in biochemical engineering/biotechnology, vol 159. Springer, Berlin

Yukawa H, Omumasaba CA, Nonaka H, Kós P, Okai N, Suzuki N, Suda M, Tsuge Y, Watanabe J, Ikeda Y, Vertès AA, Inui M (2007) Comparative analysis of the *Corynebacterium glutamicum* group and complete genome sequence of strain R. Microbiology 153:1042–1058

Zhang L, Li Y, Wang Z, Xia Y, Chen W, Tang K (2007) Recent developments and future prospects of *Vitreoscilla* hemoglobin application in metabolic engineering. Biotechnol Adv 25:123–136

Zhang B, Yu M, Zhou Y, Li Y, Ye BC (2017a) Systematic pathway engineering of *Corynebacterium glutamicum* S9114 for L-ornithine production. Microb Cell Factories 16:158

Zhang Y, Cai J, Shang X, Wang B, Liu S, Chai X, Tan T, Zhang Y, Wen T (2017b) A new genome-scale metabolic model of *Corynebacterium glutamicum* and its application. Biotechnol Biofuels 10:169

Zhang B, Ren LQ, Yu M, Zhou Y, Ye BC (2018a) Enhanced L-ornithine production by systematic manipulation of L-ornithine metabolism in engineered *Corynebacterium glutamicum* S9114. Bioresour Technol 250:60–68

Zhang B, Yu M, Wei WP, Ye BC (2018b) Optimization of L-ornithine production in recombinant *Corynebacterium glutamicum* S9114 by cg3035 overexpression and manipulating the central metabolic pathway. Microb Cell Factories 17:91

Zhang B, Yu M, Zhou Y, Ye BC (2018c) Improvement of L-ornithine production by attenuation of *argF* in engineered *Corynebacterium glutamicum* S9114. AMB Express 8:26

Zhang H, Li Y, Wang C, Wang X (2018d) Understanding the high L-valine production in *Corynebacterium glutamicum* VWB-1 using transcriptomics and proteomics. Sci Rep 8:3632

Zhang X, Zhang X, Xu G, Zhang X, Shi J, Xu Z (2018e) Integration of ARTP mutagenesis with biosensor-mediated high-throughput screening to improve L-serine yield in *Corynebacterium glutamicum*. Appl Microbiol Biotechnol 102:5939–5951

Zhao J, Hu X, Li Y, Wang X (2015) Overexpression of ribosome elongation factor G and recycling factor increases L-isoleucine production in *Corynebacterium glutamicum*. Appl Microbiol Biotechnol 99:4795–4805

Pathways at Work: Metabolic Flux Analysis of the Industrial Cell Factory *Corynebacterium glutamicum*

Judith Becker and Christoph Wittmann

Contents

Abstract The Gram-positive soil bacterium *Corynebacterium glutamicum* is a leading workhorse for the biotechnology industry. Worldwide, engineered strains of the microbe synthetize approximately 6 million tons of the amino acids L-glutamate and L-lysine. In addition to these well-known traditional fermentation products, the portfolio of *C. glutamicum* has impressively expanded over the past years and meanwhile comprises more than 70 natural and non-natural chemicals, materials, fuels, and health-care products. A major challenge to create superior strains for industrial production has always been and still is the design part, i.e. the identification of the most appropriate targets for genomic optimization into a desired phenotype. Metabolic flux analysis is about deciphering the activity of biochemical reactions and pathways of central carbon metabolism. It probably provides the most understandable insight into the core machinery of a living cell. This core part of metabolism displays the controlling centrepiece for synthetizing all the different molecules of recognized industrial value. One can therefore easily understand that

J. Becker · C. Wittmann (✉)
Institute of Systems Biotechnology, Saarland University, Saarbrücken, Germany
e-mail: christoph.wittmann@uni-saarland.de

M. Inui, K. Toyoda (eds.), *Corynebacterium glutamicum*, Microbiology Monographs 23, https://doi.org/10.1007/978-3-030-39267-3_8

metabolic flux analysis had a huge impact on the industrial career of *C. glutamicum* since its pioneering days in the 1950s, when the microbe was discovered. Initial approaches used simple isotopic tracer studies and mathematical balancing of fermentation data to assess metabolic fluxes in *C. glutamicum* for the first time. These techniques have been systematically upgraded into comprehensive technologies and enabled a number of seminal studies of fluxes in *C. glutamicum*, which provided fascinating and unexpected insights into its lifestyle and laid the foundation for metabolically engineered strains, used today in industry. Today, *C. glutamicum* probably displays the best studied microorganism on the level of metabolic fluxes. New developments promise an even brighter future.

1 Introduction

Corynebacterium glutamicum, discovered about 70 years ago, is a leading cell factory of the bio-industry. Traditionally, it has been employed as a producer of amino acids. The genetic alteration of metabolic control in the microbe generated multiple candidates for the over-production of a remarkable amino acid spectrum (Ikeda 2003; Becker and Wittmann 2012, 2015; Wendisch 2017). Today, the two major industrial products obtained by *C. glutamicum* are L-glutamate and L-lysine with an annual global market of approximately 9–10 billion dollar (Becker et al. 2018b; Eggeling and Bott 2015). In addition, strains of *C. glutamicum* were genetically modified to produce a variety of organic acids (Rohles et al. 2018; Inui et al. 2004b; Becker et al. 2015, 2018a; Tsuge et al. 2015; Okai et al. 2016; Wieschalka et al. 2012), diamines (Kind et al. 2010; Kind and Wittmann 2011; Kind et al. 2011; Becker and Wittmann 2017a; Schneider and Wendisch 2010, 2011), alcohols,(Inui et al. 2004a; Jojima et al. 2015; Smith et al. 2010; Blombach et al. 2011; Yamamoto et al. 2013) and amino-acid related compounds (Becker et al. 2013; Perez-Garcia et al. 2017; Rohles et al. 2016; Sun et al. 2016). The huge industrial relevance has stimulated research for more efficient production strains from early on. Initially, strains were developed classically, using random mutagenesis and selection. With the advent of recombinant DNA technology, strain engineering could be upgraded into a rational approach with iterative optimization of desired properties such as yield, productivity, stress tolerance, and substrate spectrum. However, the obvious complexity of the metabolism of *C. glutamicum* complicated the design of cell factories for industrial fermentation. It became clear that optimization would strongly benefit from a better knowledge about *C. glutamicum*, enabling a more straightforward prediction and evaluation of genetic targets (Becker and Wittmann 2015; Ma et al. 2017; Becker et al. 2016; Chae et al. 2017). In this regard, metabolic flux analysis, i.e. fluxomics, appeared perfect to investigate the pathways of *C. glutamicum in vivo*. Over the years, fluxomics has evolved into the most valuable guide in metabolic engineering among all omics approaches (Wittmann 2010; Becker and Wittmann 2018; Kohlstedt et al. 2010). This chapter presents the major concepts and tools of metabolic flux analysis and highlights prominent application examples for systems biology and metabolic engineering of *C. glutamicum*. The type strain *C. glutamicum* ATCC 13032 and thereof derived mutants are in the focus of most flux studies. In addition, there are some reports on fluxes in close relatives including *Brevibacterium lactofermentum* and *Microbacterium ammoniaphilum*.

2 Metabolic Network Analysis of *C. glutamicum*

The complex structure of biological networks typically impedes rational approaches for targeted optimization of cell factories for industrial application. Strategies, which appear promising on a first glance, might quickly be paralyzed by the redundancy of biochemical reactions, an undesired distribution of flux, or the sudden activation of alternative pathways upon genetic perturbation. A profound knowledge of this complex network structure is hence an essential basis for systems metabolic engineering.

2.1 Metabolic Network Reconstruction

Generally, the carbon core metabolism comprises a complex network of approximately 100–200 biochemical reactions, which are closely interconnected. The analysis of the fluxes through these reactions is primarily based on a model of the underlying reaction network. Fortunately, the extensive biochemical and physiological characterization of *C. glutamicum* during the past decades provides a rich source of information on many of the enzymes and pathways present in this organism (Becker and Wittmann 2017b; Yokota and Lindley 2005; Eikmanns 2005). Approximately 15 years ago, this knowledge was complemented and updated with information from genome sequencing and annotation (Kalinowski et al. 2003; Haberhauer et al. 2001). However, one should be aware that central metabolism is still not totally understood. We continuously discover novel features in practically all studied microbes, which have substantial impact for flux analysis, i.e. regarding enzyme specificity (Nikel et al. 2015) or even the existence of complete metabolic routes (Lange et al. 2017).

The core part of metabolism in *C. glutamicum* seems well understood although. The microbe possesses the pentose phosphate (PP) pathway, the Emden-Meyerhof-Parnas (EMP) pathway, the tricarboxylic acid (TCA) cycle, the glyoxylic acid shunt, and a rich set of enzymes, which interconnect C_3-metabolites of the EMP pathway with C_4-metabolites of the TCA cycle (Fig. 1) (Wittmann and Becker 2007; Wittmann and De Graaf 2005; Becker et al. 2016). Metabolic intermediates are withdrawn from these central metabolic pathways to serve as building blocks for biomass formation, which results in continuous anabolic fluxes in growing cells. *C. glutamicum* requires glucose 6-phosphate, fructose 6-phosphate, ribose 5-phosphate, erythrose

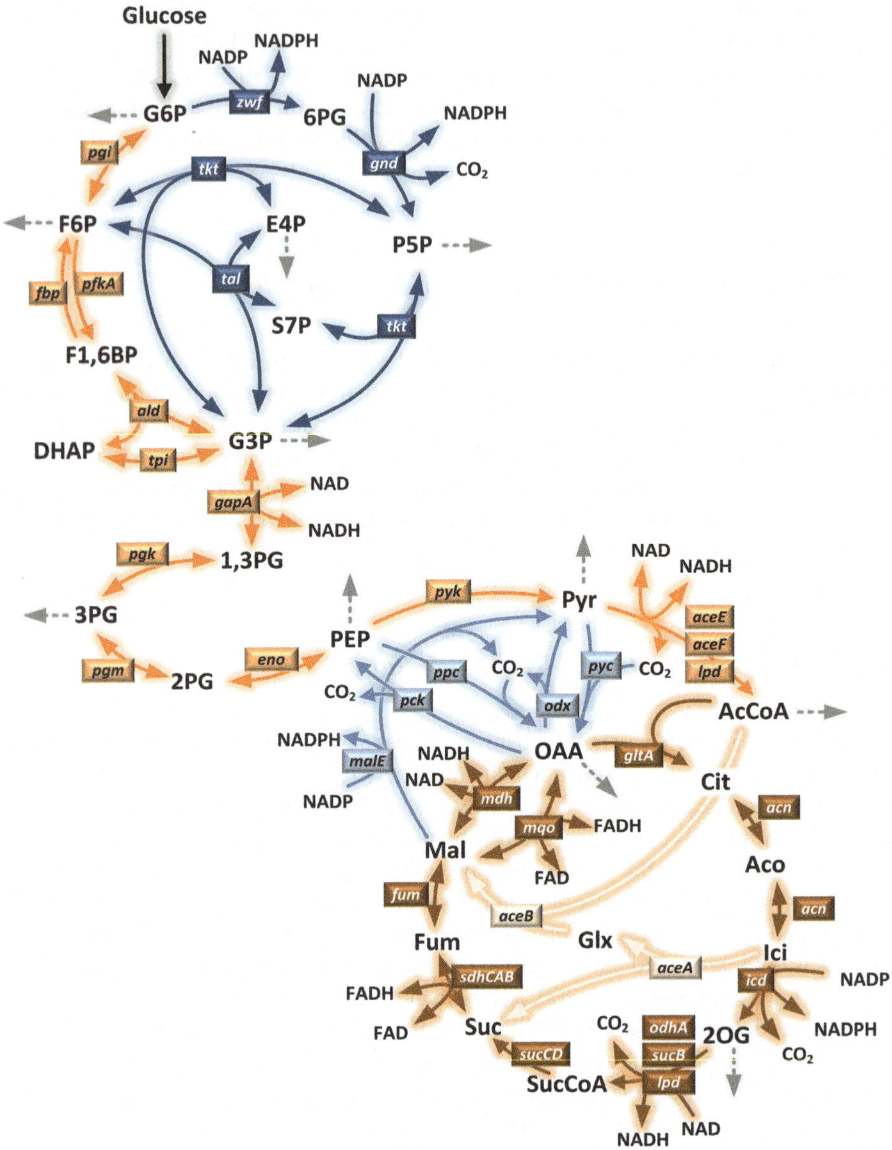

Fig. 1 Biochemical network of the central metabolism of *Corynebacterium glutamicum*

4-phosphate, glyceraldehyde 3-phosphate, pyruvate, phosphoenolpyruvate, acetyl-CoA, oxaloacetate, and 2-oxoglutarate for biomass formation (Wittmann and De Graaf 2005). Due to the structure of the cell wall, diaminopimelate, a direct precursor of L-lysine, is additionally consumed (Wehrmann et al. 1998). Beside carbon precursors, NADPH is required as reducing power for biosynthesis. Overall, 16,400 μmol NADPH is consumed to form 1 g biomass. This amount has to be generated by the

NADPH supplying enzymes, glucose 6-phosphate dehydrogenase (Ihnen and Demain 1969; Moritz et al. 2000), 6-phosphogluconate dehydrogenase (Ohnishi et al. 2005; Moritz et al. 2000), isocitrate dehydrogenase (Chen and Yang 2000) and malic enzyme (Dominguez et al. 1998; Becker et al. 2008) (Fig. 1). For most flux studies, knowledge on the cellular composition of *C. glutamicum* and the anabolic demand for the precursors is therefore important. Corresponding data have been collected in different studies for *C. glutamicum* MH20-22B in continuous culture using glucose (Marx et al. 1996), for exponentially growing *C. glutamicum* ATCC17965 using glucose (Cocaign-Bousquet et al. 1996), and for exponentially growing *C. glutamicum* BS87 (Kind et al. 2013).

In addition, reactions and pathways to secreted products have to be considered. Most important, when studying over-producers, is the biosynthetic pathway for the target product. As example, the biosynthetic pathways of L-lysine and L-glutamate start from oxaloacetate and 2-oxoglutarate, respectively, two intermediates of the carbon core network, and additionally demand for NADPH (4 NADPH per L-lysine and 1 NADPH per L-glutamate). Furthermore, *C. glutamicum* might secrete minor shares of by-products, such as trehalose, acetate, lactate, succinate, pyruvate, L-alanine, L-glutamate, and L-glycine (Krömer et al. 2004; Becker et al. 2005; Kiefer et al. 2002; Blombach et al. 2007). The observed by-product spectrum determines, which of the corresponding pathways should be taken into account.

For flux analysis, the set of biochemical reactions and pathways has to be arranged into a fully consistent network model. In such a model, the metabolites are stoichiometrically connected by exactly defined enzymatic reactions, which altogether convert substrates into intermediates, and ultimately into products (Vallino and Stephanopoulos 1993; Schilling et al. 1999; Kjeldsen and Nielsen 2009; Shinfuku et al. 2009; Melzer et al. 2009; Zhang et al. 2019; Mei et al. 2016). When designing such networks models, it is important not to base them exclusively on genomic information and stoichiometry, but also take thermodynamic considerations into account, which impact direction and reversibility of flux and constrain feasible flux states (Palsson 2000; Wang et al. 2017). Such constraints are easily implemented into the biochemical model, if the thermodynamic feasibility of the corresponding biochemical reaction is known. This allows, for instance, defining reversible and irreversible reactions in the network. Finally, such a functional biochemical network is the basic prerequisite for the determination of metabolic fluxes.

2.2 *Stoichiometric Flux Analysis*

Stoichiometric balancing, also called metabolite balancing, is a model-based approach for flux analysis. It is based on a metabolic network of the relevant reactions and applicable to cells under metabolic steady-state, e.g. growing in chemostat, balanced batch cultures, and balanced phases of fed-batch processes (Vallino and Stephanopoulos 1993; Dersch et al. 2016b). In short, a subset of the fluxes of interest is directly measured, e.g. fluxes linked to substrate uptake, biomass formation, and product synthesis (Wittmann and De Graaf 2005; von Kamp et al. 2017). Subsequently, the remaining intracellular fluxes are calculated using the stoichiometric relations of the model. The calculation is straightforward, using linear algebra. Pioneering studies compared metabolic fluxes in *C. glutamicum* during growth and L-lysine production (Vallino and Stephanopoulos 1993, 1994), as well as under different environmental conditions such as osmotic stress (Varela et al. 2003), and oxygen deprivation (Dominguez et al. 1993). Due to the nature of carbon core metabolism, stoichiometric flux models derived therefrom are usually underdetermined, which is also true for *C. glutamicum*. Consequently, additional constraints are required to achieve a unique solution of the intracellular flux distribution. These constraints can be provided by balances for ATP, NADH or NADPH, which, however, often are subjected to uncertainties and can even introduce artefacts into the flux estimation (Wittmann and De Graaf 2005; Winter and Krömer 2013; Maarleveld et al. 2013). Alternatively, the underlying network can be simplified, e.g. assuming certain pathways to be inactive (Aiba and Matsuoka 1979). Metabolic fluxes derived from stoichiometric balancing are strongly influenced by the network topology and the chosen assumptions rather than by biological data (Sauer 2006; El Massaoudi et al. 2003), so that care should be taken regarding their interpretation. In addition, important fine structures of the network of *C. glutamicum* cannot be discriminated at all, using stoichiometric balancing. As example, the two branches of L-lysine biosynthesis both exhibit the same stoichiometry (Wittmann and Becker 2007; Sonntag et al. 1993). The individual fluxes through the parallel routes are therefore not resolvable by this approach (Wittmann and De Graaf 2005). Similarly, the glyoxylate shunt, the bidirectional fluxes around the pyruvate node, and flux reversibility of individual enzymes remain inaccessible (Vallino and Stephanopoulos 1993; Wittmann and De Graaf 2005). Nevertheless, stoichiometric flux analysis was an important approach, particularly in the early days of *C. glutamicum* research and contributed strongly to our current understanding of the microbe. Moreover, increased knowledge on energy and redox metabolism (Sonderegger et al. 2004) and the consideration of transcriptomics regarding the set of expressed pathways under the studied conditions (Beckers et al. 2016) help to refine network structures and increase model quality and thereby flux accuracy.

3 State-of-Art ^{13}C Metabolic Flux Analysis

The first flux studies of *C. glutamicum* explored the microbe using isotopic tracers. These tracer substrates are enriched with ^{13}C at specific carbon atom positions (Wittmann 2007; Becker and Wittmann 2014). When fed to the studied cells, they are taken up and metabolized, whereby the ^{13}C label is distributed throughout the network. The carbon transition of each enzymatic reaction thereby specifically defined the fate of the ^{13}C label (Fig. 2). Depending on the tracer chosen, pathway specific labelling signatures are formed, which can be exploited to infer intracellular

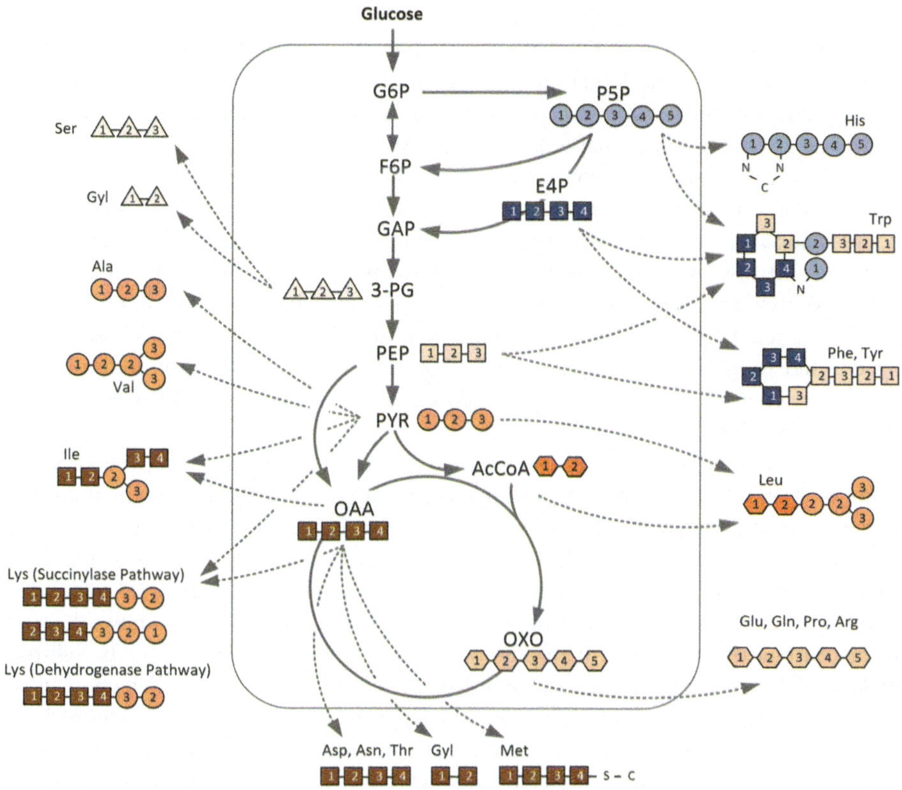

Fig. 2 Relationship between the carbon skeleton of amino acids and the carbon skeleton of their precursors withdrawn from the intermediary metabolism. Adapted from Wittmann (2007), Szyperski (1998)

fluxes. Mass spectrometry and NMR emerged as most prominent techniques to measure ^{13}C data from such isotopic experiments and provide information for flux calculation (Wittmann 2002, 2007; Kiefer et al. 2007; Massou et al. 2007). Pioneering isotope studies generated important knowledge on metabolic properties of *C. glutamicum,* such as L-histidine biosynthesis (Ishino et al. 1984), and first estimates of the fluxes through the TCA cycle or the glyoxylate shunt (Walker et al. 1982). In addition, they have been successfully used for the identification of unknown metabolites and side-products, resulting from genetic alteration of the metabolism of *C. glutamicum* (Krömer et al. 2006; Kind et al. 2010). Moreover, isotope experiments provided valuable insights into sulphur (Bolten et al. 2010) and nitrogen metabolism of *C. glutamicum* (Tesch et al. 1999), allowed to trace the carbon fate from substrates into products (Lessmeier et al. 2015; Bolten et al. 2010; Ishino et al. 1986), provided evidence for the re-uptake of secreted products (Zhao et al. 2012; Rohles et al. 2018), and led to the discovery of novel metabolic pathways (Krömer et al. 2006). Today, selected labelling data are also found useful to obtain model-independent flux estimates (Becker et al. 2010; Klingner et al. 2015; Fürch et al. 2009) and to assess net fluxes in the metabolism of *C. glutamicum* (Sato et al. 2008; Hoffmann et al. 2018).

In particular, isotopic labelling data are highly valuable to reduce the number of assumptions needed in pure stoichiometric balancing and to obtain fully determined and even overdetermined networks (Sauer 2006; Wittmann and De Graaf 2005). Over the years, this integrated concept has been developed further and further. Today, state-of-art metabolic flux analysis is a comprehensive workflow, which couples sophisticated isotopic experiments with high-resolution ^{13}C labelling analysis and software embedded modelling frameworks for flux computation (Guo et al. 2018; Wittmann 2007; Heux et al. 2017). GC-MS has evolved into the most prominent technique for the ^{13}C measurement due to its high accuracy, sensitivity, speed, and robustness (Wittmann 2002, 2007; Guo et al. 2018). In addition, also NMR has been used for this purpose (de Graaf et al. 2000; Boisseau et al. 2013; Nicolas et al. 2007).

For flux calculation, data on extracellular fluxes (substrate uptake, product secretion), anabolic fluxes, and ^{13}C labelling patterns are used to globally fit the unknown flux parameters by a computer flux model (Kiefer et al. 2004; Krömer et al. 2004; Wittmann and Heinzle 2002). Hereby, the free flux parameters of interest are varied by an optimization algorithm, until the deviation between experimental and simulated data is minimized (Wittmann 2007). Sophisticated modelling frameworks such as matrix based isotopomer mapping (Wittmann and Heinzle 1999, 2001b; Schmidt et al. 1999), cumomer balancing using the 13C-FLUX software (Wiechert et al. 2001) and the open source package OpenFLUX (Quek et al. 2009) and OpenFLUX2 (Shupletsov et al. 2014) provide convenient software solutions for flux calculation. The platforms are steadily updated and extended to allow a more user-friendly programming and visualization (Shupletsov et al. 2014; Desai and Srivastava 2018; Weitzel et al. 2013; Nöh et al. 2015; He et al. 2016). In addition, they support experimental design of tracer experiments and statistical flux evaluation (Quek et al. 2009; Millard et al. 2014; Nöh et al. 2018).

3.1 Experimental Design

The appropriate combination of tracer substrate and labelling analytes is a critical factor to obtain high quality labelling information for flux estimation (Wittmann and Heinzle 2001b; Yang et al. 2005; Nöh et al. 2018). The selection is case specific and depends on the flux parameters of interest and the studied network. Experimental design has been successfully used to evaluate the suitability of specific tracers (Wittmann and Heinzle 2001b; Yang et al. 2005; Nöh et al. 2018; Crown et al. 2016; Kohlstedt and Wittmann 2019; Millard et al. 2014) and the information content of specific mass isotopomers (Becker et al. 2008; Kohlstedt and Wittmann 2019) for a number of flux problems. Using computer simulations, suitable approaches can be identified also for *C. glutamicum*, as illustrated for different

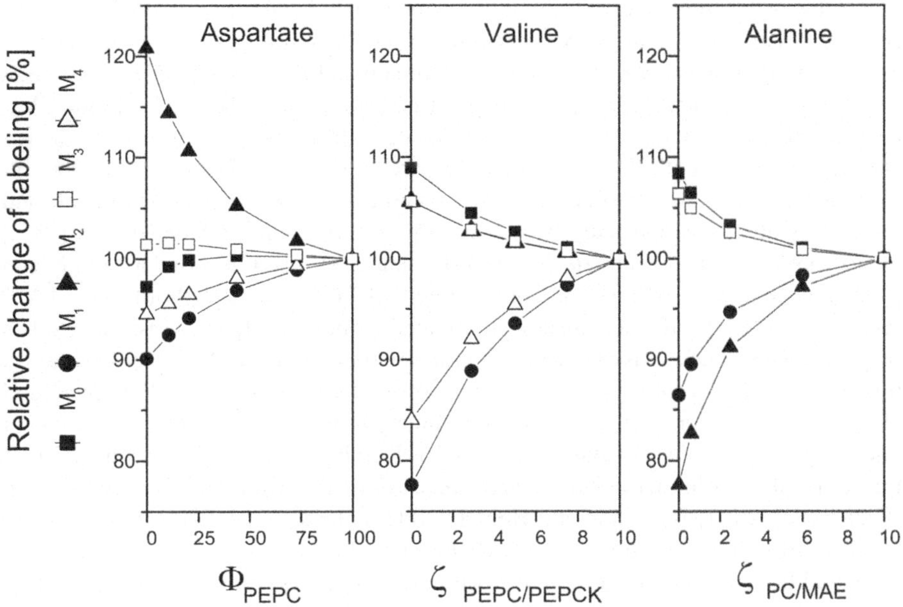

Fig. 3 Computer-based experimental design for quantification of flux parameters at the pyruvate node of *Corynebacterium glutamicum*: mass isotopomer distribution of aspartate (*m/z* 418) with varied flux partitioning between phosphoenolpyruvate carboxylase and pyruvate carboxylase (Φ_{PEPC}), mass isotopomer distribution of valine (*m/z* 288) with varied exchange flux between phosphoenolpyruvate and oxaloacetate catalyzed by phosphoenolpyruvate carboxylase and phosphoenol-pyruvate carboxykinase ($\zeta_{PEPC/PEPCK}$), mass isotopomer distribution of alanine (*m/z* 260) with varied exchange flux between pyruvate and oxaloacetate catalyzed by pyruvate carboxylase and malic enzyme ($\zeta_{PC/MalE}$). The assumed input substrate is an equimolar mixture of [$^{13}C_6$] glucose and [$^{12}C_6$] glucose. The figure is taken from Becker et al. (2008)

GC-MS fragment ions and fluxes around the complex network interconnecting glycolysis and TCA cycle (Fig. 3).

In principle, tracer experiments can be carried out on a single carbon source or a substrate mixture, using tracer substrates that are labelled at one or more positions (Becker et al. 2008; Wittmann et al. 2004a; Kiefer et al. 2004; Rados et al. 2014). The optimal labelling degree, position and/or substrate mixture of the tracer substrate can again be identified by a computer-based experimental design. Intensive computational simulations showed that specific tracers are optimal for flux analysis in *C. glutamicum* and other microbes (Wittmann and Heinzle 2001b; Wiechert and de Graaf 1997; Möllney et al. 1999; Kohlstedt and Wittmann 2019). Flux analysis with [1-^{13}C] glucose, for instance, provides labelling data that sensitively depend on the flux partitioning between the EMP and the PP pathway. The sensitivity allows a precise estimation of the corresponding *in vivo* fluxes (Wittmann 2007; Wittmann and Heinzle 2001b). On the level of individual carbon atoms, the PP pathway enzyme 6-phosphogluconate dehydrogenase exclusively releases the ^{13}C-labelled C_1 atom in the decarboxylation reaction, whereas this atom is retained in the carbon backbone of the pathway intermediates, when glucose is metabolized via the EMP pathway (Fig. 4a). Variation of the flux partitioning ratio thus strongly influences the ^{13}C enrichment in downstream metabolites (Wittmann and De Graaf 2005). At the same time, the degree of reversibility of phosphoglucoisomerase at the entry into the EMP pathway affects the labelling. The measurement of different analytes provides sensitive data to assess both flux parameters in parallel (Wittmann and Heinzle 2001b). Complementary to that, resolution of the exchange fluxes between the TCA cycle and the EMP pathway is optimally based on a mixture of naturally labelled and [$^{13}C_6$] glucose (Wittmann and Heinzle 2001b; Wittmann 2007), as C-C bond assembly and cleavage results in new labelling combinations of TCA cycle metabolites and EMP pathway metabolites that are specifically related to these reactions. Most flux studies of *C. glutamicum* hence used two parallel tracer experiments on [1-^{13}C] glucose and a mixture of naturally labelled and [$^{13}C_6$] glucose (Becker et al. 2008; Wittmann and Heinzle 2002; Kiefer et al. 2004) or a combination of [1-^{13}C] and [$^{13}C_6$] glucose in one single experiment (Iwatani et al. 2007). Hereby, parallel studies on different tracers result in a higher accuracy and precision of the metabolic flux distribution (Kiefer et al. 2004; Wittmann and Heinzle 2002; Petersen et al. 2000; Krömer et al. 2004; Kohlstedt and Wittmann 2019; Rados et al. 2014).

In addition, also other tracers have been used to address more specific questions. As stated above, *C. glutamicum* operates two parallel L-lysine biosynthetic routes. The relative contribution of each branch to the overall L-lysine flux has been determined by tracer experiments on [6-^{13}C] glucose and positional labelling analysis of L-lysine using NMR (Sonntag et al. 1993; Nicolas et al. 2008). Alternatively, the relevant labelling information to determine the L-lysine flux split was obtained in *C. glutamicum*, using GC-MS analysis of L-alanine, L-lysine and diaminopimelate from tracer studies on [4-^{13}C] glucose or [3-^{13}C] glucose (Fig. 4b) (Becker et al. 2013; Wittmann and Heinzle 2001b). In another prominent example, metabolic flux analysis of sucrose-grown *C. glutamicum* used three parallel tracer studies on [1-^{13}C glucose] sucrose, [1-^{13}C fructose] sucrose and [^{13}C fructose] sucrose (Wittmann et al. 2004a).

3.2 Labelling Analysis for Flux Estimation

Analysis of ^{13}C labelling patterns is mainly performed by GC-MS, which allows a quantification of the mass isotopomer distribution in a variety of different metabolites (Villas-Boas et al. 2005; Wittmann 2007; Roessner et al. 2000; Chen et al. 1998). Metabolites, accessible by GC-MS for flux analysis are for example amino acids (Wittmann et al. 2002; Wittmann and Heinzle 1999; Christensen et al. 2002; Christensen and Nielsen 1999), organic acids (Adler et al. 2013; Lange et al. 2017), sugars, and sugar derivatives (Kohlstedt and Wittmann 2019; Kiefer et al. 2004). Their analysis requires derivatization, GC separation, electron impact (EI) or chemical ionization (CI), mass separation, and detection (Wittmann 2007; Becker and Wittmann 2014). Due to the high energy input in the ionization process, the analytes

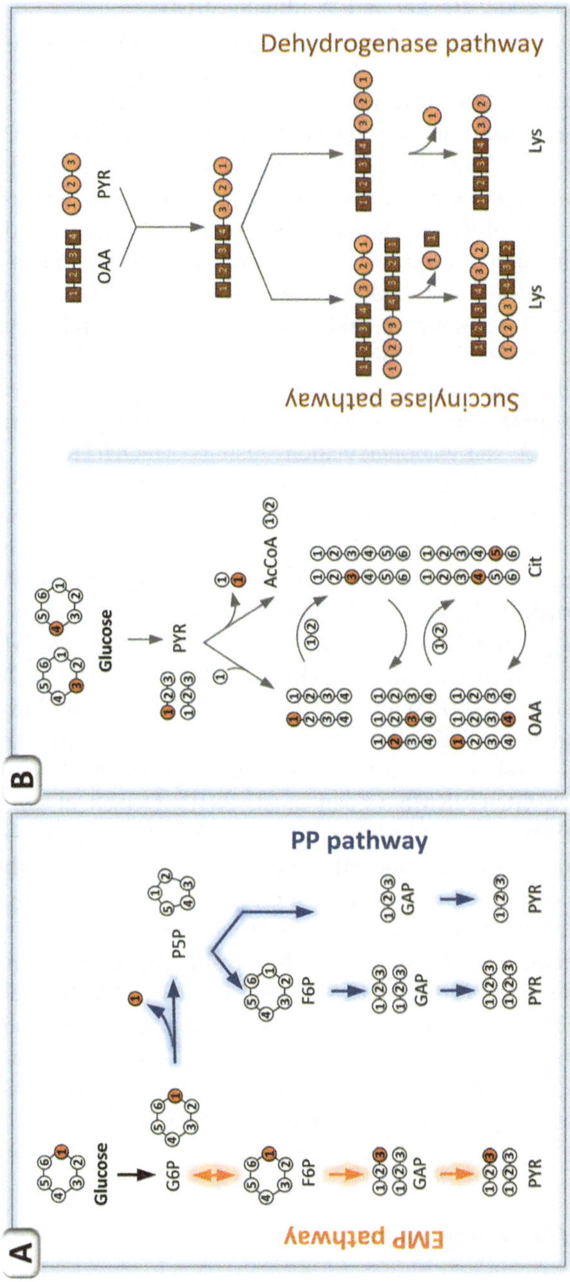

Fig. 4 (**a**) Strategy for determination of the flux partitioning ratio between glycolysis and pentose phosphate pathway using an isotope study with [1-^{13}C] glucose. (**b**) Strategy for determination of the flux partitioning between dehydrogenase branch and succinylase branch of the L-lysine biosynthetic pathway of *C. glutamicum*

are usually fragmented into specific ions. In turn, the formed fragments can be used for identification and labelling analysis (Christensen and Nielsen 1999; Wittmann 2007; Wittmann et al. 2002; Becker and Wittmann 2014).

Proteinogenic amino acids are the most common analytes used to obtain the necessary labelling information for steady-state flux analysis in growing cells (Christensen and Nielsen 1999; Dauner and Sauer 2000). A number of flux studies for *C. glutamicum* were based on these data (Wittmann and Heinzle 2002; Wittmann et al. 2004b; Krömer et al. 2004; Becker et al. 2008; Buschke et al. 2013; Kind et al. 2013). Hereby, the well-known precursor-amino acid relationship allowed to deduce the ^{13}C-labelling patterns of the respective carbon precursors in intermediary metabolism from that of the amino acids (Fig. 2) (Szyperski 1998; Wittmann 2007). The additional consideration of sugars and sugar derivates in the labelling analysis can enhance flux identifiability, resolution, and accuracy (Kohlstedt and Wittmann 2019; Adler et al. 2013; Lange et al. 2017; Becker et al. 2007). In principle, amino acids can be derivatised by silylation using reagents such as MBDSTFA (N-methyl-N-tert-butyldimethylsilyl-trifluoroacetamide) and MSTFA (N-methyl-N-(trimethylsilyl)-trifluoro acetamid), or alkylation/acylation using MeOH/TMCS (Methanol/Trimethylchlorosilane) and TMSH (Trimethylsulfonium hydroxide) for GC-MS analysis (Wittmann 2007). However, the labelling pattern of amino acids is preferentially quantified from their *t*-butyl-dimethyl-silyl (TBDMS) derivatives, which yield high quality signals for fragment ions that contain the entire carbon skeleton of the analytes and are therefore particularly useful for metabolic flux analysis. For sugars, the derivatisation typically comprises a combination of oximation and silylation using hydroxylamine and trifluoroacetamides (BSTFA, MSTFA) (Kiefer et al. 2002; Bolten et al. 2009; Wittmann et al. 2007; Wittmann 2007). In addition to GC-MS, other mass spectrometry techniques for ^{13}C labelling analysis comprise LC-MS (Kiefer et al. 2007; Oldiges et al. 2004; van Winden et al. 2005; Noack et al. 2011), CE-MS (Iwatani et al. 2007), GC-C-IRMS (Yuan et al. 2010; Dersch et al. 2016a) and MALDI-TOF MS (Wittmann and Heinzle 2001a). MS techniques provide mass isotopomer data of an investigated molecule. Orthogonal information on the positional labelling of specific carbon atoms within a molecule is obtained, using NMR and 2D-NMR, although at the cost of increased effort and reduced sensitivity (de Graaf et al. 2000; Nicolas et al. 2008; Marx et al. 1996). To some extent, positional labelling information can also be deduced from LC-MS and GC-MS data (Kappelmann et al. 2017; Becker et al. 2013). It depends on the studied metabolic flux problem, which technique and/or combinations thereof should be selected—in terms of costs and effort on one hand and flux resolution and identifiability on the other hand (Millard et al. 2017; Schwechheimer et al. 2018a, b; Sonntag et al. 1993; Rühl et al. 2012).

3.3 Technical Implementation

A spectrum of flux technologies has been developed to enable an application-oriented research of this organism, such as mutant screening at miniaturized scale, routine strain characterization and testing, bioprocess development at laboratory scale, and industrial production at large-scale (Fig. 5).

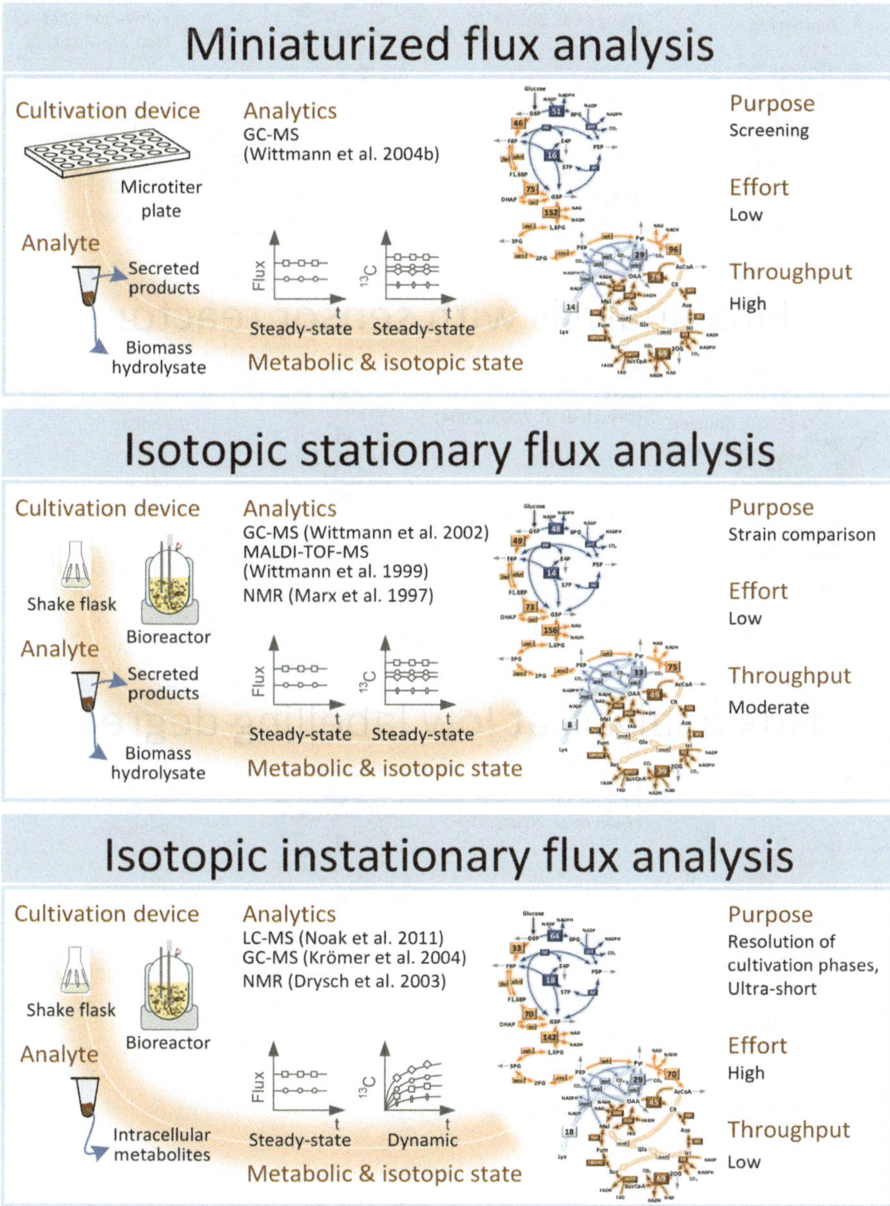

Fig. 5 Current spectrum of ^{13}C metabolic flux technologies and common application field in systems biology and systems metabolic engineering

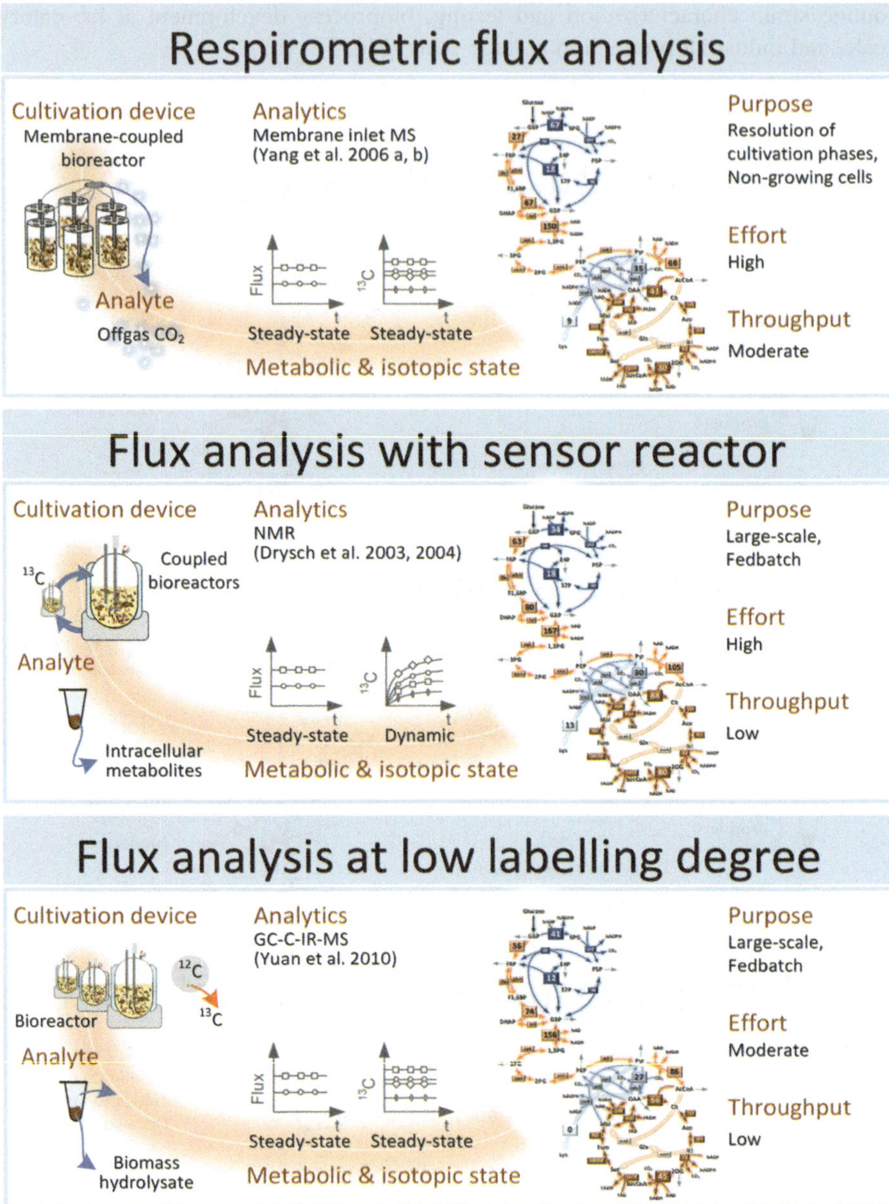

Fig. 5 (continued)

3.3.1 Routine Flux Analysis of Balanced Cultures at Lab Scale

After years of intense research and development, GC-MS-based ^{13}C metabolic flux analysis has meanwhile evolved into a routine method for determining steady-state fluxes in microbes (Sauer 2006). The approach allows high-resolution of fluxes at high precision and accuracy. The tracer studies are easily performed in shake flasks with culture volumes of 5–25 mL, but can, of course also be conducted in lab scale bioreactors. This technique is usually applied to exponentially growing cells, since these cells are in metabolic and isotopic steady-state (Becker et al. 2008; Winter and Krömer 2013), so that the amino acid labelling from the cell protein provides the necessary information for flux determination. During growth, labelled amino acids are incorporated into biomass, so that their labelling pattern is conserved in the cell protein; they are subsequently easily accessible by simple hydrolysis of the harvested biomass (Kelleher 2001). Moreover, organic acids and sugar derivatives can be analysed to provide complementary labelling data (Kohlstedt and Wittmann 2019; Wittmann and Heinzle 2002; Lange et al. 2017). It should be noticed that the verification of steady-state is an important prerequisite for appropriate analysis. Whereas isotopic steady-state can be validated by constant ^{13}C labelling patterns over time, metabolic steady-state can be inferred from constant kinetics and stoichiometry during the studies process (Rohles et al. 2016; Becker et al. 2008; Bücker et al. 2014).

This type of steady-state stationary flux analysis has proven highly value in guiding systems metabolic engineering of *C. glutamicum* (Wittmann 2007; Wittmann and Heinzle 2002; Wittmann et al. 2004a; Becker and Wittmann 2018; Becker et al. 2011). Moreover, it provided unique insights into the metabolism of the microbe as response to genetic perturbation (Bartek et al. 2011; Becker et al. 2008; Marx et al. 1997; Kim et al. 2006; Marx et al. 1999), amino acid production (Sonntag et al. 1995; Krömer et al. 2004; Shirai et al. 2006), and the use of different carbon sources, such as sucrose, fructose, xylose, mannitol, acetate and propionate (Wittmann et al. 2004a; Kiefer et al. 2004; Buschke et al. 2013; Hoffmann et al. 2018; Veit et al. 2009; Wendisch et al. 2000; Dominguez et al. 1998).

3.3.2 Miniaturized Flux Analysis for Screening Purposes

Genetic modification of microbes during strain optimization often creates a large number of mutants, which display highly interesting libraries to be screened on the flux level. To this end, approaches were developed for *C. glutamicum* (Wittmann et al. 2004b) and other microbes (Klingner et al. 2015; Berger et al. 2014; Heux et al. 2014), comprising tracer cultivation at a miniaturized scale for high-throughput (Fischer et al. 2004; Wittmann et al. 2004b; Heux et al. 2017). The established multi-well microtiter plate format allows reproducible, well-controlled cultivation in a volume of approximately 100–1000 μL with online monitoring of dissolved oxygen, offline quantification of substrates, products, and biomass, together with labelling analysis of secreted products, or optionally from hydrolysed biomass, using GC-MS (Wittmann et al. 2004b). In addition to the high number of parallel experiments, the miniaturized scale benefits from a cost advantage since it requires much lower amounts of the costly ^{13}C-labelled substrates. Using such miniaturized approaches, the fluxes through the central metabolic pathways of *C. glutamicum* can be precisely determined, as demonstrated for different L-lysine producing mutants (Wittmann et al. 2004b). In the meantime a novel platform for automated high-throughput fluxome profiling of bacteria has been established, which combines a robotic system for ^{13}C labelling experiments and sampling with NMR-based isotopic profiling including automated data interpretation (Heux et al. 2014).

3.3.3 Metabolic Flux Analysis Under Dynamic Conditions

Notably, the aforementioned techniques for flux estimation are suitable to determine steady-state fluxes. However, industrial fermentation processes are often performed as fed-batch cultures that are characterized by changing conditions. The determination of metabolic fluxes under such dynamic conditions requires specific methods. The key to resolve time-dependent changes in flux is the consideration of intracellular metabolites, which quickly adapt labelling patterns. Suitable metabolites are intracellular amino acids (Krömer et al. 2004; Drysch et al. 2004) and pathway intermediates of the EMP pathway, the PP pathway, and the TCA cycle (Kiefer et al. 2007; Drysch et al. 2004). While the amino acids can be well analysed for ^{13}C labelling by GC-MS (Wittmann et al. 2002; Krömer et al. 2004), free pathway intermediates are accessible using LC-MS (Kiefer et al. 2007; Iwatani et al. 2007; Noack et al. 2011), GC-MS (Kohlstedt et al. 2014), and NMR (Drysch et al. 2004). Previous applications investigated flux variations in response to dynamic culture conditions, such as shifts from growth to L-lysine production (Krömer et al. 2004) or from batch to fed-batch (Iwatani et al. 2007). This technique seems useful to identify metabolic bottlenecks, which specifically result from the changed flux distribution, since these would not be accessible by standard ^{13}C MFA.

3.3.4 Metabolic Flux Analysis of Non-growing Cells

Industrial fermentation processes typically support product synthesis under conditions of limited growth in order to reduce or even abolish anabolism as a competitor for carbon, energy, and reducing power. Typically, industrial production is conducted as fed-batch with an extended nutrient-limited production phase (Kind et al. 2014; Kohlstedt et al. 2018; Rohles et al. 2018). As another example, amino acid production with *C. glutamicum* traditionally uses auxotrophic mutants with decoupled phases of growth and production (Krömer et al. 2004; Vallino and Stephanopoulos 1993; Drysch et al. 2004). In addition, novel concepts for dynamic metabolic control use genetic switches and metabolic valves to change the metabolic activity from one state that optimally supports growth to another state that is more favourable for production (Lynch 2016). An elaborated respirometric approach for flux estimation has been developed to access metabolic fluxes of such producing, non-growing cells (Yang et al. 2005, 2006a, b). It is based on tracer studies with labelling measurement of carbon dioxide, which is practically always formed during metabolic activity which needs energy and redox power, even in non-growing cells. The technique is thus independent of further product formation or growth. Using a set-up with three parallel experiments on differently labelled tracer substrates, all central metabolic pathways activities of *C. glutamicum* can be resolved (Yang et al. 2006a). Another study reports on the flux estimation of non-growing succinate-producing *C. glutamicum*. A tracer combination of [1-^{13}C] glucose, [2-^{13}C] glucose, [1,2-^{13}C] glucose and [U-^{13}C] glucose combined with NMR analysis of secreted products allowed the determination of the central metabolic fluxes of non-growing *C. glutamicum* under anaerobic condition (Rados et al. 2014).

3.3.5 Metabolic Flux Analysis at Large Scale

Labelling studies for the determination of intracellular pathway activities are usually restricted to small-scale experimental set-ups due to the high cost of the required tracer substrates. However, with regard to industrial production processes it appears attractive to mitigate eventual scale effects by estimating metabolic fluxes also at large scale. This became possible with the development of a sensor-reactor approach allowing flux estimation under varying conditions representative for large-scale fermentation (El Massaoudi et al. 2003). The basic idea is to perform the tracer experiment in a small-scale bioreactor, which is operated in parallel to a large-scale production reactor. The labelling is introduced into the sensor reactor system by [$^{13}C_6$] glucose pulses; key fermentation parameters are assessed in both reactors throughout the course of the experiment to ensure process equivalence. In a case study, labelling data for flux estimation were obtained by NMR analysis of intracellular metabolites (Drysch et al. 2003). The approach was further extended to the determination of time-dependent flux changes that typically occur during fed-batch fermentation (Drysch et al. 2003, 2004).

A smart strategy for metabolic flux analysis at large scale is the use of ultrasensitive technology for labelling analysis. GC combustion isotope ratio mass spectrometry (GC-C-IRMS) can precisely estimate ^{13}C labelling at extremely low enrichment. It is approximately 100–1000-fold more sensitive than common GC-MS or LC-MS approaches (Dersch et al. 2016a). This type of measurement allows to substantially decrease the amount of the applied tracer. GC-C-IRMS allows precise metabolic flux analysis at a reduced labelling degree of the tracer substrate of only 1%, whereby the coupling to a GC enables efficient separation and subsequent analysis of amino acids (Yuan et al. 2010).

4 Metabolic Pathway Engineering in *C. glutamicum*

With regard to metabolic flux analysis, *C. glutamicum* is probably the best studies microbe (Wittmann 2010; Wittmann and De Graaf 2005; Bartek et al. 2011; Noack et al. 2011; Hoffmann et al. 2018; Rados et al. 2014). Driven by the strong industrial interest for this bacterium, a number of studies have been conducted over the past years. They provide a very detailed picture and form an excellent knowledge base for systems metabolic engineering of *C. glutamicum* (Wittmann and Becker 2007; Ikeda 2003).

4.1 Key Fluxes for the Production of L-Lysine and L-Lysine-Derived Chemicals

The feed amino acid L-lysine holds the second largest share in the amino acid business with global sales of approximately 4 billion dollars. Industrial production mainly relies on strains of *C. glutamicum,* which have been engineered by classical mutagenesis as well as metabolic engineering (Eggeling and Bott 2015; Becker et al. 2018b). In recent years, such strains became more and more attractive as chassis strains for the production of L-lysine derived platform chemicals (Becker et al. 2018b). This chemicals include diaminopentane (Kind et al. 2010; Kind and Wittmann 2011), aminovalerate (Rohles et al. 2016; Shin et al. 2016) and glutarate (Rohles et al. 2018), which serve as building blocks for the production of novel bio-based plastics with advanced material properties, and L-pipecolic acid (Perez-Garcia et al. 2017; Perez-Garcia et al. 2016), a valuable precursor for the pharmaceutical industry. With a stoichiometric demand of 1:1:4 for each molecule of L-lysine, the L-lysine biosynthetic pathway in *C. glutamicum* requires pyruvate and oxaloacetate as carbon precursors, and NADPH as reducing power (Michal 1999). Similarly, these building blocks are required for all L-lysine-derived compounds. It is obvious that optimized production requires a well-balanced supply of these building blocks.

4.1.1 Engineering of Key Fluxes for Carbon Precursor Supply

Pyruvate and oxaloacetate are both part of a network of carboxylating and decarboxylating reactions, which connect the EMP pathway and the TCA cycle. In *C. glutamicum,* this complex network comprises pyruvate carboxylase and phosphoenolpyruvate carboxylase as carboxylating enzymes, as well as malic enzyme, PEP carboxykinase and OAA decarboxylase that catalyze decarboxylation reactions from the TCA cycle towards the EMP pathway (Fig. 1) (Eikmanns 2005; Becker and Wittmann 2017b). This enzymatic set-up creates a cyclic reaction network, which controls the anaplerotic net flux into the TCA cycle and, furthermore, allows a redistribution of flux in response to metabolic or genetic disturbance (Becker et al.

2008; Petersen et al. 2000, 2001; Shirai et al. 2007; Sawada et al. 2010; Yanase et al. 2016; Netzer et al. 2004). The general role of the cycle has been attributed to energy maintenance (de Graaf et al. 2001; Sauer et al. 1997), equilibration of metabolite levels around the pyruvate node (Sauer and Eikmanns 2005), and additional NADPH supply under certain conditions (Dominguez et al. 1998; Becker et al. 2008; Netzer et al. 2004). Specifically related to the development of L-lysine overproducing strains, the cycle and its individual enzymes exhibit a key role. Using metabolic flux analysis, a close correlation between the L-lysine flux and the anaplerotic net flux was discovered for different L-lysine producers (Wittmann and De Graaf 2005; Wittmann and Heinzle 2002). This observation was used in different metabolic engineering strategies that aimed at an increased anaplerotic net flux by (1) overexpression of pyruvate carboxylase as major anaplerotic enzyme (Peters-Wendisch et al. 2001; Kind et al. 2011; Becker et al. 2011; Kind et al. 2010; Rohles et al. 2016, 2018; Buschke et al. 2011), (2) overexpression of PEP carboxylase (Sano et al. 1987), (3) deletion of the *pck* gene encoding the OAA-consuming reaction PEP carboxykinase (Petersen et al. 2001; Kind et al. 2011; Becker et al. 2011; Perez-Garcia et al. 2017, 2018; Jorge et al. 2017b; Rohles et al. 2016, 2018; Kind et al. 2010; Buschke et al. 2011) and (4) implementation of feed-back insensitive variants of pyruvate carboxylase (Ohnishi et al. 2002; Kortmann et al. 2019; Becker et al. 2011; Perez-Garcia et al. 2017, 2018; Jorge et al. 2017b; Rohles et al. 2016, 2018; Kind et al. 2010; Buschke et al. 2011; Kind et al. 2011), and PEP carboxylase (Chen et al. 2014; Yanase et al. 2016; Yokota et al. 2017).

In contrast to the anaplerotic net flux, the flux through pyruvate dehydrogenase and TCA cycle, pathways competing with L-lysine production, was found to decrease with increasing L-lysine flux (Fig. 6). This finding suggested that an attenuation of pyruvate dehydrogenase and TCA cycle could redirect the flux

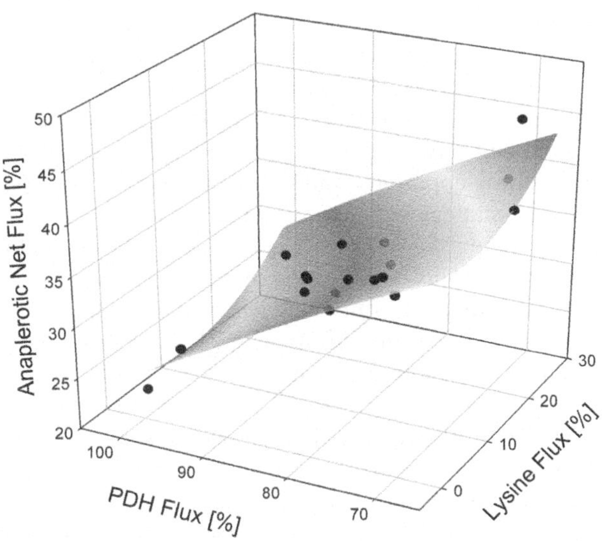

Fig. 6 Metabolic flux correlation between fluxes of L-lysine production, pyruvate dehydrogenase and anaplerotic carboxylation [%] in different strains of *Corynebacterium glutamicum*. Data represent values from 18 independent ^{13}C flux experiments taken from Becker et al. (2005, 2007, 2008); Krömer et al. (2004); Kiefer et al. (2004); Wittmann and Heinzle (2002); Kim et al. (2006); Marx et al. (1996). The correlation was obtained by paraboloid fitting of the data set

towards anaplerotic carboxylation to support the production of L-lysine and related products. Indeed, L-lysine production was successfully improved by deletion (Blombach et al. 2007) and targeted down-regulation of *pdh* (Becker et al. 2010; Buchholz et al. 2013). In addition, the down-regulation of the TCA cycle flux at the level of isocitrate dehydrogenase, citrate synthase and succinyl-CoA synthetase substantially improved production performance (Becker et al. 2009; Kind et al. 2013; van Ooyen et al. 2012; Xu et al. 2018; Yokota et al. 2017). Interestingly, perturbation of the metabolism at the level of pyruvate kinase revealed an indifferent effect on L-lysine production, depending on the genetic background of the parent strain (Becker et al. 2008; Shiio et al. 1987; Yanase et al. 2016; Yokota et al. 2017). Only in combination with the expression of a desensitized PEP carboxylase, deletion of the pyruvate kinase gene was beneficial for L-lysine production (Yanase et al. 2016; Yokota et al. 2017).

4.1.2 Key Fluxes for NADPH Supply

In addition to carbon, NADPH is required for efficient L-lysine biosynthesis and—in line with this—for all L-lysine-derived chemicals. In *C. glutamicum,* NADPH is provided by the enzymes glucose 6-phosphate dehydrogenase (G6PDH) and 6-phosphogluconate dehydrogenase, which form the oxidative part of the PP pathway (Ihnen and Demain 1969; Yokota and Lindley 2005), by the TCA cycle enzyme isocitrate dehydrogenase (Chen and Yang 2000), and by malic enzyme (Gourdon et al. 2000; Eikmanns 2005). The relevance of these enzymes for NADPH supply for L-lysine production has been elucidated to great detail by comparative flux studies (Wittmann and Heinzle 2002). The PP pathway is the most important pathway for NADPH supply in glucose-grown *C. glutamicum*. Improved production performance is accompanied by an increased carbon flux through the PP pathway and a concomitantly reduced flux through the EMP pathway (Fig. 7). The TCA cycle flux, however, decreases with increased L-lysine flux (Wittmann and Heinzle 2002; Becker et al. 2009; Wittmann and De Graaf 2005), unravelling that the cycle rather competes with production for carbon. Accordingly, manipulation of the flux partitioning ratio at the glucose 6-phosphate (G6P) branch point in favour of the PP pathway appeared promising to improve L-lysine production. This was tested by different strategies. The most radical one was a deletion of phosphoglucoisomerase (*pgi*), forcing the cell to exclusively metabolize carbon via the PP pathway (Marx et al. 2003). This modification improved L-lysine production, but severely impaired growth. Moreover, a *pgi* deficient mutant would be limited to glucose-based production, because other substrates such as sucrose, glycerol, lactate, fructose, xylose, and arabinose require an active phosphoglucoisomerase for channelling carbon into the PP pathway to form NADPH (Kiefer et al. 2004; Becker et al. 2005; Wittmann et al. 2004a; Buschke et al. 2013; Dominguez et al. 1998).

The strict regulation of the NADPH-providing enzymes G6PDH and 6-phosphogluconate dehydrogenase by NADPH and other metabolites (Moritz et al. 2000) suggested the need to mutate these enzymes in order to relieve feedback

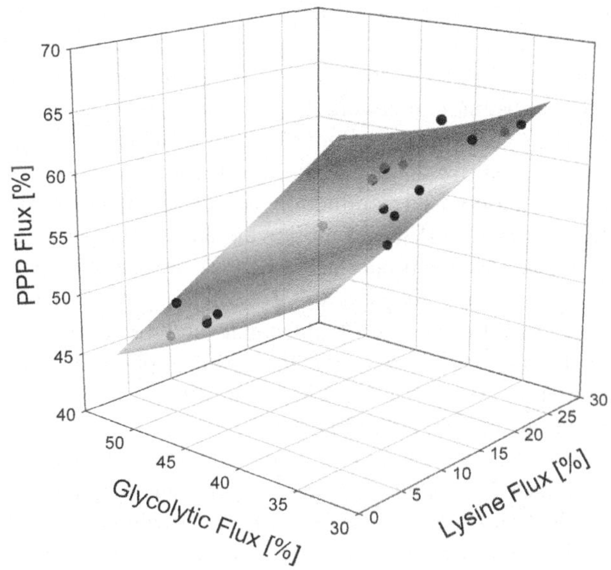

Fig. 7 Metabolic flux correlation between fluxes through L-lysine production, glycolysis and pentose phosphate pathway [%] in different strains of *Corynebacterium glutamicum*. Data represent values from 18 independent ^{13}C flux experiments taken from Becker et al. (2005, 2007, 2008); Krömer et al. (2004); Kiefer et al. (2004); Wittmann and Heinzle (2002); Kim et al. (2006); Marx et al. (1996). The correlation was obtained by paraboloid fitting of the data set

inhibition constraints. Implementation of such modified enzyme-variants in a wild type-derived L-lysine producer indeed resulted in increased L-lysine production by increased fluxes through the PP pathway (Ohnishi et al. 2005; Becker et al. 2007). Moreover, over expression of G6PDH by promoter exchange (Becker et al. 2007) and start codon exchange (Becker et al. 2010) influenced the flux split ratio at the G6P node thereby significantly improving L-lysine production. In an advanced L-lysine producer, over expression of G6PDH was realized in its complete *tkt*-operon-frame thereby additionally upregulating other enzymes of the PP pathway, including transketolase and transaldolase of PP pathway (Becker et al. 2011). Exchange of the natural *tkt*-promoter by the *sod*-promoter concomitantly increased PP pathway flux and L-lysine production further (Becker et al. 2011). This strategy was efficiently transferred to diaminopentane-producing strains with substantial improvement of glucose- and xylose-based production (Kind et al. 2014; Buschke et al. 2013).

Efficient NADPH supply in *C. glutamicum* is especially challenging, when using substrates, which only support a low flux through the oxidative PP pathway, such as fructose (Kiefer et al. 2004), sucrose (Wittmann et al. 2004a), mannitol (Hoffmann et al. 2018), and xylose (Buschke et al. 2013). To address this issue, different engineering strategies were found valuable: (1) enforced metabolic flux rerouting towards PP pathway, (2) the incorporation of alternative reactions for NADPH supply, and (3) a reduction of the NADPH demand. As example, the amplified expression of the *fbp* gene, encoding the gluconeogenetic enzyme fructose 1,6-bisphosphatase (FBPase), was highly efficient to push carbon towards the PP pathway (Becker et al. 2005, 2011). This non-obvious target was identified by metabolic flux analysis of *C. glutamicum* during growth on fructose and sucrose, highlighting this technology as a powerful tool to guide strain optimization (Kiefer et al. 2004; Wittmann et al. 2004a). Overexpression of *fbp* did not only turn out beneficial for glucose, fructose, and sucrose based production (Becker et al. 2005) but also improved production of L-lysine and L-lysine-derived compounds from other substrates including xylose (Buschke et al. 2013) and silage juice (Neuner et al. 2013). Moreover, this modification is an integral trait of hyper-producing strains, optimized for L-lysine production from mannitol (Hoffmann et al. 2018), and glucose-based production of diaminopentane (Kind et al. 2014), glutarate (Rohles et al. 2018) and aminovalerate (Rohles et al. 2016). Beside the expression level, also kinetic properties of FBPase are relevant for production. Unlike the native *C. glutamicum* enzyme, the *E. coli* enzyme variant is insensitive to metabolic regulation by fructose 1-phosphate and fructose 2,6-bisphosphate. Heterologous expression of the *E. coli fbp* gene supported L-lysine production in glucose (Xu et al. 2014), fructose (Xu et al. 2014), sucrose (Xu et al. 2014) and molasses-based media (Xu et al. 2013).

For mannitol-grown *C. glutamicum*, an additional metabolic route was established to support a better operation of the PP pathway during growth and production on this sugar alcohol (Hoffmann et al. 2018). Heterologous expression of fructokinase from either *E. coli* (*mak*) or *Clostridium acetobutylicum* (*srcK*) created a new assimilation pathway allowing higher fluxes through the NADPH-providing reactions of the PP pathway and thus improved L-lysine production (Hoffmann et al. 2018).

Instead of using the PP pathway as the natural NADPH source in *C. glutamicum*, other studies systematically manipulated the co-factor specificity of homologous enzymes or implanted heterologous enzymes in *C. glutamicum* to create novel reactions for sufficient NADPH supply. Engineering of the native NAD-dependent glyceraldehyde 3-phosphate dehydrogenase GapA towards the use of NADP as alternative co-factor substantially increased L-lysine production without a significant flux change towards the PP pathway (Bommareddy et al. 2014). Similarly, a replacement of GapA by the heterologous NADPH-generating glyceraldehyde 3-phosphate dehydrogenase GapN of *Streptococcus mutans* allowed L-lysine production independent of the PP pathway (Takeno et al. 2016). This modification was also the key to success to efficiently use mannitol as alternative carbon source for L-lysine production (Hoffmann et al. 2018). Despite a low operation of the PP pathway, sufficient NADPH supply for L-lysine production was ensured by coupling the supply of this co-factor to the high EMP pathway flux (Hoffmann et al. 2018).

As alternative strategy, enzymes were engineered for co-factor specificity to reduce the NADPH demand for L-lysine biosynthesis. A recent study reports on the mining of NADH-utilizing dehydrogenases, namely aspartate dehydrogenase from *Pseudomonas aeruginosa* (PaASPDH), aspartate-semialdehyde dehydrogenase from *Tistrella mobilis* (TmASADH), dihydrodipicolinate reductase from *Escherichia coli* (EcDHDPR), and diaminopimelate dehydrogenase from *Pseudothermotoga thermarum* (PtDAPDH), to replace the native *C. glutamicum* enzyme variants (Wu et al. 2019).

4.1.3 Systems-Wide Flux Engineering for L-Lysine Hyper-production

A seminal study used the power of ^{13}C metabolic flux analysis combined with *in silico* modelling to systematically redirect the carbon fluxes of *C. glutamicum* for optimized L-lysine production. The metabolic blueprint of the superior producer based on a differential flux map composed from (1) the *in vivo* fluxes of the wild type as quantitative measure of the pathway usage as a starting point and (2) a predicted flux scenario for optimal L-lysine production as the ultimate goal (Becker et al. 2011). By iterative implementation of twelve defined genome-based modifications, the carbon flux was stepwise primed for L-lysine production. Under industrial conditions, the final engineered strain LYS-12 produced 120 g L^{-1} L-lysine within only 30 h at a high yield of 0.55 g g^{-1}. Flux analysis guided strain engineering throughout the genealogy by unravelling the metabolic flux responses towards each of the implemented genetic perturbation. The fluxes provided a rational and quantitative measure to explain the observed changes in the production performance, and to validate the desired rerouting of metabolic fluxes, namely (1) increased L-lysine flux, (2) increased PP pathway flux, (3) increased pyruvate carboxylase flux, (4) decreased PEP carboxykinase flux and (5) decreased TCA cycle flux (Becker et al. 2011).

4.2 Key Fluxes for the Production of L-Glutamate and L-Glutamate Derived Chemicals

Similarly to L-lysine, L-glutamate overproduction is closely linked to the central metabolism. During L-glutamate production, significant amounts of the TCA cycle intermediate 2-oxoglutarate are withdrawn, resulting in the need for anaplerotic replenishment (Shirai et al. 2007; Peters-Wendisch et al. 2001; Sato et al. 2008). In order to support anaplerotic carboxylation, the central glycolytic enzyme pyruvate kinase was deleted in the wild type of *C. glutamicum*, which improved L-glutamate production under biotin-limited conditions (Sawada et al. 2010). The additional expression of a feedback deregulated PEP carboxylase was especially beneficial for L-glutamate production (Yokota et al. 2017). This is in line with previous findings, which showed that PEP carboxylase is indispensable for L-glutamate production and carries the major anaplerotic flux under these conditions (Sato et al. 2008). Studies on posttranslational modification of PEP carboxylase by acetylation provided reasonable evidence that this enzyme might indeed play a major role for carbon flux redistribution under L-glutamate-producing conditions (Nagano-Shoji et al. 2017). Interestingly, the deletion of pyruvate carboxylase resulted in an improved production of L-glutamate and the thereof derived γ-aminobutyrate (GABA) (Wang et al. 2015), likely related to the activation of PEP carboxykinase. Also targeted overexpression of the encoding *ppc* gene and deletion of the *mdh* gene, encoding the OAA consuming malate dehydrogenase, improved GABA production (Shi et al. 2017).

Regarding the L-glutamate precursor 2-oxoglutarate, the activity of isocitrate dehydrogenase (ICD) as 2-oxoglutarate-supplying reaction and the activity of the 2-oxoglutarate dehydrogenase complex (ODHC) as 2-oxoglutarate-consuming reaction have been studied. Metabolic flux analysis revealed that ICD activity does not significantly change during fermentation, while that of the ODHC significantly decreases after the induction of L-glutamate production (Shirai et al. 2005). Artificial amplification of the activities of ICD and L-glutamate dehydrogenase do not influence the fluxes at the 2-oxoglutarate branch point (Shimizu et al. 2003). In the light of this finding, the major role in flux control at the key branch point of 2-oxoglutarate can be attributed to ODHC. In this regard, production of L-glutamate and GABA was successfully improved by deletion of *odhA*, encoding a subunit of the enzyme complex (Asakura et al. 2007; Wang et al. 2015). A similar effect was attained by *odhA* antisense RNA expression that leads to a decreased specific activity of ODHC (Kim et al. 2009). A reduced ODHC activity, obtained by start codon engineering (ATG → TTG) of OdhA, also improved production of putrescine (Nguyen et al. 2015) and GABA via the putrescine route (Jorge et al. 2017a). ODHC is subject to a regulatory control mechanism involving the ODHC suppressor protein OdhI and serine/threonine protein kinase G (PknG) (Schultz et al. 2007), whereby the inhibitory effect of OdhI on ODHC can be overcome by PknG-mediated phosphorylation of OdhI at the threonine residue 14 (Schultz et al. 2007; Niebisch et al. 2006). Moreover, posttranslational L-lysine succinylation was found to regulate OdhI activity, which might be of relevance for L-glutamate production (Mizuno et al. 2016; Komine-Abe et al. 2017). Accordingly, perturbation of this regulatory mechanism by overexpression of the *odhI* gene improve the production of L-glutamate (Kim et al. 2010). Similarly, deletion of the *pknG* gene or introduction of the point mutation T14A in OdhI—thus stabilizing the inhibitory effect of OdhI on ODHC—successfully increased the production of GABA (Jorge et al. 2017a; Okai et al. 2014) and putrescine (Nguyen et al. 2015).

5 Conclusions and Future Directions

The past years have impressively demonstrated the unique value of metabolic flux analysis to understand molecular and cellular processes in *C. glutamicum* and upgrade this microbe into efficient cell factories. Future strain optimization will demand even more for efficient flux approaches in order to speed up development times and efficiently extend the application folio of the microbe to non-natural products and raw materials.

One interesting line of research is the holistic interpretation of flux data in combination with other omics data in order to decipher global regulatory mechanisms (Krömer et al. 2004; Kohlstedt et al. 2010, 2014) and to identify genes with hypothetic or unknown function. Initial efforts of multi-omics analysis of *C. glutamicum* appear promising (Krömer et al. 2004, 2008; Silberbach et al. 2005; Buschke et al. 2013; Buchinger et al. 2009). They promise highly informative data sets to elucidate regulatory mechanisms on a global level and drive ambitious multidimensional optimization (Becker et al. 2018b; Becker and Wittmann 2015, 2018). Clearly, such systematic studies will benefit from the fast progress in miniaturization and automatization of metabolic flux studies, which enable high-throughput (Wittmann et al. 2004b; Heux et al. 2014, 2017). Depending on the desired level of detailedness, metabolic flux analysis could involve systems-wide flux analysis of central metabolic pathways, targeted estimates of flux split ratios, and analysis of characteristic flux fingerprints via isotopic profiling (Heux et al. 2014, 2017; Klingner et al. 2015; Berger et al. 2014; Tang et al. 2012).

With regard to industrial implementation of *C. glutamicum*, novel approaches for ^{13}C metabolic flux analysis in complex nutrient environments are more and more deepening our understanding of industrial cell factories under production-like conditions (Schwechheimer et al. 2016, 2018b, c; Adler et al. 2013, 2014; Templeton et al. 2017). In this regard, metabolic flux analysis will remain a "bottle opener" for systems-level design of *C. glutamicum* as efficient cell factory for the production of an array of bio-based goods. Moreover, metabolic flux analysis is more and more applied to gain deeper insights in the metabolism of pathogens and their mechanism of infection and virulence (Bücker et al. 2014; Berger et al. 2014), which is also of high interest for pathogenic *Corynebacteria* and close relatives such as *Mycobacterium tuberculosis*.

References

Adler P, Bolten CJ, Dohnt K, Hansen CE, Wittmann C (2013) Core fluxome and metafluxome of lactic acid bacteria under simulated cocoa pulp fermentation conditions. Appl Environ Microbiol 79(18):5670–5681. https://doi.org/10.1128/AEM.01483-13

Adler P, Frey LJ, Berger A, Bolten CJ, Hansen CE, Wittmann C (2014) The key to acetate: metabolic fluxes of acetic acid bacteria under cocoa pulp fermentation-simulating conditions. Appl Environ Microbiol 80(15):4702–4716. https://doi.org/10.1128/AEM.01048-14

Aiba S, Matsuoka M (1979) Identification of metabolic model—citrate production from glucose by *Candida lipolytica*. Biotechnol Bioeng 21(8):1373–1386. https://doi.org/10.1002/bit.260210806

Asakura Y, Kimura E, Usuda Y, Kawahara Y, Matsui K, Osumi T, Nakamatsu T (2007) Altered metabolic flux due to deletion of *odhA* causes L-glutamate overproduction in *Corynebacterium glutamicum*. Appl Environ Microbiol 73(4):1308–1319. https://doi.org/10.1128/AEM.01867-06

Bartek T, Blombach B, Lang S, Eikmanns BJ, Wiechert W, Oldiges M, Nöh K, Noack S (2011) Comparative ^{13}C metabolic flux analysis of pyruvate dehydrogenase complex-deficient, L-valine-producing *Corynebacterium glutamicum*. Appl Environ Microbiol 77(18):6644–6652. https://doi.org/10.1128/AEM.00575-11

Becker J, Wittmann C (2012) Systems and synthetic metabolic engineering for amino acid production—the heartbeat of industrial strain development. Curr Opin Biotechnol 23(5):718–726. https://doi.org/10.1016/j.copbio.2011.12.025

Becker J, Wittmann C (2014) GC-MS-based ^{13}C metabolic flux analysis. In: Krömer JO, Nielsen LK, Blank LM (eds) Metabolic flux analysis—methods and protocols, Methods in molecular biology. Springer/Humana, New York, pp 165–174. https://doi.org/10.1007/978-1-4939-1170-7_10

Becker J, Wittmann C (2015) Advanced biotechnology: metabolically engineered cells for the bio-based production of chemicals and fuels, materials, and health-care products. Angew Chem Int Ed Engl 54:3328–3350. https://doi.org/10.1002/anie.201409033

Becker J, Wittmann C (2017a) Diamines for bio-based materials. In: Wittmann C, Liao JC (eds) Industrial biotechnology, vol 1. Advanced biotechnology, vol 4. Wiley-VCH, Weinheim, pp 393–404

Becker J, Wittmann C (2017b) Industrial microorganisms: *Corynebacterium glutamicum*. In: Wittmann C, Liao JC (eds) Industrial biotechnology, vol 1. Advanced biotechnology, vol 3a. Wiley-VCH, Weinheim, pp 183–203

Becker J, Wittmann C (2018) From systems biology to metabolically engineered cells-an omics perspective on the development of industrial microbes. Curr Opin Microbiol 45:180–188. https://doi.org/10.1016/j.mib.2018.06.001

Becker J, Klopprogge C, Zelder O, Heinzle E, Wittmann C (2005) Amplified expression of fructose 1,6-bisphosphatase in *Corynebacterium glutamicum* increases *in vivo* flux through the pentose phosphate pathway and lysine production on different carbon sources. Appl Environ Microbiol 71(12):8587–8596. https://doi.org/10.1128/AEM.71.12.8587-8596.2005

Becker J, Klopprogge C, Herold A, Zelder O, Bolten CJ, Wittmann C (2007) Metabolic flux engineering of L-lysine production in *Corynebacterium glutamicum*—over expression and modification of G6P dehydrogenase. J Biotechnol 132(2):99–109. https://doi.org/10.1016/j.jbiotec.2007.05.026

Becker J, Klopprogge C, Wittmann C (2008) Metabolic responses to pyruvate kinase deletion in lysine producing *Corynebacterium glutamicum*. Microb Cell Fact 7:8. https://doi.org/10.1186/1475-2859-7-8

Becker J, Klopprogge C, Schröder H, Wittmann C (2009) Metabolic engineering of the tricarboxylic acid cycle for improved lysine production by *Corynebacterium glutamicum*. Appl Environ Microbiol 75(24):7866–7869. https://doi.org/10.1128/AEM.01942-09

Becker J, Buschke N, Bücker R, Wittmann C (2010) Systems level engineering of *Corynebacterium glutamicum*—reprogramming translational efficiency for superior production. Eng Lif Sci 10:430–438

Becker J, Zelder O, Haefner S, Schröder H, Wittmann C (2011) From zero to hero—design-based systems metabolic engineering of *Corynebacterium glutamicum* for L-lysine production. Metab Eng 13(2):159–168. https://doi.org/10.1016/j.ymben.2011.01.003

Becker J, Schäfer R, Kohlstedt M, Harder BJ, Borchert NS, Stöveken N, Bremer E, Wittmann C (2013) Systems metabolic engineering of *Corynebacterium glutamicum* for production of the chemical chaperone ectoine. Microb Cell Fact 12:110. https://doi.org/10.1186/1475-2859-12-110

Becker J, Lange A, Fabarius J, Wittmann C (2015) Top value platform chemicals: bio-based production of organic acids. Curr Opin Biotechnol 36:168–175. https://doi.org/10.1016/j.copbio.2015.08.022

Becker J, Gießelmann G, Hoffmann SL, Wittmann C (2016) *Corynebacterium glutamicum* for sustainable bio-production: from metabolic physiology to systems metabolic engineering. In: Zhao H, Zeng AP (eds) Synthetic biology—metabolic engineering, Advances in biochemical engineering/biotechnology. Springer, Heidelberg, pp 217–263

Becker J, Kuhl M, Kohlstedt M, Starck S, Wittmann C (2018a) Metabolic engineering of *Corynebacterium glutamicum* for the production of *cis, cis*-muconic acid from lignin. Microb Cell Fact 17(1):115. https://doi.org/10.1186/s12934-018-0963-2

Becker J, Rohles CM, Wittmann C (2018b) Metabolically engineered *Corynebacterium glutamicum* for bio-based production of chemicals, fuels, materials, and healthcare products. Metab Eng 50:122–141. https://doi.org/10.1016/j.ymben.2018.07.008

Beckers V, Poblete-Castro I, Tomasch J, Wittmann C (2016) Integrated analysis of gene expression and metabolic fluxes in PHA-producing *Pseudomonas putida* grown on glycerol. Microb Cell Fact 15:73. https://doi.org/10.1186/s12934-016-0470-2

Berger A, Dohnt K, Tielen P, Jahn D, Becker J, Wittmann C (2014) Robustness and plasticity of metabolic pathway flux among uropathogenic isolates of *Pseudomonas aeruginosa*. PLoS One 9(4):e88368. https://doi.org/10.1371/journal.pone.0088368

Blombach B, Schreiner ME, Moch M, Oldiges M, Eikmanns BJ (2007) Effect of pyruvate dehydrogenase complex deficiency on L-lysine production with *Corynebacterium glutamicum*. Appl Microbiol Biotechnol 76(3):615–623. https://doi.org/10.1007/s00253-007-0904-1

Blombach B, Riester T, Wieschalka S, Ziert C, Youn JW, Wendisch VF, Eikmanns BJ (2011) *Corynebacterium glutamicum* tailored for efficient isobutanol production. Appl Environ Microbiol 77(10):3300–3310. https://doi.org/10.1128/AEM.02972-10

Boisseau R, Charrier B, Massou S, Portais JC, Akoka S, Giraudeau P (2013) Fast spatially encoded 3D NMR strategies for ^{13}C-based metabolic flux analysis. Anal Chem 85(20):9751–9757. https://doi.org/10.1021/ac402155w

Bolten CJ, Heinzle E, Müller R, Wittmann C (2009) Investigation of the central carbon metabolism of *Sorangium cellulosum*: metabolic network reconstruction and quantification of pathway fluxes. J Microbiol Biotechnol 19(1):23–36

Bolten CJ, Schröder H, Dickschat J, Wittmann C (2010) Towards methionine overproduction in *Corynebacterium glutamicum*—methanethiol and dimethyldisulfide as reduced sulfur sources. J Microbiol Biotechnol 20(8):1196–1203

Bommareddy RR, Chen Z, Rappert S, Zeng AP (2014) A de novo NADPH generation pathway for improving lysine production of *Corynebacterium glutamicum* by rational design of the coenzyme specificity of glyceraldehyde 3-phosphate dehydrogenase. Metab Eng 25:30–37. https://doi.org/10.1016/j.ymben.2014.06.005

Buchholz J, Schwentner A, Brunnenkan B, Gabris C, Grimm S, Gerstmeir R, Takors R, Eikmanns BJ, Blombach B (2013) Platform engineering of *Corynebacterium glutamicum* with reduced pyruvate dehydrogenase complex activity for improved production of L-lysine, L-valine, and 2-ketoisovalerate. Appl Environ Microbiol 79(18):5566–5575. https://doi.org/10.1128/AEM.01741-13

Buchinger S, Strösser J, Rehm N, Hänßler E, Hans S, Bathe B, Schomburg D, Krämer R, Burkovski A (2009) A combination of metabolome and transcriptome analyses reveals new targets of the *Corynebacterium glutamicum* nitrogen regulator AmtR. J Biotechnol 140(1–2):68–74. https://doi.org/10.1016/j.jbiotec.2008.10.009

Bücker R, Heroven AK, Becker J, Dersch P, Wittmann C (2014) The pyruvate—tricarboxylic acid cycle node: a focal point of virulence control in the enteric pathogen *Yersinia pseudotuberculosis*. J Biol Chem 289(43):30114–30132. https://doi.org/10.1074/jbc.M114.581348

Buschke N, Schröder H, Wittmann C (2011) Metabolic engineering of *Corynebacterium glutamicum* for production of 1,5-diaminopentane from hemicellulose. Biotechnol J 6 (3):306–317. https://doi.org/10.1002/biot.201000304

Buschke N, Becker J, Schäfer R, Kiefer P, Biedendieck R, Wittmann C (2013) Systems metabolic engineering of xylose-utilizing *Corynebacterium glutamicum* for production of 1,5-diaminopentane. Biotechnol J 8(5):557–570. https://doi.org/10.1002/biot.201200367

Chae TU, Choi SY, Kim JW, Ko YS, Lee SY (2017) Recent advances in systems metabolic engineering tools and strategies. Curr Opin Biotechnol 47:67–82. https://doi.org/10.1016/j.copbio.2017.06.007

Chen R, Yang H (2000) A highly specific monomeric isocitrate dehydrogenase from *Corynebacterium glutamicum*. Arch Biochem Biophys 383(2):238–245

Chen Z, Landman P, Colmer TD, Adams MA (1998) Simultaneous analysis of amino and organic acids in extracts of plant leaves as tert-butyldimethylsilyl derivatives by capillary gas chromatography. Anal Biochem 259(2):203–211

Chen Z, Bommareddy RR, Frank D, Rappert S, Zeng AP (2014) Deregulation of feedback inhibition of phosphoenolpyruvate carboxylase for improved lysine production in *Corynebacterium glutamicum*. Appl Environ Microbiol 80(4):1388–1393. https://doi.org/10.1128/AEM.03535-13

Christensen B, Nielsen J (1999) Isotopomer analysis using GC-MS. Metab Eng 1(4):282–290

Christensen B, Gombert AK, Nielsen J (2002) Analysis of flux estimates based on ^{13}C-labelling experiments. Eur J Biochem 269(11):2795–2800

Cocaign-Bousquet M, Guyonvarch A, Lindley ND (1996) Growth rate-dependent modulation of carbon flux through central metabolism and the kinetic consequences for glucose-limited chemostat cultures of *Corynebacterium glutamicum*. Appl Environ Microbiol 62(2):429–436

Crown SB, Long CP, Antoniewicz MR (2016) Optimal tracers for parallel labeling experiments and 13C metabolic flux analysis: a new precision and synergy scoring system. Metab Eng 38:10–18. https://doi.org/10.1016/j.ymben.2016.06.001

Dauner M, Sauer U (2000) GC-MS analysis of amino acids rapidly provides rich information for isotopomer balancing. Biotechnol Prog 16(4):642–649. https://doi.org/10.1021/bp000058h

Dersch LM, Beckers V, Rasch D, Melzer G, Bolten C, Kiep K, Becker H, Blasing OE, Fuchs R, Ehrhardt T, Wittmann C (2016a) Novel approach for high-throughput metabolic screening of whole plants by stable isotopes. Plant Physiol 171(1):25–41. https://doi.org/10.1104/pp.15.01217

Dersch LM, Beckers V, Wittmann C (2016b) Green pathways: metabolic network analysis of plant systems. Metab Eng 34:1–24. https://doi.org/10.1016/j.ymben.2015.12.001

Desai TS, Srivastava S (2018) FluxPyt: a Python-based free and open-source software for ^{13}C-metabolic flux analyses. PeerJ 6:e4716. https://doi.org/10.7717/peerj.4716

Dominguez H, Nezondet C, Lindley ND, Cocaign M (1993) Modified carbon flux during oxygen limited growth of *Corynebacterium glutamicum* and the consequences for amino acid overproduction. Biotechnol Lett 15(5):449–454

Dominguez H, Rollin C, Guyonvarch A, Guerquin-Kern JL, Cocaign-Bousquet M, Lindley ND (1998) Carbon-flux distribution in the central metabolic pathways of *Corynebacterium glutamicum* during growth on fructose. Eur J Biochem 254(1):96–102

Drysch A, El Massaoudi M, Mack C, Takors R, de Graaf AA, Sahm H (2003) Production process monitoring by serial mapping of microbial carbon flux distributions using a novel sensor reactor approach: II-^{13}C-labeling-based metabolic flux analysis and L-lysine production. Metab Eng 5 (2):96–107

Drysch A, El Massaoudi M, Wiechert W, de Graaf AA, Takors R (2004) Serial flux mapping of *Corynebacterium glutamicum* during fed-batch L-lysine production using the sensor reactor approach. Biotechnol Bioeng 85(5):497–505

Eggeling L, Bott M (2015) A giant market and a powerful metabolism: L-lysine provided by *Corynebacterium glutamicum*. Appl Microbiol Biotechnol 99(8):3387–3394. https://doi.org/10.1007/s00253-015-6508-2

Eikmanns BJ (2005) Central metabolism: tricarboxylic acid cycle and anaplerotic reactions. In: Eggeling L, Bott M (eds) Handbook of *Corynebacterium glutamicum*. CRC, Boca Raton, pp 241–276

El Massaoudi M, Spelthahn J, Drysch A, de Graaf A, Takors R (2003) Production process monitoring by serial mapping of microbial carbon flux distributions using a novel sensor reactor approach: I-Sensor reactor system. Metab Eng 5(2):86–95

Fischer E, Zamboni N, Sauer U (2004) High-throughput metabolic flux analysis based on gas chromatography-mass spectrometry derived ^{13}C constraints. Anal Biochem 325(2):308–316

Fürch T, Preusse M, Tomasch J, Zech H, Wagner-Döbler I, Rabus R, Wittmann C (2009) Metabolic fluxes in the central carbon metabolism of *Dinoroseobacter shibae* and *Phaeobacter gallaeciensis*, two members of the marine Roseobacter clade. BMC Microbiol 9:209. https://doi.org/10.1186/1471-2180-9-209

Gourdon P, Baucher MF, Lindley ND, Guyonvarch A (2000) Cloning of the malic enzyme gene from *Corynebacterium glutamicum* and role of the enzyme in lactate metabolism. Appl Environ Microbiol 66(7):2981–2987

de Graaf AA, Mahle M, Möllney M, Wiechert W, Stahmann P, Sahm H (2000) Determination of full ^{13}C isotopomer distributions for metabolic flux analysis using heteronuclear spin echo difference NMR spectroscopy. J Biotechnol 77(1):25–35 de Graaf AA, Eggeling L, Sahm H (2001) Metabolic engineering for L-lysine production by *Corynebacterium glutamicum*. Adv Biochem Eng Biotechnol 73:9–29

Guo W, Sheng J, Feng X (2018) Synergizing ^{13}C Metabolic flux analysis and metabolic engineering for biochemical production. Adv Biochem Eng Biotechnol 162:265–299. https://doi.org/10.1007/10_2017_2

Haberhauer G, Schröder H, Pompejus M, Zelder O, Kröger B (2001) *Corynebacterium glutamicum* genes encoding proteins involved in membrane synthesis and membrane transport. Patent WO 01/00805 Patent

He L, Wu SG, Zhang M, Chen Y, Tang YJ (2016) WUFlux: an open-source platform for ^{13}C metabolic flux analysis of bacterial metabolism. BMC Bioinformatics 17(1):444. https://doi.org/10.1186/s12859-016-1314-0

Heux S, Poinot J, Massou S, Sokol S, Portais JC (2014) A novel platform for automated high-throughput fluxome profiling of metabolic variants. Metab Eng 25:8–19. https://doi.org/10.1016/j.ymben.2014.06.001

Heux S, Berges C, Millard P, Portais JC, Letisse F (2017) Recent advances in high-throughput ^{13}C-fluxomics. Curr Opin Biotechnol 43:104–109. https://doi.org/10.1016/j.copbio.2016.10.010

Hoffmann SL, Jungmann L, Schiefelbein S, Peyriga L, Cahoreau E, Portais JC, Becker J, Wittmann C (2018) Lysine production from the sugar alcohol mannitol: design of the cell factory *Corynebacterium glutamicum* SEA-3 through integrated analysis and engineering of metabolic pathway fluxes. Metab Eng. https://doi.org/10.1016/j.ymben.2018.04.019

Ihnen ED, Demain AL (1969) Glucose-6-phosphate dehydrogenase and its deficiency in mutants of *Corynebacterium glutamicum*. J Bacteriol 98(3):1151–1158

Ikeda M (2003) Amino acid production processes. Adv Biochem Eng Biotechnol 79:1–35

Inui M, Kawaguchi H, Murakami S, Vertes AA, Yukawa H (2004a) Metabolic engineering of *Corynebacterium glutamicum* for fuel ethanol production under oxygen-deprivation conditions. J Mol Microbiol Biotechnol 8(4):243–254. https://doi.org/10.1159/000086705

Inui M, Murakami S, Okino S, Kawaguchi H, Vertes AA, Yukawa H (2004b) Metabolic analysis of *Corynebacterium glutamicum* during lactate and succinate productions under oxygen deprivation conditions. J Mol Microbiol Biotechnol 7(4):182–196. https://doi.org/10.1159/000079827

Ishino S, Yamaguchi K, Shirahata K, Araki K (1984) Involvement of *meso*-α,ε-diaminopimelate D-dehydrogenase in lysine biosynthesis in *Corynebacterium glutamicum*. Agric Biol Chem 48 (10):2557–2560

Ishino S, Kuga T, Yamaguchi K, Shirahata K, Araki K (1986) ^{13}C NMR-studies of histidine fermentation with a *Corynebacterium-glutamicum* mutant. Agric Biol Chem 50(2):307–310. https://doi.org/10.1080/00021369.1986.10867392

Iwatani S, Van Dien S, Shimbo K, Kubota K, Kageyama N, Iwahata D, Miyano H, Hirayama K, Usuda Y, Shimizu K, Matsui K (2007) Determination of metabolic flux changes during fed-batch cultivation from measurements of intracellular amino acids by LC-MS/MS. J Biotechnol 128 (1):93–111

Jojima T, Noburyu R, Sasaki M, Tajima T, Suda M, Yukawa H, Inui M (2015) Metabolic engineering for improved production of ethanol by *Corynebacterium glutamicum*. Appl Microbiol Biotechnol 99(3):1165–1172. https://doi.org/10.1007/s00253-014-6223-4

Jorge JM, Nguyen AQ, Perez-Garcia F, Kind S, Wendisch VF (2017a) Improved fermentative production of gamma-aminobutyric acid via the putrescine route: systems metabolic engineering for production from glucose, amino sugars, and xylose. Biotechnol Bioeng 114(4):862–873. https://doi.org/10.1002/bit.26211

Jorge JMP, Perez-Garcia F, Wendisch VF (2017b) A new metabolic route for the fermentative production of 5-aminovalerate from glucose and alternative carbon sources. Bioresour Technol 245 (Pt B):1701–1709. https://doi.org/10.1016/j.biortech.2017.04.108

Kalinowski J, Bathe B, Bartels D, Bischoff N, Bott M, Burkovski A, Dusch N, Eggeling L, Eikmanns BJ, Gaigalat L, Goesmann A, Hartmann M, Huthmacher K, Krämer R, Linke B, McHardy AC, Meyer F, Möckel B, Pfefferle W, Pühler A, Rey DA, Rückert C, Rupp O, Sahm H, Wendisch VF, Wiegrabe I, Tauch A (2003) The complete *Corynebacterium glutamicum* ATCC 13032 genome sequence and its impact on the production of L-aspartate-derived amino acids and vitamins. J Biotechnol 104(1-3):5–25

Kappelmann J, Klein B, Geilenkirchen P, Noack S (2017) Comprehensive and accurate tracking of carbon origin of LC-tandem mass spectrometry collisional fragments for ^{13}C-MFA. Anal Bioanal Chem 409(9):2309–2326. https://doi.org/10.1007/s00216-016-0174-9
Kelleher JK (2001) Flux estimation using isotopic tracers: common ground for metabolic physiology and metabolic engineering. Metab Eng 3(2):100–110
Kiefer P, Heinzle E, Wittmann C (2002) Influence of glucose, fructose and sucrose as carbon sources on kinetics and stoichiometry of lysine production by *Corynebacterium glutamicum*. J Ind Microbiol Biotechnol 28(6):338–343. https://doi.org/10.1038/sj/jim/7000252
Kiefer P, Heinzle E, Zelder O, Wittmann C (2004) Comparative metabolic flux analysis of lysine-producing *Corynebacterium glutamicum* cultured on glucose or fructose. Appl Environ Microbiol 70(1):229–239
Kiefer P, Nicolas C, Letisse F, Portais JC (2007) Determination of carbon labeling distribution of intracellular metabolites from single fragment ions by ion chromatography tandem mass spectrometry. Anal Biochem 360(2):182–188
Kim H-M, Heinzle E, Wittmann C (2006) Deregulation of aspartokinase by single nucleotide exchange leads to global flux rearrangement in the central metabolism of *Corynebacterium glutamicum*. J Microbiol Biotechnol 16(8):1174–1179
Kim J, Hirasawa T, Sato Y, Nagahisa K, Furusawa C, Shimizu H (2009) Effect of *odhA* overexpression and *odhA* antisense RNA expression on Tween-40-triggered glutamate production by *Corynebacterium glutamicum*. Appl Microbiol Biotechnol 81(6):1097–1106. https://doi.org/10.1007/s00253-008-1743-4
Kim J, Fukuda H, Hirasawa T, Nagahisa K, Nagai K, Wachi M, Shimizu H (2010) Requirement of de novo synthesis of the OdhI protein in penicillin-induced glutamate production by *Corynebacterium glutamicum*. Appl Microbiol Biotechnol 86(3):911–920. https://doi.org/10.1007/s00253-009-2360-6
Kind S, Wittmann C (2011) Bio-based production of the platform chemical 1,5-diaminopentane. Appl Microbiol Biotechnol 91(5):1287–1296. https://doi.org/10.1007/s00253-011-3457-2
Kind S, Jeong WK, Schröder H, Wittmann C (2010) Systems-wide metabolic pathway engineering in *Corynebacterium glutamicum* for bio-based production of diaminopentane. Metab Eng 12 (4):341–351. https://doi.org/10.1016/j.ymben.2010.03.005
Kind S, Kreye S, Wittmann C (2011) Metabolic engineering of cellular transport for overproduction of the platform chemical 1,5-diaminopentane in *Corynebacterium glutamicum*. Metab Eng 13 (5):617–627. https://doi.org/10.1016/j.ymben.2011.07.006
Kind S, Becker J, Wittmann C (2013) Increased lysine production by flux coupling of the tricarboxylic acid cycle and the lysine biosynthetic pathway—metabolic engineering of the availability of succinyl-CoA in *Corynebacterium glutamicum*. Metab Eng 15:184–195. https://doi.org/10.1016/j.ymben.2012.07.005
Kind S, Neubauer S, Becker J, Yamamoto M, Völkert M, Abendroth GV, Zelder O, Wittmann C (2014) From zero to hero—production of bio-based nylon from renewable resources using engineered *Corynebacterium glutamicum*. Metab Eng 25:113–123. https://doi.org/10.1016/j.ymben.2014.05.007
Kjeldsen KR, Nielsen J (2009) *In silico* genome-scale reconstruction and validation of the *Corynebacterium glutamicum* metabolic network. Biotechnol Bioeng 102(2):583–597. https://doi.org/10.1002/bit.22067
Klingner A, Bartsch A, Dogs M, Wagner-Döbler I, Jahn D, Simon M, Brinkhoff T, Becker J, Wittmann C (2015) Large-scale ^{13}C flux profiling reveals conservation of the Entner-Doudoroff pathway as a glycolytic strategy among marine bacteria that use glucose. Appl Environ Microbiol 81(7):2408–2422. https://doi.org/10.1128/AEM.03157-14
Kohlstedt M, Wittmann C (2019) GC-MS based ^{13}C metabolic flux analysis resolves the cyclic glucose metabolism of *Pseudomonas putida* KT2440 and *Pseudomonas aeruginosa* PAO1. Metab Eng 54:35–53

Kohlstedt M, Becker J, Wittmann C (2010) Metabolic fluxes and beyond-systems biology understanding and engineering of microbial metabolism. Appl Microbiol Biotechnol 88 (5):1065–1075. https://doi.org/10.1007/s00253-010-2854-2

Kohlstedt M, Sappa PK, Meyer H, Maass S, Zaprasis A, Hoffmann T, Becker J, Steil L, Hecker M, van Dijl JM, Lalk M, Mader U, Stülke J, Bremer E, Völker U, Wittmann C (2014) Adaptation of *Bacillus subtilis* carbon core metabolism to simultaneous nutrient limitation and osmotic challenge: a multi-omics perspective. Environ Microbiol 16(6):1898–1917. https://doi.org/10.1111/1462-2920.12438

Kohlstedt M, Starck S, Barton N, Stolzenberger J, Selzer M, Mehlmann K, Schneider R, Pleissner D, Rinkel J, Dickschat JS, Venus J, van Duuren JBJH, Wittmann C (2018) From lignin to nylon: cascaded chemical and biochemical conversion using metabolically engineered *Pseudomonas putida*. Metab Eng 47:279–293. https://doi.org/10.1016/j.ymben.2018.03.003

Komine-Abe A, Nagano-Shoji M, Kubo S, Kawasaki H, Yoshida M, Nishiyama M, Kosono S (2017) Effect of lysine succinylation on the regulation of 2-oxoglutarate dehydrogenase inhibitor, OdhI, involved in glutamate production in *Corynebacterium glutamicum*. Biosci Biotechnol Biochem 81(11):2130–2138. https://doi.org/10.1080/09168451.2017.1372182

Kortmann M, Mack C, Baumgart M, Bott M (2019) Pyruvate carboxylase variants enabling improved lysine production from glucose identified by biosensor-based high-throughput fluorescence-activated cell sorting screening. ACS Synth Biol 8(2):274–281. https://doi.org/10.1021/acssynbio.8b00510

Krömer JO, Sorgenfrei O, Klopprogge K, Heinzle E, Wittmann C (2004) In-depth profiling of lysine-producing *Corynebacterium glutamicum* by combined analysis of the transcriptome, metabolome, and fluxome. J Bacteriol 186(6):1769–1784

Krömer JO, Heinzle E, Schröder H, Wittmann C (2006) Accumulation of homolanthionine and activation of a novel pathway for isoleucine biosynthesis in *Corynebacterium glutamicum* McbR deletion strains. J Bacteriol 188(2):609–618. https://doi.org/10.1128/JB.188.2.609-618.2006

Krömer JO, Bolten CJ, Heinzle E, Schröder H, Wittmann C (2008) Physiological response of *Corynebacterium glutamicum* to oxidative stress induced by deletion of the transcriptional repressor McbR. Microbiology 154(Pt 12):3917–3930. https://doi.org/10.1099/mic.0.2008/021204-0

Lange A, Becker J, Schulze D, Cahoreau E, Portais JC, Haefner S, Schröder H, Krawczyk J, Zelder O, Wittmann C (2017) Bio-based succinate from sucrose: high-resolution ^{13}C metabolic flux analysis and metabolic engineering of the rumen bacterium *Basfia succiniciproducens*. Metab Eng 44:198–212. https://doi.org/10.1016/j.ymben.2017.10.003

Lessmeier L, Pfeifenschneider J, Carnicer M, Heux S, Portais JC, Wendisch VF (2015) Production of carbon-13-labeled cadaverine by engineered *Corynebacterium glutamicum* using carbon-13-labeled methanol as co-substrate. Appl Microbiol Biotechnol 99(23):10163–10176. https://doi.org/10.1007/s00253-015-6906-5

Lynch MD (2016) Into new territory: improved microbial synthesis through engineering of the essential metabolic network. Curr Opin Biotechnol 38:106–111. https://doi.org/10.1016/j.copbio.2016.01.009

Ma Q, Zhang Q, Xu Q, Zhang C, Li Y, Fan X, Xie X, Chen N (2017) Systems metabolic engineering strategies for the production of amino acids. Synth Syst Biotechnol 2(2):87–96. https://doi.org/10.1016/j.synbio.2017.07.003

Maarleveld TR, Khandelwal RA, Olivier BG, Teusink B, Bruggeman FJ (2013) Basic concepts and principles of stoichiometric modeling of metabolic networks. Biotechnol J 8(9):997–1008. https://doi.org/10.1002/biot.201200291

Marx A, de Graaf AA, Wiechert W, Eggeling L, Sahm H (1996) Determination of the fluxes in the central metabolism of *Corynebacterium glutamicum* by nuclear magnetic resonance spectroscopy combined with metabolite balancing. Biotechnol Bioeng 49(2):111–129. https://doi.org/10.1002/(SICI)1097-0290(19960120)49:2<111::AID-BIT1>3.0.CO;2-T

Marx A, Striegel K, de Graaf AA, Sahm H, Eggeling L (1997) Response of the central metabolism of *Corynebacterium glutamicum* to different flux burdens. Biotechnol Bioeng 56(2):168–180. https://doi.org/10.1002/(SICI)1097-0290(19971020)56:2<168::AID-BIT6>3.0.CO;2-N

Marx A, Eikmanns BJ, Sahm H, de Graaf AA, Eggeling L (1999) Response of the central metabolism in *Corynebacterium glutamicum* to the use of an NADH-dependent glutamate dehydrogenase. Metab Eng 1(1):35–48. https://doi.org/10.1006/mben.1998.0106

Marx A, Hans S, Möckel B, Bathe B, de Graaf AA, McCormack AC, Stapleton C, Burke K, O'Donohue M, Dunican LK (2003) Metabolic phenotype of phosphoglucose isomerase mutants of *Corynebacterium glutamicum*. J Biotechnol 104(1-3):185–197

Massou S, Nicolas C, Letisse F, Portais JC (2007) Application of 2D-TOCSY NMR to the measurement of specific ^{13}C-enrichments in complex mixtures of ^{13}C-labeled metabolites. Metab Eng 9(3):252–257. https://doi.org/10.1016/j.ymben.2007.03.001

Mei J, Xu N, Ye C, Liu L, Wu J (2016) Reconstruction and analysis of a genome-scale metabolic network of *Corynebacterium glutamicum* S9114. Gene 575(2 Pt 3):615–622. https://doi.org/10.1016/j.gene.2015.09.038

Melzer G, Esfandabadi ME, Franco-Lara E, Wittmann C (2009) Flux design: *in silico* design of cell factories based on correlation of pathway fluxes to desired properties. BMC Syst Biol 3:120. https://doi.org/10.1186/1752-0509-3-120

Michal G (1999) Biochemical pathways. Wiley, Chichester

Millard P, Sokol S, Letisse F, Portais JC (2014) IsoDesign: a software for optimizing the design of ^{13}C-metabolic flux analysis experiments. Biotechnol Bioeng 111(1):202–208. https://doi.org/10.1002/bit.24997

Millard P, Cahoreau E, Heuillet M, Portais JC, Lippens G (2017) ^{15}N-NMR-based approach for amino acids-based ^{13}C-metabolic flux analysis of metabolism. Anal Chem 89(3):2101–2106. https://doi.org/10.1021/acs.analchem.6b04767

Mizuno Y, Nagano-Shoji M, Kubo S, Kawamura Y, Yoshida A, Kawasaki H, Nishiyama M, Yoshida M, Kosono S (2016) Altered acetylation and succinylation profiles in *Corynebacterium glutamicum* in response to conditions inducing glutamate overproduction. Microbiologyopen 5 (1):152–173. https://doi.org/10.1002/mbo3.320

Möllney M, Wiechert W, Kownatzki D, de Graaf AA (1999) Bidirectional reaction steps in metabolic networks: IV. Optimal design of isotopomer labeling experiments. Biotechnol Bioeng 66(2):86–103

Moritz B, Striegel K, De Graaf AA, Sahm H (2000) Kinetic properties of the glucose-6-phosphate and 6-phosphogluconate dehydrogenases from *Corynebacterium glutamicum* and their application for predicting pentose phosphate pathway flux *in vivo*. Eur J Biochem 267 (12):3442–3452

Nagano-Shoji M, Hamamoto Y, Mizuno Y, Yamada A, Kikuchi M, Shirouzu M, Umehara T, Yoshida M, Nishiyama M, Kosono S (2017) Characterization of lysine acetylation of a phosphoenolpyruvate carboxylase involved in glutamate overproduction in *Corynebacterium glutamicum*. Mol Microbiol 104(4):677–689. https://doi.org/10.1111/mmi.13658

Netzer R, Krause M, Rittmann D, Peters-Wendisch PG, Eggeling L, Wendisch VF, Sahm H (2004) Roles of pyruvate kinase and malic enzyme in *Corynebacterium glutamicum* for growth on carbon sources requiring gluconeogenesis. Arch Microbiol 182(5):354–363. https://doi.org/10.1007/s00203-004-0710-4

Neuner A, Wagner I, Sieker T, Ulber R, Schneider K, Peifer S, Heinzle E (2013) Production of L-lysine on different silage juices using genetically engineered *Corynebacterium glutamicum*. J Biotechnol 163(2):217–224. https://doi.org/10.1016/j.jbiotec.2012.07.190

Nguyen AQ, Schneider J, Reddy GK, Wendisch VF (2015) Fermentative production of the diamine putrescine: system metabolic engineering of *Corynebacterium glutamicum*. Metabolites 5 (2):211–231. https://doi.org/10.3390/metabo5020211

Nicolas C, Kiefer P, Letisse F, Krömer J, Massou S, Soucaille P, Wittmann C, Lindley ND, Portais JC (2007) Response of the central metabolism of *Escherichia coli* to modified expression of the gene encoding the glucose-6-phosphate dehydrogenase. FEBS Lett 581(20):3771–3776. https://doi.org/10.1016/j.febslet.2007.06.066

Nicolas C, Becker J, Sanchou L, Letisse F, Wittmann C, Portais JC, Massou S (2008) Measurement of isotopic enrichments in ^{13}C-labelled molecules by 1D selective zero-quantum filtered TOCSY NMR experiments. Cr Chim 11(4–5):480–485. https://doi.org/10.1016/j.crci.2007.06.018

Niebisch A, Kabus A, Schultz C, Weil B, Bott M (2006) Corynebacterial protein kinase G controls 2-oxoglutarate dehydrogenase activity via the phosphorylation status of the OdhI protein. J Biol Chem 281(18):12300–12307. https://doi.org/10.1074/jbc.M512515200

Nikel PI, Chavarria M, Fuhrer T, Sauer U, de Lorenzo V (2015) *Pseudomonas putida* KT2440 strain metabolizes glucose through a cycle formed by enzymes of the Entner-Doudoroff, Embden-Meyerhof-Parnas, and pentose phosphate pathways. J Biol Chem 290 (43):25920–25932. https://doi.org/10.1074/jbc.M115.687749

Noack S, Nöh K, Moch M, Oldiges M, Wiechert W (2011) Stationary versus non-stationary ^{13}C-MFA: a comparison using a consistent dataset. J Biotechnol 154(2-3):179–190. https://doi.org/10.1016/j.jbiotec.2010.07.008

Nöh K, Droste P, Wiechert W (2015) Visual workflows for ^{13}C-metabolic flux analysis. Bioinformatics 31(3):346–354. https://doi.org/10.1093/bioinformatics/btu585

Nöh K, Niedenfuhr S, Beyss M, Wiechert W (2018) A Pareto approach to resolve the conflict between information gain and experimental costs: multiple-criteria design of carbon labeling experiments. PLoS Comput Biol 14(10):e1006533. https://doi.org/10.1371/journal.pcbi.1006533

Ohnishi J, Mitsuhashi S, Hayashi M, Ando S, Yokoi H, Ochiai K, Ikeda M (2002) A novel methodology employing *Corynebacterium glutamicum* genome information to generate a new L-lysine-producing mutant. Appl Microbiol Biotechnol 58(2):217–223

Ohnishi J, Katahira R, Mitsuhashi S, Kakita S, Ikeda M (2005) A novel *gnd* mutation leading to increased L-lysine production in *Corynebacterium glutamicum*. FEMS Microbiol Lett 242 (2):265–274. https://doi.org/10.1016/j.femsle.2004.11.014

Okai N, Takahashi C, Hatada K, Ogino C, Kondo A (2014) Disruption of *pknG* enhances production of gamma-aminobutyric acid by *Corynebacterium glutamicum* expressing glutamate decarboxylase. AMB Express 4:20. https://doi.org/10.1186/s13568-014-0020-4

Okai N, Miyoshi T, Takeshima Y, Kuwahara H, Ogino C, Kondo A (2016) Production of protocatechuic acid by *Corynebacterium glutamicum* expressing chorismate-pyruvate lyase from *Escherichia coli*. Appl Microbiol Biotechnol 100(1):135–145. https://doi.org/10.1007/s00253-015-6976-4

Oldiges M, Kunze M, Degenring D, Sprenger GA, Takors R (2004) Stimulation, monitoring, and analysis of pathway dynamics by metabolic profiling in the aromatic amino acid pathway. Biotechnol Prog 20(6):1623–1633

Palsson B (2000) The challenges of in silico biology. Nat Biotechnol 18(11):1147–1150

Perez-Garcia F, Peters-Wendisch P, Wendisch VF (2016) Engineering *Corynebacterium glutamicum* for fast production of L-lysine and L-pipecolic acid. Appl Microbiol Biotechnol 100(18):8075–8090. https://doi.org/10.1007/s00253-016-7682-6

Perez-Garcia F, Max Risse J, Friehs K, Wendisch VF (2017) Fermentative production of L-pipecolic acid from glucose and alternative carbon sources. Biotechnol J 12(7). https://doi.org/10.1002/biot.201600646

Perez-Garcia F, Jorge JMP, Dreyszas A, Risse JM, Wendisch VF (2018) Efficient production of the dicarboxylic acid glutarate by *Corynebacterium glutamicum* via a novel synthetic pathway. Front Microbiol 9:2589. https://doi.org/10.3389/fmicb.2018.02589

Petersen S, de Graaf AA, Eggeling L, Mollney M, Wiechert W, Sahm H (2000) *In vivo* quantification of parallel and bidirectional fluxes in the anaplerosis of *Corynebacterium glutamicum*. J Biol Chem 275(46):35932–35941. https://doi.org/10.1074/jbc.M908728199

Petersen S, Mack C, de Graaf AA, Riedel C, Eikmanns BJ, Sahm H (2001) Metabolic consequences of altered phosphoenolpyruvate carboxykinase activity in *Corynebacterium glutamicum* reveal anaplerotic regulation mechanisms *in vivo*. Metab Eng 3(4):344–361. https://doi.org/10.1006/mben.2001.0198

Peters-Wendisch PG, Schiel B, Wendisch VF, Katsoulidis E, Möckel B, Sahm H, Eikmanns BJ (2001) Pyruvate carboxylase is a major bottleneck for glutamate and lysine production by *Corynebacterium glutamicum*. J Mol Microbiol Biotechnol 3(2):295–300

Quek LE, Wittmann C, Nielsen LK, Krömer JO (2009) OpenFLUX: efficient modelling software for ^{13}C-based metabolic flux analysis. Microb Cell Fact 8:25. https://doi.org/10.1186/1475-2859-8-25

Rados D, Turner DL, Fonseca LL, Carvalho AL, Blombach B, Eikmanns BJ, Neves AR, Santos H (2014) Carbon flux analysis by ^{13}C nuclear magnetic resonance to determine the effect of CO_2 on anaerobic succinate production by *Corynebacterium glutamicum*. Appl Environ Microbiol 80(10):3015–3024. https://doi.org/10.1128/AEM.04189-13

Roessner U, Wagner C, Kopka J, Trethewey RN, Willmitzer L (2000) Technical advance: simultaneous analysis of metabolites in potato tuber by gas chromatography-mass spectrometry. Plant J 23(1):131–142

Rohles CM, Giesselmann G, Kohlstedt M, Wittmann C, Becker J (2016) Systems metabolic engineering of *Corynebacterium glutamicum* for the production of the carbon-5 platform chemicals 5-aminovalerate and glutarate. Microb Cell Fact 15(1):154. https://doi.org/10.1186/s12934-016-0553-0

Rohles CM, Gläser L, Kohlstedt M, Giesselmann G, Pearson S, del Campo A, Becker J, Wittmann C (2018) A bio-based route to the carbon-5 chemical glutaric acid and to bionylon-6,5 using metabolically engineered *Corynebacterium glutamicum*. Green Chem 20(20):4662–4674. https://doi.org/10.1039/c8gc01901k

Rühl M, Rupp B, Nöh K, Wiechert W, Sauer U, Zamboni N (2012) Collisional fragmentation of central carbon metabolites in LC-MS/MS increases precision of ^{13}C metabolic flux analysis. Biotechnol Bioeng 109(3):763–771. https://doi.org/10.1002/bit.24344

Sano K, Ito K, Miwa K, Nakamori S (1987) Amplification of the phosphoenol pyruvate carboxylase gene of *Brevibacterium lactofermentum* to improve amino acid production. Agric Biol Chem 51 (2):597–599

Sato H, Orishimo K, Shirai T, Hirasawa T, Nagahisa K, Shimizu H, Wachi M (2008) Distinct roles of two anaplerotic pathways in glutamate production induced by biotin limitation in *Corynebacterium glutamicum*. J Biosci Bioeng 106(1):51–58. https://doi.org/10.1263/jbb.106.51

Sauer U (2006) Metabolic networks in motion: ^{13}C-based flux analysis. Mol Syst Biol 2:62. https://doi.org/10.1038/msb4100109

Sauer U, Eikmanns BJ (2005) The PEP-pyruvate-oxaloacetate node as the switch point for carbon flux distribution in bacteria. FEMS Microbiol Rev 29(4):765–794. https://doi.org/10.1016/j.femsre.2004.11.002

Sauer U, Hatzimanikatis V, Bailey JE, Hochuli M, Szyperski T, Wuthrich K (1997) Metabolic fluxes in riboflavin-producing *Bacillus subtilis*. Nat Biotechnol 15(5):448–452

Sawada K, Zen-in S, Wada M, Yokota A (2010) Metabolic changes in a pyruvate kinase gene deletion mutant of *Corynebacterium glutamicum* ATCC 13032. Metab Eng 12(4):401–407. https://doi.org/10.1016/j.ymben.2010.01.004

Schilling CH, Schuster S, Palsson BO, Heinrich R (1999) Metabolic pathway analysis: basic concepts and scientific applications in the post-genomic era. Biotechnol Prog 15(3):296–303

Schmidt K, Norregaard LC, Pedersen B, Meissner A, Duus JO, Nielsen JO, Villadsen J (1999) Quantification of intracellular metabolic fluxes from fractional enrichment and 13C-13C coupling constraints on the isotopomer distribution in labeled biomass components. Metab Eng 1 (2):166–179. https://doi.org/10.1006/mben.1999.0114

Schneider J, Wendisch VF (2010) Putrescine production by engineered *Corynebacterium glutamicum*. Appl Microbiol Biotechnol 88(4):859–868. https://doi.org/10.1007/s00253-010-2778-x

Schneider J, Wendisch VF (2011) Biotechnological production of polyamines by bacteria: recent achievements and future perspectives. Appl Microbiol Biotechnol 91(1):17–30. https://doi.org/10.1007/s00253-011-3252-0

Schultz C, Niebisch A, Gebel L, Bott M (2007) Glutamate production by *Corynebacterium glutamicum*: dependence on the oxoglutarate dehydrogenase inhibitor protein OdhI and protein kinase PknG. Appl Microbiol Biotechnol 76(3):691–700. https://doi.org/10.1007/s00253-007-0933-9

Schwechheimer SK, Park EY, Revuelta JL, Becker J, Wittmann C (2016) Biotechnology of riboflavin. Appl Microbiol Biotechnol. https://doi.org/10.1007/s00253-015-7256-z

Schwechheimer SK, Becker J, Peyriga L, Portais JC, Sauer D, Müller R, Hoff B, Haefner S, Schröder H, Zelder O, Wittmann C (2018a) Improved riboflavin production with *Ashbya gossypii* from vegetable oil based on ^{13}C metabolic network analysis with combined labeling analysis by GC/MS, LC/MS, 1D, and 2D NMR. Metab Eng 47:357–373. https://doi.org/10.1016/j.ymben.2018.04.005

Schwechheimer SK, Becker J, Peyriga L, Portais JC, Wittmann C (2018b) Metabolic flux analysis in *Ashbya gossypii* using ^{13}C-labeled yeast extract: industrial riboflavin production under complex nutrient conditions. Microb Cell Fact 17(1):162. https://doi.org/10.1186/s12934-018-1003-y

Schwechheimer SK, Becker J, Wittmann C (2018c) Towards better understanding of industrial cell factories: novel approaches for ^{13}C metabolic flux analysis in complex nutrient environments. Curr Opin Biotechnol 54:128–137. https://doi.org/10.1016/j.copbio.2018.07.001

Shi F, Zhang M, Li Y (2017) Overexpression of *ppc* or deletion of *mdh* for improving production of gamma-aminobutyric acid in recombinant *Corynebacterium glutamicum*. World J Microbiol Biotechnol 33(6):122. https://doi.org/10.1007/s11274-017-2289-3

Shiio I, Yokota A, Sugimoto SI (1987) Effect of pyruvate kinase deficiency on l-lysine productivities of mutants with feedback-resistant aspartokinases. Agric Biol Chem 51(9):2485–2493

Shimizu H, Tanaka H, Nakato A, Nagahisa K, Kimura E, Shioya S (2003) Effects of the changes in enzyme activities on metabolic flux redistribution around the 2-oxoglutarate branch in glutamate production by *Corynebacterium glutamicum*. Bioprocess Biosyst Eng 25(5):291–298. https://doi.org/10.1007/s00449-002-0307-8

Shin JH, Park SH, Oh YH, Choi JW, Lee MH, Cho JS, Jeong KJ, Joo JC, Yu J, Park SJ, Lee SY (2016) Metabolic engineering of *Corynebacterium glutamicum* for enhanced production of 5-aminovaleric acid. Microb Cell Fact 15(1):174. https://doi.org/10.1186/s12934-016-0566-8

Shinfuku Y, Sorpitiporn N, Sono M, Furusawa C, Hirasawa T, Shimizu H (2009) Development and experimental verification of a genome-scale metabolic model for *Corynebacterium glutamicum*. Microb Cell Fact 8:43

Shirai T, Nakato A, Izutani N, Nagahisa K, Shioya S, Kimura E, Kawarabayasi Y, Yamagishi A, Gojobori T, Shimizu H (2005) Comparative study of flux redistribution of metabolic pathway in glutamate production by two coryneform bacteria. Metab Eng 7(2):59–69

Shirai T, Matsuzaki K, Kuzumoto M, Nagahisa K, Furusawa C, Shioya S, Shimizu H (2006) Precise metabolic flux analysis of coryneform bacteria by gas chromatography-mass spectrometry and verification by nuclear magnetic resonance. J Biosci Bioeng 102(5):413–424. https://doi.org/10.1263/jbb.102.413

Shirai T, Fujimura K, Furusawa C, Nagahisa K, Shioya S, Shimizu H (2007) Study on roles of anaplerotic pathways in glutamate overproduction of *Corynebacterium glutamicum* by metabolic flux analysis. Microb Cell Fact 6:19. https://doi.org/10.1186/1475-2859-6-19

Shupletsov MS, Golubeva LI, Rubina SS, Podvyaznikov DA, Iwatani S, Mashko SV (2014) OpenFLUX2: ^{13} C-MFA modeling software package adjusted for the comprehensive analysis of single and parallel labeling experiments. Microb Cell Fact 13(1):152. https://doi.org/10.1186/s12934-014-0152-x

Silberbach M, Schäfer M, Hüser AT, Kalinowski J, Pühler A, Krämer R, Burkovski A (2005) Adaptation of *Corynebacterium glutamicum* to ammonium limitation: a global analysis using transcriptome and proteome techniques. Appl Environ Microbiol 71(5):2391–2402. https://doi.org/10.1128/AEM.71.5.2391-2402.2005

Smith KM, Cho KM, Liao JC (2010) Engineering *Corynebacterium glutamicum* for isobutanol production. Appl Microbiol Biotechnol 87(3):1045–1055. https://doi.org/10.1007/s00253-010-2522-6

Sonderegger M, Jeppsson M, Hahn-Hagerdal B, Sauer U (2004) Molecular basis for anaerobic growth of *Saccharomyces cerevisiae* on xylose, investigated by global gene expression and metabolic flux analysis. Appl Environ Microbiol 70(4):2307–2317

Sonntag K, Eggeling L, De Graaf AA, Sahm H (1993) Flux partitioning in the split pathway of lysine synthesis in *Corynebacterium glutamicum*. Quantification by ^{13}C- and ^{1}H-NMR spectroscopy. Eur J Biochem 213(3):1325–1331

Sonntag K, Schwinde J, deGraaf AA, Marx A, Eikmanns BJ, Wiechert W, Sahm H (1995) ^{13}C NMR studies of the fluxes in the central metabolism of *Corynebacterium glutamicum* during growth and overproduction of amino acids in batch cultures. Appl Microbiol Biotechnol 44 (3–4):489–495

Sun H, Zhao D, Xiong B, Zhang C, Bi C (2016) Engineering *Corynebacterium glutamicum* for violacein hyper production. Microb Cell Fact 15(1):148. https://doi.org/10.1186/s12934-016-0545-0

Szyperski T (1998) C-13-NMR, MS and metabolic flux balancing in biotechnology research. Q Rev Biophys 31(1):41–106. https://doi.org/10.1017/S0033583598003412

Takeno S, Hori K, Ohtani S, Mimura A, Mitsuhashi S, Ikeda M (2016) l-Lysine production independent of the oxidative pentose phosphate pathway by *Corynebacterium glutamicum* with the *Streptococcus mutans gapN* gene. Metab Eng 37:1–10. https://doi.org/10.1016/j.ymben.2016.03.007

Tang JK, You L, Blankenship RE, Tang YJ (2012) Recent advances in mapping environmental microbial metabolisms through ^{13}C isotopic fingerprints. J R Soc Interface 9(76):2767–2780. https://doi.org/10.1098/rsif.2012.0396

Templeton N, Smith KD, McAtee-Pereira AG, Dorai H, Betenbaugh MJ, Lang SE, Young JD (2017) Application of ^{13}C flux analysis to identify high-productivity CHO metabolic phenotypes. Metab Eng 43(Pt B):218–225. https://doi.org/10.1016/j.ymben.2017.01.008

Tesch M, de Graaf AA, Sahm H (1999) In vivo fluxes in the ammonium-assimilatory pathways in *Corynebacterium glutamicum* studied by ^{15}N nuclear magnetic resonance. Appl Environ Microbiol 65(3):1099–1109

Tsuge Y, Yamamoto S, Kato N, Suda M, Vertes AA, Yukawa H, Inui M (2015) Overexpression of the phosphofructokinase encoding gene is crucial for achieving high production of D-lactate in *Corynebacterium glutamicum* under oxygen deprivation. Appl Microbiol Biotechnol 99 (11):4679–4689. https://doi.org/10.1007/s00253-015-6546-9

Vallino JJ, Stephanopoulos G (1993) Metabolic flux distributions in *Corynebacterium glutamicum* during growth and lysine overproduction. Biotechnol Bioeng 41(6):633–646. https://doi.org/10.1002/bit.260410606

Vallino JJ, Stephanopoulos G (1994) Carbon flux distributions at the glucose-6-phosphate branch point in *Corynebacterium glutamicum* during lysine overproduction. Biotechnol Prog 10 (3):327–334. https://doi.org/10.1021/Bp00027a014

van Ooyen J, Noack S, Bott M, Reth A, Eggeling L (2012) Improved L-lysine production with *Corynebacterium glutamicum* and systemic insight into citrate synthase flux and activity. Biotechnol Bioeng 109(8):2070–2081. https://doi.org/10.1002/bit.24486

van Winden WA, van Dam JC, Ras C, Kleijn RJ, Vinke JL, van Gulik WM, Heijnen JJ (2005) Metabolic-flux analysis of *Saccharomyces cerevisiae* CEN.PK113-7D based on mass isotopomer measurements of ^{13}C-labeled primary metabolites. FEMS Yeast Res 5 (6–7):559–568

Varela C, Agosin E, Baez M, Klapa M, Stephanopoulos G (2003) Metabolic flux redistribution in *Corynebacterium glutamicum* in response to osmotic stress. Appl Microbiol Biotechnol 60 (5):547–555

Veit A, Rittmann D, Georgi T, Youn JW, Eikmanns BJ, Wendisch VF (2009) Pathway identification combining metabolic flux and functional genomics analyses: acetate and propionate activation by *Corynebacterium glutamicum*. J Biotechnol 140(1–2):75–83. https://doi.org/10.1016/j.jbiotec.2008.12.014

Villas-Boas SG, Mas S, Akesson M, Smedsgaard J, Nielsen J (2005) Mass spectrometry in metabolome analysis. Mass Spectrom Rev 24(5):613–646

von Kamp A, Thiele S, Hädicke O, Klamt S (2017) Use of CellNetAnalyzer in biotechnology and metabolic engineering. J Biotechnol 261:221–228. https://doi.org/10.1016/j.jbiotec.2017.05.001

Walker TE, Han CH, Kollman VH, London RE, Matwiyoff NA (1982) ^{13}C nuclear magnetic resonance studies of the biosynthesis by *Microbacterium ammoniaphilum* of L-glutamate selectively enriched with carbon-13. J Biol Chem 257(3):1189–1195

Wang N, Ni Y, Shi F (2015) Deletion of *odhA* or *pyc* improves production of gamma-aminobutyric acid and its precursor L-glutamate in recombinant *Corynebacterium glutamicum*. Biotechnol Lett 37(7):1473–1481. https://doi.org/10.1007/s10529-015-1822-4

Wang L, Dash S, Ng CY, Maranas CD (2017) A review of computational tools for design and reconstruction of metabolic pathways. Synth Syst Biotechnol 2(4):243–252. https://doi.org/10.1016/j.synbio.2017.11.002

Wehrmann A, Phillipp B, Sahm H, Eggeling L (1998) Different modes of diaminopimelate synthesis and their role in cell wall integrity: a study with *Corynebacterium glutamicum*. J Bacteriol 180(12):3159–3165

Weitzel M, Nöh K, Dalman T, Niedenfuhr S, Stute B, Wiechert W (2013) 13CFLUX2—high-performance software suite for ^{13}C-metabolic flux analysis. Bioinformatics 29(1):143–145. https://doi.org/10.1093/bioinformatics/bts646

Wendisch VF (2017) Microbial production of amino acid-related compounds. Adv Biochem Eng Biotechnol 159:255–269. https://doi.org/10.1007/10_2016_34

Wendisch VF, de Graaf AA, Sahm H, Eikmanns BJ (2000) Quantitative determination of metabolic fluxes during coutilization of two carbon sources: comparative analyses with *Corynebacterium glutamicum* during growth on acetate and/or glucose. J Bacteriol 182(11):3088–3096

Wiechert W, de Graaf A (1997) Bidirectional reaction steps in metabolic networks: I. Modeling and simulation of carbon isotope labeling experiments. Biotechnol Bioeng 55(1):102–117

Wiechert W, Möllney M, Petersen S, de Graaf AA (2001) A universal framework for ^{13}C metabolic flux analysis. Metab Eng 3(3):265–283

Wieschalka S, Blombach B, Eikmanns BJ (2012) Engineering *Corynebacterium glutamicum* for the production of pyruvate. Appl Microbiol Biotechnol 94(2):449–459. https://doi.org/10.1007/s00253-011-3843-9

Winter G, Krömer JO (2013) Fluxomics—connecting 'omics analysis and phenotypes. Environ Microbiol 15(7):1901–1916. https://doi.org/10.1111/1462-2920.12064

Wittmann C (2002) Metabolic flux analysis using mass spectrometry. Adv Biochem Eng Biotechnol 74:39–64

Wittmann C (2007) Fluxome analysis using GC-MS. Microb Cell Fact 6:6. https://doi.org/10.1186/1475-2859-6-6

Wittmann C (2010) Analysis and engineering of metabolic pathway fluxes in *Corynebacterium glutamicum*. Adv Biochem Eng Biotechnol 120:21–49. https://doi.org/10.1007/10_2009_58

Wittmann C, Becker J (2007) The L-lysine story. From metabolic pathways to industrial production. In: Wendisch VF (ed) Amino acid biosynthesis—pathways, regulation and metabolic engineering, Microbiology monographs, vol 5. Springer, Berlin, pp 40–68

Wittmann C, De Graaf AA (2005) Metabolic flux analysis in *Corynebacterium glutamicum*. In: Eggeling L, Bott M (eds) Handbook of *Corynebacterium glutamicum*. CRC, Boca Raton, pp 277–304

Wittmann C, Heinzle E (1999) Mass spectrometry for metabolic flux analysis. Biotechnol Bioeng 62(6):739–750

Wittmann C, Heinzle E (2001a) Application of MALDI-TOF MS to lysine-producing *Corynebacterium glutamicum*: a novel approach for metabolic flux analysis. Eur J Biochem 268 (8):2441–2455

Wittmann C, Heinzle E (2001b) Modeling and experimental design for metabolic flux analysis of lysine-producing Corynebacteria by mass spectrometry. Metab Eng 3(2):173–191. https://doi.org/10.1006/mben.2000.0178

Wittmann C, Heinzle E (2002) Genealogy profiling through strain improvement by using metabolic network analysis: metabolic flux genealogy of several generations of lysine-producing Corynebacteria. Appl Environ Microbiol 68(12):5843–5859

Wittmann C, Hans M, Heinzle E (2002) *In vivo* analysis of intracellular amino acid labelings by GC/MS. Anal Biochem 307(2):379–382

Wittmann C, Kiefer P, Zelder O (2004a) Metabolic fluxes in *Corynebacterium glutamicum* during lysine production with sucrose as carbon source. Appl Environ Microbiol 70(12):7277–7287. https://doi.org/10.1128/AEM.70.12.7277-7287.2004

Wittmann C, Kim HM, Heinzle E (2004b) Metabolic network analysis of lysine producing *Corynebacterium glutamicum* at a miniaturized scale. Biotechnol Bioeng 87(1):1–6. https://doi.org/10.1002/bit.20103

Wittmann C, Weber J, Betiku E, Krömer J, Bohm D, Rinas U (2007) Response of fluxome and metabolome to temperature-induced recombinant protein synthesis in *Escherichia coli*. J Biotechnol 132(4):375–384. https://doi.org/10.1016/j.jbiotec.2007.07.495

Wu W, Zhang Y, Liu D, Chen Z (2019) Efficient mining of natural NADH-utilizing dehydrogenases enables systematic cofactor engineering of lysine synthesis pathway of *Corynebacterium glutamicum*. Metab Eng 52:77–86. https://doi.org/10.1016/j.ymben.2018.11.006

Xu J, Zhang J, Guo Y, Zai Y, Zhang W (2013) Improvement of cell growth and L-lysine production by genetically modified *Corynebacterium glutamicum* during growth on molasses. J Ind Microbiol Biotechnol 40(12):1423–1432. https://doi.org/10.1007/s10295-013-1329-8

Xu JZ, Zhang JL, Guo YF, Jia QD, Zhang WG (2014) Heterologous expression of *Escherichia coli* fructose-1,6-bisphosphatase in *Corynebacterium glutamicum* and evaluating the effect on cell growth and L-lysine production. Prep Biochem Biotechnol 44(5):493–509. https://doi.org/10.1080/10826068.2013.833115

Xu JZ, Wu ZH, Gao SJ, Zhang W (2018) Rational modification of tricarboxylic acid cycle for improving L-lysine production in *Corynebacterium glutamicum*. Microb Cell Fact 17(1):105. https://doi.org/10.1186/s12934-018-0958-z

Yamamoto S, Suda M, Niimi S, Inui M, Yukawa H (2013) Strain optimization for efficient isobutanol production using *Corynebacterium glutamicum* under oxygen deprivation. Biotechnol Bioeng 110(11):2938–2948. https://doi.org/10.1002/bit.24961

Yanase M, Aikoh T, Sawada K, Ogura K, Hagiwara T, Imai K, Wada M, Yokota A (2016) Pyruvate kinase deletion as an effective phenotype to enhance lysine production in *Corynebacterium glutamicum* ATCC13032: redirecting the carbon flow to a precursor metabolite. J Biosci Bioeng 122(2):160–167. https://doi.org/10.1016/j.jbiosc.2015.12.023

Yang TH, Heinzle E, Wittmann C (2005) Theoretical aspects of ^{13}C metabolic flux analysis with sole quantification of carbon dioxide labeling. Comput Biol Chem 29(2):121–133. https://doi.org/10.1016/j.compbiolchem.2005.02.005

Yang TH, Wittmann C, Heinzle E (2006a) Respirometric ^{13}C flux analysis—Part II: in vivo flux estimation of lysine-producing *Corynebacterium glutamicum*. Metab Eng 8(5):432–446. https://doi.org/10.1016/j.ymben.2006.03.002

Yang TH, Wittmann C, Heinzle E (2006b) Respirometric ^{13}C flux analysis, Part I: design, construction and validation of a novel multiple reactor system using on-line membrane inlet mass spectrometry. Metab Eng 8(5):417–431. https://doi.org/10.1016/j.ymben.2006.03.001

Yokota A, Lindley ND (2005) Central metabolism: sugar uptake and conversion. In: Eggeling L, Bott M (eds) Handbook of *Corynebacterium glutamicum*. CRC, Boca Raton, pp 215–240

Yokota A, Sawada K, Wada M (2017) Boosting anaplerotic reactions by pyruvate kinase gene deletion and phosphoenolpyruvate carboxylase desensitization for glutamic acid and lysine production in *Corynebacterium glutamicum*. Adv Biochem Eng Biotechnol 159:181–198. https://doi.org/10.1007/10_2016_31

Yuan Y, Yang TH, Heinzle E (2010) ^{13}C metabolic flux analysis for larger scale cultivation using gas chromatography-combustion-isotope ratio mass spectrometry. Metab Eng 12(4):392–400. https://doi.org/10.1016/j.ymben.2010.02.001

Zhang Y, Shang X, Wang B, Hu Q, Liu S, Wen T (2019) Reconstruction of tricarboxylic acid cycle in *Corynebacterium glutamicum* with a genome-scale metabolic network model for trans-4-hydroxyproline production. Biotechnol Bioeng 116(1):99–109. https://doi.org/10.1002/bit.26818

Zhao Z, Ding JY, Ma WH, Zhou NY, Liu SJ (2012) Identification and characterization of gamma-aminobutyric acid uptake system GabPCg (NCgl0464) in *Corynebacterium glutamicum*. Appl Environ Microbiol 78(8):2596–2601. https://doi.org/10.1128/AEM.07406-11

Amino Acid Exporters in *Corynebacterium glutamicum*

Masaaki Wachi

Contents

Abstract *Corynebacterium glutamicum* was isolated by Japanese researchers in 1957 as an L-glutamic acid-producing bacterium. L-Glutamate has a distinctive taste known as 'umami', and is widely used as a flavour enhancer. Since 1960, L-glutamate has been industrially produced by fermentation using *C. glutamicum*. *C. glutamicum* is also used for the fermentative production of other amino acids such as L-lysine and L-threonine. In the past decade, our understanding of the molecular basis of amino acid secretion has enabled the identification of *lysE*, *thrE*, and *brnFE*, i.e., the genes encoding novel carriers exporting L-lysine, L-threonine, and L-isoleucine, respectively. Although it has been suggested that the secretion of L-glutamate by *C. glutamicum* is mediated by a carrier system in the cytoplasmic membrane, no L-glutamate exporter has been identified for a long time. In 2007, it was reported that NCgl1221 gene, which encodes a homolog of the MscS protein

M. Wachi (✉)
Department of Life Science and Technology, Tokyo Institute of Technology, Midori-ku, Yokohama, Japan
e-mail: mwachi@bio.titech.ac.jp

M. Inui, K. Toyoda (eds.), *Corynebacterium glutamicum*, Microbiology Monographs 23, https://doi.org/10.1007/978-3-030-39267-3_9

(mechanosensitive channel of the small conductance), is involved in L-glutamate secretion by this bacterium. It was later designated as MscCG (mechanosensitive channel from *C. glutamicum*). Recently, it was also reported that most of *C. glutamicum* strains have one more MscS homolog named MscCG2, which functions as a minor L-glutamate exporter. These two MscS homologs were proved to function as a mechanosensitive channel by electrophysiological studies. Although the physiological significance of amino acid export systems in bacterial cells remains unclear, understanding the export processes of amino acids is important for amino acid production. The focus of this review is on the amino acid export systems of *C. glutamicum*, especially on the recently identified mechanosensitive channels NCgl1221/MscCG and MscCG2.

1 Introduction

L-Glutamate was identified as a compound responsible for a distinctive taste, called 'umami', isolated from a hot water extract of seaweed by a Japanese chemist, Kikunae Ikeda, in 1908. Umami is one of the five basic sensory qualities of taste, in addition to sweet, sour, salty, and bitter. Although it has not been recognised as a taste for very long, its corresponding taste receptor has recently been identified (Chaudhari et al. 2000). At present, it is considered that umami has the physiological function of indicating proteinaceous food. In 1957, Kinoshita and his co-workers isolated *Corynebacterium glutamicum*, originally designated *Micrococcus glutamicus*, as an L-glutamate-producing bacterium (Kinoshita et al. 1957; Udaka 1960). This was a breakthrough in the biotechnological production of amino acids by microbial fermentation. In contrast to chemical synthesis, this ensures that only the biologically active L-forms are produced. L-Glutamate is now widely used as a flavour enhancer. More than 2 million tons of monosodium glutamate is produced worldwide per year by fermentation using *C. glutamicum* or closely related species, and its demand is steadily increasing at an annual rate of 3–5% (Ajinomoto 2009).

C. glutamicum is a non-spore former Gram-positive bacterium that has an asymmetric rod shape. It often grows as V-shaped cell pairs, which result from snapping division (Fig. 1). Wild-type *C. glutamicum* releases more than 80 g/L L-glutamic acid under appropriate culture conditions. *C. glutamicum* has a unique mechanism of L-glutamate secretion. The presence of biotin, which is required by *C. glutamicum* for growth, inhibits L-glutamate production in the culture medium, whereas production is induced under biotin-limiting conditions (Shiio et al. 1962) as well as in response to fatty acid ester surfactants (Takinami et al. 1965) or penicillin (Nara et al. 1964). It is also induced by ethambutol treatment, which inhibits the formation of the mycolic acid layer of the cell wall (Radmacher et al. 2005). Since biotin limitation and the other inductive treatments cause damage to the cell surface structures of this microorganism, it has long been assumed that L-glutamate leaks through the cell membrane (Takinami et al. 1968).

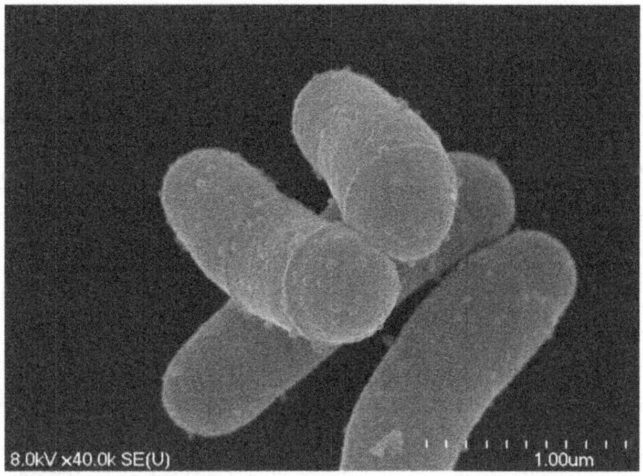

Fig. 1 Scanning electron microscopy image of *C. glutamicum* cells

The *dtsR1* (NCgl0678/cg0812) gene, which is isolated as a multicopy suppressor of a mutant hypersensitive to fatty acid ester surfactants, encodes a protein that exhibits high homology to the β subunit of acetyl-CoA carboxylase (Kimura et al. 1996). Since biotin is a cofactor of acetyl-CoA carboxylase, biotin limitation as well as treatment with fatty acid ester surfactants might affect the biotin–enzyme complex containing DtsR1. Overexpression of *cls* (NCgl2646/cg3037) encoding cardiolipin synthase or *acp* (NCgl2174/cg2473) encoding acyl carrier protein also results in trigger-independent glutamate secretion (Nampoothiri et al. 2002). Penicillin inhibits cell wall biosynthesis by binding to the penicillin-binding proteins (PBPs), which catalyze the final step of peptidoglycan synthesis. *C. glutamicum* is highly tolerant to lytic enzymes such as egg white lysozyme, which catalyzes peptidoglycan hydrolysis. Mutations in *ltsA* (NCgl2116/cg2410), which determines the lysozyme susceptibility of *C. glutamicum*, lead to glutamate production (Hirasawa et al. 2000, 2001). It was reported that the LtsA protein is involved in amidation of diaminopimelic acid residue in the peptidoglycan (Levefaudes et al. 2015). However, the connection between cell wall biosynthesis and L-glutamate secretion remains unclear.

The activity of 2-oxoglutarate dehydrogenase complex (ODHC) reportedly decreases during L-glutamate production in response to biotin limitation, fatty acid ester surfactants, and penicillin (Shingu and Terui 1971; Kawahara et al. 1997). Since ODHC is located at the point of divergence between the TCA cycle and L-glutamate biosynthesis (Fig. 2), a decrease in ODHC activity switches the metabolic flow from the TCA cycle to L-glutamate synthesis (Shimizu et al. 2003; Hasegawa et al. 2008). ODHC is generally comprised of three enzymes: 2-oxoglutarate dehydrogenase (E1o), dihydrolipoamide *S*-succinyltransferase (E2o), and dihydrolipoamide dehydrogenase (E3). Disruption of *odhA* (NCgl1084/cg1280), which encodes the E1o subunit, results in L-glutamate secretion without induction (Asakura et al. 2007). Metabolic linkage between acetyl-CoA carboxylase

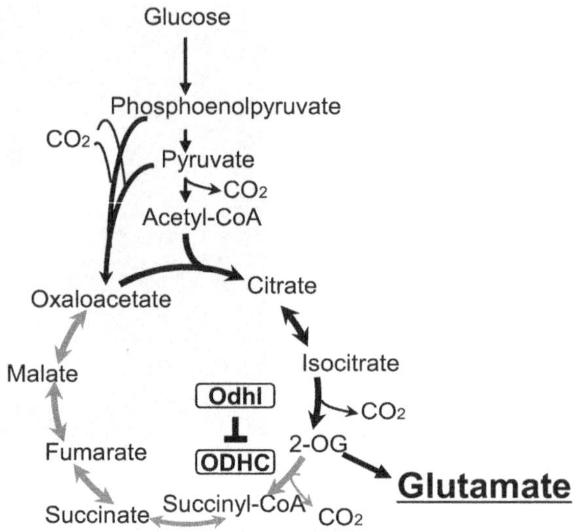

Fig. 2 Biosynthetic pathway of L-glutamate in *C. glutamicum*. L-Glutamate is synthesized from 2-oxoglutarate. 2-Oxoglutarate dehydrogenase complex (ODHC) is a key enzyme for L-glutamate synthesis. OdhI is an inhibitory protein for ODHC

and ODHC is considered to trigger L-glutamate secretion, as a *dtsR1* disruptant produces a significant amount of L-glutamate and reduces ODHC activity (Kimura 2002). Recently, a novel form of ODHC activity regulation in *C. glutamicum* was reported (Niebisch et al. 2006) in which ODHC activity is strongly inhibited by the non-phosphorylated form of the OdhI (NCgl1385/cg1630) protein, which is phosphorylated by the serine/threonine protein kinase PknG (NCgl2655/cg3046). It is also reported that OdhI protein synthesis is induced upon the induction of L-glutamate production by penicillin treatment (Kim et al. 2010).

Nevertheless, the molecular mechanism underlying the secretion of L-glutamate by *C. glutamicum* is not completely clear. Although it is suggested that the secretion of L-glutamate by *C. glutamicum* is mediated by a carrier system in the cytoplasmic membrane (Hoischen and Krämer 1989), no L-glutamate exporter has been identified for a long time.

2 Discovery of Amino Acid Exporters of *C. glutamicum*

Amino acid transport systems are ubiquitously found in eukaryotes and prokaryotes. The physiological roles of uptake systems of amino acids are obvious. Exogenous amino acids can be directly used for protein synthesis. They are also used as carbon, nitrogen, and energy sources. On the other hand, the physiological significance of amino acid export systems is still unclear. However, understanding the export processes of amino acids is important in the amino acid production by microorganisms. Krämer and his co-workers demonstrated that the excretion of L-glutamic acid, L-lysine, L-isoleucine, and L-threonine from *C. glutamicum* cells is an active process (for a review, Krämer 1994). Based on these pioneering studies, Eggeling and his co-workers successfully identified novel transporter families responsible for amino acid efflux (for reviews, Eggeling and Sahm 2003; Marin and Krämer 2007; Eggeling 2005). In particular, the cloning of the L-lysine exporter gene, namely, *lysE* (NCgl1214/cg1424) was a breakthrough in the molecular analysis of amino acid exporters.

2.1 Lysine Exporter

The lysine exporter LysE was identified through an elegant strategy devised by Eggeling and his co-workers (Vrljić et al. 1995, 1996). In contrast to the cloning of drug-resistance exporters that confer resistance to the corresponding drug, no positive selection system exists for amino acid exporters. They found that the addition of L-methionine to wild-type *C. glutamicum* results in the excretion of L-lysine; this is because L-methionine represses homoserine dehydrogenase activity in this bacterium, which results in a decrease in intracellular L-threonine levels. Since L-lysine synthesis is regulated by L-threonine, metabolic flux toward L-lysine is enhanced. An increase in the intracellular L-lysine level subsequently induces L-lysine excretion. This methionine-induced L-lysine excretion system was used to isolate mutants with defective L-lysine export. Mutagenized *C. glutamicum* cells were plated on complex medium, and clones were replica plated on two types of minimal medium plates with or without L-methionine. The ability of L-lysine excretion was then assayed by overlaying the L-lysine-requiring *C. glutamicum lysA* (NCgl1133/cg1334) mutant cells. As a result, mutants with defective L-lysine export were isolated. Using one such mutant, they succeeded in cloning *lysE* encoding a transporter responsible for L-lysine export.

LysE is a relatively small membrane protein with a molecular weight of 25,425 Da, and has five (or six) putative transmembrane segments (Vrljić et al. 1996, 1999); it may function as a dimer. The driving forces of this exporter are the electrochemical proton potential and the lysine gradient. LysE has a relatively low affinity (K_m of approximately 20 mM) for lysine (Bröer and Krämer 1991a, b). This ensures that lysine is exported only when lysine accumulates intracellularly. Since *C. glutamicum* does not possess L-lysine-degrading enzymes, the intracellular lysine concentration would increase in the absence of excretion systems. Maintaining the intracellular lysine concentration appears one of the physiological functions of this lysine excretion system when cells are grown in complex medium or in the presence of lysine-containing peptides. An important observation is that the growth of *lysE* mutants is abolished in the presence of 1 mM Lys-Ala dipeptide. In the mutant cells, lysine accumulates to an extremely high concentration of approximately 1100 mM, whereas it only reaches a concentration of 20–30 mM in wild-type cells (Vrljić et al. 1996).

LysG (NCgl1215/cg1425) is the positive regulator of *lysE* expression (Vrljić et al. 1996; Bellmann et al. 2001). In the *C. glutamicum* chromosome, the *lysG* gene encoding a LysR-type transcriptional regulator is adjacent to *lysE*. The expression of *lysE* is regulated by the intracellular concentration of L-lysine or L-arginine. LysE can mediate the export of L-arginine at a rate similar to that of L-lysine. L-Citruline and L-histidine also act as co-inducers, but these two basic amino acids are exported by different export systems. On the other hand, two other basic amino acids, L-ornithine and DL-diaminopimelate, do not act as co-inducers; LysE does not accept these two amino acids as a substrate.

2.2 Threonine Exporter

The lack of LysE exporter in *C. glutamicum* results in growth arrest in the presence of lysine-containing peptides because of extremely high accumulation of peptide-derived L-lysine to more than 1000 mM (Vrljić et al. 1996). A similar effect of impaired growth has been applied for isolating the L-threonine exporter (Simic et al. 2001). A significant growth delay was observed in the presence of 1 mM Thr-Ala or Ala-Thr, and the strongest growth reduction was observed in the presence of 1 mM Thr-Thr-Thr. The intracellular concentration of L-threonine increased by up to 130 mM in the presence of 1 mM Thr-Thr-Thr. Transposon mutagenesis was used to isolate mutant strains exhibiting increased sensitivity to Thr-containing peptides. This approach helped in identifying *thrE* (NCgl2533/cg2905) encoding an L-threonine exporter.

The gene *thrE* encodes a membrane protein with a molecular weight of 51,697 Da. ThrE has nine putative transmembrane segments with long N- and C-terminal extensions. In contrast to the lysine exporter LysE, deletion of *thrE* causes only a limited cell growth defect at high internal L-threonine concentrations. The physiological function of ThrE is still unclear.

At an intracellular L-threonine concentration of approximately 170 mM, which is achieved in the presence of 1 mM Thr-Thr-Thr, the efflux rates of L-threonine are 2.7 and 1.1 nmol min^{-1} mg^{-1} dry weight for the wild type and *thrE* deletion strain, respectively; this indicates that 59% of L-threonine export is driven by ThrE. The efflux ratio was reduced to 0.6 nmol min^{-1} mg^{-1} dry weight with CCCP addition. This suggests that passive diffusion contributes to 22% of the export, while the still-unknown carrier is expected to catalyze the remaining 19% of the export. Notably, ThrE can catalyze the export of L-serine but not of glycine.

2.3 Isoleucine Exporter

C. glutamicum excretes L-isoleucine in a process dependent on the proton motive force (Hermann and Krämer 1996). To identify the isoleucine export system, a strategy similar to that used to identify the L-threonine exporter ThrE was applied (Kennerknecht et al. 2002). *C. glutamicum* mutants sensitive to the peptide Ile-Ile were isolated by transposon mutagenesis. In one such mutant, strong peptide sensitivity resulted from insertion into a gene designated as *brnF* (NCgl0254/cg0314). The *brnF* gene encodes a membrane protein with a molecular weight of 27,333 Da with seven putative transmembrane segments. Downstream of *brnF*, *brnE* (NCgl0255/cg0315) encodes a second membrane protein with a molecular weight of 11,480 Da with four transmembrane segments. A mutant lacking both genes no longer exports L-isoleucine. BrnF and BrnE together are also responsible for the export of L-leucine and L-valine. An Lrp-like regulatory gene (NCgl0253/cg0313) is located upstream of *brnFE* and is required for the expression. L-Methionine is also a substrate of BrnFE, although it is suggested that *C. glutamicum* has at least one more methionine export system (Trötschel et al. 2005).

3 Glutamic Acid Excretion by *C. glutamicum*

Since *C. glutamicum* produces incomparably large amounts of L-glutamic acid, it was at first believed that this bacterium had an unusual biosynthetic pathway for L-glutamate. Later, it was revealed that L-glutamate is synthesised by the common pathway via glycolysis and the TCA cycle (Fig. 2) (Shingu and Terui 1971; Shiio and Ujigawa-Takeda 1980). Although glutamate excretion is suggested to be catalyzed by a specific carrier system (Hoischen and Krämer 1989, Gutmann et al. 1992), the characteristics of such a system have not been elucidated yet. *C. glutamicum* does not secrete L-glutamic acid under normal growth conditions; glutamate secretion is induced in response to biotin limitation, since this bacterium is a biotin auxotroph (Shiio et al. 1962). Fatty acid ester surfactants (Takinami et al. 1965) and penicillin (Nara et al. 1964) also induce L-glutamic acid secretion, even in the presence of biotin. Since these inducing treatments affect the integrity of the cell envelope, it has long been assumed that L-glutamate leaks through the cell envelope (Takinami et al. 1968). The history and the recent progress in studies of the mechanism of glutamate production by *C. glutamicum* are documented in review articles (Shimizu and Hirasawa 2007; Hirasawa and Wachi 2017).

3.1 Involvement of NCgl1221 Protein in Glutamate Excretion

It is reported that the activity of ODHC decreases during L-glutamate production in response to biotin limitation, fatty acid ester surfactants, and penicillin (Kawahara et al. 1997). Disruption of *odhA*, encoding a subunit of ODHC, results in L-glutamic acid secretion without induction (Asakura et al. 2007). Since ODHC is located at the point of divergence between the TCA cycle and L-glutamate biosynthesis, it is widely assumed that a change in metabolic flow from the TCA cycle to L-glutamate synthesis due to a decrease in ODHC activity induces L-glutamic acid production in this bacterium. A possible connection between ODHC activity and

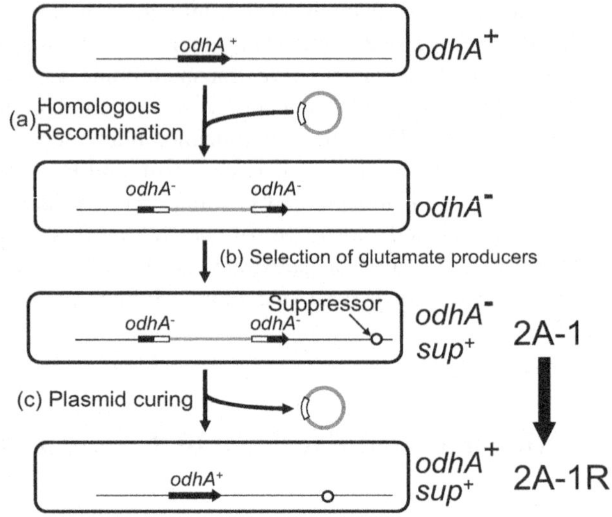

Fig. 3 Strategy for identifying a mutation responsible for L-glutamate production. (a) *odhA* disruptants were constructed by homologous recombination between the chromosomal *odhA* gene and a truncated *odhA* gene on a plasmid. (b) A clone that produced large amounts of L-glutamate in the presence of excess biotin was isolated. (c) The wild-type *odhA* gene was restored by curing the plasmid. For details, see the text

glutamate excretion is also supported by the finding of the regulatory mechanism of ODHC, including the OdhI protein, which inhibits ODHC activity (Niebisch et al. 2006) (Fig. 2). However, we and probably many other investigators in this field have also observed that some *odhA* disruptants often produce little if any L-glutamate. We speculate that additional mutations are responsible for the L-glutamate production of some *odhA* disruptants. Our approach, which led us to identify the NCgl1221 mechanosensitive channel (Nakamura et al. 2007), is briefly described below (Fig. 3).

To examine the above-mentioned hypothesis, we constructed *odhA* disruptants by homologous recombination between the chromosomal *odhA* gene and a truncated *odhA* gene on a plasmid (Fig. 3a). The resultant *odhA*-disrupted strains had extremely unstable phenotypes and formed colonies of various sizes on agar plates. Most clones produced 2-oxoglutarate, lactate, acetate, and pyruvate, but not L-glutamate; however, one clone, namely, 2A-1, produced high levels of L-glutamate in the presence of excess biotin (Fig. 3b). In order to determine whether 2A-1 had gained an additional mutation(s) that rendered it an L-glutamate producer, its wild-type *odhA* gene was restored by secondary homologous recombination (Fig. 3c).

The L-glutamate productivity of the *odhA*$^+$ revertant, namely, 2A-1R, was examined. As expected, 2A-1R retained the ability to produce L-glutamate in the presence of excess biotin. The L-glutamate produced by 2A-1R was assimilated after the glucose had been completely consumed. These findings show that *odhA* disruption does not confer the ability to produce L-glutamate efficiently. Rather, 2A-1 carries an unknown mutation(s) that independently confers the ability to produce L-glutamate (Fig. 3b, c).

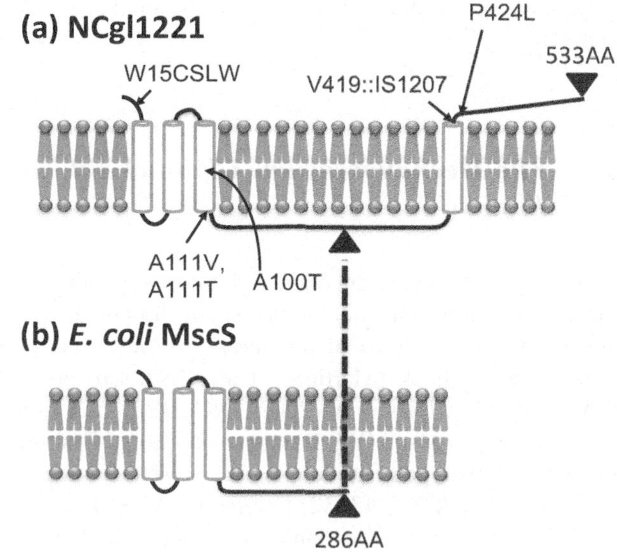

Fig. 4 Membrane topology of NCgl1221 and the *E. coli* MscS. (**a**) Membrane topology of NCgl1221 predicted by the PHD.htm program (http://www.predictprotein.org/). Arrows indicate the NCgl1221 mutations inducing constitutive L-glutamate production. (**b**) Membrane topology of the *E. coli* MscS

Further genetic analysis revealed that the coding region of the NCgl1221 (cg1434) gene of 2A-1R contains the insertion sequence IS1207 at nucleotide 1258 that results in the production of a C-terminally truncated NCgl1221 protein consisting of 423 amino acids (Fig. 4). To determine whether this mutation, termed NCgl1221 (V419::IS1207), is responsible for L-glutamate production, we constructed a derivative, designated BL1, in which the chromosomal NCgl1221 gene was replaced by the mutant gene, from the wild type. This strain exhibited remarkably high L-glutamate productivity without induction treatment, indicating that this mutation alone can lead to L-glutamate production. We also sequenced the NCgl1221 genes of other *odhA* disruptants of *C. glutamicum* that produce large amounts of L-glutamate; as expected, most had mutations in NCgl1221. The substitutions identified were Ala100→Thr, Ala111→Val, Ala111→Thr, Pro424→Leu, and Trp15→Cys-Ser-Leu-Trp (an insertion of 3 amino acids) (Fig. 4). *odhA* disruptants that produced very little or no L-glutamate carried the wild-type NCgl1221 allele. These results clearly indicate that NCgl1221 mutations cause L-glutamate overproduction on their own.

Disruption of NCgl1221 essentially abolishes L-glutamic acid secretion, causing an increase in the intracellular L-glutamic acid pool under biotin-limiting conditions. Intracellular L-glutamate concentrations in the disruptants were about 4–10 times higher than those of the wild type during the production period. Similar phenomena were observed regarding L-aspartate. Extracellular L-aspartate concentration was lower in the disruptant than in the wild type, while the reverse was true for intracellular L-aspartate. In contrast, the ratio of intra- to extracellular 2-oxoglutarate was almost identical in the wild type and in the disruptant. These results strongly promote the view that NCgl1221 encodes a major L-glutamic acid exporter.

3.2 *The Mechanosensitive Channel Activity of NCgl1221 Protein*

The NCgl1221 gene encodes a homolog of the MscS protein, which is the mechanosensitive channel of the small conductance. Mechanosensitive channels are gated by alterations of membrane tension, thus preventing cell disruption by hypo-osmotic shock (Martinac et al. 1987; Cui et al. 1995; Levina et al. 1999; Sukharev 2002; Bass et al. 2002). NCgl1221 has an N-terminal region similar to *Escherichia coli* MscS (YggB) and an additional C-terminal region unlike any known sequence. This type of protein is also present in other *Corynebacteria* spp. such as *C. glutamicum* ATCC 13032, *C. efficiens* YS314, *C. diphtheriae* NCTC 13129, and *C. jeikeium* K411 (Genbank accession numbers, NC_004369, NC_004369, NC_002935, and CR931997, respectively). NCgl1221 was previously reported to catalyze the efflux of a compatible solute, namely, betaine (Nottebrock et al. 2003).

The PHD.htm program (http://www.predictprotein.org/) predicts that NCgl1221 has three and one transmembrane segments in its N- and C-terminal domains, respectively (Fig. 4). Protein structure homology-modelling (http://swissmodel.expasy.org/) shows that the N-terminal domain of NCgl1221 forms a very similar three-dimensional structure to that of the *E. coli* MscS (Bass et al. 2002) (Fig. 5), suggesting that NCgl1221 also functions as a mechanosensitive channel. The

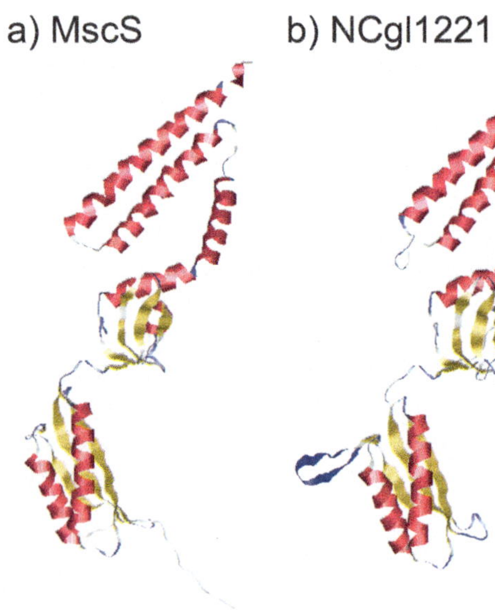

Fig. 5 Comparison of the 3D structures of the *E. coli* MscS and the N-terminal domain of NCgl1221. (**a**) *E. coli* MscS (PDB code: 2OAU) and (**b**) N-terminal domain of NCgl1221. The 3D structure of the N-terminal domain of NCgl1221 was predicted by the SWISS-MODEL protein structure homology-modelling program (http://swissmodel.expasy.org/)

W15CSLW, A100T, A111T, and A111V substitutions are located in the N-terminal domain, while the V419::IS1207 and P424L substitutions are located in the C-terminal domain (Fig. 4). The third transmembrane segment of MscS has a conserved pattern of glycine and alanine residues that are required for the conformational transition that leads to channel opening (Edwards et al. 2005). Substitutions in the N-terminal domain may affect the gating of the channel. Interestingly, the NCgl1221 (V419::IS1207) mutation, which induces constitutive L-glutamate production, results in a truncation of the C-terminal extracytoplasmic domain. This C-terminal extracytoplasmic domain may have a regulatory role in channel opening.

Electrophysiological studies performed by different research groups demonstrated that NCgl1221 indeed possesses mechanosensitive channel activity. Therefore, NCgl1221 was designated as MscCG (mechanosensitive channel from *C. glutamicum*). Using giant *E. coli* spheroplasts lacking MscS and MscL, Börngen et al. (2010) showed that NCgl1221/MscCG exhibits the typical pressure-dependent gating behavior of a stretch-activated channel with a strong rectifying activity in patch-clump analysis. The channel opened under negative pressure between 40 and 100 mmHg; the unitary conductance was 346.3 ± 22.5 pS and 99.5 ± 4.5 pS at positive and negative voltages in asymmetric spheroplast solution, respectively. These behaviors resemble those of the *E. coli* MscS. In contrast to the *E. coli* MscS, the NCgl1221/MscCG channel exhibits a slight preference for cations over anions. The conductance became saturated at KCl concentrations above 300 mM at positive but not at negative pipette voltages, which was not observed in the *E. coli* MscS. The mechanosensitive channel activity of NCgl1221/MscCG was also confirmed by patch-clump analysis using giant spheroplasts of *Bacillus subtilis* lacking mechanosensitive channels, MscL and YkuT (an MscS homolog) (Hashimoto et al. 2010). Heterologous expression of NCgl1221 suppressed cell death caused by hypoosmotic shock in a *B. subtilis* mutant lacking MscL and YkuT, indicating that NCgl1221/MscCG physiologically functions as a mechanosensitive channel at least in *B. subtilis*.

These results suggest that NCgl1221/MscCG encodes a long-lost L-glutamic acid exporter. Hashimoto et al. (2012) directly observed that glutamate is passing through the NCgl1221/MscCG channel in patch-clump assay. We proposed the following model for L-glutamate production by *C. glutamicum* (Fig. 6) (Nakamura et al. 2007): (1) Treatments leading to L-glutamate production cause changes in membrane tension by inhibiting lipid or cell wall synthesis, (2) The structure of NCgl1221/MscCG is altered by sensing changes in membrane tension, and (3) The activated NCgl1221/MscCG mechanosensitive channel catalyzes L-glutamate excretion. L-Glutamate may function as a compatible solute to prevent cells from bursting.

3.3 A Minor Glutamate Exporter MscCG2

In *C. glutamicum* ATCC 13869 strain, which was isolated by Ajinomoto Co. Ltd., and used in the study by Nakamura et al. (2007), disruption of the NCgl1221/*mscCG*

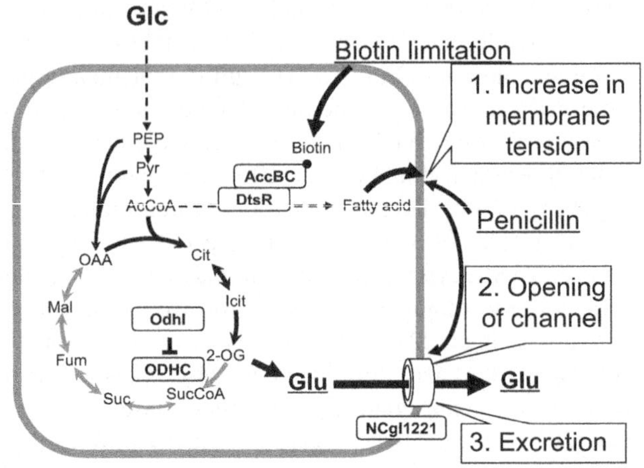

Fig. 6 A possible mechanism of L-glutamate production by *C. glutamicum.* (1) Inducing treatments alter membrane tension. (2) The NCgl1221 mechanosensitive channel opens by sensing a change in membrane tension. (3) L-Glutamate is exported as a compatible solute to prevent cell disruption. For details, see the text

gene greatly diminished the productivity of L-glutamate, but the disrupted strain still produced significant amounts (10–20 % of the wild type levels) of L-glutamate in the medium, suggesting the presence of a minor L-glutamate exporter (Nakamura et al. 2007).

The type strain of *C. gluamicum*, ATCC 13032, possesses only one MscS homolog (NCgl1221/MscCG). However, accumulation of genome sequence data from different *C. glutamicum* strains has revealed that most of *C. glutamicum* strains including ATCC 13869 have one more MscS homolog gene, named *mscCG2*. Wang et al. (2018) reported that MscCG2 also functions as an L-glutamate exporter using *C. glutamicum* Z188 strain, which is used for industrial production of L-glutamate in China. MscCG2 has three transmembrane segments in the N-terminal region and the cytoplasmic domain in the C-terminal region, which is typical for the MscS family proteins. MscCG2 shows relatively low sequence similarity with NCgl1221/MscCG (23% identity in amino acid sequence).

Single deletion of *mscCG2* did not lead to a significant decrease in L-glutamate production in Z188, probably due to the existence of NCgl1221/*mscCG*. In order to elucidate the contribution of MscCG2 in L-glutamate excretion, NCgl1221/*mscCG* single disruptant and NCgl1221/*mscCG mscCG2* double disruptant were constructed. The *ΔmscCG* single disruptant still produced lower but significant amounts of L-glutamate in the medium. However, L-glutamate production was almost eliminated in the *ΔmscCG ΔmscCG2* double disruptant. Complementation of the *mscCG2* gene restored the L-glutamate productivity. It was also demonstrated that L-glutamate excretion by MscCG2 was induced by specific treatments such as biotin limitation and penicillin addition and that a gain-of-function mutation of MscCG2 (A151V) caused constitutive production of L-glutamate. These results clearly indicate a minor but noticeable role of MscCG2 in L-glutamate excretion.

3.4 A Role of the Additional C-Terminal Domain of NCgl1221/MscCG in Glutamate Excretion

The major deference of NCgl1221/MscCG from the typical MscS family proteins is the presence of an additional long C-terminal sequence, which contains the fourth transmembrane segment and the C-terminal extracytoplasmic domain. In order to investigate the role of this additional sequence, a series of C-terminal truncation mutants of NCgl1221/MscCG were constructed and their L-glutamate production ability was examined (Yamashita et al. 2013) (Fig. 7). It was shown that the N-terminal domain of 286 amino acid long, which is homologous to the *E. coli* MscS, retained the ability of L-glutamate production in response to inducing treatments such as biotin limitation and penicillin addition. Patch-clump analysis confirmed that the N-terminal domain mediated L-glutamate secretion. It was also demonstrated that the mutant lacking the C-terminal extracytoplasmic domain produced L-glutamate without any inducing treatments (Yamashita et al. 2013). A similar observation was also done by Becker et al. (2013). They also demonstrated that the *E. coli* MscS is able to function as an L-glutamate exporter to some extent when expressed in *C. glutamicum* cells lacking NCgl1221/MscCG. It was also demonstrated that the chimeric protein between the *E. coli* MscS and the C-terminal domain of NCgl1221/MscCG showed gating hysteresis in a voltage-dependent manner like NCgl1221/MscCG (Nakayama et al. 2016). These results

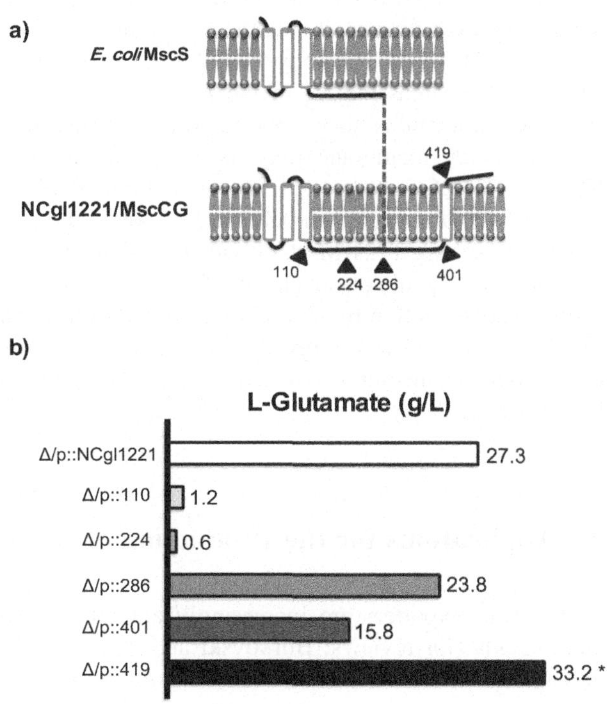

Fig. 7 L-Glutamate production of C-terminal truncation mutants of NCgl1221/MscCG. (**a**) Membrane topology of *E. coli* MscS and NCgl1221/MscCG. Arrowheads indicate the positions of truncation. (**b**) L-Glutamate production by C-terminal truncation mutants. NCgl1221/*mscCG* deletion mutant strains (Δ) carrying plasmids expressing the C-terminal truncation mutants of NCgl1221/MscCG (p::110, p::224, p::286, p::401 and p::419) were examined for their glutamate production ability. Glutamate production was induced by biotin limitation except for Δ/p::419. ∗The Δ/p::419 strain was grown in the presence of excess biotin

suggest that the N-terminal domain of NCgl1221/MscCG is necessary and sufficient for the excretion of L-glutamate in response to inducing treatments, and that the C-terminal extracytoplasmic domain has a regulatory role in L-glutamate excretion.

3.5 *Electrophysiological Analysis of NCg1221/MscCG in the* C. glutamicum *Membrane*

Early electrophysiological studies were done using *E. coli* or *B. subtilis* as an expression host of NCgl1221/MscCG (Börngen et al. 2010; Hashimoto et al. 2010) because of difficulty of giant spheroplast preparation from *C. glutamicum*. Nakayama et al. (2018) recently succeeded the preparation of giant spheroplasts from *C. glutamicum*, which enabled them to analyse the mechanosensitive channel property of NCgl1221/MscCG in the native cell membrane. Patch-clump analysis revealed the presence of three types of mechanosensitive channels, NCgl1221/MscCG, MscCG2 and MscL, in the *C. glutamicum* cell membrane. These three channels differ from each other in their conductance and mechanosensitivity. NCgl1221/MscCG has a relatively small conductance of ~340 pS compared to other two channels, an intermediate MscGC2 conductance of ~1.0 nS and a very large MscL conductance of ~5.5 nS. It was also demonstrated that membrane tension required for half activation of NCgl1221/MscCG was ~5.5 mN/m compared to ~12 mN/m for both MscCG2 and MscL. They also showed that the *C. glutamicum* membrane was much softer than the *E. coli* membrane by the micropipette aspiration technique. The areal elasticity modulus K_A of the *C. glutamicum* membrane was ~15 mN/m whereas the K_A of the *E. coli* membrane was ~44 mN/m. Molecular dynamics simulations also supported this observation. This "soft" properties of the *C. glutamicum* membrane may have a great impact on the gating behavior of NCgl1221/MscCG. NCgl1221/MscCG exhibited voltage-dependent gating hysteresis in the *C. glutamicum* membrane but not in the *E. coli* membrane. The gating hysteresis of NCgl1221/MscCG, which is characterized as different gating thresholds between opening and closing, is considered to be important for continuous L-glutamate excretion by *C. glutamicum*. Considering all these, they proposed that NCgl1221/MscCG has evolved and adapted to the mechanical properties of the *C. glutamicum* membrane, which enables the channels to specialize in export of L-glutamate.

4 Applications for the Production of Amino Acids

Amino acid exporters are increasingly attracting attention for the production of amino acids (for reviews; Burkovski and Krämer 2002; Kelle et al. 2005; Marin and Krämer 2007; Dong et al. 2011). Overexpression of *lysE* and *thrE* enhances the excretion of L-lysine and L-threonine, respectively, at least on a laboratory scale (Vrljić et al. 1996; Simic et al. 2001). Overexpression of *lysE* enhances lysine export to five times the rate of the wild-type cells. This is the first example demonstrating that the reinforcement of export systems can be a strategy for breeding industrial producer strains for primary metabolites. Overexpression of NCgl1221/MscCG also increases the production of L-glutamate, although only in response to induction (Nakamura et al. 2007). It was also demonstrated that introduction of the gain-of-function mutant NCgl1221/MscCG into the *E. coli* phenylalanine producer strain improved the productivity (Hashimoto et al. 2012). Producer strains used in the industrial production of amino acids have been generated by repeated mutagenesis and selection. It would be interesting to know whether industrial L-glutamate producer strains carry mutations in the NCgl1221/*mscCG* gene that induce the constitutive production of L-glutamate.

References

Ajinomoto (2009) Fact sheet: amino acids business. http://www.ajinomoto.co.jp/ir/pdf/fact/Aminoacids-Oct2009.pdf

Asakura Y, Kimura E, Usuda Y, Kawahara Y, Matsui K, Osumi T, Nakamatsu T (2007) Altered metabolic flux due to deletion of *odhA* causes L-glutamate overproduction in *Corynebacterium glutamicum*. Appl Environ Microbiol 73:1308–1319. https://doi.org/10.1128/AEM.01867-06

Bass RB, Strop P, Barclay M, Rees DC (2002) Crystal structure of *Escherichia coli* MscS, a voltage-modulated and mechanosensitive channel. Science 298:1582–1587. https://doi.org/10.1126/science.1077945

Becker M, Börngen K, Nomura T, Battle AR, Marin K, Martinac B, Krämer R (2013) Glutamate efflux mediated by *Corynebacterium glutamicum* MscCG, *Escherichia coli* MscS, and their derivatives. Biochim Biophys Acta 1828:1230–1240. https://doi.org/10.1016/j.bbamem.2013.01.001

Bellmann A, Vrljić M, Pátek M, Sahm H, Krämer R, Eggeling L (2001) Expression control and specificity of the basic amino acid exporter LysE of *Corynebacterium glutamicum*. Microbiology 147:1765–1774. https://doi.org/10.1099/00221287-147-7-1765

Börngen K, Battle A, Möker N, Morbach S, Marin K, Martinac B, Krämer K (2010) The properties and contribution of the *Corynebacterium glutamicum* MscS variant to fine-tuning of osmotic adaptation. Biochim Biophys Acta 1798:2141–2149. https://doi.org/10.1016/j.bbamem.2010.06.022

Bröer S, Krämer R (1991a) Lysine excretion by *Corynebacterium glutamicum*. 1. Identification of a specific secretion carrier system. Eur J Biochem 202:131–135. https://doi.org/10.1111/j.1432-1033.1991.tb16353.x

Bröer S, Krämer R (1991b) Lysine excretion by *Corynebacterium glutamicum*. 2. Energetics and mechanism of the transport system. Eur J Biochem 202:137–143. https://doi.org/10.1111/j.1432-1033.1991.tb16354.x

Burkovski A, Krämer R (2002) Bacterial amino acid transport proteins: occurrence, functions, and significance for biotechnological applications. Appl Microbiol Biotechnol 58:265–274. https://doi.org/10.1007/s00253-001-0869-4

Chaudhari N, Landin AM, Roper SD (2000) A metabotropic glutamate receptor variant functions as a taste receptor. Nat Neurosci 3:113–119

Cui C, Smith DO, Adler J (1995) Characterization of mechanosensitive channels in *Escherichia coli* cytoplasmic membrane by whole-cell patch clamp recording. J Membr Biol 144:31–42

Dong X, Quinn PJ, Wang X (2011) Metabolic engineering of *Escherichia coli* and *Corynebacterium glutamicum* for the production of L-threonine. Biotechnol Adv 29:11–23. https://doi.org/10.1016/j.biotechadv.2010.07.009

Edwards MD, Li Y, Kim S, Miller S, Bartlett W, Black S, Dennison S, Iscla I, Blount P, Bowie JU, Booth IR (2005) Pivotal role of the glycine-rich TM3 helix in gating the MscS mechanosensitive channel. Nat Struct Mol Biol 12:113–119

Eggeling L (2005) Export of amino acids and other solutes. In: Eggeling L, Bott M (eds) Handbook of *Corynebacterium glutamicum*. Taylor & Francis, Heidelberg, pp 187–214

Eggeling L, Sahm H (2003) New ubiquitous translocators: amino acid export by *Corynebacterium glutamicum* and *Escherichia coli*. Arch Microbiol 180:155–160

Gutmann M, Hoischen C, Krämer R (1992) Carrier-mediated glutamate secretion by *Corynebacterium glutamicum* under biotin limitation. Biochim Biophys Acta 1112:115–123. https://doi.org/10.1016/0005-2736(92)90261-J

Hasegawa T, Hashimoto K, Kawasaki H, Nakamatsu T (2008) Changes in enzyme activities at the pyruvate node in glutamate-overproducing *Corynebacterium glutamicum*. J Biosci Bioeng 105:12–19. https://doi.org/10.1263/jbb.105.12

Hashimoto K, Nakamura K, Kuroda T, Yabe I, Nakamatsu T, Kawasaki H (2010) The protein encoded by NCgl1221 in *Corynebacterium glutamicum* functions as a mechanosensitive channel. Biosci Biotechnol Biochem 74:2546–2549. https://doi.org/10.1271/bbb.100636

Hashimoto K, Murata J, Konishi T, Yabe I, Nakamatsu T, Kawasaki H (2012) Glutamate is excreted across the cytoplasmic membrane through the NCgl1221 channel of *Corynebacterium glutamicum* by passive diffusion. Biosci Biotechnol Biochem 76:1422–1424. https://doi.org/10.1271/bbb.100636

Hermann T, Krämer R (1996) Mechanism and regulation of isoleucine excretion in *Corynebacterium glutamicum*. Appl Environ Microbiol 62:3238–3244

Hirasawa T, Wachi M (2017) Glutamate fermentation-2: mechanism of L-glutamate overproduction in *Corynebacterium glutamicum*. Adv Biochem Eng Biotechnol 159:57–72. https://doi.org/10.1007/10_2016_26

Hirasawa T, Wachi M, Nagai K (2000) A mutation in the *Corynebacterium glutamicum ltsA* gene causes susceptibility to lysozyme, temperature-sensitive growth, and L-glutamate production. J Bacteriol 182:2696–2701. https://doi.org/10.1128/JB.182.10.2696-2701.2000

Hirasawa T, Wachi M, Nagai K (2001) L-Glutamate production by lysozyme-sensitive *Corynebacterium glutamicum ltsA* mutant strains. BMC Biotechnol 1:9. https://doi.org/10.1186/1472-6750-1-9

Hoischen C, Krämer R (1989) Evidence for an efflux carrier system involved in the secretion of glutamate by *Corynebacterium glutamicum*. Arch Microbiol 151:342–347

Kawahara Y, Takahashi-Fuke K, Shimizu E, Nakamatsu T, Nakamori S (1997) Relationship between the glutamate production and the activity of 2-oxoglutarate dehydrogenase in *Brevibacterium lactofermentum*. Biosci Biotechnol Biochem 61:1109–1112. https://doi.org/10.1271/bbb.61.1109

Kelle R, Herrmann T, Bathe B (2005) L-Lysine production. In: Eggeling L, Bott M (eds) Handbook of *Corynebacterium glutamicum*. Taylor & Francis, Heidelberg, pp 465–488

Kennerknecht N, Sahm H, Yen M-R, Pátek M, Saier MH Jr, Eggeling L (2002) Export of L-isoleucine from *Corynebacterium glutamicum*: a two-gene-encoded member of a new translocator family. J Bacteriol 184:3947–3956. https://doi.org/10.1128/JB.184.14.3947-3956.2002

Kim J, Fukuda H, Hirasawa T, Nagahisa K, Nagai K, Wachi M, Shimizu H (2010) Requirement of *de novo* synthesis of the OdhI protein in penicillin-induced glutamate production by *Corynebacterium glutamicum*. Appl Microbiol Biotechnol 86:911–920. https://doi.org/10.1007/s00253-009-2360-6

Kimura E (2002) Triggering mechanism of L-glutamate overproduction by DtsR1 in coryneform bacteria. J Biosci Bioeng 94:545–551. https://doi.org/10.1016/S1389-1723(02)80193-1

Kimura E, Abe C, Kawahara Y, Nakamatsu T (1996) Molecular cloning of a novel gene, *dtsR*, which rescues the detergent sensitivity of a mutant derived from *Brevibacterium lactofermentum*. Biosci Biotechnol Biochem 60:1565–1570. https://doi.org/10.1271/bbb.60.1565

Kinoshita S, Udaka S, Shimono M (1957) Studies on the amino acid fermentation. Part 1. Production of L-glutamic acid by various microorganisms. J Gen Appl Microbiol 3:193–205

Krämer R (1994) Secretion of amino acid by bacteria: physiology and mechanism. FEMS Microbiol Rev 13:75–94

Levefaudes M, Patin D, de Sousa-d'Auria C, Chami M, Blanot D, Hervé M, Arthur M, Houssin C, Mengin-Lecreulx D (2015) Diaminopimelic acid amidation in Corynebacteriales: new insight into the role of LtsA in peptidoglycan modification. J Biol Chem 290:13079–13094. https://doi.org/10.1074/jbc.M115.642843

Levina N, Tötemeyer S, Stokes NR, Louis P, Jones MA, Booth IR (1999) Protection of *Escherichia coli* cells against extreme turgor by activation of MscS and MscL mechanosensitive channels: identification of genes required for MscS activity. EMBO J 18:1730–1737. https://doi.org/10.1093/emboj/18.7.1730

Marin K, Krämer R (2007) Amino acid transport systems in biotechnologically relevant bacteria. Microbiol Monogr 5:289–325. https://doi.org/10.1007/7171_2006_069

Martinac B, Buechner M, Delcour AH, Adler J, Kung C (1987) Pressure-sensitive ion channel in *Escherichia coli*. Proc Natl Acad Sci USA 84:2297–2301. https://doi.org/10.1073/pnas.84.8.2297

Nakamura J, Hirano S, Ito H, Wachi M (2007) Mutations of the *Corynebacterium glutamicum* NCgl1221 gene, encoding a mechanosensitive channel homolog, induce L-glutamic acid production. Appl Environ Microbiol 73:4491–4498. https://doi.org/10.1128/AEM.02446-06

Nakayama Y, Becker M, Ebrahimian H, Konishi T, Kawasaki H, Krämer R, Martinac B (2016) The impact of the C-terminal domain on the gating properties of MscCG from *Corynebacterium glutamicum*. Biochim Biophys Acta 1858:130–138. https://doi.org/10.1016/j.bbamem.2015.10.010

Nakayama Y, Komazawa K, Bavi N, Hashimoto KI, Kawasaki H, Martinac B (2018) Evolutionary specialization of MscCG, an MscS-like mechanosensitive channel, in amino acid transport in *Corynebacterium glutamicum*. Sci Rep 8:12893. https://doi.org/10.1038/s41598-018-31219-6

Nampoothiri KM, Hoischen C, Bathe B, Möckel B, Pfefferle W, Krumbach K, Sahm H, Eggeling L (2002) Expression of genes of lipid synthesis and altered lipid composition modulates L-glutamate efflux of *Corynebacterium glutamicum*. Appl Microbiol Biotechnol 58:89–96

Nara T, Samejima H, Kinoshita S (1964) Effect of penicillin on amino acid fermentation. Agric Biol Chem 28:120–124

Niebisch A, Kabus A, Schultz C, Weil B, Bott M (2006) Corynebacterial protein kinase G controls 2-oxoglutarate dehydrogenase activity via the phosphorylation status of the OdhI protein. J Biol Chem 281:12300–12307. https://doi.org/10.1074/jbc.M512515200

Nottebrock D, Meyer U, Krämer R, Morbach S (2003) Molecular and biochemical characterization of mechanosensitive channels in *Corynebacterium glutamicum*. FEMS Microbiol Lett 218:305–309. https://doi.org/10.1111/j.1574-6968.2003.tb11533.x

Radmacher E, Stansen KC, Besra GS, Alderwick LJ, Maughan WN, Hollweg G, Sahm H, Wendisch VF, Eggeling L (2005) Ethambutol, a cell wall inhibitor of *Mycobacterium tuberculosis*, elicits L-glutamate efflux of *Corynebacterium glutamicum*. Microbiology 151:1359–1368. https://doi.org/10.1099/mic.0.27804-0

Shiio I, Ujigawa-Takeda K (1980) Presence and regulation of α-ketoglutarate dehydrogenase complex in a glutamate-producing bacterium, *Brevibacterium flavum*. Agric Biol Chem 44:1887–1890

Shiio I, Otsuka S, Takahashi M (1962) Effect of biotin on the bacterial formation of glutamic acid. I. Glutamate formation and cellular permeability of amino acids. J Biochem 51:56–62

Shimizu H, Hirasawa T (2007) Production of glutamate and glutamate-related amino acids: molecular mechanism analysis and metabolic engineering. Microbiol Monogr 5:1–38

Shimizu H, Tanaka H, Nakato A, Nagahisa K, Kimura E, Shioya S (2003) Effects of the changes in enzyme activities on metabolic flux redistribution around the 2-oxoglutarate branch in glutamate production by *Corynebacterium glutamicum*. Bioprocess Biosyst Eng 25:291–298

Shingu H, Terui G (1971) Studies on the process of glutamic acid fermentation at the enzyme level: I. On the change of α-ketoglutaric acid dehydrogenase in the course of culture. J Ferment Technol 49:400–405

Simic P, Sahm H, Eggeling L (2001) L-Threonine export: use of peptides to identify a new translocator from *Corynebacterium glutamicum*. J Bacteriol 183:5317–5324. https://doi.org/10.1128/JB.183.18.5317-5324.2001

Sukharev S (2002) Purification of the small mechanosensitive channel of *Escherichia coli* (MscS): the subunit structure, conduction, and gating characteristics in liposomes. Biophys J 83:290–298. https://doi.org/10.1016/S0006-3495(02)75169-2

Takinami K, Yoshii H, Tsuri H, Okada H (1965) Biochemical effects of fatty acid and its derivatives on L-glutamic acid fermentation. Part III. Biotin-Tween 60 relationship in the accumulation of L-glutamic acid and the growth of *Brevibacterium lactofermentum*. Agric Biol Chem 29:351–359

Takinami K, Yoshii H, Yamada Y, Okada H, Kinoshita K (1968) Control of L-glutamic acid fermentation by biotin and fatty acid. Amino Acid Nucl Acid 18:120–160

Trötschel C, Deutenberg D, Bathe B, Burkovski A, Krämer R (2005) Characterization of methionine export in *Corynebacterium glutamicum*. J Bacteriol 187:3786–3794. https://doi.org/10.1128/JB.187.11.3786-3794.2005

Udaka S (1960) Screening method for microorganisms accumulating metabolites and its use in the isolation of *Micrococcus glutamicus*. J Bacteriol 79:754–755

Vrljić M, Kronemeyer W, Sahm H, Eggeling L (1995) Unbalance of L-lysine flux in *Corynebacterium glutamicum* and its use for the isolation of excretion-deficient mutants. J Bacteriol 177:4021–4027. https://doi.org/10.1128/jb.177.14.4021-4027.1995

Vrljić M, Sahm H, Eggeling L (1996) A new type of transporter with a new type of cellular function: L-lysine export from *Corynebacterium glutamicum*. Mol Microbiol 22:815–826. https://doi.org/10.1046/j.1365-2958.1996.01527.x

Vrljić M, Garg J, Bellmann A, Wachi S, Freudl R, Malecki MJ, Sahm H, Kozina VJ, Eggeling L, Saier MH Jr (1999) The LysE superfamily: topology of the lysine exporter LysE of *Corynebacterium glutamicum*, a paradyme for a novel superfamily of transmembrane solute translocators. J Mol Microbiol Biotechnol 1:327–336

Wang Y, Cao G, Xu D, Fan L, Wu X, Ni X, Zhao S, Zheng P, Sun J, Ma Y (2018) A novel *Corynebacterium glutamicum* L-glutamate exporter, Appl Environ Microbiol. 84(6):pii: e02691-17. https://doi.org/10.1128/AEM.02691-17

Yamashita C, Hashimoto K, Kumagai K, Maeda T, Takada A, Yabe I, Kawasaki H, Wachi M (2013) L-Glutamate secretion by the N-terminal domain of the *Corynebacterium glutamicum* NCgl1221 mechanosensitive channel. Biosci Biotechnol Biochem 77:1008–1013. https://doi.org/10.1271/bbb.120988

Part IV
Metabolic Design for a Wide Variety ofProducts

Metabolic Engineering in *Corynebacterium glutamicum*

Volker F. Wendisch and Jin-Ho Lee

Contents

V. F. Wendisch (✉)
Genetics of Prokaryotes, Faculty of Biology, Bielefeld University, Bielefeld, Germany

Center for Biotechnology, Bielefeld University, Bielefeld, Germany
e-mail: volker.wendisch@uni-bielefeld.de

J.-H. Lee
Major in Food Science & Biotechnology, School of Food Biotechnology & Nutrition, Kyungsung University, Nam-gu, Busan, Republic of Korea

M. Inui, K. Toyoda (eds.), *Corynebacterium glutamicum*, Microbiology Monographs 23, https://doi.org/10.1007/978-3-030-39267-3_10

Abstract *Corynebacterium glutamicum* strain development has adopted each metabolic engineering tool immediately upon its conception and has contributed to developing these tools further. In the last more than five decades, *C. glutamicum* strains have been selected and screened after undirected mutagenesis. This approach yielded industrially relevant strains, however, it was limited by the little insight that could be gained hampering transfer to related processes. Genetic engineering, the complete genome sequence and omics tools, systems biology and synthetic biology have closed this gap and provided a profound knowledge base to develop engineering strategies driven by metabolic insight. Quite recently, serendipitous untargeted approaches came into focus again since mutants selected classically e.g. using genetically encoded biosensors or by adaptive laboratory evolution could be understood when genome resequencing was combined with genetic and biochemical experiments. We will discuss metabolic engineering from the classical, genetic engineering, systems biology eras to the era of synthetic biology for *C. glutamicum* strain development and forecast the impact of the most recent methods such as CRISPR technology and adaptive laboratory evolution.

1 Introduction

C. glutamicum is a non-pathogenic, facultatively anaerobic, Gram-positive actinobacterium. Due to its importance in the million-ton-scale amino acid production industry (Wendisch 2007), its physiology and genetics have been studied in depth (Eggeling and Bott 2005; Tatsumi and Inui 2013; Burkovski 2008). The reader is referred to other chapters of this book and excellent reviews with respect to amino acid production (Hirasawa and Shimizu 2016; Lee and Wendisch 2017b), in particular L-lysine (Koffas and Stephanopoulos 2005; Ikeda 2017) and L-glutamate production (Hirasawa et al. 2012). *C. glutamicum* strain development has previously been reviewed through the classical era (Leuchtenberger et al. 2005; Ohnishi et al. 2008), the genetic engineering era (Kirchner and Tauch 2003), the systems biology era (Wendisch 2003; Wendisch et al. 2006; Takors et al. 2007; Ma et al. 2017) and the synthetic biology era (Wendisch 2014; Lee and Wendisch 2017b). For enabling access to alternative carbon sources the reader is referred to other reviews (Zahoor et al. 2012; Wendisch et al. 2016; Choi et al. 2018). Fermentation technologies have been reviewed elsewhere (D'Este et al. 2018).

In this chapter, we provide an overview of individual metabolic engineering unit operations on the one hand. A number of successful examples of systems metabolic engineering are presented on the other hand. *C. glutamicum* has been engineered for the production of a plethora of compounds (Becker and Wittmann 2015). The product classes proteins (Freudl 2017; Liu et al. 2017), terpenoids (Heider et al. 2014a, b) and alcohols (Becker et al. 2018a, b; Eikmanns and Blombach 2014) are not covered here. Metabolic engineering advances and prospects for amino acid production have been reviewed recently (Wendisch 2019). The examples of systems metabolic engineering that are given here for illustration focus on diamines, ω-amino acids, dicarboxylic acids, aromatic compounds and alkylamines (Fig. 1).

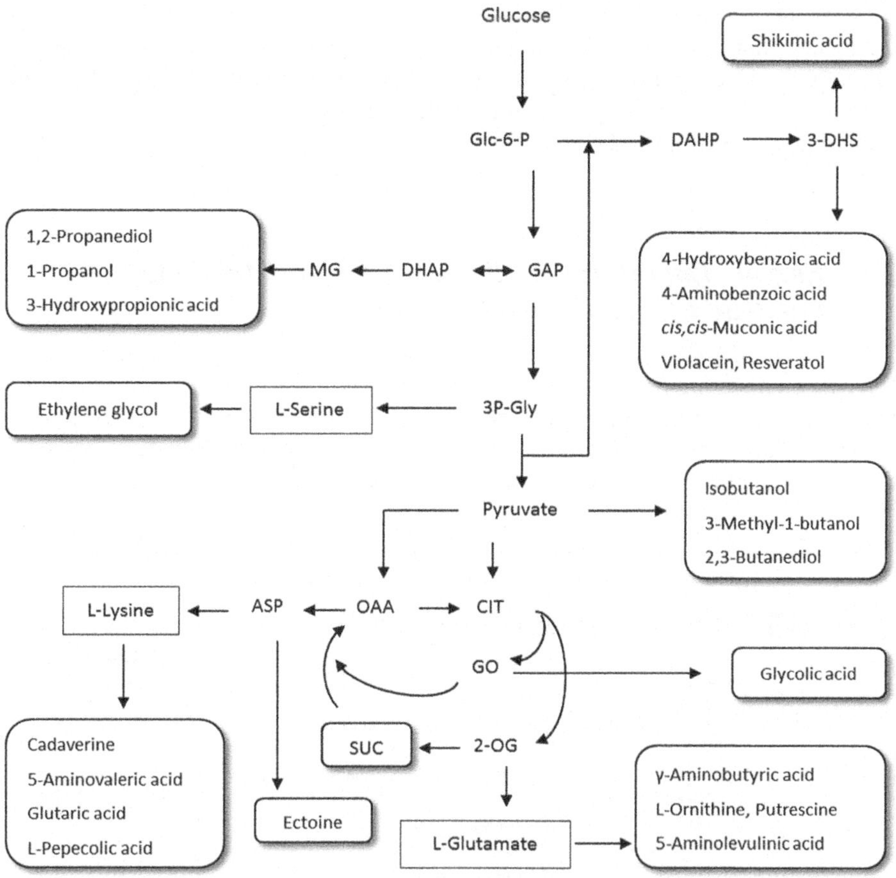

Fig. 1 Schematic representation of bio-based synthetic routes in *C. glutamicum* for the production of diamines, dicarboxylic acids, alcohols, non-proteinogenic amino acids, and aromatic compounds. Abbreviations are as follows. *ASP* aspartate, *CIT* citrate, *DAHP* 3-deoxy-D-arabinoheptulosonate-7-phosphate, *DHAP* dihydroxyacetone 3-phosphate, *3-DHS* 3-dehydroshikimate, *GAP* glyceraldehyde 3-phosphate, *3P-Gly* 3-phosphoglycerate, *GO* glyoxylate, *MG* methylglyoxyal, *OAA* oxaloacetate, *2-OG* 2-oxoglutarate, *SUC* succinate

2 Individual Strategies of Metabolic Engineering: Examples Applied to *C. glutamicum*

Systems metabolic engineering combines several individual engineering strategies for strain optimization. Each product requires a dedicated overall strategy, however, the process can be deconvoluted to recurring unit operations of metabolic engineering (Fig. 2). In the following we describe the most common of them and provide examples of their use in systems metabolic engineering. Where relevant, current technological developments are presented.

2.1 Enzyme Engineering: Alleviating Enzyme Inhibition, Catalytic Efficiency, and Cofactor Use

Enzyme engineering can be used as an important toolbox of metabolic engineering with regard to alleviating enzyme inhibition and changing catalytic efficiency or cofactor availability (Table 1) (Chen and Zeng 2016). Allosteric regulation is one of the basic mechanisms that control almost all cellular metabolisms, and thus

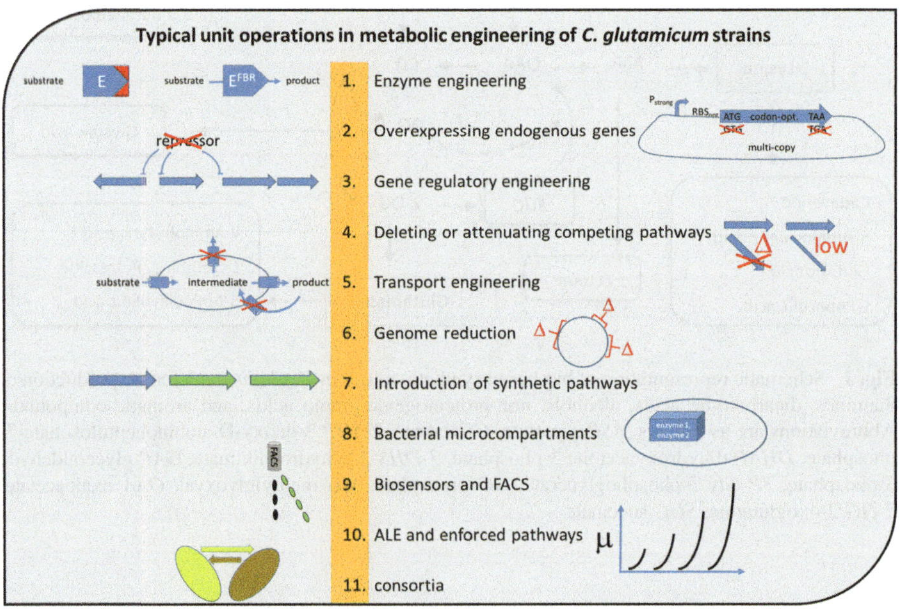

Fig. 2 Overview of the individual engineering approaches used in systems metabolic engineering. Blue arrows represent endogenous genes, green arrows represent genes of different origin. Deletions are indicated by red crosses. Feedback inhibition of an enzyme (given as block arrow) is indicated by red triangles representing ligands

Table 1 Representative examples of enzyme engineering approaches in metabolic engineering of amino acids and derivatives

Enzyme	Source	Mutations	Effect	Production	Reference
Aspartokinase	*C. glutamicum*	Q298G	Release of feedback inhibition by L-lysine and L-threonine	58 g L^{-1} L-lysine	Chen et al. (2011)
Phosphoenol-pyruvate carboxylase	*C. glutamicum*	N917G	Release of feedback inhibition by L-aspartate and malate	37% improved L-lysine production	Chen et al. (2014)
Homoserine kinase	*C. glutamicum*	A20G	Release of feedback inhibition by L-threonine	L-Threonine	Petit et al. (2018)
ATP phosphoribosyl transferase	*C. glutamicum*	N215K/L231F/T235A	Release of feedback inhibition by histidine 37-fold increase in K_i value	0.15 mM L-histidine	Zhang et al. (2012)
Homoserine dehydrogenase	*C. glutamicum*	Various mutations	Feedback inhibition by L-lysine but not by L-threonine	L-Lysine	Chen et al. (2015)
Chorismate pyruvate lyase	*E. coli*	L31A, D33A	Release of product inhibition by 4-hydroxybenzoate	55% improved 4HBA production	Han et al. (2016); Purwanto et al. (2018)
Glyceraldehyde 3-phosphate dehydrogenase	*C. glutamicum*	D35G/L36R/P192S	Similar catalytic efficiency towards NAD and NADP	60% improved L-lysine production	Bommareddy et al. (2014)
Glutamate decarboxylase	*E. coli*	Glu89Gln/*Δ452-466*	Expending the activity range to neutral pH	38 g L^{-1} of GABA	Choi et al. (2015b)
Glutamate decarboxylase	*Lactobacillus brevis*	T17I/D294G/E312S/Q346H	Expending the activity range to neutral pH	63.9% increased γ-aminobutyrate production	Thu Ho et al. (2013)
2,5-Diketo-D-gluconic acid reductase	*C. glutamicum*	R238H	7-fold increase in activity with NADH	Vitamin C	Banta et al. (2002)

deregulation of allosteric inhibition by enzyme engineering has been implemented for overproduction of amino acids, carboxylic acids, amines, and alcohols in *C. glutamicum* (Ohnishi et al. 2002; Sindelar and Wendisch 2007). Feedback-resistant enzymes, including aspartokinase (Chen et al. 2011), phosphoenolpyruvate carboxylase (Chen et al. 2014), homoserine kinase (Petit et al. 2018), and ATP phosphoribosyltransferase (Zhang et al. 2012) from *C. glutamicum*, were developed through protein engineering approaches based on statistical coupling analysis of protein sequences, structural analysis and site-specific mutation, or homology model and mutational analysis. The engineered enzymes showed dramatically decreased feedback inhibition by intermediates or end-products, and the resulting strains yielded improved L-lysine or L-histidine production. Enzymes showing a product inhibition are also considered as major targets for metabolic engineering (Purwanto et al. 2018). Chorismate-pyruvate lyase (CPL) from *E. coli* is strongly inhibited by its product, 4-hydroxybenzoate (4HBA). On the basis of a structural analysis and molecular dynamics simulations of CPL, two key residues, L31A and D33A, were identified for increasing the conformational dynamics of the flaps and thereby facilitating the product release (Han et al. 2016). *In vivo* result using *C. glutamicum* showed 55% increase of 4HBA production by introduction of the L31A mutant (Purwanto et al. 2018). Another issue is knockout or knockdown of related genes in metabolic engineering could lead to auxotrophic mutants and impaired cell growth, and thus up- and down-regulation of fine-tuned proteins are desirable for constructing well-tailored engineered strains. Homoserine dehydrogenase (HD) of *C. glutamicum* is naturally allosterically regulated by L-threonine and L-isoleucine and is a key regulation enzyme in L-lysine biosynthesis. As a proof of concept, Chen et al. (2015) attempted to design an L-lysine-sensitive HD that can use L-lysine as a signal molecule to control its activity by structural alignment and docking simulation and verified by mutagenesis experiments. The reengineered HD only responds to L-lysine inhibition but not to L-threonine.

Protein engineering can be applicable for optimizing a reconstructed artificial pathway with enhanced catalytic efficiency. In one example for γ-aminobutyrate (GABA) production in *C. glutamicum*, glutamate decarboxylases (GAD) derived from *E. coli* or lactic acid bacteria exhibit optimum activity at pH 4.0–5.0 and decreased activity at pH 7.0 (Choi et al. 2015b; Thu Ho et al. 2013). To broaden the activity range of GADs, mutant GADs from *E coli* and *Lactobacillus brevis* were obtained by protein engineering, leading to increased accumulation of GABA in *C. glutamicum* (Shi et al. 2014). Cofactors are involved in numerous intracellular reactions, so redox balance plays an important role in the biosynthesis of amino acids, alcohols, *etc* (Wang et al. 2017b). One promising approach to supply reducing power is modification of cofactor preferences by protein engineering. Bommareddy et al. (2014) succeeded in generating a *de novo* NADPH generation pathway by altering the coenzyme specificity of a native NAD-dependent glyceraldehyde 3-phosphate dehydrogenase (GAPDH) from *C. glutamicum* to NADP by rational protein engineering, yielding 60% improvement of L-lysine production in comparison to the control strain. As a reverse situation, the preference for NADPH over NADH of the *Corynebacterium* 2,5-diketo-D-gluconic acid (2,5-DKG) reductase was engineered, resulting in construction of a 2,5-DKG mutant showing a sevenfold increase in activity with NADH compared to that of the wild-type enzyme (Banta et al. 2002). This mutant 2,5-DKG reductase may be a valuable new catalyst for use in industrial production of vitamin C. To overcome a redox imbalance in L-valine production under oxygen-deprivation conditions, the coenzyme requirement for L-valine synthesis was converted from NADPH to NADH using *Lysinibacillus sphaericus* leucine dehydrogenase in place of endogenous transaminase B and by modification of nicotinamide coenzyme-binding site of acetohydroxy acid isomeroreductase (Hasegawa et al. 2012).

2.2 Genetic Engineering: Overexpression of Genes of Endogenous Pathway Enzymes

Overexpressing endogenous genes can be done by increasing the copy number using expression from a plasmid, by increasing transcription initiation using stronger and/or inducible promoters (Rytter et al. 2014), or by increasing translation efficiency using stronger ribosome binding sites (Zhang et al. 2015), using ATG instead of less preferred translational start codons and using TAA instead of less preferred translational stop codons. For example, the genome-reduced L-lysine producer strain GRLys1 has two copies of several genes relevant for L-lysine overproduction (2 copies each of *lysC*^{T311I}, *asd, dapA, dapB, ddh, lysA,* and *lysE*; Unthan et al. 2015). L-Glutamate biosynthesis was improved by modulating glutamate dehydrogenase gene *gdh* promoter activity (Asakura et al. 2007). For example, to improve provision of L-glutamate for L-ornithine production, four different −10 and/or −35 sequences of the *gdh* promoter were introduced upstream of the chromosomal *gdh* and compared. The highest specific activity for glutamate dehydrogenase was observed with “TATAAT” as −10 consensus sequence and “TTGCCA” as −35 sequence. The 4.5-fold increased glutamate dehydrogenase activity resulted in an about twofold increased L-ornithine production (Jensen et al. 2015). Replacing the less preferred translational stop codon TGA with TAA for plasmid based expression of the *Pseudomonas putida* ornithine cyclodeaminase gene *ocd* improved L-ornithine-based L-proline production fourfold (Jensen and Wendisch 2013).

2.3 Gene Regulatory Engineering: Derepression, Constant Activation and On-Demand Control

Gene regulatory engineering is often used to derepress a biosynthetic pathway (Pérez-García and Wendisch 2018). For example, L-arginine biosynthesis genes can be derepressed by deletion of the *arg* operon repressor gene *argR* (Ikeda et al. 2009). Another example is the deletion of *ltbR*, which codes for the IclR-type repressor of L-leucine and L-tryptophan biosynthesis genes (Brune et al. 2007), to derepress L-leucine biosynthesis (and tryptophan biosynthesis) genes for overproduction of L-leucine (Vogt et al. 2014) or 2-ketoisocaproate (Vogt et al. 2015; Bückle-Vallant et al. 2014). Overexpression of activator protein RamA (Auchter et al. 2011) improved dimethylitaconate production (Joo et al. 2017).

Overexpression of sigma factor genes has altered transcription globally and in the case of *sigA* overexpression improved carotenoid production (Taniguchi et al. 2017), while riboflavin overproduction resulted from *sigH* overexpression (Taniguchi and Wendisch 2015; Toyoda et al. 2015). On-demand control of biosynthetic genes has been achieved using riboswitch-mediated regulation of translational initiation of L-lysine export gene *lysE* (ON mode; Zhou and Zeng 2015a) and citrate synthase gene *gltA* (OFF mode; Zhou and Zeng 2015b).

2.4 Deleting or Attenuating Competing Pathways Incl. CRISPR Techniques

For production of 5-aminovalerate, its conversion by the GABA degradation pathway had to be avoided. To this end, either the gene *gabT* encoding GABA transaminase was deleted alone (Shin et al. 2016; Rohles et al. 2016) or the *gabTDP* operon was deleted as a whole (Jorge et al. 2017a). Two-step gene deletion by homologous recombination is firmly established in *C. glutamicum* strain development. However, metabolic engineering of *C. glutamicum* has embraced CRISPR technology as first shown for CRISPR interference and gene repression using CRISPRi/dCas9 to improve L-lysine and L-glutamate production (Cleto et al. 2016). Meanwhile, genome editing applications using CRISPR-Cpf1 (Jiang et al. 2017) and CRISPR-Cas9 (Cho et al. 2017) including multiplexing (Wang et al. 2018) have been described.

Sometimes total abrogation of a competing pathway is not optimal. For instance, putrescine production from ornithine as precursor required deletion of ornithine transcarbamoylase gene *argF*, but rendered the putrescine producer strain L-arginine auxotrophic (Schneider and Wendisch 2010). This limitation could be overcome by low-level expression of *argF* from a plasmid rendering the resulting strain bradytrophic for L-arginine (Schneider et al. 2012). Attenuation of competing pathways was also achieved by using alleles encoding less active enzymes as is the case for hom^{V59A} which led to an L-homoserine leaky phenotype with just enough flux towards L-homoserine and L-threonine avoiding L-threonine auxotrophy, but ample provision of L-aspartate semialdehyde for L-lysine production (Ikeda et al. 2006). 2-Oxoglutarate dehydrogenase has been attenuated to increase L-glutamate production (Schultz et al. 2007) or to increase provision of L-glutamate for production of L-arginine, L-ornithine, L-citrulline, putrescine, GABA, and L-proline (Park et al. 2014; Jensen et al. 2015; Nguyen et al. 2015a; Jorge et al. 2017b). To reduce flux via 2-oxoglutarate dehydrogenase in the TCA cycle either the promoter of the gene for 2-oxoglutarate dehydrogenase subunit E1 *odhA* was weakened, a less preferred translational start codon of *odhA* was used or the 2-oxoglutarate dehydrogenase inhibitory protein OdhI was modified. Changing threonine residues 14 or 15 of OdhI to alanine reduced phosphorylation of OdhI by serine/threonine protein kinases maintaining low 2-oxoglutarate dehydrogenase activity (Niebisch et al. 2006).

2.5 Transport Engineering

Transport engineering aims either to improve product export or substrate uptake or to avoid loss of intermediates by excretion (Pérez-García and Wendisch 2018). These application strategies are based on the discovery and biochemical characterization of proteins for substrate uptake and product excretion (Marin and Krämer 2007; Eggeling 2017). First, L-lysine export catabolized by LysE (Vrljic et al. 1996) is discussed in two aspects: improving production of L-lysine as final product and avoiding export of L-lysine as intermediate to improve production of L-lysine derived products. L-Lysine productivity was improved in genome reduced strains by a second chromosomally encoded copy of *lysE* (Unthan et al. 2015; Henke et al. 2018). On the other hand, deletion of *lysE* improved production of L-lysine derived products such as cadaverine (Kind et al. 2014), 5-aminovalerate (Rohles et al. 2016) and L-pipecolic acid (Pérez-García et al. 2017). As a third example, import of hexosamines by phosphoenolpyruvate:carbohydrate phosphotransferase systems (PTS) is presented. The hexosamines glucosamine and *N*-acetylglucosamine (GlcNAc) are subunits of chitin, and *N*-acetylmuramic acid (MurNAc) and GlcNAc are subunits of cell wall peptidoglycan. *C. glutamicum* imports glucosamine *via* a side activity of the glucose-specific PTS component PtsG (Uhde et al. 2013). For uptake of GlcNAc an exogenous PTS was needed: the GlcNAc-specific PTS NagE from *Corynebacterium glycinophilum* (Matano et al. 2014). Finally, uptake of MurNAc required MurNAc-specific PTS component MurP from *E. coli*. When general PTS component Crr from *E. coli* was present in addition, faster growth with MurNAc resulted (Sgobba et al. 2018a).

2.6 Genome Reduction

Most bacterial cells are highly adaptable to numerous changing environments, because they contain many genes functioning under certain conditions, but some bacteria living in restricted environments such as *Mycoplasma* possess reduced genomes (Becker et al. 2018b; Hutchison et al. 2016; Lee and Wendisch 2017b). Various bio-based compounds and proteins are now being produced from bacteria in stable bioreactor environments, in which genes encoding proteins regarding a number of metabolic, regulatory, and stress response properties are unnecessary or sometimes undesirable (Baumgart et al. 2018). Taken together, complexity of cellular network and regulation and genetic instability are another challenging field in metabolic engineering of production hosts (Mampel et al. 2013). A chassis strain harboring relevant gene sets together with essential ones which can grow fast on defined medium (Unthan et al. 2015) is considered to be a reasonable host for providing robust property and a broad range of biotechnological applications (Baumgart et al. 2018).

Mobile genetic elements, including prophages and insertion sequences elements, are the key factors in genome rearrangements and genetic instability and have important roles in the adaptation of cells to environment by inversion, recombination, and insertion (Choi et al. 2015a). Since three prophage loci exist in the genome of *C. glutamicum* ATCC 13032, the elimination of prophages sequences was carried out by Baumgart et al. (2013), leading to the construction of a prophage-free MB001 strain with a genome reduced by 6%. MB001-based strains have been extensively engineered for the production of carotenoids, e.g. lycopene, decaprenoxanthin, and astaxanthin (Heider et al. 2014a, b; Henke et al. 2016, 2017; Taniguchi et al. 2017), and amino acids, e.g. L-citrulline (Eberhardt et al. 2014), L-ornithine, L-arginine, and L-proline (Jensen et al. 2015). All genes of *C. glutamicum* were evaluated based on their essentiality, and found that 26 gene clusters deleted from the strain MB001 (*ΔCGP123ΔISCg12*) were irrelevant for the biological fitness of *C. glutamicum* on defined medium (Unthan et al. 2015). The same approach was also conducted using the prophage-free L-lysine producer (DM1933*ΔCGP123*), showing that titer, yield, and productivity were not negatively influenced by the deletions (Pérez-García et al. 2016). The resulting strain was additionally manipulated to the production of L-lysine and L-pipecolic acid (Pérez-García et al. 2016, 2017a). In particular, the engineered GRLys strain led to increased volumetric and specific productivities of L-lysine by 18–20% and 70–72%, respectively, compared to the parental strain. Another potential factor causing genetic rearrangements of genome and recombinant plasmid in *C. glutamicum* is two major IS elements (Choi et al., 2015a). To minimize the detrimental effects by these, two IS element free strains were created, and evaluated their deletion effect on the production of GFP (green fluorescence protein). The production performance of poly(3-hydroxybutyrate) and γ-aminobutyrate was also increased by 38% and 15%, respectively, in WJ008 strain (IS*Cg2*-cured) compared to that of wild-type strain. Very recently, Baumgart et al. (2018) proceeded with combinatorial deletions of irrelevant gene clusters obtained from the previous study (Unthan et al. 2015) using CR099 strain (ATCC 13032*ΔCGP123ΔISCg1ΔISCg2*), yielding a library of 28 strains. The final chassis strain C1∗ reduced the genome by 13.4% (412 deleted genes) from the wild-type strain and exhibited a similar growth behavior on defined medium.

2.7 Introduction of Synthetic Pathways by Heterologous Gene Expression

The production of diverse native and non-native bio-products is heavily explored in many microbes, which are necessary to introduce enzymes realizing artificial *de novo* biosynthetic pathways as well as optimize the system efficiency (Chen and Zeng 2016). Enzymes that provide significant variances of properties, e.g. optimum pH/temperature, substrate/cofactor preferences, enzyme regulations, and kinetics and stability, can be obtained from diverse sources. Hence, the heterologous expression of biosynthetic gene clusters enables the production of value-added natural products derived from plants and other microbes, along with a variety of compounds derived from L-amino acids, organic acids, alcohols, and so on (Kallscheuer et al. 2017; Ma et al. 2017; Oh et al. 2015).

C. glutamicum strains have been successfully engineered for the production of bio-products through glycolysis-, L-glutamate-, L-aspartate-, and aromatic amino acids-derived pathways by heterologous expression of key genes which are lacking in *C. glutamicum*, together with by modulating endogenous genes (Fig. 1) (Becker et al. 2018b). As representative examples, recombinant cells producing L-aspartate-derived bio-products, including cadaverine, 5-aminovalerate, glutarate, L-pipecolic acid, and ectoine, were developed from L-lysine-producing *C. glutamicum* by introduction of synthetic pathways (Fig. 4). Most corresponding genes not present in *C. glutamicum* are taken from other microbial sources with original or codon-optimized form. The *cadA* or *ldcC* genes encoding lysine decarboxylase isozyme from *E. coli* was employed for the synthesis of cadavarine from an L-lysine producing strain (Mimitsuka et al. 2007; Oh et al. 2015). 5-Aminovalerate (5AVA) biosynthesis was reconstructed through two alternative branch pathways from L-lysine *via* either 5-aminovaleramide or cadaverine. The former is comprised of expression of *davB*/*davA* genes encoding lysine 2-monooxygenase/5-aminovaleramidase from *Pseudomonas putida* (Shin et al. 2016), and the latter is comprised of expression of *ldcC* from *E. coli* and *patA*/*patD* genes encoding putrescine transaminase/γ-aminobutyraldehyde dehydrogenase from *E. coli* (Jorge et al. 2017a). The production of cyclic amino acid, L-pipecolic acid, was accomplished by introduction of *lysDH* from *Silicibacter pomeroyi* combined with an endogenous *proC* expression (Pérez-García et al. 2016), whereas the production of another cyclic amino acid, ectoine, was carried out by heterologous expression of ectoine operon *etcABC* from *Chromohalobacter salexigens* (Pérez-García et al. 2017b).

Although a *de novo* synthetic pathway was reconstructed by introduction of heterologous expression of genes, it can cause to unexpected physiology such as poor growth in *C. glutamicum* (Blombach et al. 2011). The reconstituted isobutanol producer consumed only small amounts of glucose due to probably a redox imbalance under oxygen depletion conditions (Blombach et al. 2011). To improve redox balance, *E. coli pntAB* encoding a membrane-bound transhydrogenase was additionally expressed. The resulting strain consumed glucose efficiently and produced 42 mM isobutanol. Similar strategies were accomplished in construction of L-lysine and succinate producers (Kabus et al. 2007; Wang et al. 2017a). The availability of NADPH pool is crucial role in L-arginine biosynthesis. Significant increase of L-arginine production in *C. glutamicum* was confirmed by expression of *ppnK* encoding NAD^+ kinase from *Corynebacterium crenatum* in combination with *E. coli pntAB* expression (Zhan et al. 2018).

2.8 Bacterial Microcompartments

Bacterial microcompartments (BMCs) are self-assembling proteinaceous organelles found in a broad range of bacteria (Lee et al. 2019). They include the anabolic carboxysomes and the catabolic metabolosomes depending on encapsulated enzymes which are surrounded by a semipermeable protein shell (Huber et al. 2017). Carboxysomes have encasing carbonic anhydrase and RuBisCO (ribulose 1,5-bisphosphate carboxylase/oxygenase) associated with the anabolic process of CO_2 fixation, enhancing carbon fixation in autotrophs. Metabolosomes found in heterotrophs are involved in the metabolism of various organic compounds, i.e. choline, ethanol, ethanolamine, fucose/rhamnose, propanediol, and amino-2-propanol utilization (Kerfeld and Erbilgin 2015). The function of BMCs allows the cell to confine toxic or volatile intermediates within a selectively permeable protein shell (Kerfeld et al. 2010), consequently leading to accelerating metabolic pathways, preventing wasteful by-reactions, and relieving cellular damage. Thus, BMCs have significant potential in application of industrial biotechnology to produce small molecules with toxic or volatile intermediates.

Application of BMC potential into metabolic engineering requires assembly of BMC shells in *C. glutamicum*. The α-carboxysomal gene cluster of *Halothiobacillus neapolitanus* was introduced into *C. glutamicum* (Baumgart et al. 2017). Though cell morphology and growth of *C. glutamicum* were slightly affected by expression of these gene clusters, BMC-like structures were observed inside cells. Moreover, encapsulation of eYFP into the carboxysomes was accomplished by fusion with the large RuBisCO subunit. This study represents that an α-carboxysomal gene cluster was first transferred to a Gram-positive model *C. glutamicum*. The Pdu operon from *Citrobacter freundii* consists of 21 genes that code for eight shell proteins (PduA, B, B′, J, K, N, U, T) along with for enzymes involved in 1,2-propanediol degradation and reactivation and recycling of cofactors. Recently, Huber et al. (2017) established *Cit. freundii* BMC in *C. glutamicum* by optimizing the expression of genetic clusters of shell components from Pdu operon. The targeted eYFP was encapsulated to the empty compartments through C-terminal fusions with synthetic scaffold interaction partners (PDZ, SH3, and GBD) as well as with a non-native C-terminal targeting peptide from *Klebsiella pneumonia* AdhDH. When PduA, one of shell proteins, was overproduced as N-terminal fusion form with the P18 and D18 encapsulation peptides from PduP and PduD, respectively, it assembled to form filamentous structures within the cytosol, generating localization of eYFP into these structures in *C. glutamicum*. These nanotube-like structures might have a great potential as scaffolds for directed cellular organization and pathway enhancement.

2.9 Biosensors and FACS

Cellular biosensors defined as sensors made by host cells that generate signals and can be recognized by the host cells to control their behavior or the behavior of heterologous pathways (Zhang and Keasling 2011). Of these, genetically encoded biosensors such as transcriptional regulators and riboswitches are sensing intracellular metabolite concentrations, regulating the expression of corresponding genes. Recently, fluorescence-activated cell sorting (FACS) technique which can detect fluorescence emitted from individual cells has been developed and applied for high-throughput screening of mutant strains. Thus, the combination of biosensor and FACS technique could greatly facilitate the implementation of systems metabolic engineering for industrial strains and processes (Ma et al. 2017).

Transcriptional regulators can be used as genetically encoded biosensors to direct gene expression responding to the intracellular metabolite concentration or to screen metabolite-accumulating mutants (Lee and Wendisch 2017b). The two amino acid-responsive transcriptional regulators, Lrp (Lange et al. 2011) and LysG (Bellmann et al. 2001), have been used to develop genetically-encoded biosensors in *C. glutamicum*. An Lrp-biosensor based on a transcriptional fusion of the *brnFE* promoter with a fluorescence reporter gene *eYFP* sensing increased intracellular concentration of branched-chain amino acids or L-methionine could be translated into fluorescence signal outputs and be able to sort and screen mutants using FACS (Mustafi et al. 2014). This system has not only applied for monitoring of live cell imaging and L-valine titer in engineered L-valine producing *C. glutamicum* (Mustafi et al. 2014) but also performed an adaptive laboratory evolution of L-valine producer (Mahr et al. 2015). An LysG-biosensor was constructed by fusion of the *lysE* promoter with a promoterless fluorescence reporter gene, thus enabling us to screen mutant libraries either expressing variants of *lysC*-encoded aspartokinase (Schendzielorz et al. 2014) or *murE*-encoded UDP-*N*-acetylmuramoy-L-alanyl-D-glutamate:meso-diaminopimelate ligase (Binder et al. 2013) for increased production of L-lysine. Another LysR-family transcriptional regulator ShiR in *C. glutamicum* senses the accumulation of shikimate and activates *shiA* gene encoding a shikimate transporter. This biosensor combined the *shiA* promoter with the *eGFP* gene, and which can be applicable for monitoring shikimate production (Liu et al. 2018). Cyclic adenosine monophosphate (cAMP) acts as an effector of transcriptional regulator GlxR in *C. glutamicum* (Kim et al. 2004). A GlxR-based cAMP biosensor was constructed by fusion of the promoter of *cg3195* gene encoding a putative flavin-containing monooxygenase with the *eyfp* gene (Schulte et al. 2017). The GlxR sensor allowed high-throughput screening of libraries for single mutant cells with an altered cAMP level.

Riboswitch is a regulatory part of an mRNA molecule that binds a metabolite and regulates the expression of downstream genes, and which is consisted of two parts: an aptamer and an expression platform (Zhou and Zeng 2015b). The structure of the expression platform is modulated by directly binding of the aptamer with metabolite, resulting in either typically turn off gene expression *via* transcription termination or turn it on (Becker et al. 2018b). A synthetic RNA device comprising a riboswitch module and a selection module for sensing a metabolite was developed in *E. coli* (Yang et al. 2013). As application, the L-lysine riboswitch modules from *E. coli* and *B. subtilis* were employed to regulate the expression of *gltA* and control TCA cycle activity in an L-lysine producing *C. glutamicum* strain, respectively (Zhou and Zeng 2015b). As an extension of this approach, a synthetic L-lysine-ON riboswitch was screened from mutant library of a natural L-lysine-OFF riboswitch from *E. coli* using *tetA*-based dual genetic selection (Zhou and Zeng 2015a). The selected L-lysine-ON riboswitch was combined with the *lysE* gene, resulting in achievement of dynamic control of L-lysine transport in an L-lysine producer. Recombinant cells harboring this system showed 89% increase of L-lysine yield compared to the control strain. Furthermore, the simultaneous introduction of the L-lysine-ON and -OFF riboswitches into the L-lysine producer achieved a 21% increase in the yield of L-lysine compared to that of cells harboring only L-lysine-OFF riboswitch.

2.10 ALE and Enforced Pathways

Gain-of-function mutations have been selected by what has recently been called Adaptive Laboratory Evolution (ALE). Survival depends on a metabolic trait that arises by mutations that are identified by genome re-sequencing, e.g. resistance to high methanol concentrations (Leßmeier and Wendisch 2015), oxidative stress (Lee et al. 2013) and heat stress (Oide et al. 2015), ability to import carbon and energy source such as dicarboxylic acids (Youn et al. 2008, 2009) and glucosamine (Uhde et al. 2013), faster catabolism of glucose (Pfeifer et al. 2017) and xylose (Radek et al. 2017). Coupled with a genetically encoded biosensor valine production was improved by ALE (Mahr et al. 2015).

Mutations have been introduced to constrain metabolism enforcing compensatory mutations. ALE identified compensatory mutations for fast glycolysis with a non-phosphorylating GAP dehydrogenase GapN accepting NADP instead of NAD (Komati Reddy et al. 2015; Takeno et al. 2016; Hoffmann et al. 2018). For example, compensatory mutations resided in non-proton-pumping NADH:ubiquinone oxidoreductase II. The identified amino acid change D213G, which was shared by related, but NADPH:ubiquinone oxidoreductase II enzymes, indeed led to NADPH oxidation. Oxidation of NADPH formed in the ATP-neutral glycolysis with NADP-dependent GapN by mutant NADPH-accepting NDH-II^{D213G} was required and, thus, by coupling to electron transport chain phosphorylation (Komati Reddy et al. 2015). Similarly, ALE identified gain-of-function mutants of ribose phosphate isomerase negative strains for methanol-dependent growth (Tuyishime et al.

2018), as has previously been shown for *E. coli* (Meyer et al. 2018). As recently established for *E. coli* (Lindner et al. 2018), ALE may offer a tool to optimize the rate of NADPH regeneration that often constraints biosynthetic pathways.

2.11 Synthetic Consortia

In nature, ecological niches are seldom dominated by a single microbial species, rather consortia prevail. Many food biotechnological processes have been established early by mankind and often involve microbial consortia, e.g. in cheese making. Metabolic engineering has focused mostly on optimization of microorganisms to be used in mono-culture. Only recently, metabolic engineering aimed at designing synthetic microbial consortia to "divide labor between two strains or species in a combined biotech process" (Sgobba and Wendisch 2020). Importantly, metabolic engineering principles are used rather than adapting natural consortia by evolutionary approaches (Sgobba and Wendisch 2020). Examples for such consortia are shown for production of fine chemicals such as resveratrol by *E. coli*-based consortia (Camacho-Zaragoza et al. 2016), paclitaxel precursors by *E. coli-S. cerevisiae* (Zhou et al. 2015b), or vitamin C production from either a recombinant *Erwinia herbicola* strain or by consortia consisting of *Gluconobacter oxydans* and *Ketogulonicigenium vulgare* (Wang et al. 2016). Besides these examples of sub-dividing a long biosynthetic pathway, other consortia contain feedstock degrading strains combined with producer strains, e.g. a co-culture of the cellulase-secreting fungus *Trichoderma reesei* and an isobutanol producing *E. coli* strain lignocellulosic feedstock-based isobutanol production (Minty et al. 2013).

Recently, synthetic *C. glutamicum-E. coli* consortia have been designed and their properties characterized to some details (Sgobba et al. 2018b). Two types of consortia were designed: one based on commensalism of an L-lysine auxotrophic, sucrose negative *E. coli* strain and an L-lysine producing *C. glutamicum* that can only catabolize the glucose moiety of sucrose due to deletion of the fructose importer gene *ptsF* and secretes fructose as by-product which in turn can be utilized by the *E. coli* strain (Sgobba et al. 2018b). Neither strain alone could grow in minimal medium with sucrose as sole carbon sources, but the consortium did. Moreover, the consortium produced a surplus of L-lysine (Sgobba et al. 2018b). The second *C. glutamicum-E. coli* consortium was designed to be mutualistic. An α-amylase secreting, L-lysine auxotrophic *E. coli* strain and a naturally amylase-negative, L-lysine overproducing *C. glutamicum* strain were combined and assayed for L-lysine production from starch. Neither strain in single culture grew with starch, whereas the consortium catabolized starch and produced L-lysine (Sgobba et al. 2018b). The process was scalable since a 25 L batch-fermentation with *C. glutamicum* strain CgLYS4(pVWEx1-crimson)(pECXT99A) and *E. coli* strain EcLYS1(pAmy)(pVWEx1-gfpuv) produced L-lysine from starch. Single cell analysis of the consortia was possible since the *E. coli* and *C. glutamicum* strains expressed different fluorescent proteins. From microscopic pictures it was evident that the *C. glutamicum* strain dominated the consortium likely indicating that starch hydrolysis requires few α-amylase secreting *E. coli* cells (Sgobba et al. 2018b). This work provided a proof-of-principle for starch based L-lysine production and it could be transferred to the starch-based production of the L-lysine derived value-added products L-pipecolic acid and cadaverine (Sgobba et al. 2018b). Performance of the consortium for starch-based production of L-lysine is superseded by using α-amylase secreting, L-lysine overproducing *C. glutamicum* strains (Seibold et al. 2006; Tsuge et al. 2013). Similarly, glucose-based production of L-lysine derived cadaverine, 5-aminovalerate, L-pipecolic acid (Mimitsuka et al. 2007; Tateno et al., 2009; Jorge et al. 2017a; Pérez-García et al. 2016, 2017a, b) leads to higher titers, yields and productivities than these first consortia. Process intensification paired with strain optimization is required for competitive biotech process based on these synthetic consortia.

Very recently, a microfluidic co-cultivation system has been designed and characterized for the analysis of growth and interactions within microbial consortia at single-cell resolution (Burmeister et al. 2018). A commensalistic consortium of an L-lysine-producing *C. glutamicum* strain and an L-lysine auxotrophic mutant *C. glutamicum* strain was co-cultivated, but physically separated. It could be shown that the auxotrophic mutant grew due to secreted L-lysine supplied by the L-lysine-producing strain (Burmeister et al. 2018). It can be expected that microfluidic co-cultivation systems will be instrumental for process intensification to obtain efficient biotech processes based on synthetic consortia.

3 Systems Metabolic Engineering of *C. glutamicum*

3.1 Systems Metabolic Engineering for Production of Diamines

Systems metabolic engineering approaches to the production of C4 and C5 diamines have been described for *C. glutamicum* (Wendisch et al. 2018a, b), while the C3 diaminopropane has only been produced by recombinant *E. coli* (Chae et al. 2017). Putrescine production by *C. glutamicum* has first been described by Schneider and Wendisch 2010 (Fig. 3). Putrescine production was improved by using a plasmid addiction system (Schneider et al. 2012), eliminating putrescine *N*-acetylation (Nguyen et al. 2015b) and systems metabolic engineering (Nguyen et al. 2015a; Jensen et al. 2015), and by repressing *carB, ilvH, ilvB* and *aroE* expression (Li et al. 2018). Putrescine has been produced from alternative feedstocks (Meiswinkel et al. 2013a, b).

In the following, the systems metabolic engineering approach to cadaverine production is described (Fig. 4). The metabolic networks of L-lysine producing *C. glutamicum* strains has been extended with L-lysine degradation pathways for production of L-lysine-derived value-added compounds such as L-PA (Pérez-García et al. 2016), 5-aminovalerate (Rohles et al. 2016; Jorge et al. 2017a) and glutarate

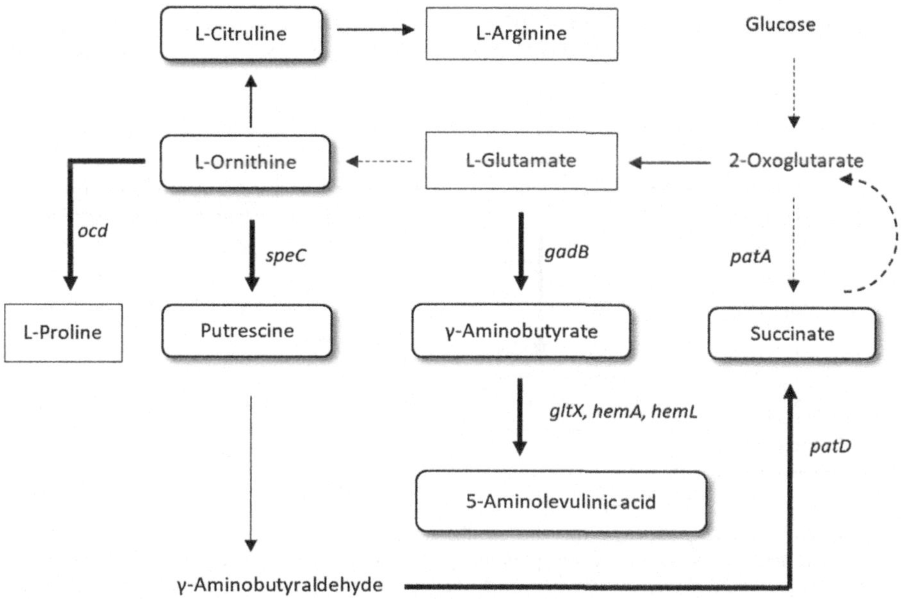

Fig. 3 Biosynthetic pathways for the synthesis of putrescine, γ-aminobutyrate (GABA), succinate, 5-aminilevulinic acid, L-citrulline and L-proline. Corresponding gene and proteins are as follows. *argF*, ornithine transcarbamoylase; *gadB*, glutamate decarboxylase from *E. coli*; *gdhA*, glutamate dehydrogenase; *ocd*, ornithine cyclodeaminase from *P. putida*; *odh*, 2-oxoglutarate dehydrogenase; *patA*, putrescine transaminase from *E. coli*; *patD*, γ-aminobutyraldehyde dehydrogenase from *E. coli*; *speC*, ornithine decarboxylase from *E. coli*. Bold line and bold dashed line mean a single-step and multi-steps, respectively, as well as heterologous gene(s) expression. Thin line means endogenous expression of gene(s)

(Rohles et al. 2016; Pérez-García et al. 2018). Of these, metabolic engineering of cadaverine production has been described first (Mimitsuka et al. 2007). The linear diamine cadaverine is a monomeric precursor of polyamides and can be polymerized with sustainably produced succinate or sebacic acid to yield nylon-5,4 and nylon-5,10, respectively (Mimitsuka et al. 2007; Kind et al. 2014; Oh et al. 2015; Matsushima et al. 2016). The advantages of converting L-lysine to cadaverine arise from the fact that a single enzyme, lysine decarboxylase, is needed and that the catalyzed reaction is a decarboxylation shifting the equilibrium to cadaverine and carbon dioxide, which leaves the fermenter in the gas phase. A biocatalytic process of cadaverine production from L-lysine using catalytically active inclusion bodies of lysine decarboxylase from *E. coli* has been described recently (Kloss et al. 2018). Either lysine decarboxylase CadA (Mimitsuka et al. 2007) or LdcC (Kind et al. 2014) from *E. coli* was used, the latter being preferred since its pH optimum of pH 7.6 is closer to the physiological intracellular pH of *C. glutamicum* (Kelle et al. 1996). Codon optimization of *ldcC* and expression from the constitutive promoter pTuf produced cadaverine with yield of 0.3 mol mol^{-1} when pyridoxal was added to

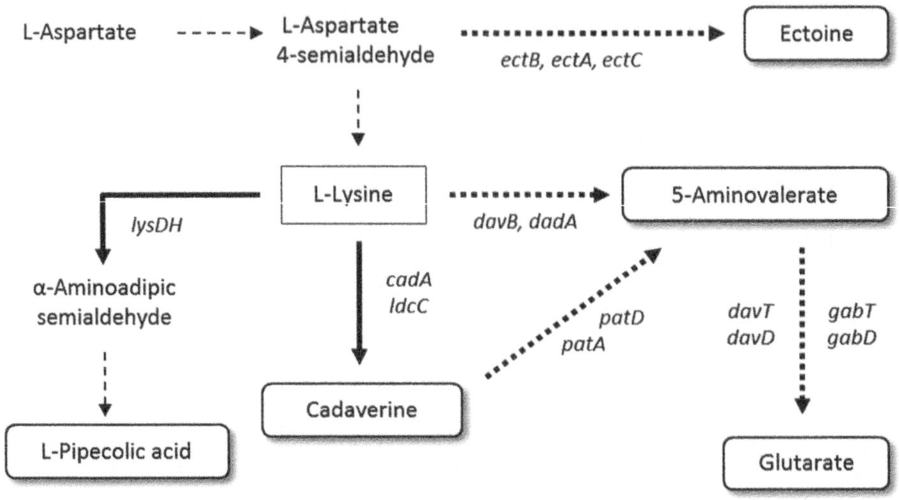

Fig. 4 Biosynthetic pathways for the synthesis of cadaverine, 5-aminovalerate, glutarate, L-pipecolic acid, and ectoine. Corresponding gene and proteins are as follows. *cadA*, lysine decarboxylase from *Escherichia coli; davA,* lysine 2-monooxygenase from *P. putida*; *davB,* 5-aminovaleramidase from *Pseudomonas putida*; *gabD,* glutarate semialdehyde dehydrogenase from *P. stutzeri*; *ectA*, L-2,4-diaminobutyric acid acetyltransferase from *Chromohalobacter* salexigens; *ectB*, L-2,4-diaminobutyric acid transaminase from *Ch. salexigens*; *ectC*, L-ectoine synthase from *Ch. Salexigens; gabT*, 5-aminovalerate transaminase from *Pseudomonas stutzeri*; *ldcC,* the second lysine decarboxylase from *E. coli*; *lysDH,* lysine dehydrogenase from *Silicibacter pomeroyi*; *patA,* putrescine transaminase from *E. coli*; *patD,* γ-aminobutyraldehyde dehydrogenase from *E. coli*. Bold line and bold dashed line mean a single-step and multi-steps, respectively, as well as heterologous gene(s) expression. Thin line means endogenous expression of gene(s)

the cultivation medium (Kind et al. 2014). The by-product *N*-acetyl-cadaverine was formed in considerable concentration due to *N*-acyltransferase SnaA, which acetylates a number of diamines including cadaverine (Nguyen et al. 2015b). Deletion of *snaA* and of the L-lysine exporter gene *lysE* (Vrljic et al. 1996) avoided formation of *N*-acetyl-cadaverine and L-lysine as by-products (Kind et al. 2014). Overexpression of *cgmA*, the gene of an L-arginine excretion protein which likely is active also with the diamines putrescine and cadaverine (Nguyen et al. 2015b; Lubitz et al. 2016) led to a strain producing cadaverine to 88 g L^{-1} in fed-batch culture without L-lysine-derived by-products. Besides glucose, production of cadaverine from alternative carbon sources has been described, e.g. from the hemicellulose pentose xylose (Buschke et al. 2011, 2013), from xylooligosaccharides due to expression of β-xylosidase *BSU17580* from *B. subtilis* as fusion to the PorH membrane anchor (Imao et al. 2017). A proof-of-concept was shown for cadaverine production by *C. glutamicum* from methanol (Leßmeier et al. 2015), but relevant production from methanol required the use of a natural methylotrophic bacterium (Naerdal et al. 2015). Starch-based cadaverine production has been achieved by synthetic consortia (Sgobba et al. 2018b).

3.2 Systems Metabolic Engineering for Production of Non-proteinogenic ω-Amino Acids

Systems metabolic engineering of *C. glutamicum* enabled production of the non-proteinogenic ω-amino acids γ-aminobutyrate (GABA) and 5-aminovalerate (5AVA). Both monomeric ω-amino acids can be lactamized (Chae et al. 2017) and converted to the polyamides nylon-4 and nylon-5, respectively.

Decarboxylation of L-glutamate yields GABA (Fig. 3), and pyridoxal phosphate-dependent glutamate decarboxylase is part of the acid stress response of enteric and lactic acid bacteria. Biotransformation of L-glutamate to GABA with a final GABA titer of around 1.1 M can be achieved with calcium alginate-entrapped *L. brevis* cells with high GAD activity (Li et al. 2010). Glucose-based GABA production became possible by systems metabolic engineering of *C. glutamicum* with plasmid-borne expression of *L. brevis*-derived glutamate decarboxylase gene (Shi and Li 2011). Alternatively, glutamate decarboxylase from *E. coli* was used (Takahashi et al. 2012) or two *Lactobacillus* enzymes co-produced (Shi et al. 2013). Manipulation of the 2-oxoglutarate dehydrogenase complex activity by deletion of the gene for serine/threonine protein kinase led to GABA production of about 31 g L^{-1} (Okai et al. 2014). To accommodate the discrepancy between optimal pH of the glutamate decarboxylase enzymes and the intracellular physiological pH of *C. glutamicum* glutamate decarboxylase enzymes were engineered (Shi et al. 2014) and/or a two-phase process was used: first, cell growth and L-glutamate production at neutral pH followed by GABA synthesis during stationary phase at acidic pH (Choi et al. 2015b). Gene expression was improved by altering ribosome binding and promoter sequences (Shi et al. 2018) and strain optimization made use of IS element deletions and combinatorial gene deletions (Choi et al. 2015a, 2017).

A novel pathway for GABA production has been established (Jorge et al. 2016, 2017b). In this pathway (Fig. 3), putrescine is converted to γ-aminobutyraldehyde by putrescine transaminase (PatA) followed by oxidation of γ-aminobutyraldehyde to GABA by γ-aminobutyraldehyde dehydrogenase (PatD) (Jorge et al. 2016). GABA production *via* the putrescine-based pathway led to GABA titers of about 63 g L^{-1} in fed-batch cultivation at the highest volumetric productivity for fermentative GABA production reported to date (1.34 g L^{-1} h^{-1}; Jorge et al. 2017b). GABA was produced by *C. glutamicum* from agro-wastes such as empty fruit bunches (Baritugo et al. 2018) and from carbon sources that do not have competing uses in the food or feed industries such as xylose, glucosamine, and *N*-acetyl-glucosamine (Jorge et al. 2017b).

Two independent pathways for 5AVA production by *C. glutamicum* have been established (Fig. 4) and both have been extended to glutarate production (Kim et al. 2019: Pérez-García et al. 2018). Since 5AVA is an intermediate of the so-called 5AVA pathway of L-lysine degradation in *Pseudomonas putida* (Park et al. 2014), genes for lysine 2-monooxygenase (DavB) and aminovaleramidase (DavA) were expressed in L-lysine producing *C. glutamicum* (Buschke et al. 2011). 5AVA production was increased upon reducing formation of glutarate as by-product by deletion of the GABA transaminase gene *gabT* (Shin et al. 2016; Rohles et al. 2016) and by deletion of *lysE* to avoid loss of L-lysine (Rohles et al. 2016). A titer of 28 g L^{-1} of 5AVA and a volumetric productivity of 0.9 g L^{-1} h^{-1} resulted (Rohles et al. 2016). Since this pathway involved a monooxygenase and since oxygen availability often limits aerobic fermentation processes, an alternative three-step pathway comprising lysine decarboxylase (LdcC), putrescine transaminase (PatA), and γ-aminobutyraldehyde dehydrogenase (PatD) from *E. coli* was developed (Jorge et al. 2017a). Very efficient 5AVA production in shake flasks to a 5AVA titer of 5.1 g L^{-1} with a volumetric productivity of 0.12 g L^{-1} h^{-1} was achieved (Jorge et al. 2017a). 5AVA was produced from alternative carbon sources, namely from *Miscanthus* hydrolysate (Joo et al. 2017) and starch, glucosamine, xylose and arabinose (Jorge et al. 2017a).

3.3 Systems Metabolic Engineering for Production of Dicarboxylic Acids

Dicarboxylic acids, including succinate, glutarate, and *cis,cis*-muconate (ccMA), can be used as most important bio-based building block chemicals having numerous potential applications (Ahn et al. 2016; Becker et al. 2018a; Wendisch et al. 2018a, b). Recent achievement in the production of dicarboxylic acids using recombinant *C. glutamicum* strains are highlighted in the following.

Succinate, a C4 dicarboxylic acid, is one of the most important platform chemicals and is used as surfactants, chelators, an additive food and pharmaceutical industries (Ahn et al. 2016). Cell growth of *C. glutamicum* is arrested in anaerobic condition, but the cells maintain the capability to metabolize glucose to organic acids, such as lactate, acetate, and succinate (Okino et al. 2008). With these features, bi-phasic culture process has been performed by achieving high cell density in aerobic growth phase and then succinate production in the second anaerobic production phase through reductive TCA cycle (Ahn et al. 2016; Chung et al. 2017). Some succinate production was conducted under aerobic conditions through oxidative TCA cycle and the glyoxylate shunt (Litsanov et al. 2012; Zhu et al. 2013). A variety of metabolic engineering strategies have been implemented to develop succinate producing *C. glutamicum*. These include eliminating the pathways for byproducts accumulation such as acetate and lactate (Litsanov et al. 2012; Zhu et al. 2013), increasing the anaplerotic pathways to improve the carbon flux to oxaloacetate *via* PEP or pyruvate (Chung et al. 2017; Jojima et al. 2016; Litsanov et al. 2012; Zhu et al. 2013), increasing glyoxylate shunt enzymes (Zhu et al. 2014), increasing the NADH redox potential (Xu et al. 2016; Zhou et al. 2015a), optimization of production process (Jojima et al. 2016), and utilization of renewable biomass from lignocellulosic hydrolysate (Mao et al. 2018). Moreover, the end-product succinate was known to inhibit the rates of glucose consumption and succinate production (Chung et al. 2017). Overexpression of NCgl0275 gene, which was downregulated by the addition of succinate, was found to suppress succinate inhibition, resulting in 37.7% increase in succinate production. Finally, the S071 strain *(ΔldhΔpta-ackAΔactAΔpoxBpyc*^{P458S}*Δpck_*P_{tuf}*::Ms.pckA*P_{tuf}*::ppcΔptsG)* overexpressing NCgl0275 gene was able to produce 152.2 g L^{-1} succinate with a yield of 1.67 mol mol^{-1} glucose using fed-batch fermentation, in which cells were initially aerobically grown during the growth stage and anaerobically cultivated during the production stage (Chung et al. 2017).

Glutarate, a C5 dicarboxylic acid, can be used as a building block for bio-based polymers such as polyesters, polyamides, and polyols (Becker et al. 2018b). Additionally, glutarate is converted to 1,5-pentanediol by hydrogenation, which is used as a plasticizer and precursor of polyesters. Two alternative routes to produce glutarate from L-lysine *via* 5AVA has been described (Kim et al. 2019; Pérez-García et al. 2018). The first route was accomplished by expressing *davB, davA, davT*, and *davD* encoding L-lysine 2-monooxygenase, 5-aminovaleramidase, 5-aminovalerate transaminase, and glutarate semialdehyde dehydrogenase, respectively, from *P. putida* in L-lysine producer (Kim et al. 2019). Final engineered strain could produce 24.5 g L^{-1} glutarate from glucose *via* L-lysine and 5AVA route on fed-batch fermentation. Second route comprised five enzymatic steps converting L-lysine to glutarate *via* cadaverine and 5AVA, and which consisted of L-lysine decarboxylase, putrescine transaminase, and γ-aminobutyraldehyde dehydrogenase encoded by *ldcA*, *patD*, and *patA*, respectively, from *E. coli* and GABA/5AVA amino transferase and succinate/glutarate semialdehyde dehydrogenase encoded by *gabT* and *gabD*, respectively, from *Pseudomonas stutzeri* (Pérez-García et al. 2018). Glutarate production by the final strain with additional deletion of *gdh* was achieved by fed-batch fermentation with a titer of 25 g L^{-1} and an overall yield of 0.17 g of succinate per g of glucose.

ccMA, a C6 unsaturated dicarboxylic acid, has highly attractive owing to its extensive industrial applications in chemicals, pharmaceuticals, functional resins, and agrochemicals as well as a precursor of adipic acid and terephthalic acid (Wendisch et al. 2018a, b). In *C. glutamicum*, ccMA production was attempted by converting 3-dehydroshikimate via endogenous 3-dehydroshikimate dehydratase QsuB, heterologous protocatechuate decarboxylase YBD and endogenous ring-opening catechol 1,2-dioxygenase CatA (Fig. 5) (Shin et al. 2018). By adopting non-PTS glucose uptake system, the final strain *(ΔaroEΔpcaGΔpcaHΔcatBΔptsIΔiolR)* was constructed and produced 4.5 g L^{-1} of ccMA with a yield of 0.22 mol mol^{-1} glucose. Lignin valorization to targeted chemicals is highly challenging, but attractive and important for the chemical industry (Wendisch et al. 2018a, b). Becker et al. (2018a) engineered *C. glutamicum* for high-level production of bio-based ccMA from lignin-derived aromatics and lignin hydrolysates. This system realized by stepwise metabolic engineering of *C. glutamicum* to convert the aromatics benzoic acid, catechol, and phenol into ccMA. Biotransformation of catechol into ccMA was carried out in a fed-batch process, yielding 85 g L^{-1} ccMA. Moreover, this system successfully converted lignin hydrolysates to ccMA with 1.8 g L^{-1} titer.

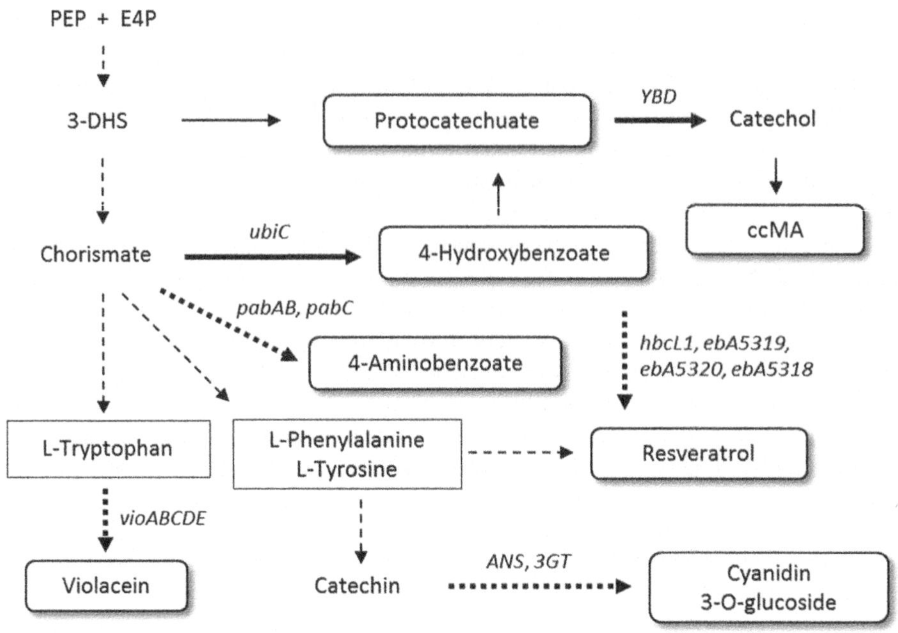

Fig. 5 Biosynthetic pathways for aromatics and derivatives. Abbreviations are as follows. *ccMA cis, cis*-muconic acid, *DAHP* 3-deoxy-D-arabinoheptulosonate-7-phosphate, *3-DHS* 3-dehydroshikimate, *E4P* erythrose 4-phosphate, *PEP* phosphoenolpyruvate. Corresponding genes and proteins are as follows. *ubiC*, chorismate-pyruvate lyase from *E. coli* or *Providencia rustigianii*; *YBD*, protocatechuate decarboxylase; *pabAB*, aminodeoxychorismate synthase from *Corynebacterium callunae*; *pabC*, 4-amino-4-deoxychorismate lyase from *Xenorhabdus bovienii*; *ANS*, anthocyanidin synthase from *Petunia hybrida*; *3GT*, 3-*O*-glucosyltransferase from *Arabidopsis thaliana; hbcL1*, *ebA5319*, *ebA5320*, and *ebA5318* from *Aromatoleum aromaticum* encoding 4-hydroxybenzoate:CoA ligase, β-ketothiolase, 3-hydroxyacyl-CoA dehydrogenase, and enoyl-CoA hydratase, respectively. Bold line and bold dashed line mean a single-step and multi-steps, respectively, as well as heterologous gene(s) expression. Thin line means endogenous expression of gene(s)

3.4 Systems Metabolic Engineering for Production of Aromatic Compounds

Aromatics represent a diverse class of chemicals with significance implications in the chemical, pharmaceutical, and food industries (Averesch and Krömer 2018). The shikimate pathway intermediates and derivatives enrich a wide potential for bio-replacements of fossil fuel-derived aromatics as well as naturally produced secondary metabolites. *C. glutamicum* has been as a prominent host strain for the production of aromatic amino acids including halogenated L-tryptophan (Veldmann et al. 2019a, b), 4-hydroxybenzoate (4HBA), 4-aminobenzoate (4ABA), and protocatechuate along with value-added natural compounds from plants and other microbes e.g. violacein, anthocyanin, and resveratrol (Fig. 5) (Averesch and Krömer 2018; Lee and Wendisch 2017a).

4HBA is a base material for antibacterial parabens used as food preservatives, a major building block for liquid crystal polymer, and a valuable platform intermediate for the production of commodity and fine chemicals (Lee and Wendisch 2017a). Kitade et al. (2018) and Purwanto et al. (2018) obtained *C. glutamicum* strains producing 36.7 g L^{-1} and 19 g L^{-1} of 4HBA, respectively, on a growth-arrested bioprocess and a fed-batch fermentation. Basically, both strains were constructed by heterologous production of feedback-resistant CPL from *Providencia rustigianii* and *E. coli*, respectively, and expression and deletion of endogenous key genes involved in the shikimate pathways. 4ABA is used in dyes, feed additives, and pharmaceuticals and has the potential to serve as a building block for aromatic polymers (Koma et al. 2014). Overexpression of *pabAB* from *Corynebacterium callunae* and *pabC* from *Xenorhabdus bovienii* were combined with a strain overexpressing the genes of the shikimate pathway (Kubota et al. 2016). The resulting strain produced 43 g L^{-1} of 4ABA titer. Violacein is a bis-indole pigment with anti-tumoral, anti-bacterial, and anti-oxidant activities (Sun et al. 2016). The *vioABCDE* operon from *Chromobacterium violaceum* was overexpressed in *C. glutamicum* to have obtained 532 mg L^{-1} violacein from glucose *via* L-tryptophan. Plant polyphenols, e.g. anthocyanin and resveratrol, have potential applications as nutraceuticals or pharmaceuticals (Kallscheuer et al. 2017). Recombinant *C. glutamicum* that could produce cyanidin 3-*O*-glucoside (C3G), one of anthocyanins, from catechin was organized by co-expression of *ANS* encoding anthocyanidin synthase from *Petunia hybrida* and *3GT* encoding 3-*O*-glucosyltransferase from *Arabidopsis thaliana* (Zha et al. 2018). Ammonia lyases connecting endogenous aromatic amino acid biosynthesis with phenylpropanoid synthesis proved to be the limiting step during microbial polyphenol production (Kallscheuer et al. 2017). To solve this limitation, Kallscheuer et al. (2017) recruited a reversal reaction of CoA-dependent, β-oxidative phenylpropanoid degradation pathway starting from 4-hydroxybenzoate. The synthetic pathway consisted of *hbcL1*, *ebA5319*, *ebA5320*, and *ebA5318* from *Aromatoleum aromaticum* encoding 4-hydroxybenzoate:CoA ligase, β-ketothiolase, 3-hydroxyacyl-CoA dehydrogenase, and enoyl-CoA hydratase, respectively, as well as *Petroselinum crispum 4cl* encoding 4-coumarate:CoA ligase and *Arachis hypogaea sts* encoding stilbene synthase. Consequently, 5 mg L^{-1} of resveratrol was accumulated from 4-hydroxybenzoate in a recombinant strain.

3.5 Systems Metabolic Engineering for Production of Alkylamines

In nature, *N*-alkylated amino acids are present in diverse biological molecules in bacteria, archaea and eukaryotes, for example in destruxin B and cyclosporine A (Wu et al. 2013). *N*-Alkylation of amino acids renders them more lipophilic.

Peptides with *N*-alkylated amino acids exhibit increased membrane permeability. A number of peptide-based drugs contain *N*-alkylated amino acids such as *N*-methylated tubulysin which showed higher antimitotic activity as compared to tubulysin (Patterson et al. 2008). In addition to a higher membrane permeability these peptidomimetics are less prone to proteolytic hydrolysis which may result in better pharmacokinetics (Di Gioia et al. 2016). Chemical synthesis of *N*-methylated amino acids is possible, but characterized by incomplete stereoselectivity and low yields.

Recently, two fermentative routes to *N*-methylated amino acids have been described (Mindt et al. 2018a, b). The fermentative production of about 18 g L^{-1} *N*-methylglutamate from glycerol and monomethylamine by recombinant *P. putida* expressing two genes of the catabolic monomethylamine pathway from methylotrophic *Methylobacterium extorquens* has been described (Mindt et al. 2018b). A potentially more versatile route involving the imine reductase (IRED) DpkA from *Pseudomonas putida* has been developed for *C. glutamicum* and about 32 g L^{-1} *N*-methyl-L-alanine were produced in fed-batch fermentation from glucose and monomethylamine (Mindt et al. 2018a). DpkA is part of the catabolic D-lysine degradation pathway in *P. putida* reducing piperideine-2-carboxylate to L-pipecolic acid, however, this IRED also methylaminates pyruvate to yield *N*-methyl-L-alanine (Mihara et al. 2005; Muramatsu et al. 2005). For efficient fermentative *N*-methyl-L-alanine production, *dpkA* had to be expressed in a pyruvate overproducing *C. glutamicum* strain. *C. glutamicum* strain ELB-P overproduced pyruvate due to systems metabolic engineering (Wieschalka et al. 2012). A number of genes were deleted to prevent conversion of pyruvate to acetyl-CoA (Δ*aceE* encoding the E1p subunit pyruvate dehydrogenase), acetate (Δ*pqo* encoding pyruvate-quinone oxidoreductase), L-lactate (Δ*ldhA* encoding fermentative NAD-dependent L-lactic acid dehydrogenase), branched-chain amino acids (Δ*ilvN* encoding the C-terminal regulatory domain of acetohydroxyacid synthase), and to L-alanine (Δ*alaT*Δ*avtA* encoding alanine aminotransferase and valine-pyruvate aminotransferase, respectively). Due to these deletions *C. glutamicum* ELB-P overproduces pyruvate from glucose (Wieschalka et al. 2012). As consequence of the *aceE* deletion, *C. glutamicum* ELB-P requires acetate for biomass formation and, thus, pyruvate production from glucose can be decoupled from growth with acetate (Wieschalka et al. 2012). Heterologous expression of *dpkA* in *C. glutamicum* ELB-P allowed for production of *N*-methyl-L-alanine, however, the ratios of the medium components glucose and acetate as carbon and energy sources, ammonium sulfate and urea as nitrogen sources and monomethylamine as substrate for *N*-alkylation had to be optimized (Mindt et al. 2018a). Transfer from shake flasks to 4 L bioreactor cultivation in fed-batch enabled production to a titer of about 31.7 g L^{-1} *N*-methyl-L-alanine with a product yield of 0.71 g *N*-methyl-L-alanine per g of glucose and a volumetric productivity of 0.35 g L^{-1} h^{-1}. In addition, sustainable production of *N*-methyl-L-alanine could be also demonstrated from starch and the lignocellulosic pentose sugars xylose and arabinose (Mindt et al. 2018a).

The substrate promiscuity of the IRED DpkA from *P. putida* provides the potential to produce a wider spectrum of *N*-alkylated amino acids (Mihara et al. 2005; Muramatsu et al. 2005), whereas the catabolic monomethylammonium pathway from *M. extorquens* is characterized by high substrate specificity (Gruffaz et al. 2014). Metabolic engineering to tap on the potential of substrate promiscuity of DpkA with respect to use other *N*-alkyl donors such as monoethylamine or keto acids such as glyoxylate for fermentative production a broader range of *N*-alkylated amino acids has just started (Mindt et al. 2019a, b).

4 Perspective

The number of articles on *C. glutamicum* and its metabolic engineering increases year by year. One of the strengths of this bacterium is its safe use in food and feed biotech industry since more than five decades. Unlike *E. coli* it lacks for example endotoxins.

C. glutamicum strains are excellent amino acid producers and fermentative L-glutamate and L-lysine production has reached production volumes of about 6 million metric tons per year. Metabolic engineering has diversified the product spectrum of *C. glutamicum* and, typically, amino acid derived products such as diamines or amino acid precursors such as keto acids can be produced with very good titers, yields and productivities. The advent of new technologies such CRISPR interference and genome editing will accelerate *C. glutamicum* strain development in particular when multiplexing can be realized. As in the past, fast and potent metabolic engineering tools will be applied to *C. glutamicum* as soon as they arise. Thus, we can expect a wider range of amino acid related products being produced by *C. glutamicum* strains engineered using the newest suite of metabolic engineering tools.

We can also foreshadow that sustainable production from second generation feedstocks without competing uses in human and animal nutrition will play more important roles and access to more complex polymeric feedstocks will become possible. Design and realization of microbial consortia containing *C. glutamicum* strains are in its infancy, but this approach may prove useful in the biorefinery concept. However, dividing "labour" between different specialized strains or species has to outcompete approaches aimed at multi-potent strains able to convert all relevant feedstocks to the desired product.

Metabolic engineering has developed in waves and we can expect similar future developments. After classical mutagenesis combined with selection and screening was successful although treating the cell as a black box, genetic engineering, omics-based systems biology and genome-scale modeling deepened our insight into the cell's physiology. Currently, synthetic biology approaches are combined with adaptive laboratory evolution. Typically improved traits can be obtained, but insight into causality is often lacking. Future research will have to conceptualize the underlying molecular causes and principles to drive metabolic engineering of *C. glutamicum* to the next level.

Acknowledgements Both authors are particularly grateful to funding by the German-Korean MOBKOR program jointly funded by the National Research Foundation of Korea (NRF-2016K1A3A1A04940618) and the German Federal Ministry of Education and Research. VFW acknowledges funding by the EU and the state Nordrhein-Westfalen in the ERDF.NRW project "CKB-CLIB Kompetenzcluster Biotechnologie", and by ERACoBiotech and FNR-BMELV in the project INDIE. JH LEE acknowledges funding by Basic Science Research Program through the National Research Foundation of Korea (NRF-2018R1D1A1B07047207) and the BB21+ Project in 2018.

References

Ahn JH, Jang YS, Lee SY (2016) Production of succinic acid by metabolically engineered microorganisms. Curr Opin Biotechnol 42:54–66

Asakura Y, Kimura E, Usuda Y, Kawahara Y, Matsui K, Osumi T, Nakamatsu T (2007) Altered metabolic flux due to deletion of *odhA* causes L-glutamate overproduction in *Corynebacterium glutamicum*. Appl Environ Microbiol 73:1308–1319

Auchter M, Cramer A, Hüser A, Rückert C, Emer D, Schwarz P, Arndt A, Lange C, Kalinowski J, Wendisch VF, Eikmanns BJ (2011) RamA and RamB are global transcriptional regulators in *Corynebacterium glutamicum* and control genes for enzymes of the central metabolism. J Biotechnol 154:126–139

Averesch NJH, Krömer JO (2018) Metabolic engineering of the shikimate pathway for production of aromatics and derived compounds—present and future strain construction strategies. Front Bioeng Biotechnol 26:32

Banta S, Swanson BA, Wu S, Jarnagin A, Anderson S (2002) Alteration of the specificity of the cofactor-binding pocket of *Corynebacterium* 2,5-diketo-D-gluconic acid reductase A. Protein Eng 15:131–140

Baritugo KA, Kim HT, David Y, Khang TU, Hyun SM, Kang KH, Yu JH, Choi JH, Song JJ, Joo JC, Park SJ (2018) Enhanced production of gamma-aminobutyrate (GABA) in recombinant *Corynebacterium glutamicum* strains from empty fruit bunch biosugar solution. Microb Cell Fact 17:129

Baumgart M, Unthan S, Rückert C, Sivalingam J, Grünberger A, Kalinowski J, Bott M, Noack S, Frunzke J (2013) Construction of a prophage-free variant of *Corynebacterium glutamicum* ATCC 13032 for use as a platform strain for basic research and industrial biotechnology. Appl Environ Microbiol 79:6006–6015

Baumgart M, Huber I, Abdollahzadeh I, Gensch T, Frunzke J (2017) Heterologous expression of the *Halothiobacillus neapolitanus* carboxysomal gene cluster in *Corynebacterium glutamicum*. J Biotechnol 258:126–135

Baumgart M, Unthan S, Kloß R, Radek A, Polen T, Tenhaef N, Müller MF, Küberl A, Siebert D, Brühl N, Marin K, Hans S, Krämer R, Bott M, Kalinowski J, Wiechert W, Seibold G, Frunzke J, Rückert C, Wendisch VF, Noack S (2018) *Corynebacterium glutamicum* chassis C1∗ building and testing a novel platform host for synthetic biology and industrial biotechnology. ACS Synth Biol 7:132–144

Becker J, Wittmann C (2015) Advanced biotechnology: metabolically engineered cells for the bio-based production of chemicals and fuels, materials, and health-care products. Angew Chem Int Ed Engl 54:3328–3350

Becker J, Kuhl M, Kohlstedt M, Starck S, Wittmann C (2018a) Metabolic engineering of *Corynebacterium glutamicum* for the production of *cis, cis*-muconic acid from lignin. Microb Cell Fact 17:115

Becker J, Rohles CM, Wittmann C (2018b) Metabolically engineered *Corynebacterium glutamicum* for bio-based production of chemicals, fuels, materials, and healthcare products. Metab Eng 50:122–141

Bellmann A, Vrljić M, Pátek M, Sahm H, Krämer R, Eggeling L (2001) Expression control and specificity of the basic amino acid exporter LysE of *Corynebacterium glutamicum*. Microbiology 147:1765–1774

Binder S, Siedler S, Marienhagen J, Bott M, Eggeling L (2013) Recombineering in *Corynebacterium glutamicum* combined with optical nanosensors: a general strategy for fast producer strain generation. Nucleic Acids Res 41:6360–6369

Blombach B, Riester T, Wieschalka S, Ziert C, Youn JW, Wendisch VF, Eikmanns BJ (2011) *Corynebacterium glutamicum* tailored for efficient isobutanol production. Appl Environ Microbiol 77:3300–3310

Bommareddy RR, Chen Z, Rappert S, Zeng AP (2014) A de novo NADPH generation pathway for improving lysine production of *Corynebacterium glutamicum* by rational design of the coenzyme specificity of glyceraldehyde 3-phosphate dehydrogenase. Metab Eng 25:30–37

Brune I, Jochmann N, Brinkrolf K, Hüser AT, Gerstmeir R, Eikmanns BJ, Kalinowski J, Pühler A, Tauch A (2007) The IclR-type transcriptional repressor LtbR regulates the expression of leucine and tryptophan biosynthesis genes in the amino acid producer *Corynebacterium glutamicum*. J Bacteriol 189:2720–2733

Bückle-Vallant V, Krause FS, Messerschmidt S, Eikmanns BJ (2014) Metabolic engineering of *Corynebacterium glutamicum* for 2-ketoisocaproate production. Appl Microbiol Biotechnol 98:297–311

Burkovski A (2008) Corynebacteria: genomics and molecular biology. Caister Academic, Wymondham, UK

Burmeister A, Hilgers F, Langner A, Westerwalbesloh C, Kerkhoff Y, Tenhaef N, Drepper T, Kohlheyer D, von Lieres E, Noack S, Grünberger A (2018) A microfluidic co-cultivation platform to investigate microbial interactions at defined microenvironments. Lab Chip 19:98–110

Buschke N, Schröder H, Wittmann C (2011) Metabolic engineering of *Corynebacterium glutamicum* for production of 1,5-diaminopentane from hemicellulose. Biotechnol J 6:306–317

Buschke N, Becker J, Schäfer R, Kiefer P, Biedendieck R, Wittmann C (2013) Systems metabolic engineering of xylose-utilizing *Corynebacterium glutamicum* for production of 1,5-diaminopentane. Biotechnol J 8:557–570

Camacho-Zaragoza JM, Hernandez-Chavez G, Moreno-Avitia F, Ramirez-Iniguez R, Martinez A, Bolivar F, Gosset G (2016) Engineering of a microbial coculture of *Escherichia coli* strains for the biosynthesis of resveratrol. Microb Cell Fact 15:163

Chae TU, Ko YS, Hwang KS, Lee SY (2017) Metabolic engineering of *Escherichia coli* for the production of four-, five- and six-carbon lactams. Metab Eng 41:82–91

Chen Z, Zeng AP (2016) Protein engineering approaches to chemical biotechnology. Curr Opin Biotechnol 42:198–205

Chen Z, Meyer W, Rappert S, Sun J, Zeng AP (2011) Coevolutionary analysis enabled rational deregulation of allosteric enzyme inhibition in *Corynebacterium glutamicum* for lysine production. Appl Environ Microbiol 77:4352–4360

Chen Z, Bommareddy RR, Frank D, Rappert S, Zeng AP (2014) Deregulation of feedback inhibition of phosphoenolpyruvate carboxylase for improved lysine production in *Corynebacterium glutamicum*. Appl Environ Microbiol 80:1388–1393

Chen Z, Rappert S, Zeng AP (2015) Rational design of allosteric regulation of homoserine dehydrogenase by a nonnatural inhibitor L-lysine. ACS Synth Biol 4:126–131

Cho JS, Choi KR, Prabowo CPS, Shin JH, Yang D, Jang J, Lee SY (2017) CRISPR/Cas9-coupled recombineering for metabolic engineering of *Corynebacterium glutamicum*. Metab Eng 42:157–167

Choi JW, Yim SS, Kim MJ, Jeong KJ (2015a) Enhanced production of recombinant proteins with *Corynebacterium glutamicum* by deletion of insertion sequences (IS elements). Microb Cell Fact 14:207

Choi JW, Yim SS, Lee SH, Kang TJ, Park SJ, Jeong KJ (2015b) Enhanced production of gamma-aminobutyrate (GABA) in recombinant *Corynebacterium glutamicum* by expressing glutamate decarboxylase active in expanded pH range. Microb Cell Fact 14:21

Choi JW, Jeon EJ, Jeong KJ (2018) Recent advances in engineering *Corynebacterium glutamicum* for utilization of hemicellulosic biomass. Curr Opin Biotechnol 57:17–24

Chung SC, Park JS, Yun J, Park JH (2017) Improvement of succinate production by release of end-product inhibition in *Corynebacterium glutamicum*. Metab Eng 40:157–164

Cleto S, Jensen JV, Wendisch VF, Lu TK (2016) *Corynebacterium glutamicum* metabolic engineering with CRISPR interference (CRISPRi). ACS Synth Biol 5:375–385

D'Este M, Alvarado-Morales M, Angelidaki I (2018) Amino acids production focusing on fermentation technologies—A review. Biotechnol Adv 36:14–25

Di Gioia ML, Leggio A, Malagrinò F, Romio E, Siciliano C, Liguori A (2016) *N*-methylated α-amino acids and peptides: synthesis and biological activity. Mini Rev Med Chem 16:683–690

Eberhardt D, Jensen JV, Wendisch VF (2014) L-citrulline production by metabolically engineered *Corynebacterium glutamicum* from glucose and alternative carbon sources. AMB Express 4:85

Eggeling L (2017) Exporters for production of amino acids and other small molecules. Adv Biochem Eng Biotechnol 159:199–225

Eggeling L, Bott M (2005) Handbook of *Corynebacterium glutamicum*. CRC, Boca Raton, FL

Eikmanns BJ, Blombach B (2014) The pyruvate dehydrogenase complex of *Corynebacterium glutamicum*: an attractive target for metabolic engineering. J Biotechnol 192:339–345

Freudl R (2017) Beyond amino acids: use of the *Corynebacterium glutamicum* cell factory for the secretion of heterologous proteins. J Biotechnol 258:101–109

Gruffaz C, Muller EE, Louhichi-Jelail Y, Nelli YR, Guichard G, Bringel F (2014) Genes of the *N*-methylglutamate pathway are essential for growth of *Methylobacterium extorquens* DM4 with monomethylamine. Appl Environ Microbiol 80:3541–3550

Han SS, Kyeong HH, Choi JM, Sohn YK, Lee JH, Kim HS (2016) Engineering of the conformational dynamics of an enzyme for relieving the product inhibition. ACS Catalysis 6:8440–8445

Hasegawa S, Uematsu K, Natsuma Y, Suda M, Hiraga K, Jojima T, Inui M, Yukawa H (2012) Improvement of the redox balance increases L-valine production by *Corynebacterium glutamicum* under oxygen deprivation conditions. Appl Environ Microbiol 78:865–875

Heider SA, Wolf N, Hofemeier A, Peters-Wendisch P, Wendisch VF (2014a) Optimization of the IPP precursor supply for the production of lycopene, decaprenoxanthin and astaxanthin by *Corynebacterium glutamicum*. Front Bioeng Biotechnol 2:28

Heider SA, Peters-Wendisch P, Wendisch VF, Beekwilder J, Brautaset T (2014b) Metabolic engineering for the microbial production of carotenoids and related products with a focus on the rare C50 carotenoids. Appl Microbiol Biotechnol 98:4355–4368

Henke NA, Heider SA, Peters-Wendisch P, Wendisch VF (2016) Production of the marine carotenoid astaxanthin by metabolically engineered *Corynebacterium glutamicum*. Mar Drugs 14(7)

Henke NA, Heider SAE, Hannibal S, Wendisch VF, Peters-Wendisch P (2017) Isoprenoid pyrophosphate-dependent transcriptional regulation of carotenogenesis in *Corynebacterium glutamicum*. Front Microbiol 8:633

Henke NA, Wiebe D, Pérez-García F, Peters-Wendisch P, Wendisch VF (2018) Coproduction of cell-bound and secreted value-added compounds: simultaneous production of carotenoids and amino acids by *Corynebacterium glutamicum*. Bioresour Technol 247:744–752

Hirasawa T, Shimizu H (2016) Recent advances in amino acid production by microbial cells. Curr Opin Biotechnol 42:133–146

Hirasawa T, Kim J, Shirai T, Furusawa C, Shimizu H (2012) Molecular mechanisms and metabolic engineering of glutamate overproduction in *Corynebacterium glutamicum*. Subcell Biochem 64:261–281

Hoffmann SL, Jungmann L, Schiefelbein S, Peyriga L, Cahoreau E, Portais JC, Becker J, Wittmann C (2018) Lysine production from the sugar alcohol mannitol: design of the cell factory *Corynebacterium glutamicum* SEA-3 through integrated analysis and engineering of metabolic pathway fluxes. Metab Eng 47:475–487

Huber I, Palmer DJ, Ludwig KN, Brown IR, Warren MJ, Frunzke J (2017) Construction of recombinant Pdu metabolosome shells for small molecule production in *Corynebacterium glutamicum*. ACS Synth Biol 6:2145–2156

Hutchison CA 3rd, Chuang RY, Noskov VN, Assad-Garcia N, Deerinck TJ, Ellisman MH, Gill J, Kannan K, Karas BJ, Ma L, Pelletier JF, Qi ZQ, Richter RA, Strychalski EA, Sun L, Suzuki Y, Tsvetanova B, Wise KS, Smith HO, Glass JI, Merryman C, Gibson DG, Venter JC (2016) Design and synthesis of a minimal bacterial genome. Science 351(6280):aad6253

Ikeda M (2017) Lysine fermentation: history and genome breeding. Adv Biochem Eng Biotechnol 159:73–102

Ikeda M, Ohnishi J, Hayashi M, Mitsuhashi S (2006) A genome-based approach to create a minimally mutated *Corynebacterium glutamicum* strain for efficient L-lysine production. J Ind Microbiol Biotechnol 33:610–615

Ikeda M, Mitsuhashi S, Tanaka K, Hayashi M (2009) Reengineering of a *Corynebacterium glutamicum* L-arginine and L-citrulline producer. Appl Environ Microbiol 75:1635–1641

Imao K, Konishi R, Kishida M, Hirata Y, Segawa S, Adachi N, Matsuura R, Tsuge Y, Matsumoto T, Tanaka T, Kondo A (2017) 1,5-Diaminopentane production from xylooligosaccharides using metabolically engineered *Corynebacterium glutamicum* displaying beta-xylosidase on the cell surface. Bioresour Technol 245:1684–1691

Jensen JV, Wendisch VF (2013) Ornithine cyclodeaminase-based proline production by *Corynebacterium glutamicum*. Microb Cell Fact 12:63

Jensen JV, Eberhardt D, Wendisch VF (2015) Modular pathway engineering of *Corynebacterium glutamicum* for production of the glutamate-derived compounds ornithine, proline, putrescine, citrulline, and arginine. J Biotechnol 214:85–94

Jiang Y, Qian F, Yang J, Liu Y, Dong F, Xu C, Sun B, Chen B, Xu X, Li Y, Wang R, Yang S (2017) CRISPR-Cpf1 assisted genome editing of Corynebacterium glutamicum. Nat Commun 8:15179

Jojima T, Noburyu R, Suda M, Okino S, Yukawa H, Inui M (2016) Improving process yield in succinic acid production by cell recycling of recombinant *Corynebacterium glutamicum*. Fermentation 2:5

Joo YC, You SK, Shin SK, Ko YJ, Jung KH, Sim SA, Han SO (2017) Bio-based production of dimethyl itaconate from rice wine waste-derived itaconic acid. Biotechnol J 12:11

Jorge JM, Leggewie C, Wendisch VF (2016) A new metabolic route for the production of gamma-aminobutyric acid by *Corynebacterium glutamicum* from glucose. Amino Acids 48:2519–2531

Jorge JMP, Pérez-García F, Wendisch VF (2017a) A new metabolic route for the fermentative production of 5-aminovalerate from glucose and alternative carbon sources. Bioresour Technol 245:1701–1709

Jorge JMP, Nguyen AQ, Pérez-García F, Kind S, Wendisch VF (2017b) Improved fermentative production of gamma-aminobutyric acid via the putrescine route: systems metabolic engineering for production from glucose, amino sugars, and xylose. Biotechnol Bioeng 114:862–873

Kabus A, Georgi T, Wendisch VF, Bott M (2007) Expression of the *Escherichia coli pntAB* genes encoding a membrane-bound transhydrogenase in *Corynebacterium glutamicum* improves L-lysine formation. Appl Microbiol Biotechnol 75:47–53

Kallscheuer N, Vogt M, Marienhagen J (2017) A novel synthetic pathway enables microbial production of polyphenols independent from the endogenous aromatic amino acid metabolism. ACS Synth Biol 6:410–415

Kelle R, Laufer B, Brunzema C, Weuster-Botz D, Krämer R, Wandrey C (1996) Reaction engineering analysis of L-lysine transport by *Corynebacterium glutamicum*. Biotechnol Bioeng 51:40–50

Kerfeld CA, Erbilgin O (2015) Bacterial microcompartments and the modular construction of microbial metabolism. Trends Microbiol 23:22–34

Kerfeld CA, Heinhorst S, Cannon GC (2010) Bacterial microcompartments. Annu Rev Microbiol 64:391–408

Kim HJ, Kim TH, Kim Y, Lee HS (2004) Identification and characterization of *glxR*, a gene involved in regulation of glyoxylate bypass in *Corynebacterium glutamicum*. J Bacteriol 186:3453–3460

Kim HT, Khang TU, Baritugo KA, Hyun SM, Kang KH, Jung SH, Song BK, Park K, Oh MK, Kim GB, Kim HU, Lee SY, Park SJ, Joo JC (2019) Metabolic engineering of *Corynebacterium glutamicum* for the production of glutaric acid, a C5 dicarboxylic acid platform chemical. Metab Eng 51:99–109

Kind S, Neubauer S, Becker J, Yamamoto M, Völkert M, Abendroth G, Zelder O, Wittmann C (2014) From zero to hero—production of bio-based nylon from renewable resources using engineered *Corynebacterium glutamicum*. Metab Eng 25:113–123

Kirchner O, Tauch A (2003) Tools for genetic engineering in the amino acid-producing bacterium *Corynebacterium glutamicum*. J Biotechnol 104:287–299

Kitade Y, Hashimoto R, Suda M, Hiraga K, Inui M (2018) Production of 4-hydroxybenzoic acid by an aerobic growth-arrested bioprocess using metabolically engineered *Corynebacterium glutamicum*. Appl Environ Microbiol 84(6):pii: e02587-17

Kloss R, Limberg MH, Mackfeld U, Hahn D, Grünberger A, Jäger VD, Krauss U, Oldiges M, Pohl M (2018) Catalytically active inclusion bodies of L-lysine decarboxylase from *E. coli* for 1,5-diaminopentane production. Sci Rep 8:5856

Koffas M, Stephanopoulos G (2005) Strain improvement by metabolic engineering: lysine production as a case study for systems biology. Curr Opin Biotechnol 16:361–366

Koma D, Yamanaka H, Moriyoshi K, Sakai K, Masuda T, Sato Y, Toida K, Ohmoto T (2014) Production of *p*-aminobenzoic acid by metabolically engineered *Escherichia coli*. Biosci Biotechnol Biochem 78:350–357

Komati Reddy G, Lindner SN, Wendisch VF (2015) Metabolic engineering of an ATP-neutral Embden-Meyerhof-Parnas pathway in *Corynebacterium glutamicum*: growth restoration by an adaptive point mutation in NADH dehydrogenase. Appl Environ Microbiol 81:1996–2005

Kubota T, Watanabe A, Suda M, Kogure T, Hiraga K, Inui M (2016) Production of *para*-aminobenzoate by genetically engineered *Corynebacterium glutamicum* and non-biological formation of an *N*-glucosyl byproduct. Metab Eng 38:322–330

Lange C, Mustafi N, Frunzke J, Kennerknecht N, Wessel M, Bott M, Wendisch VF (2011) Lrp of *Corynebacterium glutamicum* controls expression of the brnFE operon encoding the export system for L-methionine and branched-chain amino acids. J Biotechnol 158:231–241

Lee JH, Wendisch VF (2017a) Biotechnological production of aromatic compounds of the extended shikimate pathway from renewable biomass. J Biotechnol 257:211–221

Lee JH, Wendisch VF (2017b) Production of amino acids—genetic and metabolic engineering approaches. Bioresour Technol 245:1575–1587

Lee JY, Seo J, Kim ES, Lee HS, Kim P (2013) Adaptive evolution of *Corynebacterium glutamicum* resistant to oxidative stress and its global gene expression profiling. Biotechnol Lett 35:709–717

Lee MJ, Palmer DJ, Warren MJ (2019) Biotechnological advances in bacterial microcompartment technology. Trends Biotechnol 37(3):325–336

Leßmeier L, Wendisch VF (2015) Identification of two mutations increasing the methanol tolerance of *Corynebacterium glutamicum*. BMC Microbiol 15:216

Leßmeier L, Pfeifenschneider J, Carnicer M, Heux S, Portais JC, Wendisch VF (2015) Production of carbo*N*-13-labeled cadaverine by engineered *Corynebacterium glutamicum* using carbo*N*-13-labeled methanol as co-substrate. Appl Microbiol Biotechnol 99:10163–10176

Leuchtenberger W, Huthmacher K, Drauz K (2005) Biotechnological production of amino acids and derivatives: current status and prospects. Appl Microbiol Biotechnol 69:1–8

Li H, Qiu T, Huang G, Cao Y (2010) Production of gamma-aminobutyric acid by *Lactobacillus brevis* NCL912 using fed-batch fermentation. Microb Cell Fact 9:85

Li Z, Shen YP, Jiang XL, Feng LS, Liu JZ (2018) Metabolic evolution and a comparative omics analysis of *Corynebacterium glutamicum* for putrescine production. J Ind Microbiol Biotechnol 45:123–139

Lindner SN, Calzad Iacute Az Ramirez L, Krüsemann J, Yishai O, Belkhelfa S, He H, Bouzon M, Döring V, Bar-Even A (2018) NADPH-auxotrophic *E. coli*: a sensor strain for testing in vivo regeneration of NADPH. ACS Synth Biol 7:2742–2749

Litsanov B, Brocker M, Bott M (2012) Toward homosuccinate fermentation: metabolic engineering of *Corynebacterium glutamicum* for anaerobic production of succinate from glucose and formate. Appl Environ Microbiol 78:3325–3337

Liu X, Zhang W, Zhao Z, Dai X, Yang Y, Bai Z (2017) Protein secretion in *Corynebacterium glutamicum*. Crit Rev Biotechnol. 37:541–551

Liu C, Zhang B, Liu YM, Yang KQ, Liu SJ (2018) New intracellular shikimic acid biosensor for monitoring shikimate synthesis in *Corynebacterium glutamicum*. ACS Synth Biol 7:591–601

Lubitz D, Jorge JM, Pérez-García F, Taniguchi H, Wendisch VF (2016) Roles of export genes *cgmA* and *lysE* for the production of L-arginine and L-citrulline by *Corynebacterium glutamicum*. Appl Microbiol Biotechnol 100:8465–8474

Ma Q, Zhang Q, Xu Q, Zhang C, Li Y, Fan X, Xie X, Chen N (2017) Systems metabolic engineering strategies for the production of amino acids. Synth Syst Biotechnol 2:87–96

Mahr R, Gätgens C, Gätgens J, Polen T, Kalinowski J, Frunzke J (2015) Biosensor-driven adaptive laboratory evolution of L-valine production in *Corynebacterium glutamicum*. Metab Eng 32:184–194

Mampel J, Buescher JM, Meurer G, Eck J (2013) Coping with complexity in metabolic engineering. Trends Biotechnol 31:52–60

Mao Y, Li G, Chang Z, Tao R, Cui Z, Wang Z, Tang YJ, Chen T, Zhao X (2018) Metabolic engineering of *Corynebacterium glutamicum* for efficient production of succinate from lignocellulosic hydrolysate. Biotechnol Biofuels 11:95

Marin K, Krämer R (2007) Amino acid transport systems in biotechnologically relevant bacteria. In: Wendisch VF (ed) Amino acid biosynthesis—pathways, regulation and metabolic engineering. Springer, Heidelberg, pp 289–326

Matano C, Uhde A, Youn JW, Maeda T, Clermont L, Marin K, Krämer R, Wendisch VF, Seibold GM (2014) Engineering of *Corynebacterium glutamicum* for growth and L-lysine and lycopene production from *N*-acetyl-glucosamine. Appl Microbiol Biotechnol 98:5633–5643

Matsushima Y, Hirasawa T, Shimizu H (2016) Enhancement of 1,5-diaminopentane production in a recombinant strain of *Corynebacterium glutamicum* by Tween 40 addition. J Gen Appl Microbiol 62:42–45

Meiswinkel TM, Gopinath V, Lindner SN, Nampoothiri KM, Wendisch VF (2013a) Accelerated pentose utilization by *Corynebacterium glutamicum* for accelerated production of lysine, glutamate, ornithine and putrescine. Microb Biotechnol 6:131–140

Meiswinkel TM, Rittmann D, Lindner SN, Wendisch VF (2013b) Crude glycerol-based production of amino acids and putrescine by *Corynebacterium glutamicum*. Bioresour Technol 145:254–258

Meyer F, Keller P, Hartl J, Gröninger OG, Kiefer P, Vorholt JA (2018) Methanol-essential growth of *Escherichia coli*. Nat Commun 9:1508

Mihara H, Muramatsu H, Kakutani R, Yasuda M, Ueda M, Kurihara T, Esaki N (2005) *N*-methyl-l-amino acid dehydrogenase from *Pseudomonas putida*. A novel member of an unusual NAD(P)-dependent oxidoreductase superfamily. FEBS J 272:117–123

Mimitsuka T, Sawai H, Hatsu M, Yamada K (2007) Metabolic engineering of *Corynebacterium glutamicum* for cadaverine fermentation. Biosci Biotechnol Biochem 71:2130–2135

Mindt M, Risse JM, Gruß H, Sewald N, Eikmanns BJ, Wendisch VF (2018a) One-step process for production of *N*-methylated amino acids from sugars and methylamine using recombinant *Corynebacterium glutamicum* as biocatalyst. Sci Rep 8:12895

Mindt M, Walter T, Risse JM, Wendisch VF (2018b) Fermentative production of *N*-methylglutamate from glycerol by recombinant *Pseudomonas putida*. Front Bioeng Biotechnol 6:159

Mindt M, Heuser M, Wendisch VF (2019a) Xylose as preferred substrate for sarcosine production by recombinant *Corynebacterium glutamicum*. Bioresour Technol 281:135–142

Mindt M, Hannibal S, Heuser M, Risse JM, Sasikumar K, Nampoothiri KM, Wendisch VF (2019b) Fermentative production of *N*-alkylated glycine derivatives by recombinant *Corynebacterium glutamicum* using a mutant of imine reductase DpkA from *Pseudomonas putida*. Front Bioeng Biotechnol 7:232

Minty JJ, Singer ME, Scholz SA, Bae CH, Ahn JH, Foster CE, Liao JC, Lin XN (2013) Design and characterization of synthetic fungal-bacterial consortia for direct production of isobutanol from cellulosic biomass. Proc Natl Acad Sci USA 110:14592–14597

Muramatsu H, Mihara H, Kakutani R, Yasuda M, Ueda M, Kurihara T, Esaki N (2005) The putative malate/lactate dehydrogenase from Pseudomonas putida is an NADPH-dependent Δ1-piperideine-2-carboxylate/Δ1-pyrroline-2-carboxylate reductase involved in the catabolism of d-lysine and d-proline. J Biol Chem 280:5329–5335

Mustafi N, Grünberger A, Mahr R, Helfrich S, Nöh K, Blombach B, Kohlheyer D, Frunzke J (2014) Application of a genetically encoded biosensor for live cell imaging of L-valine production in pyruvate dehydrogenase complex-deficient *Corynebacterium glutamicum* strains. PLoS One 9: e85731

Naerdal I, Pfeifenschneider J, Brautaset T, Wendisch VF (2015) Methanol-based cadaverine production by genetically engineered *Bacillus methanolicus* strains. Microb Biotechnol 8:342–350

Nguyen AQD, Schneider J, Reddy GK, Wendisch VF (2015a) Fermentative production of the diamine putrescine: system metabolic engineering of *Corynebacterium glutamicum*. Metabolites 5:211–231

Nguyen AQ, Schneider J, Wendisch VF (2015b) Elimination of polyamine *N*-acetylation and regulatory engineering improved putrescine production by *Corynebacterium glutamicum*. J Biotechnol 201:75–85

Niebisch A, Kabus A, Schultz C, Weil B, Bott M (2006) Corynebacterial protein kinase G controls 2-oxoglutarate dehydrogenase activity via the phosphorylation status of the OdhI protein. J Biol Chem 281:12300–12307

Oh YH, Choi JW, Kim EY, Song BK, Jeong KJ, Park K, Kim IK, Woo HM, Lee SH, Park SJ (2015) Construction of synthetic promoter-based expression cassettes for the production of cadaverine in recombinant *Corynebacterium glutamicum*. Appl Biochem Biotechnol 176:2065–2075

Ohnishi J, Mitsuhashi S, Hayashi M, Ando S, Yokoi H, Ochiai K, Ikeda M (2002) A novel methodology employing *Corynebacterium glutamicum* genome information to generate a new L-lysine-producing mutant. Appl Microbiol Biotechnol 58:217–223

Ohnishi J, Mizoguchi H, Takeno S, Ikeda M (2008) Characterization of mutations induced by *N*-methyl-N′-nitro-*N*-nitrosoguanidine in an industrial *Corynebacterium glutamicum* strain. Mutat Res 649:239–244

Oide S, Gunji W, Moteki Y, Yamamoto S, Suda M, Jojima T, Yukawa H, Inui M (2015) Thermal and solvent stress cross-tolerance conferred to *Corynebacterium glutamicum* by adaptive laboratory evolution. Appl Environ Microbiol 81:2284–2298

Okai N, Takahashi C, Hatada K, Ogino C, Kondo A (2014) Disruption of *pknG* enhances production of gamma-aminobutyric acid by *Corynebacterium glutamicum* expressing glutamate decarboxylase. AMB Express 4:20

Okino S, Noburyu R, Suda M, Jojima T, Inui M, Yukawa H (2008) An efficient succinic acid production process in a metabolically engineered *Corynebacterium glutamicum* strain. Appl Microbiol Biotechnol 81:459–464

Park SH, Kim HU, Kim TY, Park JS, Kim SS, Lee SY (2014) Metabolic engineering of *Corynebacterium glutamicum* for L-arginine production. Nat Commun 5:4618

Patterson AW, Peltier HM, Ellman JA (2008) Expedient synthesis of *N*-methyl tubulysin analogues with high cytotoxicity. J Org Chem 73:4362–4369

Pérez-García F, Wendisch VF (2018) Transport and metabolic engineering of the cell factory *Corynebacterium glutamicum*. FEMS Microbiol Lett 365:16

Pérez-García F, Peters-Wendisch P, Wendisch VF (2016) Engineering *Corynebacterium glutamicum* for fast production of L-lysine and L-pipecolic acid. Appl Microbiol Biotechnol 100:8075–8090

Pérez-García F, Max Risse J, Friehs K, Wendisch VF (2017a) Fermentative production of L-pipecolic acid from glucose and alternative carbon sources. Biotechnol J 12(7)

Pérez-García F, Ziert C, Risse JM, Wendisch VF (2017b) Improved fermentative production of the compatible solute ectoine by *Corynebacterium glutamicum* from glucose and alternative carbon sources. J Biotechnol 258:59–68

Pérez-García F, Jorge JMP, Dreyszas A, Risse JM, Wendisch VF (2018) Efficient production of the dicarboxylic acid glutarate by *Corynebacterium glutamicum* via a novel synthetic pathway. Front Microbiol 9:2589

Petit C, Kim Y, Lee SK, Brown J, Larsen E, Ronning DR, Suh JW, Kang CM (2018) Reduction of feedback inhibition in homoserine kinase (ThrB) of *Corynebacterium glutamicum* enhances L-threonine biosynthesis. ACS Omega 3:1178–1186

Pfeifer E, Gätgens C, Polen T, Frunzke J (2017) Adaptive laboratory evolution of *Corynebacterium glutamicum* towards higher growth rates on glucose minimal medium. Sci Rep 7:16780

Purwanto HS, Kang MS, Ferrer L, Han SS, Lee JY, Kim HS, Lee JH (2018) Rational engineering of the shikimate and related pathways in *Corynebacterium glutamicum* for 4-hydroxybenzoate production. J Biotechnol 282:92–100

Radek A, Tenhaef N, Müller MF, Brüsseler C, Wiechert W, Marienhagen J, Polen T, Noack S (2017) Miniaturized and automated adaptive laboratory evolution: evolving *Corynebacterium glutamicum* towards an improved d-xylose utilization. Bioresour Technol 245:1377–1385

Rohles CM, Gießelmann G, Kohlstedt M, Wittmann C, Becker J (2016) Systems metabolic engineering of *Corynebacterium glutamicum* for the production of the carbo*N*-5 platform chemicals 5-aminovalerate and glutarate. Microb Cell Factories 15:154

Rytter JV, Helmark S, Chen J, Lezyk MJ, Solem C, Jensen PR (2014) Synthetic promoter libraries for *Corynebacterium glutamicum*. Appl Microbiol Biotechnol 98:2617–2623

Schendzielorz G, Dippong M, Grünberger A, Kohlheyer D, Yoshida A, Binder S, Nishiyama C, Nishiyama M, Bott M, Eggeling L (2014) Taking control over control: use of product sensing in single cells to remove flux control at key enzymes in biosynthesis pathways. ACS Synth Biol 3:21–29

Schneider J, Wendisch VF (2010) Putrescine production by engineered *Corynebacterium glutamicum*. Appl Microbiol Biotechnol 88:859–868

Schneider J, Eberhardt D, Wendisch VF (2012) Improving putrescine production by *Corynebacterium glutamicum* by fine-tuning ornithine transcarbamoylase activity using a plasmid addiction system. Appl Microbiol Biotechnol 95:169–178

Schulte J, Baumgart M, Bott M (2017) Development of a single-cell GlxR-based cAMP biosensor for *Corynebacterium glutamicum*. J Biotechnol 258:33–40

Schultz C, Niebisch A, Gebel L, Bott M (2007) Glutamate production by *Corynebacterium glutamicum*: dependence on the oxoglutarate dehydrogenase inhibitor protein OdhI and protein kinase PknG. Appl Microbiol Biotechnol 76:691–700

Seibold G, Auchter M, Berens S, Kalinowski J, Eikmanns BJ (2006) Utilization of soluble starch by a recombinant *Corynebacterium glutamicum* strain: growth and lysine production. J. Biotechnol 124:381–391

Sgobba E, Wendisch VF (2020) Synthetic microbial consortia for small molecule production. Curr Opin Biotechnol 62:72–79

Sgobba E, Blöbaum L, Wendisch VF (2018a) Production of food and feed additives from non-food-competing feedstocks: valorizing *N*-acetylmuramic acid for amino acid and carotenoid fermentation with *Corynebacterium glutamicum*. Front Microbiol 9:2046

Sgobba E, Stumpf AK, Vortmann M, Jagmann N, Krehenbrink M, Dirks-Hofmeister ME, Moerschbacher B, Philipp B, Wendisch VF (2018b) Synthetic *Escherichia coli-Corynebacterium glutamicum* consortia for l-lysine production from starch and sucrose. Bioresour Technol 260:302–310

Shi F, Li Y (2011) Synthesis of γ-aminobutyric acid by expressing *Lactobacillus brevis*-derived glutamate decarboxylase in the *Corynebacterium glutamicum* strain ATCC 13032. Biotechnol Lett 33:2469–2474

Shi F, Jiang J, Li Y, Li Y, Xie Y (2013) Enhancement of γ-aminobutyric acid production in recombinant *Corynebacterium glutamicum* by co-expressing two glutamate decarboxylase genes from *Lactobacillus brevis*. J Ind Microbiol Biotechnol 40:1285–1296

Shi F, Xie Y, Jiang J, Wang N, Li Y, Wang X (2014) Directed evolution and mutagenesis of glutamate decarboxylase from *Lactobacillus brevis* Lb85 to broaden the range of its activity toward a near-neutral pH. Enzyme Microb Technol 61-62:35–43

Shi F, Luan M, Li Y (2018) Ribosomal binding site sequences and promoters for expressing glutamate decarboxylase and producing γ-aminobutyrate in *Corynebacterium glutamicum*. AMB Express 8:61

Shin JH, Park SH, Oh YH, Choi JW, Lee MH, Cho JS, Jeong KJ, Joo JC, Yu J, Park SJ, Lee SY (2016) Metabolic engineering of *Corynebacterium glutamicum* for enhanced production of 5-aminovaleric acid. Microb Cell Fact 15:174

Shin WS, Lee D, Lee SJ, Chun GT, Choi SS, Kim ES, Kim S (2018) Characterization of a non-phosphotransferase system for *cis,cis*-muconic acid production *in Corynebacterium glutamicum*. Biochem Biophys Res Commun 499:279–284

Sindelar G, Wendisch VF (2007) Improving lysine production by *Corynebacterium glutamicum* through DNA microarray-based identification of novel target genes. Appl Microbiol Biotechnol 74:677–689

Sun H, Zhao D, Xiong B, Zhang C, Bi C (2016) Engineering *Corynebacterium glutamicum* for violacein hyper production. Microb Cell Fact 15:148

Takahashi C, Shirakawa J, Tsuchidate T, Okai N, Hatada K, Nakayama H, Tateno T, Ogino C, Kondo A (2012) Robust production of gamma-amino butyric acid using recombinant *Corynebacterium glutamicum* expressing glutamate decarboxylase from *Escherichia coli*. Enzyme Microb Technol 51:171–176

Takeno S, Hori K, Ohtani S, Mimura A, Mitsuhashi S, Ikeda M (2016) l-Lysine production independent of the oxidative pentose phosphate pathway by *Corynebacterium glutamicum* with the *Streptococcus mutans gapN* gene. Metab Eng 37:1–10

Takors R, Bathe B, Rieping M, Hans S, Kelle R, Huthmacher K (2007) Systems biology for industrial strains and fermentation processes-example: amino acids. J Biotechnol 129:181–190

Taniguchi H, Wendisch VF (2015) Exploring the role of sigma factor gene expression on production by *Corynebacterium glutamicum*: sigma factor H and FMN as example. Front Microbiol 6:740

Taniguchi H, Henke NA, Heider SAE, Wendisch VF (2017) Overexpression of the primary sigma factor gene *sigA* improved carotenoid production by *Corynebacterium glutamicum*: application to production of β-carotene and the non-native linear C50 carotenoid bisanhydrobacterioruberin. Metab Eng Commun 4:1–11

Tateno T, Okada Y, Tsuchidate T, Tanaka T, Fukuda H, Kondo A (2009) Direct production of cadaverine from soluble starch using *Corynebacterium glutamicum* coexpressing alpha-amylase and lysine decarboxylase. Appl Microbiol Biotechnol 82:115–121

Tatsumi N, Inui M (2013) *Corynebacterium glutamicum*—biology and biotechnology. Springer, Heidelberg

Thu Ho NA, Hou CY, Kim WH, Kang TJ (2013) Expanding the active pH range of *Escherichia coli* glutamate decarboxylase by breaking the cooperativeness. J Biosci Bioeng 115:154–158

Toyoda K, Teramoto H, Yukawa H, Inui M (2015) Expanding the regulatory network governed by the extracytoplasmic function sigma factor σH in *Corynebacterium glutamicum*. J Bacteriol 197:483–496

Tsuge Y, Tateno T, Sasaki K, Hasunuma T, Tanaka T, Kondo A (2013) Direct production of organic acids from starch by cell surface-engineered *Corynebacterium glutamicum* in anaerobic conditions. AMB Express 3:72

Tuyishime P, Wang Y, Fan L, Zhang Q, Li Q, Zheng P, Sun J, Ma Y (2018) Engineering *Corynebacterium glutamicum* for methanol-dependent growth and glutamate production. Metab Eng 49:220–231

Uhde A, Youn JW, Maeda T, Clermont L, Matano C, Kramer R et al (2013) Glucosamine as carbon source for amino acid-producing Corynebacterium glutamicum. Appl Microbiol. Biotechnol 97:1679–1687

Unthan S, Baumgart M, Radek A, Herbst M, Siebert D, Brühl N, Bartsch A, Bott M, Wiechert W, Marin K, Hans S, Krämer R, Seibold G, Frunzke J, Kalinowski J, Rückert C, Wendisch VF, Noack S (2015) Chassis organism from *Corynebacterium glutamicum*-a top-down approach to identify and delete irrelevant gene clusters. Biotechnol J 10:290–301

Veldmann KH, Minges H, Sewald N, Lee JH, Wendisch VF (2019a) Metabolic engineering of *Corynebacterium glutamicum* for the fermentative production of halogenated tryptophan. J Biotechnol 291:7–216

Veldmann KH, Dachwitz S, Risse JM, Lee JH, Sewald N, Wendisch VF (2019b) Bromination of L-tryptophan in a fermentative process with *Corynebacterium glutamicum*. Front Bioeng Biotechnol 7:219

Vogt M, Haas S, Klaffl S, Polen T, Eggeling L, van Ooyen J, Bott M (2014) Pushing product formation to its limit: metabolic engineering of *Corynebacterium glutamicum* for L-leucine overproduction. Metab Eng 22:40–52

Vogt M, Haas S, Polen T, van Ooyen J, Bott M (2015) Production of 2-ketoisocaproate with *Corynebacterium glutamicum* strains devoid of plasmids and heterologous genes. Microb Biotechnol 8:351–360

Vrljic M, Sahm H, Eggeling L (1996) A new type of transporter with a new type of cellular function: L-lysine export from *Corynebacterium glutamicum*. Mol Microbiol 22:815–826

Wang EX, Ding MZ, Ma Q, Dong XT, Yuan YJ (2016) Reorganization of a synthetic microbial consortium for one-step vitamin C fermentation. Microb Cell Fact 15:21

Wang C, Zhou Z, Cai H, Chen Z, Xu H (2017a) Redirecting carbon flux through *pgi*-deficient and heterologous transhydrogenase toward efficient succinate production in *Corynebacterium glutamicum*. J Ind Microbiol Biotechnol 44:1115–1126

Wang M, Chen B, Fang Y, Tan T (2017b) Cofactor engineering for more efficient production of chemicals and biofuels. Biotechnol Adv 35:1032–1039

Wang Y, Liu Y, Liu J, Guo Y, Fan L, Ni X, Zheng X, Wang M, Zheng P, Sun J, Ma Y (2018) MACBETH: multiplex automated *Corynebacterium glutamicum* base editing method. Metab Eng 47:200–210

Wendisch VF (2003) Genome-wide expression analysis in Corynebacterium glutamicum using DNA microarrays. J Biotechnol 104:273–285

Wendisch VF (2007) Amino acid biosynthesis—pathways, regulation and metabolic engineering. Springer, Heidelberg

Wendisch VF (2014) Microbial production of amino acids and derived chemicals: synthetic biology approaches to strain development. Curr Opin Biotechnol 30:51–58

Wendisch VF (2019) Metabolic engineering advances and prospects for amino acid production. Metab Eng. https://doi.org/10.1016/j.ymben.2019.03.008

Wendisch VF, Bott M, Kalinowski J, Oldiges M, Wiechert W (2006) Emerging *Corynebacterium glutamicum* systems biology. J Biotechnol 124:74

Wendisch VF, Brito LF, Gil Lopez M, Hennig G, Pfeifenschneider J, Sgobba E, Veldmann KH (2016) The flexible feedstock concept in Industrial Biotechnology: metabolic engineering of *Escherichia coli, Corynebacterium glutamicum, Pseudomonas,* Bacillus and yeast strains for access to alternative carbon sources. J Biotechnol 234:139–157

Wendisch VF, Kim Y, Lee JH (2018a) Chemicals from lignin: recent depolymerization techniques and upgrading extended pathways. Curr Opin Green Sustain Chem 14:33–39

Wendisch VF, Mindt M, Pérez-García F (2018b) Biotechnological production of mono- and diamines using bacteria: recent progress, applications, and perspectives. Appl Microbiol Biotechnol 102:3583–3594

Wieschalka S, Blombach B, Eikmanns BJ (2012) Engineering *Corynebacterium glutamicum* for the production of pyruvate. Appl Microbiol Biotechnol 94:449–459

Wu CC, Chen TH, Liu BL, Wu LC, Chen YC, Tzeng YM, Hsu SL (2013) Destruxin B isolated from entomopathogenic fungus *Metarhizium anisopliae* induces apoptosis via a Bcl-2 family-dependent mitochondrial pathway in human nonsmall cell lung cancer cells. Evid Based Complement Altern Med 2013:548929

Xu H, Zhou Z, Wang C, Chen Z, Cai H (2016) Enhanced succinic acid production in *Corynebacterium glutamicum* with increasing the available NADH supply and glucose consumption rate by decreasing H(+)-ATPase activity. Biotechnol Lett 38:1181–1186

Yang J, Seo SW, Jang S, Shin SI, Lim CH, Roh TY, Jung GY (2013) Synthetic RNA devices to expedite the evolution of metabolite-producing microbes. Nat Commun 4:1413

Youn JW, Jolkver E, Kramer R, Marin K, Wendisch VF (2008) Identification and characterization of the dicarboxylate uptake system DccT in *Corynebacterium glutamicum*. J Bacteriol 190:6458–6466

Youn JW, Jolkver E, Kramer R, Marin K, Wendisch VF (2009) Characterization of the dicarboxylate transporter DctA in *Corynebacterium glutamicum*. J Bacteriol 191:5480–5488

Zahoor A, Lindner SN, Wendisch VF (2012) Metabolic engineering of *Corynebacterium glutamicum* aimed at alternative carbon sources and new products. Comput Struct Biotechnol J 3:e201210004

Zha J, Zang Y, Mattozzi M, Plassmeier J, Gupta M, Wu X, Clarkson S, Koffas MAG (2018) Metabolic engineering of *Corynebacterium glutamicum* for anthocyanin production. Microb Cell Fact 17:143

Zhan M, Kan B, Dong J, Xu G, Han R, Ni Y (2018) Metabolic engineering of *Corynebacterium glutamicum* for improved L-arginine synthesis by enhancing NADPH supply. J Ind Microbiol Biotechnol. https://doi.org/10.1007/s10295-018-2103-8

Zhang F, Keasling J (2011) Biosensors and their applications in microbial metabolic engineering. Trends Microbiol 19:323–329

Zhang Y, Shang X, Deng A, Chai X, Lai S, Zhang G, Wen T (2012) Genetic and biochemical characterization of *Corynebacterium glutamicum* ATP phosphoribosyltransferase and its three mutants resistant to feedback inhibition by histidine. Biochimie 94:829–838

Zhang B, Zhou N, Liu YM, Liu C, Lou CB, Jiang CY, Liu SJ (2015) Ribosome binding site libraries and pathway modules for shikimic acid synthesis with *Corynebacterium glutamicum*. Microb Cell Fact 14:71

Zhou LB, Zeng AP (2015a) Engineering a lysine-ON riboswitch for metabolic control of lysine production in *Corynebacterium glutamicum*. ACS Synth Biol 4:1335–1340

Zhou LB, Zeng AP (2015b) Exploring lysine riboswitch for metabolic flux control and improvement of L-lysine synthesis in *Corynebacterium glutamicum*. ACS Synth Biol 4:729–734

Zhou Z, Wang C, Chen Y, Zhang K, Xu H, Cai H, Chen Z (2015a) Increasing available NADH supply during succinic acid production by *Corynebacterium glutamicum*. Biotechnol Prog 31:12–19

Zhou K, Qiao K, Edgar S, Stephanopoulos G (2015b) Distributing a metabolic pathway among a microbial consortium enhances production of natural products. Nat Biotechnol 33:377–383

Zhu N, Xia H, Wang Z, Zhao X, Chen T (2013) Engineering of acetate recycling and citrate synthase to improve aerobic succinate production in *Corynebacterium glutamicum*. PLoS One 8:e60659

Zhu N, Xia H, Yang J, Zhao X, Chen T (2014) Improved succinate production in *Corynebacterium glutamicum* by engineering glyoxylate pathway and succinate export system. Biotechnol Lett 36:553–560

Aromatic Compound Catabolism in *Corynebacterium glutamicum*

Yukihiro Kitade, Kazumi Hiraga, and Masayuki Inui

Contents

Abstract Many kinds of aromatic compounds can be assimilated by *Corynebacterium glutamicum*. Genome sequencing has enabled the elucidation of genetic and biochemical identification, and the transcriptional regulation of the degradation pathway genes. This chapter is divided into two sections, summarized based on current knowledge; Sect. 1, basic research on functionally identified or putative uptake systems and degradation pathways of aromatic compounds in *C. glutamicum*, and Sect. 2, applied research on strain construction for production of various aromatic compounds using aromatic catabolic genes.

Y. Kitade · K. Hiraga
Research Institute of Innovative Technology for the Earth (RITE), Kizugawa, Kyoto, Japan

M. Inui (✉)
Research Institute of Innovative Technology for the Earth (RITE), Kizugawa, Kyoto, Japan

Graduate School of Biological Sciences, Nara Institute of Science and Technology, Ikoma, Nara, Japan
e-mail: inui@rite.or.jp

M. Inui, K. Toyoda (eds.), *Corynebacterium glutamicum*, Microbiology Monographs 23, https://doi.org/10.1007/978-3-030-39267-3_11

1 Introduction

Corynebacterium has the ability to metabolize various aromatic compounds and use them as carbon sources for growth. Using aromatic compound-degrading *Corynebacterium* strains, biochemical studies were performed in the 1990s, and several possible pathways for aromatic degradation were proposed (Sikkema and de Bont 1993; Romanov and Hausinger 1996; Itoh et al. 1996). Understanding of aromatic catabolic pathways by *C. glutamicum* has drastically advanced by the discovery of its genome data after 2003 (Ikeda and Nakagawa 2003; Kalinowski et al. 2003). The first genetic and biochemical identification of an aromatic degradation gene from this bacterium was carried out by Shen et al. (2004). Transcriptional regulation of catabolic pathways for aromatic compounds was reported by Brinkrolf et al. (2006). The gene encodes a catechol 1,2-dioxygenase (i.e. CatA) in strain ATCC 13032 and the functional homologs were also found in many other strains (e.g., R, ATCC 14067, K051, S9114). Since the pioneer work by Shen et al. (2004), various findings on degradation and assimilation of aromatic compounds by *C. glutamicum* during the first 10 years have been summarized at a molecular level (Shen et al. 2012, 2015). To date, five central aromatic intermediates namely, protocatechuate (PCA), catechol (CA), hydroxyquinol, phenylacetyl-CoA, and gentisate, for direct ring-cleavage have been verified in *C. glutamicum* strains. In addition, recent studies on the aromatic catabolic pathways by *C. glutamicum* were briefly described in Lee and Wendisch (2017). In this chapter, we will summarize the knowledge on catabolism of aromatic and related compounds by *C. glutamicum* and include recent advances in both basic and applied research.

2 Degradation of Aromatic Compounds in *C. glutamicum*

2.1 Uptake and Transport of Aromatic Compounds

Several functionally identified or putative *C. glutamicum* transporters that are specific for aromatic compounds have been reported as described below.

4-Hydroxybenzoate (4-HBA) Transporter: *PcaK*
Genome-wide investigation of aromatic acid transporters in *C. glutamicum* was performed by Chaudhry et al. (2007). They demonstrated that PcaK encoded by *pcaK* (cg1225, cgR_1159) is a 4-HBA transporter of *C. glutamicum* by phenotypic characterization of genetically disrupted and complemented strains as well as resting cell assays for the uptake of 4-HBA. In addition, *pcaK* is reportedly involved in the transport of PCA, by phenotypic characterization of genetically disrupted and complemented strains while resting cell assays for the uptake of PCA have not been performed. PcaK has a 31% and 30% amino acid identity to 4-hydroxybenzoate and protocatechuate transporters of *Chromobacterium violaceum* ATCC 12472 and *Burkholderia pseudomallei*, respectively.

Vanillate Transporter: *VanK*

Merkens et al. (2005) reported that *vanK* encodes a putative protocatechuate transporter. Subsequently, Chaudhry et al. (2007) demonstrated that VanK encoded by *vanK* (cg2618, cgR_2267) is a vanillate (4-hydroxy-3-methoxybenzoate) transporter of *C. glutamicum* by phenotypic characterization of genetically disrupted and complemented strains as well as resting cell assays for the uptake of vanillate. VanK has a 45% and 42% amino acid identity to the vanillate transport protein of *Pseudomonas putida* and *Acinetobacter* sp. ADP1, respectively.

PhdT

C. glutamicum grows on phenylpropanoids as the sole carbon source. Recently, the *phd* gene cluster responsible for phenylpropanoid utilization was identified by Kallscheuer et al. (2016a). The putative transporter gene *phdT* (cg0340, cgR_0369) is one of the six adjacent genes, which showed enhanced transcript levels in the presence of phenylpropanoids. Although deletion of *phdT* led to some reduction in the growth rate in the cultivation experiments with phenylpropanoids, the gene was not essential for growth. The uptake mechanism of phenylpropanoids remains unknown. Other transporter candidates for phenylpropanoids could be found in the future.

Shikimate Transporter: *ShiA*

C. glutamicum grows well with either quinate (QA) or shikimate as the sole carbon source. Although the catabolic route for degradation of QA and shikimate has already been characterized, the gene involved in shikimate uptake has not been experimentally identified. It is presumed that QsuA encoded by *qsuA* (cg0501, cgR_0492) is a QA/shikimate transporter because *qsuA* is part of an operon of QA/shikimate utilization genes (Teramoto et al. 2009; Kubota et al. 2014). However, the function of QsuA is still unknown. Recently, our group demonstrated that ShiA, which belongs to the major facilitator superfamily, is the primary shikimate transporter of *C. glutamicum* by both shikimate consumption and accumulation assays as well as fluorescent imaging with mTFP1-tagged ShiA (Kubota et al. 2015). The ShiA encoded by *shiA* (cgR_2523) has a 36% amino acid identity to *E. coli* shikimate transporter. Disruption of *shiA* resulted in loss of the ability to grow on shikimate and to consume extracellular shikimate. In addition, it was reported that ShiR encoded by *shiR* (cgR_2524), which is located immediately upstream of *shiA*, acts as an activator of the gene. The gene homolog of *shiR* (cg2899) was found in the genome of *C. glutamicum* ATCC 13032, while *shiA* was absent in the strain ATCC 13032.

Benzoate Transporters: *BenK* and *BenE*

Chaudhry et al. (2007) demonstrated that BenK encoded by *benK* (cg2642, cgR_2290) and BenE encoded by *benE* (cg2643, cgR_2291) are involved in benzoate transport in *C. glutamicum* by phenotypic characterization of genetically disrupted and complemented strains as well as resting cell assays for the uptake of benzoate. A double-knockout mutant of *benK* and *benE* lost the ability to grow on benzoate, while the double-knockout mutants complemented by either one of these genes exhibited normal growth. In addition, the double mutant strain lost the ability to take up benzoate, while every single mutant strain retained it, and the uptake rate for benzoate in the *benE* mutant was higher than the rate in the *benK* one. In membrane proteomic analysis, BenK was identified as one of the 24 proteins significantly upregulated during growth on benzoate in comparison to glucose (Haußmann et al. 2009). In addition, it was confirmed that both BenK and BenE were expressed during growth on benzoate at the protein level. Wang et al. (2011) identified three critical residues in the hydrophilic cytoplasmic loops (Gly-80, Asp-84 and Asp-312) and four in the hydrophobic transmembrane regions (Asp-35, Arg-119, Glu-139 and Arg-386) of BenK for benzoate transport by site-directed mutagenesis.

PaaT

Chen et al. (2012) identified the genes involved in phenylacetic acid (PAA) catabolism. These genes are organized as a continuous *paa* gene cluster in *C. glutamicum* AS 1.542. The putative transporter gene (*paaT*) is part of the gene cluster, which comprises 15 genes, and its transcription is regulated by the PAA catabolic pathway-specific repressor PaaR, which is encoded in the cluster. However, gene function analysis with the disruption of *paaT* has not been performed.

Gentisate Transporter: *GenK*

Shen et al. (2005b) reported that NCgl2922 encodes a putative transporter protein and is essential to gentisate (2,5-dihydroxybenzoate) assimilation. Subsequently, NCgl2922 was designated as *genK* by Chaudhry et al. (2007). Several years later, Xu et al. (2012) demonstrated that GenK is an active gentisate transporter in *C. glutamicum* ATCC 13032 by uptake assays with [^{14}C]-labeled substrates. In the uptake assays, gentisate was transported, but the transport of 3-hydroxybenzoate and benzoate was not detected. In addition, competition and inhibition experiments revealed that gentisate and 3-hydroxybenzoate inhibited gentisate uptake activity by 94%, and benzoate, catechol, and protocatechuate inhibited the activity by about 40–60%.

2.2 Degradation Pathways for Aromatic Compounds

Several individual peripheral and four common degradation pathways for aromatic compounds have been reported in *C. glutamicum* (Fig. 1). The β-ketoadipate pathway has PCA and CA branches and is involved in the degradation of various aromatic compounds such as 4-HBA, vanillate, 4-cresol (4-methylphenol), phenylpropanoids, hydroaromatics, benzoate, and phenol. In addition, it has been reported that degradation of four central aromatic intermediates such as PCA, CA, hydroxyquinol, and phenylacetyl-CoA gives rise to acetyl-CoA and succinyl-CoA, whereas degradation of other central aromatic intermediates such as gentisate yields fumarate and pyruvate (Shen et al. 2012; Lee and Wendisch 2017).

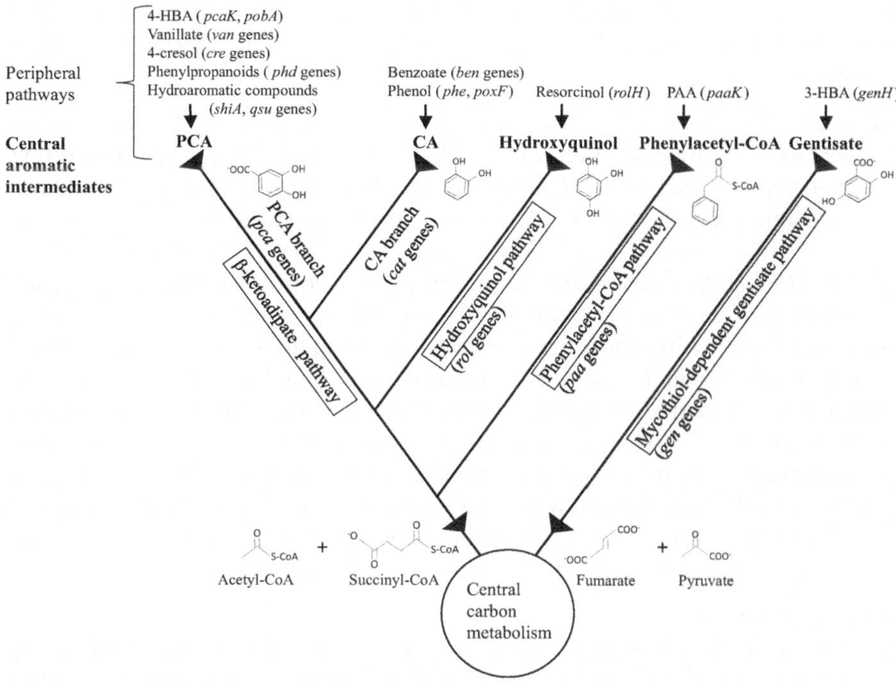

Fig. 1 Schematic representation of catabolic pathways for aromatic compounds in *Corynebacterium glutamicum*. *4-HBA* 4-hydroxybenzoate, *PAA* phenylacetic acid, *3-HBA* 3-hydroxybenzoate, *PCA* protocatechuate, *CA* catechol

2.2.1 PCA and CA Branches of the β-Ketoadipate Pathway

4-HBA and Vanillate

4-HBA is catabolized via the PCA branch of the β-ketoadipate pathway (Fig. 1). The gene *pobA* (cg1226, cgR_1160) encoding 4-HBA hydroxylase, involved in the conversion of 4-HBA into PCA in *C. glutamicum,* was identified by Huang et al. (2008). The *pobA* deletion mutant lost the ability to grow on 4-HBA as sole carbon source, and complementation with intact *pobA* restored growth on 4-HBA. The gene was expressed in *E. coli*, and the gene product was purified and biochemically characterized, which revealed that PobA from *C. glutamicum* is unique in its preference for NADPH instead of NADH as a cofactor.

Vanillate is also catabolized via the PCA branch of the β-ketoadipate pathway (Fig. 1). The vanillate-*O*-demethylase, converting vanillate to PCA, is encoded by the *vanAB* (cg2616-2617, cgR_2265-2266). Merkens et al. (2005) verified that the transcription of *vanABK* operon is induced by vanillate, and the *vanA* disruptant is unable to grow on vanillate. Recently, it was demonstrated that the *vanABK* operon is regulated by a PadR-type repressor, VanR, which is encoded by *vanR* (cg2615, cgR_2264), by Morabbi Heravi et al. (2015). In addition, it was revealed that vanillate is the only effector of VanR. Analyses of interaction between the promoter region of the *vanABK* operon and the VanR were performed, and it has been proposed that the VanR-DNA complex contains two VanR dimers at the VanR operator.

4-Cresol

C. glutamicum grows on 4-cresol as sole carbon source. Shen and Liu (2005) revealed that 4-cresol is catabolized via the PCA branch of the β-ketoadipate pathway, although peripheral pathway to the common intermediate for ring-cleavage was unknown (Fig. 1). In later years, Li et al. (2014) identified the *creABCDEFGHIJR* gene cluster (designated for 4-cresol, NCgl0531-0521), and six (*creD*, *creE*, *creG*, *creH*, *creI*, and *creJ*) of the *cre* genes were experimentally confirmed to be involved in 4-cresol catabolism. In addition, the *creG* was heterologously expressed in *E. coli* and purified. It was found that the purified CreG catalyzed the oxidation of 4-hydroxybenzyl alcohol into 4-hydroxybenzaldehyde in the presence of NAD^+ as an electron acceptor. Recently, all biodegradation enzymes encoded by the *cre* gene cluster of *C. glutamicum* were cloned, expressed, and functionally characterized using *in vitro* enzymatic assays (Du et al. 2016). It should be noted that a unique phosphorylation reaction mediated by CreHI was identified as a novel initiating step of 4-cresol biodegradation.

Phenylpropanoids

C. glutamicum grows on phenylpropanoids such as *p*-coumaric acid, ferulic acid, caffeic acid, 3-(4-hydroxyphenyl) propionic acid (4-HPP) as sole carbon source. Kallscheuer et al. (2016a) identified the *phd* (phenylpropanoid degradation) gene cluster responsible for phenylpropanoid utilization in *C. glutamicum* by analysis of single-gene deletion mutants, complementation studies, and determination of accumulating pathway intermediates. In addition, it was shown that the *phd* gene cluster is transcriptionally controlled by a MarR-type repressor encoded by cg0343 (*phdR*). This study suggested that these four phenylpropanoids described above are subsequently catabolized via the PCA branch of the β-ketoadipate pathway (Fig. 1), and a CoA-dependent, β-oxidative deacetylation pathway for degradation of phenylpropanoids in *C. glutamicum* was proposed.

Hydroaromatic Compounds

Our group found that *C. glutamicum* grows with either QA or shikimate as the sole carbon source and that the *qsuD* encoding QA/shikimate dehydrogenase is essential for QA/shikimate utilization (Teramoto et al. 2009). Moreover, it was revealed that transcription of *qsuA*, *qsuB*, *qsuC* and *qsuD* genes is induced in the presence of QA or shikimate, and the *qsuR*, which is located immediately upstream of the *qsuA* in the opposite direction, encodes a LysR-type transcriptional activator of the operon. These hydroaromatic compounds are catabolized via the PCA branch of the β-ketoadipate pathway (Fig. 1). It was presumed that not only QsuD (QA/shikimate dehydrogenase) but also two enzymes, QsuC (3-dehydroquinate dehydratase) and QsuB (3-dehydroshikimate dehydratase), or one enzyme QsuB are involved in the peripheral pathway of QA or shikimate catabolism, respectively. Several years later, the QsuD was characterized biochemically. Our group demonstrated that the catalytic reaction of the QsuD is highly susceptible to pH, and reduces 3-dehydroquinate (DHQ) using NADH and oxidizes QA using NAD^+ as a cofactor (Kubota et al. 2013). Recently, our group verified that the QsuR activates the *qsu* operon in the presence of QA or shikimate, the QsuR tetramer binds to a promoter region upstream of the operon, and that chorismate is a direct effector of QsuR (Kubota et al. 2014).

Benzoate and Phenol

Shen et al. (2004) reported that *C. glutamicum* uses benzoate and phenol as sole carbon sources for growth. The key enzymes involved in the metabolism of benzoate and phenol were characterized and it was presumed that degradation of benzoate and phenol in *C. glutamicum* occurs through a CA branch of the β-ketoadipate pathway, although peripheral pathways to the common intermediate for ring-cleavage were unknown (Fig. 1). Shen et al. (2005a) confirmed that NCgl2319 (*catA*) encodes CatA (catechol 1,2-dioxygenase) by gene disruption; a deletion mutant of *catA* did not grow with phenol as the carbon source. Although the *benABCDRKE* cluster (cg2637-2643, cgR_2285-2291), which encodes putative proteins responsible for the conversion of benzoate into CA, was identified by genome analysis of *C. glutamicum*, gene functional analyses have not been performed except for *benK* and *benE*, both of which encode benzoate transporters as described above. It has been assumed that a putative LuxR-type regulator, BenR encoded by *benR* (cg2641) positively regulates the *ben* and/or *cat* sub-clusters (Brinkrolf et al. 2006). However, the role of BenR in benzoate catabolism has not been studied and even the target genes of BenR have not been identified experimentally. Haußmann et al. (2009) verified that two putative benzoate degrading enzymes, BenA (cg2637) and BenD (cg2640), and the putative LuxR-type transcriptional regulator of the *ben* gene cluster, BenR (cg2641), are induced during growth on benzoate by comparison of the membrane proteomes of glucose- and benzoate-grown *C. glutamicum* cells.

Shen et al. (2015) proposed that phenol is converted to CA by two putative phenol monooxygenases encoded by NCgl2588 and NCgl0050 based on new annotations for genes involved in aromatic compound degradation with *C. glutamicum* genomes. The gene homologs of NCgl2588 from strain ATCC 13032 were found in genomes of *C. glutamicum* R, ATCC 14067, K051, and S9114 while the gene homolog of NCgl0050 was found only in the genome of *C. glutamicum* K051. In strain R, the gene homolog of NCgl2588 has been described as *poxF* (cgR_2584, Kitade et al. 2018). Recently, genetic and biochemical characterization of phenol hydroxylase (Phe, NCgl2588) was performed by Xiao et al. (2015). This study revealed that Phe from *C. glutamicum* mainly functions in the peripheral pathway of phenol catabolism (Fig. 1) since the deletion mutant significantly reduced its ability to grow with phenol as the sole carbon source. Gene functional analysis of NCgl0050 has not been performed.

2.2.2 Hydroxyquinol Pathway

C. glutamicum grows on resorcinol (1,3-dihydroxybenzene) as the sole source of carbon. Resorcinol is catabolized via the hydroxyquinol pathway (Fig. 1). Shen et al. (2005a) identified two gene clusters, NCgl1110-1113 and NCgl2950-2953, for degradation of resorcinol by genome data mining and by experimental analysis. Two hydroxyquinol 1,2-dioxygenase genes (NCgl1113 and NCgl2951) were functionally identified. Huang et al. (2006) verified that the NCgl1111-encoded resorcinol hydroxylase is involved in the peripheral pathway of resorcinol catabolism. Although it was revealed that the gene cluster NCgl2950-2953 is not essential for resorcinol catabolism, both NCgl2952 and NCgl1112 encode maleylacetate reductases. It was confirmed that when the double mutant Δ(NCgl2950-2953)ΔNCgl1112 was complemented with the NCgl2952, the ability to grow on resorcinol was restored. In addition, it was revealed that both NCgl1110 and NCgl2950 encode putative TetR family repressors, but only NCgl1110 is transcribed and functional. Chaudhry et al. (2007) reported that the NCgl2953, which showed sequence similarity with the permease of the major facilitator superfamily (MFS), is not involved in resorcinol and hydroxyquinol transport, but its function has not been identified. The gene cluster NCgl1110-1113 was named the *rol* (designated for resorcinol) gene cluster *rolRHMD* by Li et al. (2012). This study revealed that RolR negatively regulates the transcription of *rolHMD* and of its own gene, and specifically binds to a 29-bp sequence (*rolO*) located at the intergenic region between *rolR* and *rolHMD*.

2.2.3 Phenylacetyl-CoA Pathway

Chen et al. (2012) reported that *C. glutamicum* strain AS 1.542, which is isolated from winery soil in southeast China, could grow on PAA as sole carbon source, but the type strain ATCC 13032 cannot utilize PAA. The *paa* gene cluster composed of 15 genes was identified by genome sequencing of *C. glutamicum* AS 1.542, while such gene cluster was absent in the strain ATCC 13032. In addition, *paa* gene clusters with the same genetic organization to that of the strain AS 1.542 were found in the genomes of *C. glutamicum* R and *C. efficiens* YS-314. Strain ATCC 13032 gained the ability to grow on PAA as a sole carbon source by the introduction of *paa* gene cluster of the strain AS 1.542, confirming the role of the gene cluster in PAA degradation. Further experiments to characterize the PAA catabolic pathway genes using each gene disruption of the gene cluster have not been performed except for *paaR*, which encodes a TetR-type transcriptional repressor of the cluster (Fig. 1).

2.2.4 Mycothiol-Dependent Gentisate Pathway for 3-HBA and Gentisate Catabolism

Shen et al. (2005b) experimentally verified that the gene cluster NCgl2918-2923 (cgR_2908-2913) is involved in 3-HBA and gentisate catabolism, and NCgl2918 (cgR_2908) encodes a glutathione-independent maleylpyruvate isomerase, which converts 3-maleylpyruvate into 3-fumarylpyruvate during the catabolism. Moreover, Feng et al. (2006) found that the maleylpyruvate isomerase (GenM) needed mycothiol as a cofactor, and demonstrated that mycothiol is essential for the growth of *C. glutamicum* with 3-HBA or gentisate as carbon sources. 3-HBA 6-hydroxylase (GenH) encoded by NCgl2923 (cgR_2913, *genH*) in *C. glutamicum* was biochemically characterized by Yang et al. (2010) and shown to be involved in the peripheral pathway of 3-HBA catabolism (Fig. 1). Chao and Zhou (2013) demonstrated that GenR (NCgl2921, cgR_2911), an IclR-type regulator, activates the transcription of *genKH* (NCgl2922-2923, cgR_2912-2913) and *genDFM* (NCgl2920-2918, cgR_2910-2908) operons for 3-HBA and gentisate catabolism, and represses its own expression. Recently, it was revealed that not only GenR but also global regulator GlxR is involved in this pathway, and GlxR would contribute to the synchronization of expression of the *gen* operons in collaboration with the specific regulator GenR. Chao and Zhou (2014) demonstrated that GlxR binds to three sites in the promoter regions of *gen* operons, and represses the transcription.

3 Application of Aromatic Catabolic Genes to Strain Construction for Production of Various Aromatic Compounds

Several studies reported on the bioproduction of aromatic compounds (Averesch and Krömer 2018; Kallscheuer and Marienhagen 2018). *C. glutamicum* is an excellent host for bioproduction of aromatic compounds because of its high productivity, and highly resistant properties against the compounds. Molecular breeding of *C. glutamicum* for production of aromatic compounds is described in the other chapter of this book. Catabolic genes are disrupted for strain construction of aromatic compounds. In this chapter, modification of some catabolic pathways in *C. glutamicum* for aromatic compounds production is reviewed.

3.1 qsu *Genes and* pobA

The *qsu* gene cluster is involved in QA/shikimate utilization. In order to construct a shikimate-overproducing strain, Kogure et al. (2016) disrupted the *qsuB*, encoding 3-dehydroshikimate (DHS) dehydratase, which catalyzes the conversion of DHS into PCA, and the *qsuD*, encoding QA/shikimate dehydrogenase, which preferably catalyzes the conversion of shikimate into DHS as well as the interconversion of QA and DHQ (Table 1). *qsuB* and *qsuD* were also disrupted in 4-HBA-producing strains because both 4-HBA and shikimate are produced via the shikimate pathway (Kitade et al. 2018; Syukur Purwanto et al. 2018).

Since *pobA* is involved in the degradation of 4-HBA in *C. glutamicum*, it is necessary to disrupt it in strain construction for production of 4-HBA (Table 1).

3.2 cat, pca, *and* van *Genes*

The *cat* gene cluster is involved in the conversion of CA into β-ketoadipate. Becker et al. (2018) disrupted *catB*, encoding muconate cycloisomerase, to block *cis*, *cis*-muconic acid (MA) degradation, and overexpressed *catA* to overcome the bottleneck at the level of CatA for construction of MA producing strain from catechol (Table 1).

The *pca* gene cluster is involved in PCA catabolism. Lee et al. (2018) disrupted not only *catB* as described above but also the *pcaG/H*, encoding PCA dioxygenase alpha/beta subunits, to block PCA degradation because the authors redesigned *C. glutamicum* to accomplish the conversion of PCA to CA by expressing a codon-optimized heterologous PCA decarboxylase gene from *Klebsiella pneumoniae*. Kallscheuer and Marienhagen (2018) deleted the *pcaFDO* – *pcaCBGH* genes to block PCA degradation, and overexpressed *qsuB* for construction of producing platform of hydroxybenzoic acids, including PCA (Table 1).

Okai et al. (2017) modified a PCA-producing strain of *C. glutamicum* ATCC 21420 by overexpressing *vanAB*, encoding the vanillate *O*-demethylase subunits A and B, to utilize efficiently the lignin-derived phenolic compound ferulic acid as a start material. Genomic analysis revealed mutations in the *pca* gene cluster in the strain, which are the most probable causes of PCA production without deletion of the gene cluster (Table 1).

3.3 gen *Genes*

The *gen* gene cluster is involved in 3-HBA and gentisate catabolism. Kallscheuer and Marienhagen (2018) deleted *genDFM* – *genR* – *genKH* to block 3-HBA degradation for construction of 3-HBA producing strain (Table 1).

3.4 phd *Genes*

The *phd* gene cluster is involved in phenylpropanoid utilization. Phenylpropanoids such as cinnamic acid, *p*-coumaric acid, and caffeic acid can be produced by

Table 1 Utilization of aromatic catabolic genes for strain construction of aromatic and related compounds production

Target compound	Feedstock	Knocked out gene in catabolic pathway	Overexpressed gene in catabolic pathway	Titer (g/L)	Fermentation mode	Reference
Shikimate	Glucose	*qsuB*, *qsuD*	–	141.2	Jar fermenter	Kogure et al. (2016)
4-HBA	Glucose	*qsuB*, *qsuD*, *pobA*	–	36.6	Jar fermenter	Kitade et al. (2018)
	Glucose	*qsuB*, *pobA*	–	3.3	Flask	Kallscheuer and Marienhagen (2018)
	Glucose	*qsuABD*, *pobA*	–	19.0	Jar fermenter	Syukur Purwanto et al. (2018)
MA	Glucose	*catB*, *pcaGH*	*qsuB*	4.5	Culture tube	Shin et al. (2018)
	Catechol	*catB*	*catA*	85	Jar fermenter	Becker et al. (2018)
	Glucose	*catB*, *pcaGH*	–	53.8	Jar fermenter	Lee et al. (2018)
PCA	Ferulic acid	–	*vanAB*	1.1	Jar fermenter	Okai et al. (2017)
	Glucose	*pcaFDO* – *pcaCBGH*	*qsuB*	2.0	Flask	Kallscheuer and Marienhagen (2018)
3-HBA	Glucose	*genDFM* – *genR* – *genKH*	–	0.3	Flask	Kallscheuer and Marienhagen (2018)
Resveratrol	Glucose	*phdBCDE*, *qsuB*	–	0.06	Flask	Kallscheuer et al. (2016b)

4-HBA 4-hydroxybenzoate, *MA cis*, *cis*-muconic acid, *PCA* protocatechuate, *3-HBA* 3-hydroxybenzoate

extending synthetic pathway of aromatic amino acids phenylalanine or tyrosine, and also used as intermediates of polyphenol production (e.g., (2S)-flavanones and stilbenes). Kallscheuer et al. (2016b) blocked phenylpropanoid degradation by deletion of the *phd* cluster to construct a strain for the production of the stilbene resveratrol. *qsuB* in the strain was also deleted to reduce PCA accumulation (Table 1).

4 Conclusion and Future Directions

Several transporters (PcaK, VanK, ShiA, BenK, BenE, GenK), which were involved in aromatic compounds catabolism were identified functionally in *C. glutamicum*. However, uptake mechanisms of phenylpropanoids, PAA, 4-cresol, phenol, and resorcinol should be elucidated genetically and biochemically in the future.

A limited number of *pca* genes such as *pcaHG* and *pcaO* were functionally identified (Zhao et al. 2010; Shen et al. 2015), but more than half of the genes from the *pca* gene cluster involved in PCA branch of the β-ketoadipate pathway have not been identified functionally to date. Fundamental studies are needed to understand the catabolic pathway precisely. The *van*, *cre*, *phd*, and *qsu* gene clusters, as well as *pobA* involved in peripheral pathways of the PCA branch of the β-ketoadipate pathway, have already been functionally identified. The *cat* gene cluster involved in the CA branch of the β-ketoadipate pathway has not been functionally identified except for *catA*. In addition, functional identifications of the *ben* gene cluster and the phenol degradation gene involved in peripheral pathways of the CA branch have not been reported. Further research is needed for elucidating the mechanism of benzoate and phenol catabolism. The *rol* gene cluster involved in hydroxyquinol pathway and the *gen* gene cluster involved in gentisate one have already been identified functionally. The *paa* gene cluster involved in phenylacetyl-CoA pathway exists within genomes of some strains of *C. glutamicum* excluding ATCC 13032. The *C. glutamicum* strain ATCC 13032 gained function to catabolize PAA by introducing its gene cluster from strain AS 1.542. However, most of the *paa* genes have not been identified functionally to date. Further research, including biochemical analyses, is needed to elucidate the mechanism of PAA catabolism.

Several aromatic catabolic genes, including *qsuB*, *qsuD*, *pobA*, *catA*, *catB*, *pcaGH*, and *vanAB*, have been utilized for strain construction of aromatic compounds production by disruption and/or overexpression. Since microbial production of aromatic compounds requires relatively long biosynthetic pathways, glucose or various metabolic intermediates of target compounds may be used as start materials for bioconversion. Recently two new approaches for strain construction of hydroaromatic compound (i.e., shikimate) production were developed in *C. glutamicum*. In order to monitor the shikimate production of different *C. glutamicum* strains, one approach used the *shiA* promoter region of *C. glutamicum* as a shikimic acid biosensor (Liu et al. 2018). The other approach used the CRISPRi system to regulate gene expression in *C. glutamicum* at transcriptional level for adjustment of metabolic flux involving shikimate synthetic pathway, although aromatic catabolic genes were not used as a critical target (Zhang et al. 2016). These approaches may accelerate various applied researches using aromatic catabolic genes in *C. glutamicum.*

In general, lignin-derived phenolic compounds (*ex.* ferulic acid, vanillin, vanillate, 4-hydroxybenzoate, phenol) and furfurals produced in the lignocellulosic hydrolysates inhibit microbial fermentation and result in a severe reduction of the desired products (Klinke et al. 2004). Our research group showed that *C. glutamicum* is resistant to pretreatment-derived inhibitors of lignocellulosic biomass such as 4-hydroxybenzaldehyde, furfural, and hydroxymethylfurfural under growth-arrested conditions (Sakai et al. 2007). Aromatic compound degradation ability and cell wall/membrane structure may be involved in the resistant mechanisms of *C. glutamicum.* Thus, *C. glutamicum* will be an excellent platform for efficient bioconversion of lignocellulosic feedstocks into high-value products by engineered cells.

References

Averesch NJH, Krömer JO (2018) Metabolic engineering of the shikimate pathway for production of aromatics and derived compounds-present and future strain construction strategies. Front Bioeng Biotechnol 6:32

Becker J, Kuhl M, Kohlstedt M, Starck S, Wittmann C (2018) Metabolic engineering of *Corynebacterium glutamicum* for the production of *cis, cis*-muconic acid from lignin. Microb Cell Fact 17:115

Brinkrolf K, Brune I, Tauch A (2006) Transcriptional regulation of catabolic pathways for aromatic compounds in *Corynebacterium glutamicum.* Genet Mol Res 5:773–789

Chao H, Zhou NY (2013) GenR, an IclR-type regulator, activates and represses the transcription of *gen* genes involved in 3-hydroxybenzoate and gentisate catabolism in *Corynebacterium glutamicum.* J Bacteriol 195:1598–1609

Chao H, Zhou NY (2014) Involvement of the global regulator GlxR in 3-hydroxybenzoate and gentisate utilization by *Corynebacterium glutamicum.* Appl Environ Microbiol 80:4215–4225

Chaudhry MT, Huang Y, Shen XH, Poetsch A, Jiang CY, Liu SJ (2007) Genome-wide investigation of aromatic acid transporters in *Corynebacterium glutamicum.* Microbiology 153:857–865

Chen X, Kohl TA, Rückert C, Rodionov DA, Li LH, Ding JY, Kalinowski J, Liu SJ (2012) Phenylacetic acid catabolism and its transcriptional regulation in *Corynebacterium glutamicum.* Appl Environ Microbiol 78:5796–5804

Du L, Ma L, Qi F, Zheng X, Jiang C, Li A, Wan X, Liu SJ (2016) Characterization of a unique pathway for 4-cresol catabolism initiated by phosphorylation in *Corynebacterium glutamicum.* J Biol Chem 291:6583–6594

Feng J, Che Y, Milse J, Yin YJ, Liu L, Rückert C, Shen XH, Qi SW, Kalinowski J, Liu SJ (2006) The gene *ncgl2918* encodes a novel maleylpyruvate isomerase that needs mycothiol as cofactor and links mycothiol biosynthesis and gentisate assimilation in *Corynebacterium glutamicum.* J Biol Chem 281:10778–10785

Haußmann U, Qi SW, Wolters D, Rögner M, Liu SJ, Poetsch A (2009) Physiological adaptation of *Corynebacterium glutamicum* to benzoate as alternative carbon source—a membrane proteome-centric view. Proteomics 9:3635–3651

Huang Y, Zhao KX, Shen XH, Chaudhry MT, Jiang CY, Liu SJ (2006) Genetic characterization of the resorcinol catabolic pathway in *Corynebacterium glutamicum.* Appl Environ Microbiol 72:7238–7245

Huang Y, Zhao KX, Shen XH, Jiang CY, Liu SJ (2008) Genetic and biochemical characterization of a 4-hydroxybenzoate hydroxylase from *Corynebacterium glutamicum*. Appl Microbiol Biotechnol 78:75–73

Ikeda M, Nakagawa S (2003) The *Corynebacterium glutamicum* genome: features and impacts on biotechnological processes. Appl Microbiol Biotechnol 62:99–109

Itoh N, Yoshida K, Okada K (1996) Isolation and identification of styrene-degrading *Corynebacterium* strains, and their styrene metabolism. Biosci Biotechnol Biochem 60:1826–1830

Kalinowski J, Bathe B, Bartels D, Bischoff N, Bott M, Burkovski A, Dusch N, Eggeling L, Eikmanns BJ, Gaigalat L, Goesmann A, Hartmann M, Huthmacher K, Krämer R, Linke B, McHardy AC, Meyer F, Möckel B, Pfefferle W, Pühler A, Rey DA, Rückert C, Rupp O, Sahm H, Wendisch VF, Wiegräbe I, Tauch A (2003) The complete *Corynebacterium glutamicum* ATCC13032 genome sequence and its impact on the production of L-aspartate-derived amino acids and vitamins. J Biotechnol 104:5–25

Kallscheuer N, Marienhagen J (2018) *Corynebacterium glutamicum* as platform for the production of hydroxybenzoic acids. Microb Cell Fact 17:70

Kallscheuer N, Vogt M, Kappelmann J, Krumbach K, Noack S, Bott M, Marienhagen J (2016a) Identification of the *phd* gene cluster responsible for phenylpropanoid utilization in *Corynebacterium glutamicum*. Appl Microbiol Biotechnol 100:1871–1881

Kallscheuer N, Vogt M, Stenzel A, Gätgens J, Bott M, Marienhagen J (2016b) Construction of a *Corynebacterium glutamicum* platform strain for the production of stilbenes and (2S)-flavanones. Metab Eng 38:47–55

Kitade Y, Hashimoto R, Suda M, Hiraga K, Inui M (2018) Production of 4-hydroxybenzoic acid by an aerobic growth-arrested bioprocess using metabolically engineered *Corynebacterium glutamicum*. Appl Environ Microbiol 84:e02587–17

Klinke HB, Thomsen AB, Ahring BK (2004) Inhibition of ethanol-producing yeast and bacteria by degradation products produced during pre-treatment of biomass. Appl Microbiol Biotechnol 66:10–26

Kogure T, Kubota T, Suda M, Hiraga K, Inui M (2016) Metabolic engineering of *Corynebacterium glutamicum* for shikimate overproduction by growth-arrested cell reaction. Metab Eng 38:204–216

Kubota T, Tanaka Y, Hiraga K, Inui M, Yukawa H (2013) Characterization of shikimate dehydrogenase homologues of *Corynebacterium glutamicum*. Appl Microbiol Biotechnol 97:8139–8149

Kubota T, Tanaka Y, Takemoto N, Watanabe A, Hiraga K, Inui M, Yukawa H (2014) Chorismate-dependent transcriptional regulation of quinate/shikimate utilization genes by LysR-type transcriptional regulator QsuR in *Corynebacterium glutamicum*: carbon flow control at metabolic point. Mol Microbiol 92:356–368

Kubota T, Tanaka Y, Takemoto N, Hiraga K, Yukawa H, Inui M (2015) Identification and expression analysis of a gene encoding a shikimate transporter of *Corynebacterium glutamicum*. Microbiology 161:254–263

Lee JH, Wendisch VF (2017) Biotechnological production of aromatic compounds of the extended shikimate pathway from renewable biomass. J Biotechnol 257:211–221

Lee HN, Shin WS, Seo SY, Choi SS, Song JS, Kim JY, Park JH, Lee D, Kim SY, Lee SJ, Chun GT, Kim ES (2018) *Corynebacterium* cell factory design and culture process optimization for muconic acid biosynthesis. Sci Rep 8:18041

Li T, Zhao K, Huang Y, Li D, Jiang CY, Zhou N, Fan Z, Liu SJ (2012) The TetR-type transcriptional repressor RolR from *Corynebacterium glutamicum* regulates resorcinol catabolism by binding to a unique operator, *rolO*. Appl Environ Microbiol 78:6009–6016

Li T, Chen X, Chaudhry MT, Zhang B, Jiang CY, Liu SJ (2014) Genetic characterization of 4-cresol catabolism in *Corynebacterium glutamicum*. J Biotechnol 192:355–365

Liu C, Zhang B, Liu YM, Yang KQ, Liu SJ (2018) New intracellular shikimic acid biosensor for monitoring shikimate synthesis in *Corynebacterium glutamicum*. ACS Synth Biol 7:591–601

Merkens H, Beckers G, Wirtz A, Burkovski A (2005) Vanillate metabolism in *Corynebacterium glutamicum*. Curr Microbiol 51:59–65

Morabbi Heravi K, Lange J, Watzlawick H, Kalinowski J, Altenbuchner J (2015) Transcriptional regulation of the vanillate utilization genes (*vanABK* operon) of *Corynebacterium glutamicum* by VanR, a PadR-like repressor. J Bacteriol 197:959–972

Okai N, Masuda T, Takeshima Y, Tanaka K, Yoshida K, Miyamoto M, Ogino C, Kondo A (2017) Biotransformation of ferulic acid to protocatechuic acid by *Corynebacterium glutamicum* ATCC 21420 engineered to express vanillate *O*-demethylase. AMB Expr 7:130

Romanov V, Hausinger RP (1996) NADPH-dependent reductive *ortho* dehalogenation of 2,4-dichlorobenzoic acid in *Corynebacterium sepedonicum* KZ-4 and Coryneform bacterium strain NTB-1 via 2,4-dichlorobenzoyl coenzyme A. J Bacteriol 178:2656–2661

Sakai S, Tsuchida Y, Okino S, Ichihashi O, Kawaguchi H, Watanabe T, Inui M, Yukawa H (2007) Effect of lignocellulose-derived inhibitors on growth of and ethanol production by growth-arrested *Corynebacterium glutamicum* R. Appl Environ Microbiol 73:2349–2353

Shen X, Liu S (2005) Key enzymes of the protocatechuate branch of the β-ketoadipate pathway for aromatic degradation in *Corynebacterium glutamicum*. Sci China C Life Sci 48:241–249

Shen XH, Liu ZP, Liu SJ (2004) Functional identification of the gene locus (*ncg*l2319) and characterization of catechol 1,2-dioxygenase in *Corynebacterium glutamicum*. Biotechnol Lett 26:575–580

Shen XH, Huang Y, Liu SJ (2005a) Genomic analysis and identification of catabolic pathways for aromatic compounds in *Corynebacterium sepedonicum*. Microbes Environ 20:160–167

Shen XH, Jiang CY, Huang Y, Liu ZP, Liu SJ (2005b) Functional identification of novel genes involved in the glutathione-independent gentisate pathway in *Corynebacterium glutamicum*. Appl Environ Microbiol 71:3442–3452

Shen XH, Zhou NY, Liu SJ (2012) Degradation and assimilation of aromatic compounds by *Corynebacterium glutamicum*: another potential for applications for this bacterium? Appl Microbiol Biotechnol 95:77–89

Shen XH, Li T, Xu Y, Zhou NY, Liu SJ (2015) Transport, degradation and assimilation of aromatic compounds and their regulation in *Corynebacterium glutamicum*. In: Burkovski A (ed) *Corynebacterium glutamicum*. From systems biology to biotechnological applications. Caister Academic, Norfolk, pp 83–109

Shin WS, Lee D, Lee SJ, Chun GT, Choi SS, Kim ES, Kim S (2018) Characterization of a non-phosphotransferase system for *cis,cis*-muconic acid production in *Corynebacterium glutamicum*. Biochem Biophys Res Commun 499:279–284

Sikkema J, de Bont JAM (1993) Metabolism of tetralin (1,2,3,4-tetrahydronaphthalene) in *Corynebacterium* sp. Strain C125. Appl Environ Microbiol 59:567–572

Syukur Purwanto H, Kang MS, Ferrer L, Han SS, Lee JY, Kim HS, Lee JH (2018) Rational engineering of the shikimate and related pathways in *Corynebacterium glutamicum* for 4-hydroxybenzoate production. J Biotechnol 282:92–100

Teramoto H, Inui M, Yukawa H (2009) Regulation of expression of genes involved in quinate and shikimate utilization in *Corynebacterium glutamicum*. Appl Environ Microbiol 75:3461–3468

Wang SH, Xu Y, Liu SJ, Zhou NY (2011) Conserved residues in the aromatic acid/H^+ symporter family are important for benzoate uptake by NCgl2325 in *Corynebacterium glutamicum*. Int Biodeterior Biodegrad 65:527–532

Xiao X, Si M, Yang Z, Zhang Y, Guan J, Chaudhry MT, Wang Y, Shen X (2015) Molecular characterization of a eukaryotic-like phenol hydroxylase from *Corynebacterium glutamicum*. J Gen Appl Microbiol 61:99–107

Xu Y, Wang SH, Chao HJ, Liu SJ, Zhou NY (2012) Biochemical and molecular characterization of the gentisate transporter GenK in *Corynebacterium glutamicum*. PLoS One 7:e38701

Yang YF, Zhang JJ, Wang SH, Zhou NY (2010) Purification and characterization of the *ncgl2923*-encoded 3-hydroxybenzoate 6-hydoxylase from *Corynebacterium glutamicum*. J Basic Microbiol 50:599–604

Zhang B, Liu ZQ, Liu C, Zheng YG (2016) Application of CRISPRi in *Corynebacterium glutamicum* for shikimic acid production. Biotechnol Lett 38:2153–2161

Zhao KX, Huang Y, Chen X, Wang NX, Liu SJ (2010) PcaO positively regulates *pcaHG* of the β-ketoadipate pathway in *Corynebacterium glutamicum*. J Bacteriol 192:1565–1572

Aromatic Compound Production by *Corynebacterium glutamicum*

Takahisa Kogure, Takeshi Kubota, and Masayuki Inui

Contents

T. Kogure
Research Institute of Innovative Technology for the Earth (RITE), Kizugawa, Kyoto, Japan

T. Kubota
Graduate School of Biological Sciences, Nara Institute of Science and Technology, Nara, Japan

M. Inui (✉)
Research Institute of Innovative Technology for the Earth (RITE), Kizugawa, Kyoto, Japan

Graduate School of Biological Sciences, Nara Institute of Science and Technology, Nara, Japan
e-mail: inui@rite.or.jp; mmg-lab@rite.or.jp

M. Inui, K. Toyoda (eds.), *Corynebacterium glutamicum*, Microbiology Monographs 23, https://doi.org/10.1007/978-3-030-39267-3_12

Abstract Aromatic compounds represent important bulk and fine chemicals with numerous applications in the food, feed, cosmetics, pharmaceutical, and chemical industries. These are currently produced from non-renewable petroleum resources via chemical conversion with high environmental load or via inefficient extraction from natural plant resources. To cope with such problems, microbial production of aromatic compounds from abundant and renewable sugar feedstocks has recently received great attention as an eco-friendly and efficient alternative to producing aromatic compounds. In this context, the product spectrum of *Corynebacterium glutamicum* has recently extended to industrially valuable aromatic compounds ranging from various benzene/benzoate derivatives to more intricate plant polyphenols, some of which could be produced from sugars with markedly high productivities. This chapter summarizes recent advances in the metabolic engineering of *C. glutamicum* for the production of valuable aromatic and related compounds.

1 Introduction

Aromatic compounds, unsaturated chemical compounds typically containing one or more planar rings, have recently gained more and more attention. They serve a vast market in the chemical industry and have numerous industrial applications as food, food additives, nutraceuticals, pharmaceuticals, and the building blocks for the synthesis of polymer materials like functional plastics and fibers (Averesch and Krömer 2018; Krömer et al. 2013). Currently, most aromatic compounds that have relatively low molecular weight are produced by chemical conversion from petroleum-based feedstocks (e.g., benzene, toluene, xylene), which are not sustainable and rely on the extensive use of energy and harmful solvents, resulting in high CO_2 emissions. From the environmental point of view, biotechnological production of aromatic compounds from renewable feedstocks in an eco-friendly manner and by utilizing the ability of microorganisms to synthesize a broad range of different aromatic compounds has received much attention as a promising alternative (Averesch and Krömer 2018; Koma et al. 2012; Lee and Wendisch 2017; Noda and Kondo 2017; Wang et al. 2017). On the other hand, diverse and structurally intricate aromatic compounds like plant polyphenols, such as stilbenoids and flavonoids, are produced as natural secondary metabolites by plants and several fungi. They exhibit various human and animal health-promoting activities, and thus, their medicinal effects have been extensively studied in recent years (Scalbert et al. 2005b; Suastegui and Shao 2016). They are currently prepared by chemical synthesis or extraction from natural plant producers, but their productivities and yields are limited. As an alternative technology to address such an issue, fermentative production of those valuable natural compounds directly from abundant and inexpensive sugar feedstocks also came into focus (Chae et al. 2017; Jiang and Zhang 2016; Lee and Wendisch 2017; Lee et al. 2012; Suastegui and Shao 2016; Vargas-Tah et al. 2015; Wang et al. 2017; Wendisch et al. 2016).

Corynebacterium glutamicum has a long history as an industrial amino acid producer. Aromatic amino acid production by this organism has also been studied for many years (Ikeda 2006). However, this microbe's potential to produce other aromatic compounds of industrial significance had almost not been explored until recently. This could be partly attributed to the intrinsic ability of *C. glutamicum* to degrade and assimilate various aromatic compounds such as phenol and hydroxybenzoate via the catabolic pathway network, which is in contrast to the fact that aromatic amino acids are not further metabolized (Brinkrolf et al. 2006; Du et al. 2016; Kallscheuer et al. 2016a; Merkens et al. 2005; Shen and Liu 2005; Shen et al. 2012). However, the presence of such multiple metabolic pathways for various aromatic compounds would also contribute to the creation of aromatic producers since interception of these pathways at a selected metabolic step would lead to the accumulation of aromatic intermediates such as protocatechuate (PCA) relatively easily via endogenous metabolism. The presence of natural aromatic intermediates is also useful to produce derived aromatic compounds of interest if desired (Fig. 2). Moreover, it has been shown that *C. glutamicum* exhibits higher resistance to toxic aromatic compounds compared to other representative industrial microbes such as *Escherichia coli*, which would be advantageous for fermentative aromatics production (Kitade et al. 2018; Kubota et al. 2016; Liu et al. 2013). Our recent studies demonstrated that metabolically engineered strains of *C. glutamicum* have a marked ability to produce shikimate and aromatic compounds derived thereof (Kitade et al. 2018; Kogure et al. 2016; Kubota et al. 2016; Kogure and Inui 2018). The product portfolio of this bacterium has also been extended to valuable natural chemicals such as plant polyphenols by combining endogenous aromatic amino acid biosynthetic pathways with the plant-derived secondary metabolic pathways (Kallscheuer et al. 2016b, 2017). In this chapter, we summarized recent advances in metabolic engineering of *C. glutamicum,* focusing on the production of valuable aromatic and related compounds.

2 The Shikimate Pathway in *C. glutamicum* as the Common Pathway of Aromatics Biosynthesis

Most of the aromatic compounds, including aromatic amino acids and aromatic vitamins, are biosynthesized via the shikimate pathway, also known as the common pathway of aromatics biosynthesis, present in plants, fungi, and bacteria (Bentley 1990; Herrmann 1995). In this pathway, phosphoenolpyruvate (PEP) and erythrose-4-phosphate (E4P) generated in the central carbon metabolism are condensed into 3-deoxy-D-arabino-heptulosonate-7-phosphate (DAHP) and then converted via six sequential reactions to chorismate, a key branching point metabolite, from which the pathway branches off to the synthesis of aromatic amino acids, folate, ubiquinone, menaquinone, and other secondary metabolites like enterobactin (Fig. 1). Since the extension of the shikimate pathway via combining with heterologous metabolic

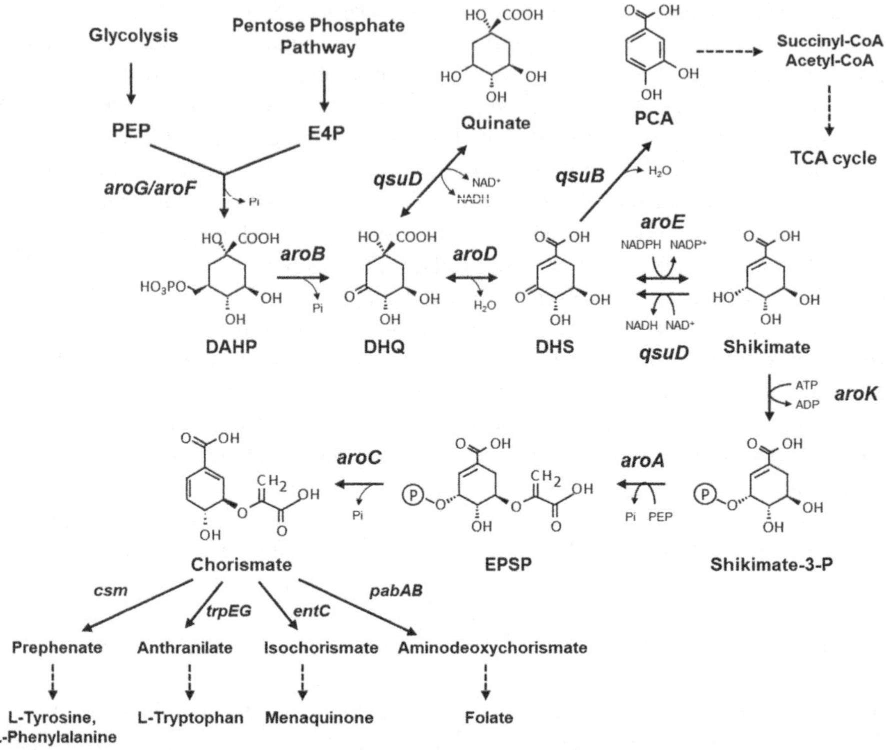

Fig. 1 The shikimate pathway in *C. glutamicum*. The genes (indicated in italics) encode the following enzymes: *aroG* and *aroF*, DAHP synthase; *aroB*, DHQ synthase; *aroD*, DHQ dehydratase; *aroE*, shikimate dehydrogenase; *aroK*, shikimate kinase; *aroA*, 5-enolpyruvyl shikimate-3-phosphate synthase; *aroC*, chorismite synthase; *qsuB*, DHS dehydratase; *qsuD*, quinate/shikimate dehydrogenase. Dashed arrows represent two or more catalytic steps. PEP, phosphoenolpyruvate; E4P, erythrose-4-phosphate; DAHP, 3-deoxy-D-arabino-heptulosonate-7-phosphate; DHQ, 3-dehydroquinate; DHS, 3-dehydroshikimate; PCA, protocatechuate; EPSP, 5-enolpyruvylshikimate-3-phosphate

enzymes enables production of a wide range of valuable aromatic and related compounds with diverse applications, metabolic engineering focusing on this pathway has gained much attention in the past decade (Averesch and Krömer 2018; Jiang and Zhang 2016; Koma et al. 2012; Lee and Wendisch 2017; Noda et al. 2016; Rodriguez et al. 2014). DAHP synthase (DS) represents the first and rate-limiting enzyme to regulate the carbon flow toward the shikimate pathway. Therefore, deregulation and enhancement of the DS activity through the overexpression of feedback inhibition-resistant DS forms are crucial to enhance the carbon flux directed to the shikimate pathway (Frost and Draths 1995; Ikeda 2006; Krämer et al. 2003), which results in the improvement of microbial aromatics production. *E. coli* has three DSs including, AroG, AroF, and AroH, each of which is subjected to the specific feedback inhibition by phenylalanine (Phe), tyrosine (Tyr), or tryptophan (Trp), respectively (Umbarger 1978). On the other hand, *C. glutamicum* ATCC 13032 has two DSs encoded by the *aroF* (NCgl0950) and *aroG* (NCgl2098) genes, of which the latter was shown to represent the active DS form in this organism, since deletion of *aroG* but not *aroF* caused auxotrophy for aromatic amino acids (Liu et al. 2008). Notably, AroG was not subject to the feedback inhibition by Tyr and/or Phe, a property inconsistent with the formerly accepted properties for AroG-like DS found in *C. glutamicum* subsp. flavum, which was shown to be strongly and synergistically inhibited by Phe and Tyr (Ikeda 2006; Shiio et al. 1974). It was shown that AroG forms a complex with the branch-point enzyme chorismate mutase (CM, NCgl0819), the first enzyme for Phe and Tyr biosynthesis catalyzing the conversion of chorismate into prephenate, and this interaction leads to substantial enhancement and regulation of DS activity by chorismate and prephenate (Li et al. 2009, 2013). The feedback inhibition insensitivity of *C. glutamicum* AroG seems to be useful for aromatics production. However, feedback-resistant mutant DS forms derived from *E. coli* ($AroF^{FBR}$ and $AroG^{FBR}$ and $AroH^{FBR}$) were shown to be superior with respect to DS activity to *C. glutamicum* AroG and accordingly they have been successfully employed to engineer several aromatics producer strains of *C. glutamicum* (Kallscheuer and Marienhagen 2018; Kitade et al. 2018; Kogure et al. 2016; Kubota et al. 2016; Syukur Purwanto et al. 2018).

While little is known about the regulation of the shikimate pathway genes in *C. glutamicum*, it was shown that the *qsuABCD* gene products, which constitute the shikimate/quinate utilization pathway that shares the route with the shikimate pathway (Fig. 1), are transcriptionally activated by the regulator QsuR in chorismate dependent manner (Kubota et al. 2014; Teramoto et al. 2009). The presence of this catabolic pathway necessitates its inactivation for efficient production of shikimate and derived aromatic compounds in this microbe (Kogure et al. 2016).

3 Production of Shikimate

Shikimate, a six-carbon hydroaromatic compound as the fourth intermediate in the shikimate pathway, attracts attention as a key building block for the synthesis of anti-influenza drug oseltamivir (Tamiful®) and many other pharmaceuticals (Martinez et al. 2015; Rawat et al. 2013; Tripathi et al. 2015). It also has potential applications in cosmetics such as a skin whitening agent attributed to its innate biological activity. While shikimate is currently produced by low-yield and costly isolation from the seed of the Chinese star anise of *Illicium* plants (Bochkov et al. 2012; Ghosh et al. 2012; Rawat et al. 2013), its more efficient microbial production from abundant sugars represents an attractive alternative to meet its increasing demand (Bilal et al. 2018; Gu et al. 2017; Krämer et al. 2003; Martinez et al. 2015). Metabolic engineering for shikimate production have been mainly attempted on *E. coli*, which represents classical and key approaches for aromatics production: (1) inactivation of the product-consuming activities (*aroK* and *aroL*-encoded shikimate kinase in *E. coli*), (2) enhancement of the carbon flux through the shikimate pathway by overexpressing a feedback-resistant form of DAHP synthase as well as other rate-limiting enzymes like *aroB*-encoded 3-dehydroquinate (DHQ) synthase and *aroE*-encoded shikimate dehydrogenase, (3) improvement of cellular PEP availability, which can be achieved via following approaches: (a) use of a non-PTS (PEP:sugar phosphotransferase system) glucose uptake route comprised of a native galactose permease (GalP) or a heterologous glucose facilitator (Glf) from *Zymomonas mobilis* and ATP-dependent glucokinase, instead of native PEP-consuming PTS, (b) overexpression of PEP synthase (encoded by *ppsA*) to reconvert pyruvate to PEP, and (c) inactivation of PEP-consuming enzymes such as pyruvate kinases (encoded by *pykA/pykF*) and PEP carboxylase, and (4) improvement of the E4P availability via overexpression of transketolase (*tkt*) and transaldolase (*tal*) (Draths et al. 1999; Escalante et al. 2010; Frost and Draths 1995; Knop et al. 2001; Krämer et al. 2003; Licona-Cassani et al. 2013; Rodriguez et al. 2013). The combination of these approaches has been examined in *E. coli* strains, resulting in the production of 87 g/L shikimate with a yield of 36% from glucose in fed-batch fermentation (Chandran et al. 2003). In addition to such intuitive engineering approaches, omics analysis-based system biology technologies such as transcriptome, metabolome, and fluxome have also contributed to the identification of novel genetic targets to develop shikimate overproducer (Bilal et al. 2018; Cortes-Tolalpa et al. 2014; Gu et al. 2017; Martinez et al. 2015).

Recently, shikimate overproduction has been achieved by engineering *C. glutamicum* (Fig. 2) (Kogure et al. 2016). In the engineered strain, $aroG^{S180F}$ encoding a feedback-resistant and high-activity DS from *E. coli* as well as endogenous *aroB*, *aroD*, *aroE* and *tkt-tal* were overexpressed, whereas *aroK* (shikimate kinase), *qsuB* (3-dehydroshikimate (DHS) dehydratase) and *qsuD* (quinate/shikimate dehydrogenase), which are involved in the consumption of shikimate and its precursor (Kubota et al. 2014; Teramoto et al. 2009), were disrupted. Additionally, as a strategy to improve the precursor PEP supply, endogenous *myo*-inositol permease IolT1 and glucokinases (Glk1, Glk2, and Ppgk) were overexpressed as a PEP-independent glucose uptake route (Ikeda et al. 2011; Lindner et al. 2011), whereas PTS was inactivated. However, in the resulting non-PTS strain of the shikimate producer, significant overflow metabolism toward dihydroxyacetone (DHA) synthesis from the glycolytic pathway was observed. To cope up with this byproduct formation, *gapA* encoding rate-limiting glycolytic enzyme glyceraldehyde-3-phosphate dehydrogenase (GAPDH) was overexpressed and *hdpA* encoding dihydroxyacetone phosphate phosphatase involved in DHA synthesis (Jojima et al. 2012) was deleted (Fig. 2). This approach successfully eliminated DHA formation and significantly enhanced glucose consumption and concomitant shikimate production, resulting in production of 141 g/L shikimate from glucose with a yield of 51% (mol/mol) after 48 h in a fermenter-controlled reaction employing growth-arrested cells, which represented the highest productivity for microbial shikimate production reported so far (Table 1) (Kogure et al. 2016). Another advantage of the obtained strain is the simultaneous utilization ability of C5/C6 mixed sugars (glucose, xylose, and arabinose) due to the engineering of the pentose utilization pathway of the host *C. glutamicum* strain (Sasaki et al. 2009).

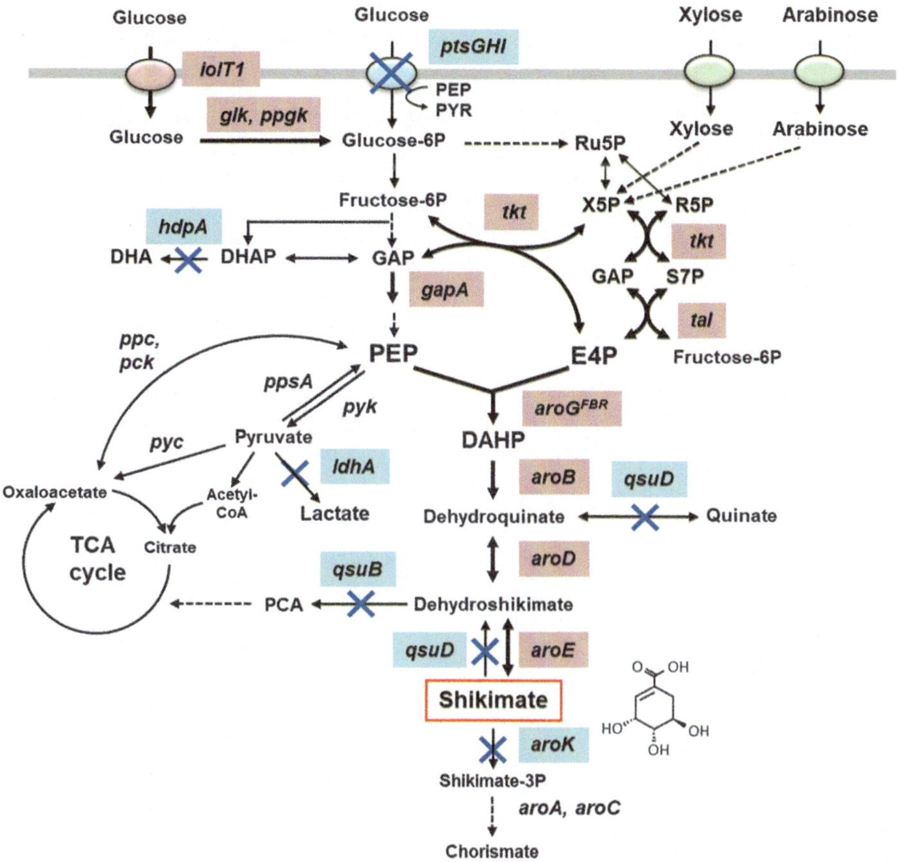

Fig. 2 Biosynthetic pathways and their engineering for shikimate production in *C. glutamicum*. Thick arrows with gene names highlighted in red indicate catalytic steps for which corresponding genes (indicated in italics) were overexpressed; Catalytic steps for which corresponding genes were deleted are indicated with crosses with gene names highlighted in blue. Dashed arrows represent two or more catalytic steps. The genes encode the following enzymes: *ptsGHI*, glucose-specific sugar:phosphoenolpyruvate phosphotransferase; *iolT1*, *myo*-inositol permease; *glk*, glucokinase; *ppgk*, polyphosphate glucokinase; *tkt*, transketolase; *tal*, transaldolase; *gapA*, glyceraldehyde-3-phosphate dehydrogenase; *hdpA*, dihydroxyacetone phosphate phosphatase; *ldhA*, L-lactate dehydrogenase: *aroG*, DAHP synthase; *aroB*, DHQ synthase; *aroD*, DHQ dehydratase; *aroE*, shikimate dehydrogenase; *aroK*, shikimate kinase; *aroA*, 5-enolpyruvylshikimate-3-phosphate synthase; *aroC*, chorismate synthase; *qsuB*, DHS dehydratase; *qsuD*, quinate/shikimate dehydrogenase. *pyk*, pyruvate kinase; *ppc*, PEP carboxylase; *pck*, PEP carboxykinase; *pyc*, pyruvate carboxylase. GAP, glyceraldehyde-3-phosphate; DHAP, 1,3-dihydroxyacetone phosphate; DHA, 1,3-dihydroxyacetone; PEP, phosphoenolpyruvate; Ru5P, ribulose-5-phosphate; R5P, ribose-5-phosphate; X5P, xylulose-5-phosphate; S7P, sedoheptulose-7-phosphate; E4P, erythrose-4-phosphate; DAHP, 3-deoxy-D-arabino-heptulosonate-7-phosphate; PCA, protocatechuate

Such a feature enables efficient shikimate production even from lignocellulosic feedstock-derived mixed sugars (Table 1) (Kogure et al. 2016). These results

Table 1 Summary of studies on aromatic and related compounds production by engineered *C. glutamicum*

Target compound	Host strain	Carbon source	Genes overexpressed[a]	Genes knocked out	Titer (g/L)	Yield[b] (%)	Time (h)	Culture style	Reference
Shikimate	*C. glutamicum* R	Glucose	*aroB, aroD, aroE, iolT1, glk1, glk2, ppgk, tkt, tal, gapA* *aroG*^{S180F} (*E. coli*)	*ldhA, aroK, qusB, qsuD, ptsH, hdpA*	141	51	48	Fermenter	Kogure et al. (2016)
	C. glutamicum Res167	Sucrose	*aroG, aroB, aroD, aroE* *mgsA* (*E. coli*), DKFPS, ADTHS, DHQS (*Methanosarcina mazei*)	*aroK*	4.7	ND	ND	Flask	Zhang et al. (2015a)
	C. glutamicum Res167	Sucrose	*aroG, aroB, aroD, aroE*	*aroK*	11.3	24	ND	Fermenter	Zhang et al. (2015b)
	C. glutamicum Res167	Sucrose	*aroG, aroB, aroD, aroE, tkt*	*aroK, pyk*, ncgl1856	23.8	ND	ND	Fermenter	Zhang et al. (2016)
PCA	*C. glutamicum* ATCC 21420	Glucose	*ubiC* (*E. coli*)		1.1	1.1	96	Fermenter	Okai et al. (2016)
	C. glutamicum ATCC 21420	Ferulic acid	*vanAB* (*Corynebacterium efficiens*)		1.1	43.2	12	Fermenter	Okai et al. (2017)
	C. glutamicum MB001	Glucose	*qsuB* , *tkt*, P_{O6}-*iolT1* *aroF*∗ (*E. coli*)	cg0344-47, cg2625-40, cg1226, cg0502, cg3349-54, P_{dapA}-*gltA*	2	ND	48	Flask	Kallscheuer and Marienhagen (2018)
Salicylate (2-HBA)	*C. glutamicum* MB001	Glucose	*Irp9* (*Yercinia enterocolitica*) *aroF*∗ (*E. coli*)	*cg0344-47, cg2625-40, cg1226, cg0502, cg3349-54*	0.01	ND	48	Flask	Kallscheuer and Marienhagen (2018)
3-HBA	*C. glutamicum* MB001	Glucose	*tkt*, P_{O6}-*iolT1*, *hyg5* (*Streptomyces hygroscopicus*), *aroF*∗ (*E. coli*)	cg0344-47, cg2625-40, cg1226, cg0502, cg3349-54, P_{dapA}-*gltA*	0.3	ND	48	Flask	Kallscheuer and Marienhagen (2018)

4-HBA	*C. glutamicum* R	Glucose	*tkt-tal, aroCKB, aroD, aroA, aroE* *aroG*^{S180F} (*E. coli*), *ubiC* (*Providencia rustigianii*)	*ldhA, qsuB, qsuD, pobA, poxF, pyk, hdpA*	36.6	41	24	Fermenter	Kitade et al. (2018)
	C. glutamicum MB001	Glucose	*tkt*, P_{O6}-*iolT1*, *ubiC* (*E. coli*), *aroH* (*E. coli*)	cg0344-47, cg2625-40, cg1226, cg0502, cg3349-54, P_{dapA}-*gltA*	3.3	ND	48	Flask	Kallscheuer and Marienhagen (2018)
	C. glutamicum ATCC 13032	Glucose	*ubiC*^{FBR}, *aroF*^{FBR} (*aroG*^{FBR}) (*E. coli*), *aroE, aroCKB, aroK* (*Methanocaldococcus jannaschii*)	*trpE, csm, pobA, qsuABD*	19	9.7	65	Fermenter	Syukur Purwanto et al. (2018)
3, 4-AHBA	*C. glutamicum* ATCC 21799	Sweet sorghum juice	*griH, griI* (*Streptomyces griseus*)		1	ND	72	Flask	Kawaguchi et al. (2015)
4-ABA	*C. glutamicum* R	Glucose	*aroCKB, aroD, aroA, aroE* *aroG*^{S180F}, *pabAB* (*Corynebacterium callunae*), *pabC* (*Xenorhabdus bovienii*)	*ldhA*	43	20	48	Fermenter	Kubota et al. (2016)
MA	*C. glutamicum* ATCC 13032	Glucose	*aroY/kpdBD*	*aroE, pcaGH, catB*	54	ND	168	Fermenter	Lee et al. (2018)
	C. glutamicum ATCC 13032	Glucose	*qsuB, YBD*	*aroE, pcaGH, catB ptsI, iolR*	4.5	22	72	Tube	Shin et al. (2018)
	C. glutamicum ATCC 13032	Catechol	*catB*	*catA*	85	100	60	Fermenter	Becker et al. (2018)
		Softwood lignin (catechol and phenol)			1.8	100	27	Fermenter	Becker et al. (2018)

ND not described
[a]The source organisms of heterologous genes are described in parentheses
[b]The yield is based on mol/mol of consumed sugar

highlighted the superior potential of *C. glutamicum* to produce shikimate and derived aromatic compounds.

Zhang et al. (2015a, b, 2016) also engineered *C. glutamicum* for shikimate production by applying synthetic biology strategies. They enforced the native shikimate pathway by implementing expression modules for the pathway genes (*aroG*, *aroB*, *aroD*, and *aroE*), where the translation of respective genes was regulated with different strengths of the ribosome-binding site (RBS) in a shikimate kinase (*aroK*) deficient host strain (Zhang et al. 2015b). The resulting strain with most effective genetic module produced 11.3 g/L of shikimate from sucrose with the yield of 24% (mol/mol) in the fed-batch fermentation (Table 1) (Zhang et al. 2015b). In a following study, they showed the availability of archaeal shikimate synthesis pathway (DKFP pathway) to enhance shikimate production in *C. glutamicum* (Zhang et al. 2015a). In that pathway, a glycolytic intermediate dihydroxyacetone phosphate (DHAP) is converted into DHQ via four reaction steps catalyzed by methylglyoxal synthase, 6-deoxy-5-ketofructose 1-phosphate (DKFP) synthase, 2-amino-3,7-dideoxy-D-threo-hept-6-ulosonate (ADTH) synthase, and DHQ synthase. Then, DHQ could be converted to shikimate by endogenous shikimate pathway. By introducing the archaeal DKFP pathway from *Methanosarcina mazei* into the previously engineered shikimate producer, the resulting hybrid strain harboring both archaeal and bacterial shikimate pathway produced 4.7 g/L of shikimate in flask culture, a 27% higher titer compared to that obtained by the strain harboring only native shikimate pathway (3.7 g/L), showing an positive contribution of archaeal pathway on shikimate production (Table 1) (Zhang et al. 2015a). The exploitation of archaeal shikimate pathway has an advantage that it does not depend on PEP for shikimate synthesis and therefore does not compete for the precursor with canonical shikimate pathway. Zhang et al. also applied CRISPR-dCas9 system to manipulate the transcription level of multiple genes related to shikimate production, resulting in the production of 23.8 g/L of shikimate from sucrose in the fermenter (Table 1) (Zhang et al. 2016).

4 Production of Aromatics and Related Compounds Derived from the Shikimate Pathway

Several industrially important aromatic compounds can be derived from the shikimate pathway via conversion of the branch point metabolites DHS and/or chorismate. These include hydroxybenzene like phenol and catechol, hydroxybenzoates (HBA) like 4-HBA, salicylate (2-HBA), 3-HBA, protocatechuate, as well as aminobenzoates (ABA) like 4-ABA (Fig. 3) (Kogure and Inui 2018; Koma et al. 2012; Lee and Wendisch 2017; Noda and Kondo 2017). These aromatic compounds represent typical bulk petrochemical products, and due to recent growing interest in the environment, their environmentally friendly and sustainable bioproduction has attracted attention. Their microbial production has

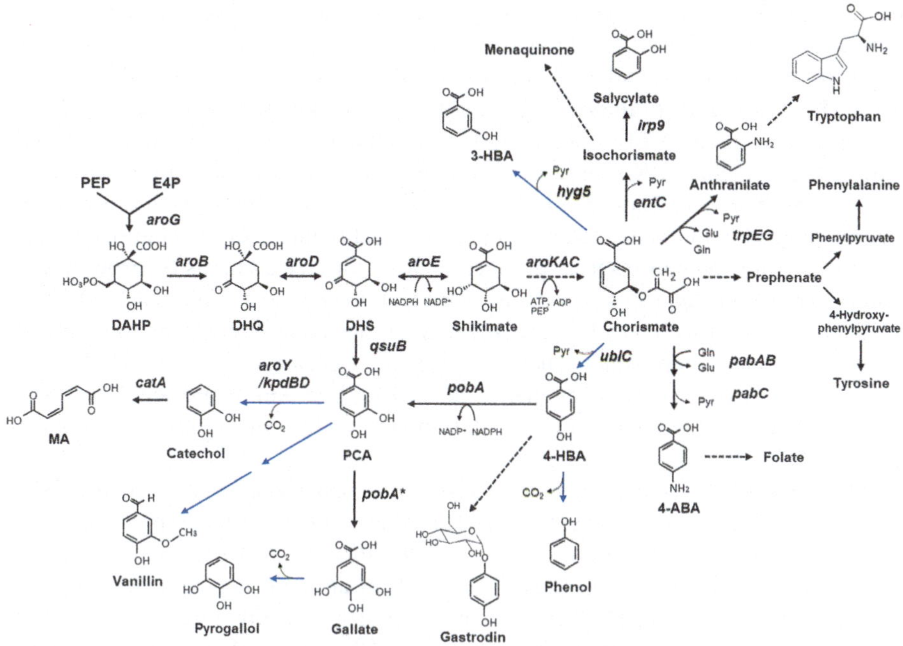

Fig. 3 Metabolic pathways for synthesis of aromatic compounds in *C. glutamicum.* Catalytic steps that can be enabled by recruiting heterologous enzymes are indicated with blue arrows (native pathways are indicated with black arrows). Genes encoding enzymes for corresponding catalytic steps are indicated in italics. *pobA*, 4-HBA 3-hydroxylase; *pobA*∗, mutated *pobA*; *aroY/kpdBD*, PCA decarboxylase and subunits; *catA*, catechol 1,2-dioxygenase; *ubiC*, chorismate pyruvate lyase; *irp9*, isochorismate synthase/isochorismate pyruvate lyase (salicylate synthase); *hyg5*, chorismatase (3-HBA synthase); *trpEG*, anthranilate synthase; *pabAB*, 4-amino-4-deoxychorismate (ADC) synthase, *pabC*, ADC lyase

been widely researched based on the engineering of several microbes including *E. coli*, *S. cerevisiae* and the solvent-tolerant bacterium *P. putida* (Gottardi et al. 2017; Lee and Wendisch 2017; Noda and Kondo 2017; Noda et al. 2016), however, most of their production levels remained low. On the other hand, our recent studies demonstrated the high potential of *C. glutamicum* as a promising host bacteria to overproduce aromatics such as 4-HBA and 4-ABA from sugar substrates at a significantly high titer applicable to the commercialization of their bioproduction (Kitade et al. 2018; Kubota et al. 2016). Higher tolerance of *C. glutamicum* to the aromatics toxicity would be an advantage for toxic aromatics production.

4.1 Protocatechuate

Protocatechuate (3, 4-dihydroxybenzoate; PCA) is a naturally occurring aromatic compound found in many medicinal plants. It has been shown that this compound has numerous pharmacological activities including antioxidant, antibacterial, and anti-inflammatory activities, which provides its potential applications in pharmaceuticals, functional foods, and cosmetics (Kakkar and Bais 2014; Krzysztoforska et al. 2017). PCA also attracts attention as a building block monomer or a precursor for the bio-based production of adipate and terephthalate via *cis*, *cis*-muconic acid (MA) (Fig. 3). Since PCA is currently obtained by costly extraction from plant sources (Kakkar and Bais 2014; Krzysztoforska et al. 2017), its more efficient microbial production is desired to enhance competitiveness for its commercial use. Several marine and soil bacteria produce PCA in the free form, as the iron-binding component of the siderophore such as petrobactin (Williams et al. 2012). Notably, *C. glutamicum* forms it as a key intermediate of the catabolic pathway to assimilate several aromatic compounds such as ferulate, caffeate, *p*-coumarate, 4-cresol, shikimate, and quinate (Kallscheuer et al. 2016a; Lee and Wendisch 2017; Shen et al. 2012). These facts indicate the innate ability of several microbes to produce PCA.

Microbial PCA production from glucose was first reported as an intermediate toward catechol production by *E. coli*. This was enabled by the expression of the *aroZ*-encoded DHS dehydratase from *Klebsiella pneumoniae* to convert the shikimate pathway intermediate DHS to PCA in an *aroE*-deficient host strain to accumulate the precursor DHS (Draths and Frost 1995). Further optimization of the host metabolism via the overexpression of $aroG^{FBR}$, *tkt*, *aroB* and *ppsA* resulted in production of 41 g/L PCA with a yield of 26% (mol/mol) from glucose, although PCA was not produced as an intended end product (Li et al. 2005).

Later, Okai et al. reported PCA production by engineering a phenylalanine-producer strain of *C. glutamicum* by employing an alternative biosynthesis route branched from chorismate and forms 4-HBA as a direct precursor of PCA (Okai et al. 2016). This was achieved via the overexpression of *E. coli ubiC*-encoded chorismate-pyruvate lyase (UbiC), a missing enzyme in *C. glutamicum* to convert chorismate into 4-HBA, which was further converted to PCA by endogenous *pobA*-encoded 4-HBA hydroxylase (Fig. 3). The resulting strain produced 1.1 g/L PCA from glucose after 96 h in the fermenter (Table 1) (Okai et al. 2016). More recently, Kallscheuer and Marienhagen also constructed *C. glutamicum* strains producing PCA and other hydroxybenzoates (salicylate (2-HBA), 3-HBA, and 4-HBA) based on the host strain $DelAro^5$, where endogenous multiple catabolic pathways for aromatics utilization were abolished (Kallscheuer and Marienhagen 2018). To enable PCA production, they overexpressed endogenous *qsuB*-encoded DHS dehydratase along with the *E. coli aroF*-encoded feedback-resistant DS. Additionally, host metabolism was engineered to increase precursor availability by overexpression of the *iolT1*-encoded PEP-independent glucose permease and *tkt* as well as by downregulation of the *gltA*-encoded citrate synthase by promoter replacement. The constructed strain produced 2.0 g/L (13.0 mM) of PCA in shake flasks (Table 1) (Kallscheuer and Marienhagen 2018).

Notably, PCA represents a pivotal precursor for the biosynthesis of other valuable aromatic compounds such as catechol, MA, gallate, pyrogallol, and vanillin (Fig. 3). Decarboxylation of PCA leads to catechol, and its oxidative ring cleavage by catechol 1,2-dioxygenase generates a dicarboxylic acid MA as the biosynthetic precursor of adipate and terephthalate. An important flavor compound vanillin can be produced from PCA via introducing heterologous two-step reactions catalyzed by catechol *O*-methyltransferase and aromatic carboxylate reductase (Hansen et al. 2009). It has been reported that PCA can be converted to gallate, a compound with strong anti-bacterial and anti-oxidant activities, by employing a mutated PobA (Y385F/T294A), which displayed much higher activity toward PCA than the wild-type form, in an engineered *E. coli* (Chen et al. 2017). Gallate could be further converted into its glycosylated product β-glucogallin, which has strong anti-oxidative and ultraviolet-photoprotective activity, in engineered *E. coli* overexpressing glucosyltransferase from *Vitis vinifera* (European grape) (De Bruyn et al. 2015).

4.2 cis, cis-*Muconic Acid*

***cis*, *cis*-Muconic acid** (MA) is a naturally occurring dicarboxylic acid attracting attention as a biological precursor of high-demand platform chemicals adipate and terephthalate, representing a constituent monomer of nylon-6,6 fiber, and polyethylene terephthalate (PET), respectively. Biotechnological production of MA offers a promising alternative to its conventional petrochemical synthesis (Sengupta et al. 2015; Polen et al. 2013; Xie et al. 2014). The non-natural biosynthetic route to produce MA from glucose was first established in the Frost laboratory using *E. coli* based on the expression of DHS dehydratase (AroZ) and PCA decarboxylase (AroY) from *Klebsiella pneumoniae* and catechol 1,2-dioxygenase (CatA) from *Acinetobacter calcoaceticus* to convert the shikimate pathway intermediate DHS to MA via PCA and catechol (Draths and Frost 1994). With the modification of the central carbon and the shikimate pathway metabolism to improve precursor availability and the flux through the shikimate pathway by inactivation of *aroE* and overexpression of $aroF^{FBR}$, *ppsA*, *aroB* and *tktA*, the resulting strain produced 36.8 g/L of MA from glucose with a yield of 22% (mol/mol) within 48 h in fermenter (Niu et al. 2002). However, in this strain, the inactivation of AroE (shikimate dehydrogenase) caused the auxotrophy for aromatic amino acids and aromatic vitamins, resulting in a drawback because of production cost-efficiency.

De novo MA production from glucose (and glycerol) via four different pathways diverging from chorismate proceeds via different benzoic acid intermediates, namely, anthranilate, salicylate, 2,3-dihydroxybenzoate, or 4-HBA, which have also been engineered in *E. coli*. However, MA production achieved employing these alternative pathways suffered from low product titers, (i.e., 0.39 g/L, 1.5 g/L, 0.6 g/L, and 0.17 g/L, respectively) (Jung et al. 2015; Lin et al. 2014; Sengupta et al. 2015; Sun et al. 2013, 2014; Weber et al., 2012; Xie et al., 2014).

Recently, *C. glutamicum* strains overproducing MA from glucose via DHS, PCA, and catechol were engineered (Lee et al. 2018; Shin et al. 2018). *C. glutamicum* is advantageous in designing metabolic pathway for MA production: it requires only a single heterologous decarboxylase gene to convert PCA to CA to form MA, whereas three heterologous genes are required in *E. coli*. However, this microbe also has catabolic pathways for MA and its precursors DHS and PCA, which need to be blocked in the engineered strain by deleting the *aroE*-encoded shikimate dehydrogenase, *pcaGH*-encoded PCA 3,4-dioxygenase, and *catB*-encoded chloromuconate cycloisomerase. To achieve the conversion of PCA to catechol, a step that is absent in *C. glutamicum*, a codon-optimized genes for PCA decarboxylase (*aroY*/*kpdC*) and associated subunit (*kpdB* and *kpdD*) from *K. pneumoniae* were overexpressed as a single operon under the strong *sod* promoter in the host strain deficient in the catabolic pathway. The resulting strain, when grown in the optimized culture medium, produced 38 g/L and 54 g/L MA from glucose in 7 and 50 L fed-batch fermentations, representing the highest titer of MA production yet reported (Lee et al. 2018). In another study, the same group examined engineering strategies to increase intracellular PEP availability to improve MA production. They deleted *ptsI* and *iolR* (a transcriptional repressor of *iolT1*) genes to enhance IolT1-mediated glucose uptake in PTS-deficient host strain to save PEP precursor (Ikeda et al. 2011). This engineering, along with the overexpression of *qsuB*-encoded DHS dehydratase, improved MA production to 4.5 g/L at a yield of 22% (mol/mol) in flask culture (Table 1).

MA production from lignin-derived aromatic compounds represents an alternative approach for sustainable MA production. In this context, MA production from *p*-coumarate at a titer of 13.5 g/L and 15.6 g/L was recently achieved by engineering the aromatics utilizing pathways of *Pseudomonas putida* strains (Johnson et al. 2016). In a very recent study, Becker et al. reported engineered *C. glutamicum* to overproduce MA from aromatics and lignin hydrolysates (Becker et al. 2018). They found that in a strain with deletion of *catB* encoding muconate cycloisomerase in the catechol branch of the β-ketoadipate pathway, catechol, phenol, and benzoate were completely converted into MA when glucose was used as the growth substrate. Overexpression of the endogenous *catA* encoding catechol 1,2-dioxygenase in the *catB*-deleted strain resulted in the accumulation of 85 g/L MA from catechol within 60 h in the fermenter reaction, when catechol was fed pulsed-wise manner to circumvent its toxicity (Table 1) (Becker et al. 2018). This strain could also produce 1.8 g/L MA directly from softwood lignin hydrolysate by completely converting catechol and phenol present in the hydrolysate, into MA (Table 1) (Becker et al. 2018).

4.3 4-Hydroxybenzoate

4-Hydroxybenzoate (4-HBA) is a valuable compound used as a building block to produce high-performance liquid crystal polymers and parabens as food preservatives (Barker and Frost 2001; Wang et al. 2018). 4-HBA also attracts attention as a biosynthetic precursor for microbial production of other commodity and/or fine aromatic chemicals such as phenol (Kim et al. 2014), PCA (Okai et al. 2016), and gastrodin (4-hydroxymethyl phenyl-β-D-glucopyranoside) (Bai et al. 2016), a herbal plant-derived analgesic component (Fig. 3) (Lee and Wendisch 2017; Wang et al. 2018). Although 4-HBA is currently synthesized from benzene via cumene and phenol, it also represents a natural product synthesized during the ubiquinone biosynthesis pathway in several microbes. In this context, microbial 4-HBA production has been explored in *Klebsiella pneumoniae*, *E. coli* and *P. putida* (Barker and Frost 2001; Meijnen et al. 2011; Muller et al. 1995; Verhoef et al. 2007). However, the production in these microbes was limited as represented by 12 g/L at a yield of 13% (mol/mol) by *E. coli* (Barker and Frost 2001) or 90 mg/L at a yield of 0.8% (mol/mol) by *S. cerevisiae* (Krömer et al. 2013).

Three research groups recently reported metabolic engineering of *C. glutamicum* for 4-HBA production (Kallscheuer and Marienhagen 2018; Kitade et al. 2018; Syukur Purwanto et al. 2018). As *C. glutamicum* does not possess endogenous chorismate-pyruvate lyase (UbiC) activity catalyzing the reaction yielding 4-HBA from chorismate, its biosynthesis in this microbe was enabled by expressing heterologous *ubiC* genes (Fig. 3) (Siebert et al. 1994). We created 4-HBA producer by overexpressing the *ubiC* gene from the intestinal bacterium *Providencia rustigianii*, which showed the highest resistance to the feedback inhibition by 4-HBA and was the most effective for 4-HBA production among other screened genes (Kitade et al. 2018). To enhance 4-HBA production, host metabolism was engineered by overexpressing the shikimate pathway and the pentose phosphate pathway genes (*aroG*^{S180F} from *E. coli* encoding feedback-resistant DS, *aroD*, *aroA*, *aroE*, and *tkt-tal*) as well as by deleting the byproduct forming pathways (*ldhA*, *qsuB*, *qsuD*, *hdpA*, *pyk* and *pobA*). The resulting strain produced 36.6 g/L of 4-HBA from glucose with a yield of 41% (mol/mol) after 24 h in an aerobic growth-arrested bioprocess, representing the highest titer of microbial 4-HBA production reported so far (Table 1) (Kitade et al. 2018). On the other hand, Kallscheuer and Marienhagen engineered 4-HBA-producing *C. glutamicum,* based on the expression of the *aroH* and *ubiC* genes from *E. coli* in a platform strain to produce hydroxybenzoates, where central carbon metabolism was altered to improve the supply of the shikimate pathway precursors PEP and E4P, and catabolic pathways for aromatics were deleted (using the strain DelAro^5 as a parent strain) (Kallscheuer and Marienhagen 2018). The engineered strain produced 3.3 g/L 4-HBA with 10.8% yield in the shake flask (Table 1) (Kallscheuer and Marienhagen 2018). More recently, Syukur Purwanto et al. also reported other 4-HBA producing *C. glutamicum* strains (Syukur Purwanto et al. 2018). They expressed feedback-resistant forms of UbiC and DS (encoded by *ubiC*^{PR} and *aroF*^{FBR}/*aroG*^{FBR} from *E. coli*) and blocked the carbon flux into aromatic amino acid biosynthesis and 4-HBA catabolism by deleting the *trpE*, *csm*, and *pobA* genes encoding anthranilate synthase I, chorismate mutase, and 4-HBA hydroxylase, respectively. Additionally, shikimate pathway genes *aroE* and *aroCKB* were overexpressed, whereas the *qsuABD* genes involved in the assimilation of the pathway intermediates, i.e., shikimate and DHS, were deleted. Notably, the introduction of *aroK* from *Methanocaldococcus jannaschii* encoding shikimate-resistant shikimate kinase was shown to be effective in reducing the accumulation of pathway intermediates shikimate and 3-DHS. These metabolic alterations resulted in the production of 19 g/L 4-HBA with 9.7% yield from glucose after 65 h in a 5 L bioreactor (Table 1) (Syukur Purwanto et al. 2018). However, in the resulting strain, the blockage of aromatic amino acid biosynthetic pathway to accumulate precursor chorismate caused the auxotrophy for aromatic amino acids and 4-ABA, which would be a drawback with regard to the cost efficiency of the biotechnological 4-HBA production.

4.4 Salicylate and 3-Hydroxybenzoate

Salicylate (2-HBA) is a natural metabolite serving as a phytohormone in plants and a biosynthetic precursor of siderophore in some microbes. It has application as pharmaceuticals such as aspirin (acetylsalicylate) and lamivudine (an anti-HIV drug) as well as cosmetics products. Salicylate is biosynthesized from chorismate via isochorismate by consecutive reactions by isochorismate synthase (ICS) and isochorismate pyruvate lyase (IPL). Its microbial production was first achieved in *E. coli* as a biological precursor of 4-hydroxycoumarin and MA by Lin et al. (2013, 2014). They achieved salicylate production via extending shikimate pathway by introducing ICS (EntC from *E. coli*) and IPL (PchB from *Pseudomonas fluorescens*) to convert chorismate into salicylate, which was coupled with the enhanced chorismate synthesis due to disruption of *pheA* and *tyrA* as well as overexpression of *aroG*^{FBR}, *aroL*, *ppsA*, and *tktA*. These engineering led to the production of 1.2 g/L salicylate from glycerol and glucose (Lin et al. 2014). Later on, Noda et al. constructed salicylate-producing *E. coli* using the same pathway consisting of *E. coli menF* and *P. aeruginosa pchB* for ICS and IPL activities, respectively (Noda et al. 2016). In this study, the central metabolism of the host strain was further engineered to enhance precursor (PEP) availability by the replacement of endogenous PTS by galactose permease and glucokinase (GalP-Glk) derived from *E. coli*, deletion of pyruvate kinase genes (*pykA* and *pykF*), as well as the blockage of the carbon flux into phenylalanine and tyrosine biosynthesis by deleting *pheA* and *tyrA*. The created strain produced 11.5 g/L salicylate after 48 h with a yield of 41.1% (mol/mol) from glucose in fermenter cultivation. In the engineered strain, pyruvate produced in the salicylate-forming reaction would enable cell growth even in the absence of both PTS and pyruvate kinase, which represent major source to produce pyruvate. (Noda et al. 2016).

In *C. glutamicum*, Kallscheuer et al. recently engineered strains for the production of salicylate and 3-HBA using the same platform strain for HBA production mentioned in the above section for PCA and 4-HBA production (Kallscheuer and Marienhagen 2018). The producer strains for salicylate and 3-HBA were created by overexpressing AroF (feedback-resistant DS from *E. coli*) and either heterologous genes *irp9* (encoding isochorismate synthase/isochorismate pyruvate lyase (salicylate synthase)) or *hyg5* (encoding 3-HBA synthase), respectively (Fig. 3). However, the product titers of the optimized strains remained very low (0.01 g/L salicylate and

0.3 g/L 3-HBA in shaking flasks) (Table 1), presumably due to the limiting activities of introduced heterologous enzymes (Kallscheuer and Marienhagen 2018).

4.5 4-Aminobenzoate

4-Aminobenzoate (4-ABA) is a precursor compound for microbial folate biosynthesis in the pathway branched from chorismate. Biotechnologically, it has a great potential to serve as a building block of biopolymers such as engineering plastics. Currently, 4-ABA is produced by chemical conversions from petroleum-derived toluene, and its bioproduction from biomass resources would contribute to addressing environmental issues. Its biosynthesis from chorismate routes via 4-amino-4-deoxychorismate (ADC) by consecutive reactions catalyzed by ADC synthase, encoded by *pabA* and *pabB*, and ADC lyase encoded by *pabC* (Fig. 3). Microbial 4-ABA production from glucose was first achieved in *S. cerevisiae*, where 4-ABA synthase gene ABZ1 was overexpressed along with the deletion of the branched pathways from chorismate toward aromatic amino acid synthesis, although the titer and yield were limited (0.25 mM and <1%) (Krömer et al. 2013). 4-ABA-producing *E. coli* strain was also engineered by introducing *pabAB* from *C. efficiens*, native *pabC*, and $aroF^{FBR}$ encoding feedback-resistant DS (Koma et al. 2014). The generated strain produced 4.8 g/L 4-ABA from glucose with a yield of 21% in fed-batch cultivation (Koma et al. 2014). Noda et al. also demonstrated the 4-ABA production in engineered *E. coli* by introducing *E. coli pabA*, *pabB*, and *pabC* into the platform strain with enhanced central carbon metabolism to chorismate, which resulted in production of 2.9 g/L 4-ABA after 48 h in the test-tube culture (Noda et al. 2016).

Metabolic engineering of *C. glutamicum* for 4-ABA production was first reported by our research group (Kubota et al. 2016). We screened heterologous *pabABC* genes and found that the combination of the *pabAB* from *Corynebacterium callunae* and *the pabC* from *Xenorhabdus bovienii* conferred the highest 4-ABA production. Overexpression of these genes in a host strain with co-overexpression of the shikimate pathway genes resulted in production of 43 g/L of 4-ABA from glucose after 48 h with a 20% yield (mol/mol) in a fermenter, which represented the highest reported titer for microbial 4-ABA production (Table 1) (Kubota et al. 2016).

4.6 3-Amino-4-Hydroxybenzoate

3-Amino-4-hydroxybenzoate (3,4-AHBA) is a natural aromatic compound synthesized as a metabolic intermediate of the biosynthesis of several secondary metabolites produced by *Streptomyces* including grixazone, yellow pigments produced under phosphate limitation by *Streptomyces griseus* (Suzuki et al. 2006). 3,4-AHBA is a valuable aromatic compound serving as a precursor for the synthesis of polybenzoxazole, a thermostable bioplastic (Kawaguchi et al. 2015). Notably, biosynthesis of 3,4-AHBA does not depend on the multi-step shikimate pathway as the source of most aromatic compounds but relies on only two enzymes, GriH and GriI, which catalyze 3,4-AHBA synthesis from C3 and C4 primary metabolites, namely, dihydroxyacetone phosphate and L-aspartate-4-semialdehyde, respectively (Suzuki et al. 2006). GriI catalyzes an aldol condensation reaction between L-aspartate-4-semialdehyde and dihydroxyacetone phosphate, whereas GriH converts the resulting C7 metabolite into 3,4-AHBA. Thus, this pathway represents a simple alternative route for microbial aromatic ring formation (Suzuki et al. 2006). Exploiting this unique enzyme system, Kawaguchi et al. constructed an engineered *C. glutamicum* producing 3,4-AHBA by expressing the *griH* and *griI* genes from *S. griseus* in a lysine-producing strain *C. glutamicum* ATCC 21799. The constructed strain produced 1.0 g/L of 3,4-AHBA from sweet sorghum juice, a potentially suitable feedstock for biochemical production (Table 1) (Kawaguchi et al. 2015).

5 Production of Polyphenols

Plant polyphenols constitute a large group of aromatic compounds of the secondary plant metabolites (Scalbert et al. 2005a). They include naringenin, kaempferol, resveratrol, pterostilbene derived from *p*-coumaric acid, piceatannol, eriodictyol, quercetin derived from caffeic acid, and violacein derived from tryptophan. They exhibit diverse human health-promoting activities such as antioxidant, anti-inflammatory, anti-microbial, anti-cancer or anti-diabetic activities and thus have potential applications as food supplements, pharmaceuticals, and cosmetic ingredients. Because the content of polyphenols in plants is very low, traditional plant extraction becomes highly costly, which promoted extensive researches on microbiological production of various plant polyphenols (Table 2). In this section, recent approaches for polyphenol production by *C. glutamicum* or other microorganisms are introduced (Fig. 4).

5.1 Naringenin

Plants produce naringenin as a secondary metabolite. It has biological activities such as antioxidant, anti-cancer, anti-obesity, and anti-inflammatory activities (Patel et al. 2018). In the biosynthetic pathway, naringenin is produced from tyrosine via *p*-coumaric acid. *p*-Coumaric acid is converted to *p*-coumaroyl-CoA by 4-coumarate: CoA ligase (4CL); then, naringenin chalcone is produced using malonyl-CoA in the fatty acid synthesis pathway. After the ring formation reaction, naringenin is produced. Ganesan et al. successfully produced naringenin from glucose using an *E. coli*-*E. coli* co-culture system in which an *E. coli* strain produced tyrosine and *p*-coumaric acid and the other strain consumed them as substrates and converted

Table 2 Summary of studies on plant polyphenol production by engineered microbes

Target compound	Host strain	Carbon source	Titer (mg/L)	Time (h)	Culture style	References
Naringenin	*E. coli* (coculture)	Glucose	41.5	36	Flask	Ganesan et al. (2017)
	S. cerevisiae	Glucose	109	38	Fermenter	Koopman et al. (2012)
	C. glutamicum	Glucose	32	36	Flask	Kallscheuer et al. (2016b)
Kaempferol	*S. cerevisiae*	Glucose with 1 mM *p*-coumarate	66.3	40	Flask	Duan et al. (2017)
	C. glutamicum	*p*-Coumaric acid	23	140	Flask	Kallscheuer et al. (2017)
Resveratrol	*S. cerevisiae*	Glucose	416	100	Fermenter	Li et al. (2015)
	E. coli	Glucose	4.6	ND	Flask	Liu et al. (2016)
	C. glutamicum	Glucose	59	ND	Flask	Kallscheuer et al. (2016b)
Pterostilbene	*E. coli*	Glucose with 1 mM methionine	33.6	72	Flask	Heo et al. (2017)
	C. glutamicum	*p*-Coumaric acid	42	140	Flask	Kallscheuer et al. (2017)
Piceatannol	*E. coli*	*p*-Coumaric acid	124	72	Flask	Shrestha et al. (2018)
	C. glutamicum	Caffeic acid	56	24	Flask	Kallscheuer et al. (2016b)
Eriodictyol	*E. coli*	Tyrosine	107	48	Flask	Zhu et al. (2014)
	C. glutamicum	Caffeic acid	37	ND	Flask	Kallscheuer et al. (2016b)
Quercetin	*C. glutamicum*	Caffeic acid	10	140	Flask	Kallscheuer et al. (2017)
Violacein	*E. coli*	Glucose	1.75	48	Fermenter	Fang et al. (2015)
	E. coli	Glucose	4.45	48	Fermenter	Zhou et al. (2018)
	C. glutamicum	Glucose	5.4	115	Fermenter	Sun et al. (2016)

ND not described

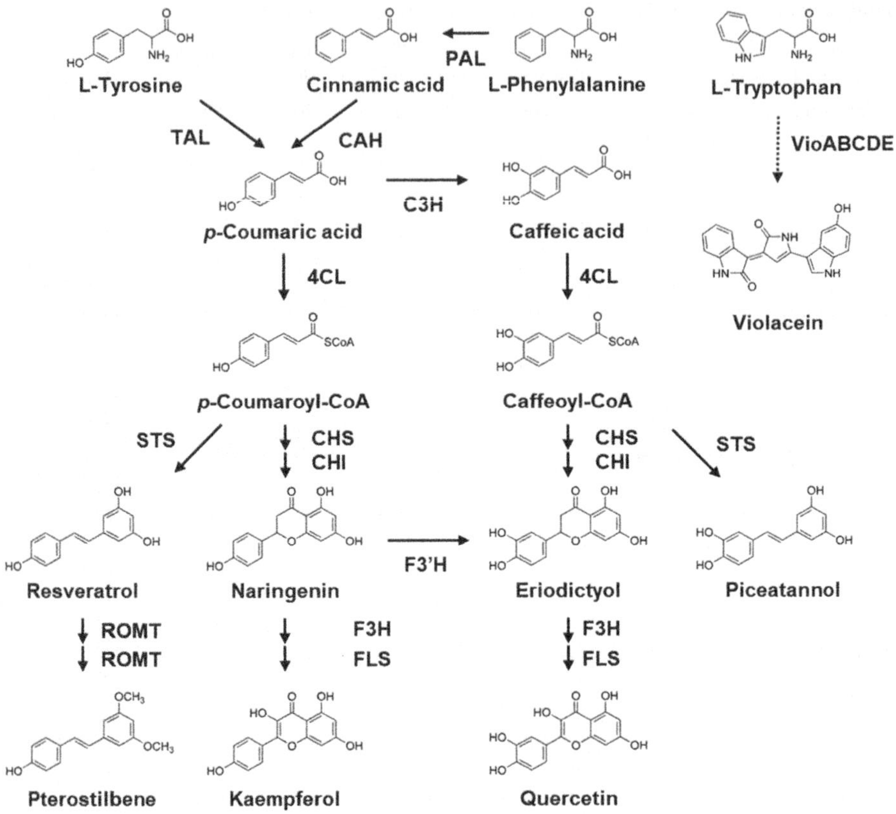

Fig. 4 Typical pathways for synthesis of plant polyphenols from aromatic amino acids. *TAL* tyrosine ammonia lyase, *PAL* phenylalanine ammonia lyase, *CAH* cinnamic acid hydroxylase, *C3H* *p*-coumarate 3-hydroxylase, *4CL* 4-coumarate:CoA ligase, *STS* stilbene synthase, *CHS* chalcone synthase, *CHI* chalcone isomerase, *ROMT* resveratrol *O*-methyltransferase, *F3′H* flavonoid 3′-hydroxylase, *F3H* flavanone 3-hydroxylase, *FLS* flavonol synthase

them to naringenin. The best strain ratio of 1:5 produced 41.5 mg/L of naringenin at 36 h of cultivation (Table 2) (Ganesan et al. 2017).

S. cerevisiae was engineered using six genes of *Arabidopsis thaliana* to produce naringenin. The genes were phenylalanine ammonia lyase (PAL), cinnamate 4-hydroxylase (C4H), cytochrome P450 reductase (CPR), 4CL, chalcone synthase (CHS), and chalcone isomerase (CHI). To further optimize host metabolism, DS was overexpressed, and phenylpyruvate decarboxylase was deleted in the *S. cerevisiae*. Upon cultivation in a fermenter, the resulting stain accumulated over 400 μM (about 109 mg/L) of naringenin in glucose medium after 38 h (Table 2) (Koopman et al. 2012).

Kallscheuer et al. confirmed that engineered *C. glutamicum* is also suitable as a platform to produce plant polyphenols (Kallscheuer et al. 2016b). They found

C. glutamicum can grow on phenylpropanoids as sole carbon and energy source (Kallscheuer et al. 2016a). Deletion of genes involved in phenylpropanoid degradation was a key step for the construction of polyphenol producers. Subsequent overexpression of plant-derived genes encoding a 4CL of *Petroselinum crispum*, a CHS and a CHI of *Petunia* × *hybrida* enabled formation of 32 mg/L naringenin from glucose after 36 h cultivation (Table 2) (Kallscheuer et al. 2016b).

5.2 *Kaempferol*

Kaempferol has broad bioactivity like naringenin and is found in many dietary plants. Kaempferol is converted from naringenin by two reactions, hydroxylation of naringenin by flavanone 3-hydroxylase (F3H) and dehydration of dihydrokaempferol by flavonol synthase (FLS). Duan et al. engineered *S. cerevisiae* to produce kaempferol. A strain expressing genes for kaempferol biosynthesis pathway derived from plants (*A. thaliana* or *Populus deltoides*) produced approximately 7 mg/L of kaempferol from glucose. Then, a strain further expressing genes in the acetyl-CoA biosynthetic pathway was constructed. The strain produced 66 mg/L of kaempferol from glucose supplemented with 1 mM *p*-coumaric acid after 40 h (Table 2) (Duan et al. 2017).

A *C. glutamicum* engineered for naringenin production was further engineered to produce kaempferol by introducing F3H from *P. hybrida* and FLS from *P. deltoides*. The strain converted *p*-coumaric acid to kaempferol via naringenin. The titer was 23 mg/L after 140 h in medium supplemented with 820 mg/L *p*-coumaric acid (Table 2) (Kallscheuer et al. 2017).

5.3 *Resveratrol*

A stilbene, resveratrol, is a polyphenol compound found in a few plant species and used as a food supplement and cosmetic ingredient. Because of its wide range of beneficial properties, microbial production of resveratrol has been extensively investigated. *p*-Coumaric acid is a key intermediate for the production of resveratrol. After the formation of *p*-coumaroyl-CoA by 4CL, stilbene synthase (STS) catalyzes to form resveratrol using three molecules of malonyl-CoA.

An engineered *S. cerevisiae* produced a high concentration of resveratrol directly from glucose. Heterologous enzymes expressed in the strain were tyrosine ammonia lyase (TAL) from *Herpetosiphon aurantiacus*, 4CL from *A. thaliana*, STS from *Vitis vinifera*, the feedback-insensitive allele of DS and chorismite mutase, and acetyl-CoA carboxylase to improve the supply of the precursor malonyl-CoA. Fed-batch fermentation of this strain with glucose resulted in a resveratrol titer of 416 mg/L (Table 2). It is noteworthy that when the feeding source was replaced by ethanol, the same strain produced 531 mg/L of resveratrol (Li et al. 2015).

Liu et al. engineered *E. coli* to produce resveratrol. They introduced genes of TAL from *Rhodotorula glutinis*, 4CL from *P. crispum*, and STS from *V. vinifera* into the loci of genes *tyrR* and *trpED* in the chromosome. TyrR is a transcriptional regulator and showed that inactivation of *tyrR* enhances expression of genes relevant to tyrosine biosynthesis. Blocking of tryptophan biosynthesis by the inactivation of *trpED* is also effective for tyrosine accumulation. The recombinant strain produced 4.6 mg/L of resveratrol from glucose after 48 h of cultivation (Table 2) (Liu et al. 2016).

In *C. glutamicum*, the platform strain DelAro[5] constructed by Kallscheuer et al. described above was also efficient for the production of resveratrol. High titer of 158 mg/L of resveratrol using *C. glutamicum* was reported, but the production depended on the addition of *p*-coumaric acid. For the de novo production without substrate supplementation, the platform strain was further engineered to overexpress 4CL derived from *P. crispum* and STS derived from *Arachis hypogaea*. The strain produced 59 mg/L of resveratrol directly from glucose as a substrate (Table 2) (Kallscheuer et al. 2016b).

The titers of resveratrol produced by *E. coli* or *C. glutamicum* were lower when compared to that in *S. cerevisiae*, perhaps because the eukaryote *S. cerevisiae* can express the plant genes better than prokaryotes.

5.4 Pterostilbene

Resveratrol is relatively sensitive to light and oxygen exposure. To increase stability, the substitution of a hydroxy group with a methoxy group is effective. Several plants have a gene characterized as possible resveratrol *O*-methyltransferase (ROMT). Two methylation cycles of resveratrol by ROMT yield a dimethoxy derivative, pterostilbene. The *O*-methyltransferase (OMT) family needs *S*-adenosylmethionine (SAM) to catalyze the transfer of a methyl group. To increase intracellular SAM pool, methionine supplementation in the medium is known to be effective.

To construct a pterostilbene producer, an engineered tyrosine producer *E. coli* was further engineered by introducing genes of TAL from *Saccharothrix espanaensis*, 4CL from *Nicotiana tabacum*, and STS from *V. vinifera*. Then a novel ROMT from *A. thaliana* was introduced to yield a pterostilbene producer. The titer of the pterostilbene was 33.6 mg/L after 72 h of cultivation in a minimal medium containing glucose supplemented with 1 mM methionine (Table 2) (Heo et al. 2017).

Expression of an OMT in a resveratrol-producing *C. glutamicum* strain allowed synthesis of 42 mg/L of pterostilbene from *p*-coumaric acid after 140 h of cultivation (Table 2). Notably, increasing the solubility of the OMT by expressing it as a fusion protein with the maltose-binding protein of *E. coli* was required to obtain this titer (Kallscheuer et al. 2017).

5.5 *Piceatannol*

Piceatannol is a 3′-hydroxylated derivative of resveratrol found in various plants. Piceatannol is reported to have a variety of roles in inhibiting cancer, inflammation, proliferation, invasion, migration, and adipogenesis (Kershaw and Kim 2017; Piotrowska et al. 2012).

A resveratrol-producing *E. coli* strain was modified to produce piceatannol by overexpressing 4-hydroxyphenylacetate hydroxylase (HpaBC). This strain produced 124 mg/L piceatannol from *p*-coumaric acid in medium supplemented with disodium malonate and sodium acetate (Table 2) (Shrestha et al. 2018).

A *C. glutamicum* strain expressing plant derived 4CL and STS produced 56 mg/L piceatannol from corresponding phenylpropanoid caffeic acid (Table 2) (Kallscheuer et al. 2016b).

5.6 *Eriodictyol*

Eriodictyol, a flavonoid found in citrus fruits, possesses anti-inflammatory, antioxidative, and antidiabetic activities. It is also known as a bitter taste blocker.

Zhu et al. reported that engineered *E. coli* produced 107 mg/L of eriodictyol from tyrosine (Table 2) (Zhu et al. 2014). To achieve this, they constructed a naringenin biosynthesis route in *E. coli* by expressing TAL from *R. glutinis*, 4CL from *P. crispum*, CHI from *Medicago sativa*, and CHS from *P. X hybrida*. To improve malonyl-CoA availability, acetyl-CoA carboxylase from *C. glutamicum* was further expressed. Naringenin was converted to eriodictyol by flavonoid 3′-hydroxylase (F3′H), which was expressed as a functional fusion protein of truncated F3′H and truncated P450 reductase to improve its solubility.

Kallscheuer et al. engineered *C. glutamicum* to produce eriodictyol from caffeic acid via caffeoyl-CoA. Overexpression of plant-derived genes encoding 4CL, CHS, and CHI enabled the formation of corresponding (2S)-flavanones eriodictyol. The strain produced 37 mg/L of eriodictyol in the presence of 5 mM caffeic acid (Table 2) (Kallscheuer et al. 2016b).

5.7 *Quercetin*

Quercetin is converted from eriodictyol by F3H and FLS. It is found in many plants, especially in oak bark, tea leaves, and onion skins. It is also known as a yellow dye.

Kalluscheuer et al. converted caffeic acid to quercetin in the engineered *C. glutamicum,* which had the same genotype as the strain used for kaempferol production. The titer was 10 mg/L after 140 h in medium supplemented with 900 mg/L caffeic acid (Table 2) (Kallscheuer et al. 2017).

5.8 Violacein

Violacein is a natural bisindole compound derived from the condensation of two molecules of tryptophan. Since violacein and deoxyviolacein have similar biological properties, they are mainly produced in the form of a mixture as crude violacein. Crude violacein is known to possess potential pharmaceutical application due to its antibacterial, antitumoral, antiviral, and antioxidant activities (Choi et al. 2015). Violacein biosynthesis after tryptophan is catalyzed by enzymes VioA, VioB, VioE, VioD, and VioC successively. Several bacteria such as *Duganella* sp. B2, *Chromobacterium violaceum*, and *Janthinobacterium lividum* were reported to naturally produce violacein as their secondary metabolites (Durán et al. 2012).

Fang et al. reported an engineered *E. coli* for the production of violacein. A tryptophan producer was constructed through the introduction of deregulated *trpE*^{fbr} and *trpD* genes by deleting three genes, *trpR*, *pheA*, and *tnaA*. Then, a *vio* gene cluster from *Duganella* sp. B2 (Jiang et al. 2010) was introduced into the engineered strain. The strain produced 1.75 g/L crude violacein from glucose after 48 h (Table 2) (Fang et al. 2015).

The same group further tuned the *E. coli* host strain to improve the tryptophan production capability by overexpressing *aroG*^{fbr} and *serA*^{fb} genes. This strain that overexpressed the *vio* gene cluster from plasmid produced 4.45 g/L crude violacein in a 5-L bioreactor from glucose (Table 2) (Zhou et al. 2018).

Since *C. glutamicum* ATCC 21850 is an established tryptophan producer (Heery and Dunican 1993), it is a suitable platform to produce crude violacein. Sun et al. transformed the host strain with a plasmid which harbors the *vioABCDE* operon from *C. violaceum*. The operon had an IPTG-inducible promoter and a strong *C. glutamicum* ribosome-binding site. The resulting strain produced 5.4 g/L of violacein in fed-batch fermentation in a 3-L bioreactor, representing the highest reported titer and productivity (Table 2) (Sun et al. 2016).

6 Concluding Remarks and Perspectives

To date, a variety of aromatic compounds have been produced by genetically engineered microorganisms. Model organisms *E. coli* and *S. cerevisiae* are often used probably due to their easier culture, well-known genetics, and easy genetic manipulation. However, regarding their productivity, most of the bio-based aromatic compounds are still not applicable to commercial production or even at the stage of confirming the proof of concept. In this regard, *C. glutamicum* has some advantages compared to the model organisms in experimental handling and has achieved markedly high productivity in several value-added aromatic chemicals from renewable sugars at a level amenable to commercialization. Therefore, it can serve as a promising platform for aromatic chemical bioproduction as an eco-friendly bioprocess. Since the productivity of several aromatics and structurally intricate compounds such as plant polyphenols is still relatively low, effective approaches have been desired. A well-known bottleneck point for polyphenol production is the incorporation step of three malonyl-CoA. The low intracellular pool of malonyl-CoA should be managed. Functional expression of eukaryotic genes in a prokaryote is also important. This is partly solved by codon optimization based on the *C. glutamicum* codon usage database and commercially available gene synthesis services. To further improve the productivity, strategies for the rational metabolic engineering based on advanced computer-assisted tools as well as omics-based analysis would be needed. Such approaches in a synergistic way would contribute to the development of engineered microbes for overproducing aromatic compounds with the most appropriate metabolic pathways from genome-wide perspectives.

References

Averesch NJH, Krömer JO (2018) Metabolic engineering of the shikimate pathway for production of aromatics and derived compounds-present and future strain construction strategies. Front Bioeng Biotechnol 6:32. https://doi.org/10.3389/fbioe.2018.00032

Bai Y, Yin H, Bi H, Zhuang Y, Liu T, Ma Y (2016) De novo biosynthesis of Gastrodin in *Escherichia coli*. Metab Eng 35:138–147. https://doi.org/10.1016/j.ymben.2016.01.002

Barker JL, Frost JW (2001) Microbial synthesis of *p*-hydroxybenzoic acid from glucose. Biotechnol Bioeng 76:376–390

Becker J, Kuhl M, Kohlstedt M, Starck S, Wittmann C (2018) Metabolic engineering of *Corynebacterium glutamicum* for the production of *cis*, *cis*-muconic acid from lignin. Microb Cell Fact 17:115. https://doi.org/10.1186/s12934-018-0963-2

Bentley R (1990) The shikimate pathway—a metabolic tree with many branches. Crit Rev Biochem Mol Biol 25:307–384. https://doi.org/10.3109/10409239009090615

Bilal M, Wang S, Iqbal HMN, Zhao Y, Hu H, Wang W, Zhang X (2018) Metabolic engineering strategies for enhanced shikimate biosynthesis: current scenario and future developments. Appl Microbiol Biotechnol 102:7759–7773. https://doi.org/10.1007/s00253-018-9222-z

Bochkov DV, Sysolyatin SV, Kalashnikov AI, Surmacheva IA (2012) Shikimic acid: review of its analytical, isolation, and purification techniques from plant and microbial sources. J Chem Biol 5:5–17. https://doi.org/10.1007/s12154-011-0064-8

Brinkrolf K, Brune I, Tauch A (2006) Transcriptional regulation of catabolic pathways for aromatic compounds in *Corynebacterium glutamicum*. Genet Mol Res 5:773–789

Chae TU, Choi SY, Kim JW, Ko YS, Lee SY (2017) Recent advances in systems metabolic engineering tools and strategies. Curr Opin Biotechnol 47:67–82. https://doi.org/10.1016/j.copbio.2017.06.007

Chandran SS, Yi J, Draths KM, von Daeniken R, Weber W, Frost JW (2003) Phosphoenolpyruvate availability and the biosynthesis of shikimic acid. Biotechnol Prog 19:808–814. https://doi.org/10.1021/bp025769p

Chen Z, Shen X, Wang J, Yuan Q, Yan Y (2017) Rational engineering of *p*-hydroxybenzoate hydroxylase to enable efficient gallic acid synthesis via a novel artificial biosynthetic pathway. Biotechnol Bioeng 114:2571–2580. https://doi.org/10.1002/bit.26364

Choi SY, Kim S, Lyuck S, Kim SB, Mitchell RJ (2015) High-level production of violacein by the newly isolated *Duganella violaceinigra* str. NI28 and its impact on *Staphylococcus aureus*. Sci Rep 5:15598. https://doi.org/10.1038/srep15598

Cortes-Tolalpa L, Gutierrez-Rios RM, Martinez LM, de Anda R, Gosset G, Bolivar F, Escalante A (2014) Global transcriptomic analysis of an engineered *Escherichia coli* strain lacking the phosphoenolpyruvate: carbohydrate phosphotransferase system during shikimic acid production in rich culture medium. Microb Cell Fact 13:28. https://doi.org/10.1186/1475-2859-13-28

De Bruyn F, De Paepe B, Maertens J, Beauprez J, De Cocker P, Mincke S, Stevens C, De Mey M (2015) Development of an in vivo glucosylation platform by coupling production to growth: production of phenolic glucosides by a glycosyltransferase of *Vitis vinifera*. Biotechnol Bioeng 112:1594–1603. https://doi.org/10.1002/bit.25570

Draths KM, Frost JW (1994) Environmentally compatible synthesis of adipic acid from D-glucose. J Am Chem Soc 116:399–400

Draths KM, Frost JW (1995) Environmentally compatible synthesis of catechol from D-glucose. J Am Chem Soc 117:2395–2400

Draths KM, Knop DR, Frost JW (1999) Shikimic acid and quinic acid: replacing isolation from plant sources with recombinant microbial biocatalysis. J Am Chem Soc 121:1603–1604

Du L, Ma L, Qi F, Zheng X, Jiang C, Li A, Wan X, Liu SJ, Li S (2016) Characterization of a unique pathway for 4-cresol catabolism initiated by phosphorylation in *Corynebacterium glutamicum*. J Biol Chem 291:6583–6594. https://doi.org/10.1074/jbc.M115.695320

Duan L, Ding W, Liu X, Cheng X, Cai J, Hua E, Jiang H (2017) Biosynthesis and engineering of kaempferol in *Saccharomyces cerevisiae*. Microb Cell Fact 16:165. https://doi.org/10.1186/s12934-017-0774-x

Durán M, Ponezi AN, Faljoni-Alario A, Teixeira MFS, Justo GZ, Durán N (2012) Potential applications of violacein: a microbial pigment. Med Chem Res 21:1524–1532

Escalante A, Calderon R, Valdivia A, de Anda R, Hernandez G, Ramirez OT, Gosset G, Bolivar F (2010) Metabolic engineering for the production of shikimic acid in an evolved *Escherichia coli* strain lacking the phosphoenolpyruvate: carbohydrate phosphotransferase system. Microb Cell Fact 9:21. https://doi.org/10.1186/1475-2859-9-21

Fang MY, Zhang C, Yang S, Cui JY, Jiang PX, Lou K, Wachi M, Xing XH (2015) High crude violacein production from glucose by *Escherichia coli* engineered with interactive control of tryptophan pathway and violacein biosynthetic pathway. Microb Cell Fact 14:8. https://doi.org/10.1186/s12934-015-0192-x

Frost JW, Draths KM (1995) Biocatalytic syntheses of aromatics from D-glucose: renewable microbial sources of aromatic compounds. Annu Rev Microbiol 49:557–579. https://doi.org/10.1146/annurev.mi.49.100195.003013

Ganesan V, Li Z, Wang X, Zhang H (2017) Heterologous biosynthesis of natural product naringenin by co-culture engineering. Synth Syst Biotechnol 2:236–242. https://doi.org/10.1016/j.synbio.2017.08.003

Ghosh S, Chisti Y, Banerjee UC (2012) Production of shikimic acid. Biotechnol Adv 30:1425–1431. https://doi.org/10.1016/j.biotechadv.2012.03.001

Gottardi M, Reifenrath M, Boles E, Tripp J (2017) Pathway engineering for the production of heterologous aromatic chemicals and their derivatives in *Saccharomyces cerevisiae*: bioconversion from glucose. FEMS Yeast Res 17. https://doi.org/10.1093/femsyr/fox035

Gu P, Fan X, Liang Q, Qi Q, Li Q (2017) Novel technologies combined with traditional metabolic engineering strategies facilitate the construction of shikimate-producing *Escherichia coli*. Microb Cell Fact 16:167. https://doi.org/10.1186/s12934-017-0773-y

Hansen EH, Moller BL, Kock GR, Bunner CM, Kristensen C, Jensen OR, Okkels FT, Olsen CE, Motawia MS, Hansen J (2009) De novo biosynthesis of vanillin in fission yeast (*Schizosaccharomyces pombe*) and baker's yeast (*Saccharomyces cerevisiae*). Appl Environ Microbiol 75:2765–2774. https://doi.org/10.1128/AEM.02681-08

Heery DM, Dunican LK (1993) Cloning of the *trp* gene cluster from a tryptophan-hyperproducing strain of *Corynebacterium glutamicum*: identification of a mutation in the *trp* leader sequence. Appl Environ Microbiol 59:791–799

Heo KT, Kang SY, Hong YS (2017) De novo biosynthesis of pterostilbene in an *Escherichia coli* strain using a new resveratrol *O*-methyltransferase from *Arabidopsis*. Microb Cell Fact 16:30. https://doi.org/10.1186/s12934-017-0644-6

Herrmann KM (1995) The shikimate pathway: early steps in the biosynthesis of aromatic compounds. Plant Cell 7:907–919. https://doi.org/10.1105/tpc.7.7.907

Ikeda M (2006) Towards bacterial strains overproducing L-tryptophan and other aromatics by metabolic engineering. Appl Microbiol Biotechnol 69:615–626. https://doi.org/10.1007/s00253-005-0252-y

Ikeda M, Mizuno Y, Awane S, Hayashi M, Mitsuhashi S, Takeno S (2011) Identification and application of a different glucose uptake system that functions as an alternative to the phosphotransferase system in *Corynebacterium glutamicum*. Appl Microbiol Biotechnol 90:1443–1451. https://doi.org/10.1007/s00253-011-3210-x

Jiang M, Zhang H (2016) Engineering the shikimate pathway for biosynthesis of molecules with pharmaceutical activities in *E. coli*. Curr Opin Biotechnol 42:1–6. https://doi.org/10.1016/j.copbio.2016.01.016

Jiang PX, Wang HS, Zhang C, Lou K, Xing XH (2010) Reconstruction of the violacein biosynthetic pathway from *Duganella* sp. B2 in different heterologous hosts. Appl Microbiol Biotechnol 86:1077–1088. https://doi.org/10.1007/s00253-009-2375-z

Johnson CW, Salvachua D, Khanna P, Smith H, Peterson DJ, Beckham GT (2016) Enhancing muconic acid production from glucose and lignin-derived aromatic compounds via increased protocatechuate decarboxylase activity. Metab Eng Commun 3:111–119. https://doi.org/10.1016/j.meteno.2016.04.002

Jojima T, Igari T, Gunji W, Suda M, Inui M, Yukawa H (2012) Identification of a HAD superfamily phosphatase, HdpA, involved in 1,3-dihydroxyacetone production during sugar catabolism in *Corynebacterium glutamicum*. FEBS Lett 586:4228–4232. https://doi.org/10.1016/j.febslet.2012.10.028

Jung HM, Jung MY, Oh MK (2015) Metabolic engineering of Klebsiella pneumoniae for the production of *cis,cis*-muconic acid. Appl Microbiol Biotechnol 99:5217–5225. https://doi.org/10.1007/s00253-015-6442-3

Kakkar S, Bais S (2014) A review on protocatechuic Acid and its pharmacological potential. ISRN Pharmacol 2014:952943. https://doi.org/10.1155/2014/952943

Kallscheuer N, Marienhagen J (2018) *Corynebacterium glutamicum* as platform for the production of hydroxybenzoic acids. Microb Cell Fact 17:70. https://doi.org/10.1186/s12934-018-0923-x

Kallscheuer N, Vogt M, Kappelmann J, Krumbach K, Noack S, Bott M, Marienhagen J (2016a) Identification of the *phd* gene cluster responsible for phenylpropanoid utilization in *Corynebacterium glutamicum*. Appl Microbiol Biotechnol 100:1871–1881. https://doi.org/10.1007/s00253-015-7165-1

Kallscheuer N, Vogt M, Stenzel A, Gätgens J, Bott M, Marienhagen J (2016b) Construction of a *Corynebacterium glutamicum* platform strain for the production of stilbenes and (2S)-flavanones. Metab Eng 38:47–55. https://doi.org/10.1016/j.ymben.2016.06.003

Kallscheuer N, Vogt M, Bott M, Marienhagen J (2017) Functional expression of plant-derived *O*-methyltransferase, flavanone 3-hydroxylase, and flavonol synthase in *Corynebacterium glutamicum* for production of pterostilbene, kaempferol, and quercetin. J Biotechnol 258:190–196. https://doi.org/10.1016/j.jbiotec.2017.01.006

Kawaguchi H, Sasaki K, Uematsu K, Tsuge Y, Teramura H, Okai N, Nakamura-Tsuruta S, Katsuyama Y, Sugai Y, Ohnishi Y, Hirano K, Sazuka T, Ogino C, Kondo A (2015) 3-Amino-4-hydroxybenzoic acid production from sweet sorghum juice by recombinant *Corynebacterium glutamicum*. Bioresour Technol 198:410–417. https://doi.org/10.1016/j.biortech.2015.09.024

Kershaw J, Kim KH (2017) The therapeutic potential of piceatannol, a natural stilbene, in metabolic diseases: a review. J Med Food 20:427–438. https://doi.org/10.1089/jmf.2017.3916

Kim B, Park H, Na D, Lee SY (2014) Metabolic engineering of *Escherichia coli* for the production of phenol from glucose. Biotechnol J 9:621–629. https://doi.org/10.1002/biot.201300263

Kitade Y, Hashimoto R, Suda M, Hiraga K, Inui M (2018) Production of 4-hydroxybenzoic acid by an aerobic growth-arrested bioprocess using metabolically engineered *Corynebacterium glutamicum*. Appl Environ Microbiol 84:pii: e02587-17. https://doi.org/10.1128/AEM.02587-17

Knop DR, Draths KM, Chandran SS, Barker JL, von Daeniken R, Weber W, Frost JW (2001) Hydroaromatic equilibration during biosynthesis of shikimic acid. J Am Chem Soc 123:10173–10182

Kogure T, Inui M (2018) Recent advances in metabolic engineering of *Corynebacterium glutamicum* for bioproduction of value-added aromatic chemicals and natural products. Appl Microbiol Biotechnol 102:8685–8705. https://doi.org/10.1007/s00253-018-9289-6

Kogure T, Kubota T, Suda M, Hiraga K, Inui M (2016) Metabolic engineering of *Corynebacterium glutamicum* for shikimate overproduction by growth-arrested cell reaction. Metab Eng 38:204–216. https://doi.org/10.1016/j.ymben.2016.08.005

Koma D, Yamanaka H, Moriyoshi K, Ohmoto T, Sakai K (2012) Production of aromatic compounds by metabolically engineered *Escherichia coli* with an expanded shikimate pathway. Appl Environ Microbiol 78:6203–6216. https://doi.org/10.1128/AEM.01148-12

Koma D, Yamanaka H, Moriyoshi K, Sakai K, Masuda T, Sato Y, Toida K, Ohmoto T (2014) Production of *p*-aminobenzoic acid by metabolically engineered *Escherichia coli*. Biosci Biotechnol Biochem 78:350–357. https://doi.org/10.1080/09168451.2014.878222

Koopman F, Beekwilder J, Crimi B, van Houwelingen A, Hall RD, Bosch D, van Maris AJ, Pronk JT, Daran JM (2012) De novo production of the flavonoid naringenin in engineered *Saccharomyces cerevisiae*. Microb Cell Fact 11:155. https://doi.org/10.1186/1475-2859-11-155

Krämer M, Bongaerts J, Bovenberg R, Kremer S, Muller U, Orf S, Wubbolts M, Raeven L (2003) Metabolic engineering for microbial production of shikimic acid. Metab Eng 5:277–283

Krömer JO, Nunez-Bernal D, Averesch NJ, Hampe J, Varela J, Varela C (2013) Production of aromatics in *Saccharomyces cerevisiae*—a feasibility study. J Biotechnol 163:184–193. https://doi.org/10.1016/j.jbiotec.2012.04.014

Krzysztoforska K, Mirowska-Guzel D, Widy-Tyszkiewicz E (2017) Pharmacological effects of protocatechuic acid and its therapeutic potential in neurodegenerative diseases: review on the basis of in vitro and in vivo studies in rodents and humans. Nutr Neurosci 1–11. https://doi.org/10.1080/1028415X.2017.1354543

Kubota T, Tanaka Y, Takemoto N, Watanabe A, Hiraga K, Inui M, Yukawa H (2014) Chorismate-dependent transcriptional regulation of quinate/shikimate utilization genes by LysR-type transcriptional regulator QsuR in *Corynebacterium glutamicum*: carbon flow control at metabolic branch point. Mol Microbiol 92:356–368. https://doi.org/10.1111/mmi.12560

Kubota T, Watanabe A, Suda M, Kogure T, Hiraga K, Inui M (2016) Production of para-aminobenzoate by genetically engineered *Corynebacterium glutamicum* and non-biological formation of an N-glucosyl byproduct. Metab Eng 38:322–330. https://doi.org/10.1016/j.ymben.2016.07.010

Lee JH, Wendisch VF (2017) Biotechnological production of aromatic compounds of the extended shikimate pathway from renewable biomass. J Biotechnol 257:211–221. https://doi.org/10.1016/j.jbiotec.2016.11.016

Lee JW, Na D, Park JM, Lee J, Choi S, Lee SY (2012) Systems metabolic engineering of microorganisms for natural and non-natural chemicals. Nat Chem Biol 8:536–546. https://doi.org/10.1038/nchembio.970

Lee HN, Shin WS, Seo SY, Choi SS, Song JS, Kim JY, Park JH, Lee D, Kim SY, Lee SJ, Chun GT, Kim ES (2018) Corynebacterium cell factory design and culture process optimization for muconic acid biosynthesis. Sci Rep 8:18041. https://doi.org/10.1038/s41598-018-36320-4

Li W, Xie D, Frost JW (2005) Benzene-free synthesis of catechol: interfacing microbial and chemical catalysis. J Am Chem Soc 127:2874–2882. https://doi.org/10.1021/ja045148n

Li PP, Liu YJ, Liu SJ (2009) Genetic and biochemical identification of the chorismate mutase from *Corynebacterium glutamicum*. Microbiology 155:3382–3391. https://doi.org/10.1099/mic.0.029819-0

Li PP, Li DF, Liu D, Liu YM, Liu C, Liu SJ (2013) Interaction between DAHP synthase and chorismate mutase endows new regulation on DAHP synthase activity in *Corynebacterium glutamicum*. Appl Microbiol Biotechnol 97:10373–10380. https://doi.org/10.1007/s00253-013-4806-0

Li M, Kildegaard KR, Chen Y, Rodriguez A, Borodina I, Nielsen J (2015) De novo production of resveratrol from glucose or ethanol by engineered *Saccharomyces cerevisiae*. Metab Eng 32:1–11. https://doi.org/10.1016/j.ymben.2015.08.007

Licona-Cassani C, Lara AR, Cabrera-Valladares N, Escalante A, Hernandez-Chavez G, Martinez A, Bolivar F, Gosset G (2013) Inactivation of pyruvate kinase or the phosphoenolpyruvate: sugar phosphotransferase system increases shikimic and dehydroshikimic acid yields from glucose in *Bacillus subtilis*. J Mol Microbiol Biotechnol 24:37–45. https://doi.org/10.1159/000355264

Lin Y, Shen X, Yuan Q, Yan Y (2013) Microbial biosynthesis of the anticoagulant precursor 4-hydroxycoumarin. Nat Commun 4:2603. https://doi.org/10.1038/ncomms3603

Lin Y, Sun X, Yuan Q, Yan Y (2014) Extending shikimate pathway for the production of muconic acid and its precursor salicylic acid in *Escherichia coli*. Metab Eng 23:62–69. https://doi.org/10.1016/j.ymben.2014.02.009

Lindner SN, Seibold GM, Henrich A, Krämer R, Wendisch VF (2011) Phosphotransferase system-independent glucose utilization in *Corynebacterium glutamicum* by inositol permeases and glucokinases. Appl Environ Microbiol 77:3571–3581. https://doi.org/10.1128/AEM.02713-10

Liu YJ, Li PP, Zhao KX, Wang BJ, Jiang CY, Drake HL, Liu SJ (2008) *Corynebacterium glutamicum* contains 3-deoxy-D-arabino-heptulosonate 7-phosphate synthases that display novel biochemical features. Appl Environ Microbiol 74:5497–5503. https://doi.org/10.1128/AEM.00262-08

Liu YB, Long MX, Yin YJ, Si MR, Zhang L, Lu ZQ, Wang Y, Shen XH (2013) Physiological roles of mycothiol in detoxification and tolerance to multiple poisonous chemicals in *Corynebacterium glutamicum*. Arch Microbiol 195:419–429. https://doi.org/10.1007/s00203-013-0889-3

Liu X, Lin J, Hu H, Zhou B, Zhu B (2016) De novo biosynthesis of resveratrol by site-specific integration of heterologous genes in *Escherichia coli*. FEMS Microbiol Lett 363(8):pii: fnw061. https://doi.org/10.1093/femsle/fnw061

Martinez JA, Bolivar F, Escalante A (2015) Shikimic acid production in *Escherichia coli*: from classical metabolic engineering strategies to omics applied to improve its production. Front Bioeng Biotechnol 3:145. https://doi.org/10.3389/fbioe.2015.00145

Meijnen JP, Verhoef S, Briedjlal AA, de Winde JH, Ruijssenaars HJ (2011) Improved *p*-hydroxybenzoate production by engineered *Pseudomonas putida* S12 by using a mixed-substrate feeding strategy. Appl Microbiol Biotechnol 90:885–893. https://doi.org/10.1007/s00253-011-3089-6

Merkens H, Beckers G, Wirtz A, Burkovski A (2005) Vanillate metabolism in *Corynebacterium glutamicum*. Curr Microbiol 51:59–65. https://doi.org/10.1007/s00284-005-4531-8

Muller R, Wagener A, Schmidt K, Leistner E (1995) Microbial production of specifically ring-13C-labelled 4-hydroxybenzoic acid. Appl Microbiol Biotechnol 43:985–988

Niu W, Draths KM, Frost JW (2002) Benzene-free synthesis of adipic acid. Biotechnol Prog 18:201–211. https://doi.org/10.1021/bp010179x

Noda S, Kondo A (2017) Recent advances in microbial production of aromatic chemicals and derivatives. Trends Biotechnol 35:785–796. https://doi.org/10.1016/j.tibtech.2017.05.006

Noda S, Shirai T, Oyama S, Kondo A (2016) Metabolic design of a platform *Escherichia coli* strain producing various chorismate derivatives. Metab Eng 33:119–129. https://doi.org/10.1016/j.ymben.2015.11.007

Okai N, Miyoshi T, Takeshima Y, Kuwahara H, Ogino C, Kondo A (2016) Production of protocatechuic acid by *Corynebacterium glutamicum* expressing chorismate-pyruvate lyase from *Escherichia coli*. Appl Microbiol Biotechnol 100:135–145. https://doi.org/10.1007/s00253-015-6976-4

Okai N, Masuda T, Takeshima Y, Tanaka K, Yoshida K-i, Miyamoto M, Ogino C, Kondo A (2017) Biotransformation of ferulic acid to protocatechuic acid by Corynebacterium glutamicum ATCC 21420 engineered to express vanillate O-demethylase. AMB Express 7(1):130

Patel K, Singh GK, Patel DK (2018) A review on pharmacological and analytical aspects of naringenin. Chin J Integr Med 24:551–560. https://doi.org/10.1007/s11655-014-1960-x

Piotrowska H, Kucinska M, Murias M (2012) Biological activity of piceatannol: leaving the shadow of resveratrol. Mutat Res 750:60–82. https://doi.org/10.1016/j.mrrev.2011.11.001

Polen T, Spelberg M, Bott M (2013) Toward biotechnological production of adipic acid and precursors from biorenewables. J Biotechnol 167:75–84. https://doi.org/10.1016/j.jbiotec.2012.07.008

Rawat G, Tripathi P, Saxena RK (2013) Expanding horizons of shikimic acid. Recent progresses in production and its endless frontiers in application and market trends. Appl Microbiol Biotechnol 97:4277–4287. https://doi.org/10.1007/s00253-013-4840-y

Rodriguez A, Martinez JA, Baez-Viveros JL, Flores N, Hernandez-Chavez G, Ramirez OT, Gosset G, Bolivar F (2013) Constitutive expression of selected genes from the pentose phosphate and aromatic pathways increases the shikimic acid yield in high-glucose batch cultures of an *Escherichia coli* strain lacking PTS and *pykF*. Microb Cell Fact 12:86. https://doi.org/10.1186/1475-2859-12-86

Rodriguez A, Martinez JA, Flores N, Escalante A, Gosset G, Bolivar F (2014) Engineering *Escherichia coli* to overproduce aromatic amino acids and derived compounds. Microb Cell Fact 13:126. https://doi.org/10.1186/s12934-014-0126-z

Sasaki M, Jojima T, Kawaguchi H, Inui M, Yukawa H (2009) Engineering of pentose transport in *Corynebacterium glutamicum* to improve simultaneous utilization of mixed sugars. Appl Microbiol Biotechnol 85:105–115. https://doi.org/10.1007/s00253-009-2065-x

Scalbert A, Johnson IT, Saltmarsh M (2005a) Polyphenols: antioxidants and beyond. Am J Clin Nutr 81:215S–217S. https://doi.org/10.1093/ajcn/81.1.215S

Scalbert A, Manach C, Morand C, Remesy C, Jimenez L (2005b) Dietary polyphenols and the prevention of diseases. Crit Rev Food Sci Nutr 45:287–306. https://doi.org/10.1080/1040869059096

Sengupta S, Jonnalagadda S, Goonewardena L, Juturu V (2015) Metabolic engineering of a novel muconic acid biosynthesis pathway via 4-hydroxybenzoic acid in *Escherichia coli*. Appl Environ Microbiol 81:8037–8043. https://doi.org/10.1128/AEM.01386-15

Shen X, Liu S (2005) Key enzymes of the protocatechuate branch of the beta-ketoadipate pathway for aromatic degradation in *Corynebacterium glutamicum*. Sci China C Life Sci 48:241–249

Shen XH, Zhou NY, Liu SJ (2012) Degradation and assimilation of aromatic compounds by *Corynebacterium glutamicum*: another potential for applications for this bacterium? Appl Microbiol Biotechnol 95:77–89. https://doi.org/10.1007/s00253-012-4139-4

Shiio I, Sugimoto S, Miyajima R (1974) Regulation of 3-deoxy-D-arabino-heptulosonate-7-phosphate synthetase in *Brevibacterium flavum*. J Biochem 75:987–997

Shin WS, Lee D, Lee SJ, Chun GT, Choi SS, Kim ES, Kim S (2018) Characterization of a non-phosphotransferase system for *cis,cis*-muconic acid production in *Corynebacterium glutamicum*. Biochem Biophys Res Commun 499:279–284. https://doi.org/10.1016/j.bbrc.2018.03.146

Shrestha A, Pandey RP, Pokhrel AR, Dhakal D, Chu LL, Sohng JK (2018) Modular pathway engineering for resveratrol and piceatannol production in engineered *Escherichia coli*. Appl Microbiol Biotechnol 102:9691–9706. https://doi.org/10.1007/s00253-018-9323-8

Siebert M, Severin K, Heide L (1994) Formation of 4-hydroxybenzoate in *Escherichia coli*: characterization of the *ubiC* gene and its encoded enzyme chorismate pyruvate-lyase. Microbiology 140(Pt 4):897–904. https://doi.org/10.1099/00221287-140-4-897

Suastegui M, Shao Z (2016) Yeast factories for the production of aromatic compounds: from building blocks to plant secondary metabolites. J Ind Microbiol Biotechnol 43:1611–1624. https://doi.org/10.1007/s10295-016-1824-9

Sun X, Lin Y, Huang Q, Yuan Q, Yan Y (2013) A novel muconic acid biosynthesis approach by shunting tryptophan biosynthesis via anthranilate. Appl Environ Microbiol 79:4024–4030. https://doi.org/10.1128/AEM.00859-13

Sun X, Lin Y, Yuan Q, Yan Y (2014) Biological production of muconic acid via a prokaryotic 2,3-dihydroxybenzoic acid decarboxylase. ChemSusChem 7:2478–2481. https://doi.org/10.1002/cssc.201402092

Sun H, Zhao D, Xiong B, Zhang C, Bi C (2016) Engineering *Corynebacterium glutamicum* for violacein hyper production. Microb Cell Fact 15:148. https://doi.org/10.1186/s12934-016-0545-0

Suzuki H, Ohnishi Y, Furusho Y, Sakuda S, Horinouchi S (2006) Novel benzene ring biosynthesis from C(3) and C(4) primary metabolites by two enzymes. J Biol Chem 281:36944–36951. https://doi.org/10.1074/jbc.M608103200

Syukur Purwanto H, Kang MS, Ferrer L, Han SS, Lee JY, Kim HS, Lee JH (2018) Rational engineering of the shikimate and related pathways in *Corynebacterium glutamicum* for 4-hydroxybenzoate production. J Biotechnol 282:92–100. https://doi.org/10.1016/j.jbiotec.2018.07.016

Teramoto H, Inui M, Yukawa H (2009) Regulation of expression of genes involved in quinate and shikimate utilization in *Corynebacterium glutamicum*. Appl Environ Microbiol 75:3461–3468. https://doi.org/10.1128/AEM.00163-09

Tripathi P, Rawat G, Yadav S, Saxena RK (2015) Shikimic acid, a base compound for the formulation of swine/avian flu drug: statistical optimization, fed-batch and scale up studies along with its application as an antibacterial agent. Antonie Van Leeuwenhoek 107:419–431. https://doi.org/10.1007/s10482-014-0340-z

Umbarger HE (1978) Amino acid biosynthesis and its regulation. Annu Rev Biochem 47:532–606. https://doi.org/10.1146/annurev.bi.47.070178.002533

Vargas-Tah A, Martinez LM, Hernandez-Chavez G, Rocha M, Martinez A, Bolivar F, Gosset G (2015) Production of cinnamic and *p*-hydroxycinnamic acid from sugar mixtures with engineered *Escherichia coli*. Microb Cell Fact 14:6. https://doi.org/10.1186/s12934-014-0185-1

Verhoef S, Ruijssenaars HJ, de Bont JA, Wery J (2007) Bioproduction of *p*-hydroxybenzoate from renewable feedstock by solvent-tolerant *Pseudomonas putida* S12. J Biotechnol 132:49–56. https://doi.org/10.1016/j.jbiotec.2007.08.031

Wang J, Shen X, Rey J, Yuan Q, Yan Y (2017) Recent advances in microbial production of aromatic natural products and their derivatives. Appl Microbiol Biotechnol 102:47–61. https://doi.org/10.1007/s00253-017-8599-4

Wang S, Bilal M, Hu H, Wang W, Zhang X (2018) 4-Hydroxybenzoic acid-a versatile platform intermediate for value-added compounds. Appl Microbiol Biotechnol 102:3561–3571. https://doi.org/10.1007/s00253-018-8815-x

Weber C, Bruckner C, Weinreb S, Lehr C, Essl C, Boles E (2012) Biosynthesis of *cis,cis*-muconic acid and its aromatic precursors, catechol and protocatechuic acid, from renewable feedstocks by *Saccharomyces cerevisiae*. Appl Environ Microbiol 78:8421–8430. https://doi.org/10.1128/AEM.01983-12

Wendisch VF, Brito LF, Gil Lopez M, Hennig G, Pfeifenschneider J, Sgobba E, Veldmann KH (2016) The flexible feedstock concept in industrial biotechnology: metabolic engineering of *Escherichia coli*, *Corynebacterium glutamicum*, *Pseudomonas*, *Bacillus* and yeast strains for access to alternative carbon sources. J Biotechnol 234:139–157. https://doi.org/10.1016/j.jbiotec.2016.07.022

Williams KM, Martin WE, Smith J, Williams BS, Garner BL (2012) Production of protocatechuic acid in *Bacillus Thuringiensis* ATCC33679. Int J Mol Sci 13:3765–3772. https://doi.org/10.3390/ijms13033765

Xie NZ, Liang H, Huang RB, Xu P (2014) Biotechnological production of muconic acid: current status and future prospects. Biotechnol Adv 32:615–622. https://doi.org/10.1016/j.biotechadv.2014.04.001

Zhang B, Jiang CY, Liu YM, Liu C, Liu SJ (2015a) Engineering of a hybrid route to enhance shikimic acid production in *Corynebacterium glutamicum*. Biotechnol Lett 37:1861–1868. https://doi.org/10.1007/s10529-015-1852-y

Zhang B, Zhou N, Liu YM, Liu C, Lou CB, Jiang CY, Liu SJ (2015b) Ribosome binding site libraries and pathway modules for shikimic acid synthesis with *Corynebacterium glutamicum*. Microb Cell Fact 14:71. https://doi.org/10.1186/s12934-015-0254-0

Zhang B, Liu ZQ, Liu C, Zheng YG (2016) Application of CRISPRi in *Corynebacterium glutamicum* for shikimic acid production. Biotechnol Lett 38:2153–2161. https://doi.org/10.1007/s10529-016-2207-z

Zhou Y, Fang MY, Li G, Zhang C, Xing XH (2018) Enhanced production of crude violacein from glucose in *Escherichia coli* by overexpression of rate-limiting key enzyme(s) involved in violacein biosynthesis. Appl Biochem Biotechnol 186:909–916. https://doi.org/10.1007/s12010-018-2787-2

Zhu S, Wu J, Du G, Zhou J, Chen J (2014) Efficient synthesis of eriodictyol from L-tyrosine in *Escherichia coli*. Appl Environ Microbiol 80:3072–3080. https://doi.org/10.1128/AEM.03986-13

Index

M. Inui, K. Toyoda (eds.), *Corynebacterium glutamicum*, Microbiology Monographs 23, https://doi.org/10.1007/978-3-030-39267-3

E

F

G

H

I

K

L

M

Q

R

The manufacturer's authorised representative in the EU is Springer Nature Customer Service Centre GmbH, Europaplatz 3, 69115 Heidelberg, Germany. If you have any concerns regarding our products, please contact ProductSafety@springernature.com

Printed and bound by CPI Group (UK) Ltd, Croydon, CR0 4YY
22/09/2026
02220956-0010